The Ecology of FYNBOS

Nutrients, Fire and Diversity

Edited by:
R M Cowling

1992
Cape Town
Oxford University Press

OXFORD UNIVERSITY PRESS

Walton Street, Oxford OX2 6DP, United Kingdom

Oxford New York Toronto
Delhi Bombay Calcutta Madras Karachi
Petaling Jaya Singapore Hong Kong Tokyo
Nairobi Dar es Salaam Cape Town
Melbourne Auckland

and associated companies in
Berlin Ibadan

The Ecology of Fynbos
Nutrients, Fire and Diversity
ISBN 0 19 570661 7

First published 1992

Published by Oxford University Press Southern Africa,
Harrington House, Barrack Street, Cape Town, 8001, South Africa

Set in 10 on 11 pt Garamond by Clyson
Reproduction by Photoprint
Printed and bound by Clyson

CONTENTS

DEDICATION

To Margaret Rutherford Bryan Levyns BA (Hons), DSc, FRSSA (1890–1975) who with dignity and integrity rose above the prejudice of her time, to make one of the greatest contributions to our knowledge of Cape flora and vegetation.

ACKNOWLEDGEMENTS

The chapters in this book are based on invited papers delivered at a colloquium held at Franschhoek from 30 June to 1 July 1990. I wish to thank all the authors and acknowledge their willingness to meet deadlines, even when goaded to breaking point by Shirley Pierce and myself. Shirley played a major role in producing this book by co-ordinating all aspects of editorial work prior to printing, and critically reading many of the manuscripts. Wendy Paisley was responsible for subediting, a thankless task admirably performed.

The editing of this book was greatly facilitated by the constructive and prompt comments of the reviewers. Many thanks to John Beard, David Bell, Charlie Boucher, Martin Cody, Peter Goldblatt, Richard Groves, Peter Grubb, Tony Hall, Rob Hengeveld, Craig Hilton-Taylor, Jonathan Majer, Eugene Moll, Tim O'Connor, Norman Pammenter, Nigel Penn, Bob Scholes, Judy Seely, Armin Seydack, Michael Usher, Abrahaam van Wyk, Rob Whelan, and Mark Williamson.

A special thanks to Colin Paterson-Jones for selecting slides and providing the cover photographs. Wendy Powell, Glenda Younge, Graham Arbuckle, and Tony Mays of Oxford University Press are warmly thanked for their patience and support.

Finally, I wish to acknowledge a grant for publication costs from the Department of Environment Affairs obtained on the initiative of the Foundation for Research Development, Pretoria who also provided administrative support.

CONTRIBUTORS

Nicky Allsopp, Department of Botany, University of Cape Town, Rondebosch 7700

William Bond, Department of Botany, University of Cape Town, Rondebosch 7700

Richard Cowling, Department of Botany, University of Cape Town, Rondebosch 7700

Hilary Deacon, Department of Archaeology, University of Stellenbosch, Stellenbosch 7600

Freddy Ellis, Faculty of Forestry, University of Stellenbosch, Stellenbosch 7600

Peter Grubb, Department of Botany, Cambridge University, Cambridge, UK

Pat Holmes, Department of Botany, University of Cape Town, Rondebosch 7700

Brian Huntley, National Botanical Institute at Kirstenbosch, Private Bag X 7, Claremont 7735

Steve Johnson, Department of Botany, University of Cape Town, Rondebosch 7700

Mark Jury, Department of Oceanography, University of Cape Town, Rondebosch 7700

Jon Keeley, Department of Biology, Occidental College, Los Angeles, California, 90041, USA

Dave le Maitre, CSIR, Division of Forest Science and Technology, Jonkershoek Forestry Research Station, Private Bag X 5011, Stellenbosch 7600

Owen Lewis, Department of Botany, University of Cape Town, Rondebosch 7700

Peter Linder, Department of Botany, University of Cape Town, Rondebosch 7700

Ian Macdonald, FitzPatrick Institute, University of Cape Town, Rondebosch 7700

Mike Meadows, Department of Environmental and Geographical Science, University of Cape Town, Rondebosch 7700

Jeremy Midgley, CSIR, Division of Forest Science and Technology, Saasveld Forestry Research Station, Private Bag X 6515, George 6530

Tony Rebelo, National Botanical Institute at Kirstenbosch, Private Bag X 7, Claremont 7735

Mike Richards, Department of Botany, University of Cape Town, Rondebosch 7700

Dave Richardson, CSIR, Division of Forest Science and Technology, Jonkershoek Forestry Research Station, Private Bag X 5011, Stellenbosch 7600

Willy Stock, Department of Botany, University of Cape Town, Rondebosch 7700

Brian van Wilgen, CSIR, Division of Forest Science and Technology, Jonkershoek Forestry Research Station, Private Bag X 5011, Stellenbosch 7600

PREFACE

This book is primarily about fynbos. But what is fynbos? The considerable confusion that surrounds the definition of fynbos (Moll and Jarman 1984; Campbell 1985; Boucher 1987) is characteristic of the typological controversies generated by phytogeographers working in the southern African region. Given the astonishing diversity of biomes, communities, and plant species in the subcontinent (Cowling et al. 1989), this confusion is perhaps understandable. However, since the term fynbos is used in this volume in the context of a vegetation type, a biome, and a flora, it seems appropriate to provide some clarity at the outset.

Dutch settlers referred to the predominant vegetation of the south-western Cape as 'fijnbosch'. Taylor (1978) suggested that this term evoked the ubiquitous fine- or small-leaved component of the vegetation. Boucher (1987) cites sources which link the term to the vegetation's potential as a forestry resource i.e. timber too slender (fine) for harvesting. Despite attempts to introduce less parochial terms such as macchia (Acocks 1953), fynbos has become embedded in the biological literature, and generally this colloquialism has been enthusiastically assimilated by scientists and laypeople alike.

Fynbos is the dominant vegetation type and contributes most of the species to the flora of the Cape Floristic Region (Bond and Goldblatt 1984; Cowling and Holmes this volume). The Cape Floristic Region is a biogeographically complex zone including elements from two floras, five phytochoria, and five biomes (see chart). Some chapters in this book refer to the Cape Floristic Region and include material on non-fynbos vegetation types and biota. The scope of other chapters is limited to the fynbos biome (Rutherford and Westfall 1986): they thus deal mainly with the fire-prone fynbos and renoster shrubland which are sometimes collectively referred to as 'Cape shrublands'. A few chapters focus only on fynbos either as a vegetation type or a biota. Throughout this volume, I have attempted to standardize biogeographic and vegetation concepts and terms (see Cowling and Holmes this volume for details) but some confusion will probably slip through the gaps.

Why a book on fynbos? It has been some 15 years since the launch of the Fynbos Biome Project (Kruger 1978). This co-operative and interdisciplinary programme provided the stimulus for a small and isolated scientific community to carry out some interesting research on a fascinating biota (Huntley this volume). Some of this research has had an impact beyond the boundaries of the Cape Floristic Region. Despite the deepening academic boycott and the recent demise of the Fynbos Biome Project, the questions being asked by researchers are more interesting than ever. It seemed appropriate, therefore, to capture some of the excitement of this era in a book.

This volume reviews and synthesizes research largely co-ordinated within the Fynbos Biome Project. With a few exceptions, the contributors comprise the younger corps of researchers who started their careers during the course of the project. As is evident in the reference lists, most of these researchers have worked collaboratively at some stage. It is important, however, to acknowledge the his-

Chart showing floristic units represented in the Cape Floristic Region.

Floral kingdom	Cape	Palaeotropical				Takhtajan (1986)
Phytochorion	Capensis	Karoo-Namib	Afromontane	Tongaland-Pondoland	Kalahari-Highveld	Werger (1978)
Biome	Fynbos	Succulent karoo	Forest	Savanna	Grassland	Rutherford and Westfall (1986)
Vegetation	Cape fynbos Renoster shrubland	Karroid shrubland	Forest	Thicket	Grassland and grassy shrubland	Campbell (1985)

torical roots of this productive and interesting phase in fynbos research. People like R S Adamson, J P H Acocks, M R B Levyns, R Marloth, and more recently, Peter Goldblatt, Tony Hall, Fred Kruger, Owen Lewis, Derek Mitchell, Eugene Moll, John Rourke, Roy Siegfried, and Hugh Taylor have provided the inspiration and leadership essential for a successful research programme. In the pre-boycott era scientists from abroad, including Ray Specht, Martin Cody, Malcolm Gill, Richard Groves, Phil Miller, and Byron Lamont, made an important impression on the fynbos research community.

Not all of the chapters in this volume have adopted a review format. Many present new approaches to understanding long-established patterns, while others include previously unpublished data which often challenge established interpretations. I have also allowed reference to submitted manuscripts in the hope that these will appear in the literature at a later stage. Thus, the book incorporates research which was current at the time of publication.

The book is divided into three substantive parts. The section on biogeography includes chapters on the contemporary and historical selective regimes, plant ecological biogeography, the patterns and evolution of plant species diversity and endemism, and the historical biogeography of the Cape flora. Collectively these chapters cover a broad range of traditions in biogeography, both in terms of approach and temporal scale (cf. Myers and Giller 1988). A serious limitation is the absence of a contribution on the biogeography of the fynbos fauna. What emerges from this section is that the massive speciation within relatively few fynbos genera accompanied the extinction of a phylogenetically more diverse flora as climatic conditions deteriorated towards the end of the Tertiary. The peculiar and diverse soils of the Cape Floristic Region, together with fire and relatively moderate Pleistocene climatic fluctuations, have promoted rapid turnover of a limited number of lineages with a predictable suite of biological traits.

The five chapters on fynbos ecology reflect the strengths in fynbos research and by no means provide a comprehensive account. Nonetheless, they straddle the organism, community, and ecosystem levels of organization. Exploiting the opportunities afforded by fynbos, the contributions address some important debates regarding plant life history evolution, coevolution, coexistence of ecologically equivalent species, and stress ecology. As indicated in the title of this book, nutrient poverty and recurrent fire emerge as the major variables driving ecological and evolutionary processes in fynbos.

A major strength of the Fynbos Biome Project was that it fostered research for management (Kruger 1978; Huntley this volume). It also provided a forum for the communication of research results, ensuring their rapid incorporation into management programmes. Thus, a section of this volume is devoted to the management of fynbos. The component chapters indicate that fynbos ecosystems have been managed for upward of 100 000 years but only in the last 300 years has the human impact resulted in large-scale transformations. The ease with which alien trees and shrubs invade fynbos is an unusual and fascinating feature of its ecology, warranting special emphasis in a chapter on invasive biota. This and other transformations have severely impacted the fynbos biota, particularly on the coastal forelands. Although the basis for an optimal reserve network is beginning to emerge, implementation of these recommendations are highly problematic. Similarly, the chapter on management shows that while research has provided management guidelines for a wide range of objectives, many programmes cannot be implemented for lack of funds. Like many other biodiversity 'hot spots', the fynbos is under siege.

Since the contributions to this volume are by a small group of Cape biologists who are on personal terms and hail from a limited number of institutes, it seemed important to get an outside perspective. Fortunately Peter Grubb and Jon Keeley, both of whom have recently experienced fynbos for the first time, agreed to write a foreword and a final chapter, respectively.

To those familiar with them, fynbos landscapes have special qualities of perspective, light, and texture that cannot be communicated in a scientific text. Therefore, I have included in the book a photographic essay which can stand alone but also provides complementary material for most of the chapters. It is highly likely and, indeed, intended that all who should pick up this book will peruse this section.

Finally, it is my wish and certainly that of the other contributors, that this book stimulates new and interesting research on fynbos by scientists from home and abroad. In contrast to many tropical centres of diversity, the fynbos flora is relatively well known and a more than adequate infrastructure means that most fynbos regions are highly accessible for research. Moreover, the tractable dynamics of fynbos and the presence in a landscape of many closely related congeners provide excellent opportunities for research. In order to maintain pressure for the preservation of fynbos, it is important that interest in both its cultural and scientific value is sustained.

R M Cowling
Cape Town, July 1991

REFERENCES

ACOCKS J P H (1953). Veld types of South Africa. *Memoirs of the Botanical Society of South Africa* **28**, 1–192.

BOND P and GOLDBLATT P (1984). Plants of the Cape Flora. A descriptive catalogue. *Journal of South African Botany Supplementary Volume* **13**, 1–455.

BOUCHER C (1987). A phytosociological study of transects through the Western Cape coastal foreland, South Africa. Ph.D. thesis, University of Stellenbosch.

CAMPBELL B M (1985). A classification of the mountains of the Fynbos Biome. *Memoirs of the Botanical Survey of South Africa* **50**, 1–115.

COWLING R M, GIBBS RUSSELL G E, HOFFMAN M T and HILTON-TAYLOR C (1989). Patterns of plant species diversity in southern Africa. In *Biotic diversity in southern Africa. Concepts and conservation.* (ed Huntley B J) Oxford University Press, Cape Town, 19–50.

KRUGER F J (1978). A description of the Fynbos Biome Project. *South African National Scientific Programmes Report* **28**, CSIR, Pretoria.

MOLL E J and JARMAN M L (1984). Clarification of the term fynbos. *South African Journal of Science* **80**, 351–2.

MYERS A A and GILLER P S (eds) (1988). *Analytical biogeography.* Chapman and Hall, London and New York.

RUTHERFORD M C and WESTFALL R H (1986). Biomes of southern Africa — an objective categorization. *Memoirs of the Botanical Survey of South Africa* **54**, 1–98.

TAKHTAJAN A (1986). *Floristic regions of the world.* University of California Press, Berkeley.

TAYLOR H C (1978). Capensis. In *Biogeography and ecology of southern Africa.* (ed Werger M J A) Junk, The Hague, 171–230.

WERGER M J A (1978). Biogeographical division of southern Africa. In *Biogeography and ecology of southern Africa.* (ed Werger M J A) Junk, The Hague, 145–70.

FOREWORD

The average well-informed ecologist anywhere in the world knows that the heath-like fynbos is remarkable for the species richness of its flora and the nutrient paucity of its soils, and thinks of it as being the dominant vegetation-type of one of the five areas of the world with a mediterranean-type climate. In fact, that perception is not quite right. At the present time, the incidence of frequent fire rather than a mediterranean-type climate is the key environmental factor which is coupled with nutrient paucity. The fires do not merely initiate phases of regeneration, as in so many vegetation types, but are vital for the persistence of the fynbos flora and vegetation in the landscape. A large part of the fynbos, it is true, experiences a mediterranean-type climate, but that part merges into a second large part where the rainfall in summer is no less than that in winter. Moreover, in the area with a mediterranean-type climate, one finds in places where fire penetrates only rarely — along gullies and on large block screes — forest with a structure and physiognomy like that typical of the Mediterranean Basin or central Chile. Trees are constantly spreading into fynbos and being destroyed by fire. Where the summers are much less dry, forest covers larger areas, but lightning strikes are much more frequent in this region and fynbos is certainly an important part of the natural landscape. Hence the title of this book puts a proper emphasis on fire as the leading determinant of the system, a determinant that has had its effect over the last few million years through an interaction with nutrient paucity, and has led to the development of an extraordinary diversity of plants. The exact nature of the changing climate under which fynbos evolved is still not certain, but it is unlikely that at any time the whole of the fynbos region was characterized by winter rainfall.

South Africa is the ideal region in which to study the interactive effects of fire and nutrient paucity along a gradient of greater or lesser summer drought because the gradient in climate and vegetation is continuous along the east-west axis of the major mountain ranges. In Australia the range of heathland climate is greater, with a distinct summer maximum in rainfall in the wallum heaths of Queensland, but the relevant areas are highly disjunct. The summer wet wallum heaths do grade into the summer-dry kwongan heaths of western Australia via the heathlands of New South Wales, Victoria, and south Australia, but there are gaps of many hundreds of kilometres between the relevant areas.

This book not only revises the perspective in which fynbos is seen, but summarizes the results of exciting new kinds of research carried out in the last decade with particular vigour in the fynbos region. We are presented, for example, with much more detailed information on the patterns of diversity in the plant communities than is available for any other warm temperate region, and this information is invaluable in providing a well-documented contrast with the patterns seen in tropical lowland rain forest — the vegetation-type best known of all for its species richness.

We are also given an account of the fossil evidence which suggests that 'firestick farming' may have increased the incidence of fire by an order of magnitude for at least 100 000 years.

A new emphasis is seen in the studies of fynbos dynamics. In place of the classical emphases on the description of succession, temporal changes in productivity, and the impact of different fire frequencies, we see attention being focused on the effects of fire at different times of the year and under different conditions, on the implications of short-lived seed banks, on the interrelations between plants and the animals that disperse or eat their seeds, and on the significance of the patchwork of fires in the landscape — one area burning one year, and adjacent areas in different years. All these more recent approaches are proving vital in understanding the maintenance of the species richness of the system, and in providing a reliable basis for making recommendations on management for both the conservation of the system and the sustained harvesting of commercially valuable flowers.

The unpalatability of the fynbos plants, with their notoriously low nutrient content, is being approached in a new light, as a testing ground for generalizing ideas about the evolution of defences against herbivores. So, also, the possible role of nutrient status in moulding many features of plant reproduction is being explored.

Fynbos is under very serious threat from introduced species. Recent research has isolated the biological properties which make some species so much more invasive than others, and has shown that certain of the invading plants not only dominate the canopy space but enrich the soil nutrient supply, an impact which may constitute an even more serious threat to the persistence of a diverse flora in the long run.

Many important unknowns remain. For example, have at least some of the fynbos plants been selected to drive soil fertility downward in the inter-fire periods and reduce the chances of their being taken over by trees, or is the low fertility of the soil simply a function of its parent materials and persistent losses of nutrient capital in frequent fires? How does 2–20% of leaf area get eaten, as in most plant communities, when the biomass of animals present is so low? Exactly how did the extraordinary species richness in certain genera evolve, and how is it that certain genera have speciated so much more than others?

Although such key questions remain unanswered, this book represents a very significant step forward in understanding fynbos, and in reorienting research in fire-prone and nutrient-poor systems. I commend it to all with an interest in the central problem of explaining patterns in biodiversity, and their maintenance.

Peter J Grubb
Cambridge, May 1991

1

The Fynbos Biome Project

B J Huntley

The early 1970s witnessed an awakening of terrestrial ecological research in South Africa. Despite the early and important contributions of Adamson, Bews, Phillips, Smuts, Acocks, and others, South African ecology had, for the most part, followed an erratic and uncharted path. No serious attempt had been made to draw together a co-operative team of researchers until the National Programme for Ecosystem Research was initiated by the Council for Scientific and Industrial Research (CSIR) in 1974. This multi-organizational, interdisciplinary programme followed in the wake of the biome studies of the International Biological Programme (IBP), inheriting both strengths and weaknesses from the IBP tradition (Huntley 1987).

The strong influence of agricultural and wildlife conservation lobbies in the national programme resulted in an early emphasis in savanna, grassland, and karoo biomes and the neglect of indigenous forest and fynbos systems. However, a strong body of young biologists was emerging in the Cape, moving rapidly ahead of the thin spread of phytosociological and taxonomic studies that had characterized work on fynbos through the 1950s and 1960s. Rising concern for the conservation of the extreme plant diversity of the Cape Floristic Region — over 8 500 species in less than 90 000 km² (Bond and Goldblatt 1984; Cowling et al. this volume) — was based on the alarming rate of habitat loss, the impact of invasive woody plants, and the poorly understood influence of changing fire regimes throughout the fynbos biome (Rebelo this volume; Van Wilgen et al. this volume). By mid-1977 the Fynbos Biome Project (FBP) had been established and developed within a few years into one of the most successful co-operative research ventures of its kind. A chronology of the FBP's initial development is provided in Table 1.1.

The overall objective of the FBP was 'to provide sound scientific knowledge of the structure and functioning of constituent ecosystems as a basis for the conservation and management of the fynbos biome'. This objec-

TABLE 1.1 Profile of the initial development of the Fynbos Biome Project.

Event	Date
First mention of the need at Committee for Terrestrial Ecosystems	April 1976
First planning meeting	June 1976
First meeting of project steering committee	March 1977
Reconnaissance survey for research site	Dec 1977
Allocation of first funds for research	Jan 1978
Publication of project description	June 1978
Appointment of project liaison officer	Jan 1979
First annual research meeting	Jan 1979
Lease of Pella study site	June 1979
Publication of preliminary synthesis	Dec 1979
International Mediterranean Ecosystems Conference	Sept 1980
Appointment of project co-ordinator	July 1981
Preparation of fynbos bibliography	Aug 1981

tive, and the approaches to its achievement, was spelt out in a descriptive document that resulted from a series of planning workshops convened within the first year of the project (Kruger 1978).

The document included an overview of the ecological characteristics of and environmental problems in fynbos, and the objectives, research strategy, and phasing of the project. In particular, it identified key scientific questions relating to each of the 12 main thrusts of the programme — providing a research agenda for the decade-long study, planned to run from 1977 to 1986.

The scientific results of the FBP have been published widely in the local and international media, with some of the major findings being synthesized in this volume. The several hundred papers that resulted from the study are a measure of its success. But the wider impacts of the project, on the development of a scientific 'invisible college' within the fynbos biome, on influencing land management practices, and on stimulating public awareness of the values and conservation needs of the biome, although profound, are more difficult to quantify. Like the International Biological Programme (Younes 1991), the FBP's influence was both conceptual and sociological.

The success of the project can be ascribed to many factors, most of them novel at the time to the South African research community. The termination of co-operative scientific programmes in the late 1980s led to much of this tradition being lost. The FBP's role in the renaissance of terrestrial ecology in this country therefore deserves consideration in this record of fynbos research.

The primary characteristic of the FBP was that it was a co-operative venture of academics, managers, and administrators. Each was seen as an equal partner in the setting of objectives, processes, and priorities. Participants were drawn from six universities, three museums, five government and provincial departments, and three statutory councils. While the CSIR provided funds for 42 university-based projects (at a total cost of only R1.29 million) most of the research effort came from workers in participating organizations, each covering their own costs.

Contrary to the recommendation of senior academics consulted at the initiation of the FBP, a 'principal investigator' was not appointed to lead the project. A successful alternative was found in the provision of a low-profile secretariat, which facilitated meetings, discussion groups, symposia, and the rapid publication of material arising from such deliberations. A remarkable level of consensus was attained in all decision making, despite widely divergent viewpoints held on many topics at the start of discussions. This consensus was possible because the project was not designed around a single central paradigm or hypothesis. Thinking and action converged around a series of loosely knit themes, retaining a freshness which did not remain trapped in the dogma of the 1970s. Thus, the stultifying effect of an ecosystem analysis modelled on the IBP soon gave way to community and population level studies.

Much if not most of the project's intellectual, and physical, energy was generated by a rapidly expanding corps of 'Young Turks' in the fynbos community. The spirit of open debate, where mutual trust and a genuine desire to share knowledge and experience replaced the social and academic hierarchies of the prevailing autocracy, contributed significantly to the rapid maturity of the new research cohort. The energetic exchange of views not only influenced the way team members thought, but also the way the research programme was conducted.

With few exceptions, students at South African universities had been cut off from modern trends in ecology long before the academic boycott of the 1980s. This intellectual isolation was rapidly broken down as the FBP progressed and as participants were exposed to the results of the Chilean/Californian convergence project, and to important debates regarding the determinants of species richness and equilibrium/non-equilibrium models of community structure. Enthusiastic attempts were made to test hypotheses arising from these debates in fynbos ecosystems. The development of a framework of theoretical and evolutionary ecology with which to underpin problem-oriented studies proved a source of considerable stimulation to young students drawn into a project which was seen to be both exciting and relevant.

The growth of the FBP paralleled the development of MEDECOS, the 'Club Med' of mediterranean ecology. Fortuitously, the third MEDECOS meeting was held in the Cape in 1980 (Kruger et al. 1983), before the academic boycott made visits by most foreign academics difficult. Thus at an early stage in the FBP, strong personal and professional links were established with colleagues such as Cody, Miller, Mooney, Oechel and Westman in California, Fuentes and Montenegro in Chile, Gill, Groves, Lamont, and Specht in Australia, and Orshan in the Mediterranean Basin. Over the years, such interaction has continued through study visits, post-doctorals, joint publications, and correspondence.

The transfer of ecological knowledge into management practice was remarkably successful in the fynbos compared with experience in other South African biomes. More than elsewhere, land managers in the fynbos have been remarkably receptive to new ideas and proposals relating to the control of alien invasive plants, to fire management, and to the conservation of endangered species. Representatives from all the major land use agencies — in agriculture, forestry, and conservation — regularly and actively participated in workshops, field studies, and synthesis exercises. That the recommendations from many of these studies have not been fully implemented has usually been due to logistic constraints rather than to conflicting opinion.

Workshops soon became one of the primary activities of the FBP, providing forums for the ongoing exchange and debate of ideas and information throughout the project. Annual research meetings drew up to 250 participants, although the core group of the FBP was in the order of 100 active researchers and managers. Most of the 45 workshops, symposia, and seminars convened by the FBP were limited from 10 to 30 participants, usually focusing on a topic of common interest which had served as the theme of a study group. Over the lifetime of the FBP, 15 study groups were formed, undertook co-operative projects, debated their findings, and published syntheses in a series of project reports (Table 1.2).

The study group activities reflected the open and opportunistic philosophy that characterized the administration of the project. Good research was supported whenever and wherever a 'critical mass' of interest and energy was available. But the programme was not lacking in structure. The dominant characteristics of fynbos — an extremely diverse mix of species-rich, fire-prone communities on nutrient-poor soils — provide a fairly predictable research agenda. Thus vegetation typology (Campbell et al. 1981; Moll et al. 1984), the role of nutrients (Day 1983), and of fire (Kruger and Bigalke 1984) enjoyed emphasis from the start. The conservation and management of fynbos were also given prominence, both in terms of identifying threatened species (Hall and Veldhuis 1985) and habitats (Jarman 1986) and in relation to the impacts of alien invasive woody plants (Macdonald and Jarman 1984). A particularly important contribution to the success of the project was that of the palaeoecology and soils group, which provided a strong historic perspective to the interpretation of the patterns, structure, and functioning of fynbos ecosystems (Deacon et al. 1983; Deacon et al. this volume).

A dedicated secretariat, that not only convened the meetings but also saw manuscripts through the press, removed an enormous workload from the research group and thus contributed significantly to the extremely high rate of publications emerging from the project. Two 'in-house' publications — the *South African National Scientific Programmes Reports* and the *Occasional Reports* of Ecosystems Programmes — further facilitated rapid communication between workers in the FBP.

Participation in international programmes, such as those of SCOPE (Scientific Committee on Problems of the Environment), proved an invaluable stimulus to research in the fynbos. The 'Ecological Effects of Fire' (Booysen and Tainton 1984) and the 'Ecology of Biological Invasions' (Macdonald et al. 1987) projects resulted in dozens of publications and two synthesis volumes. Strong links with IUCN (International Union for the Conservation of Nature and Natural Resources) also promoted the review of South African experience in habitat conservation (Siegfried and Davies 1982) and in the patterns and problems of biotic diversity (Huntley 1989) — themes for two major syntheses drawing on the results of the FBP.

Not the least important factor contributing to the success of the project was the tremendous love of and identity with fynbos displayed by the people of this spectacularly

TABLE 1.2 Study group activities in the Fynbos Biome Project, 1978–1988.

Study group	Convener	1978	1979	1980	1981	1982	1983	1984	1985	1986	1987	1988	Products
Soils, geomorphology & palaeoecology	H J Deacon		▬	▬	▬	▬	▬						SANSPR No. 75[1]
Biogeography	H J Deacon and E J Moll									▬	▬	▬	
Vegetation mapping & classification	M L Jarman			▬	▬	▬	▬	▬	▬				SANSPR Nos. 52, 57, and 83. Fynbos map
Community ecology	G J Breytenbach and W J Bond				▬	▬	▬	▬	▬	▬	▬	▬	Occasional Report No. 12
Fire ecology	B W van Wilgen	▬	▬	▬	▬	▬	▬	▬	▬				
Competition	F Pressinger						▬	▬					
Pollination biology	A G Rebelo						▬	▬	▬	▬	▬		SANSPR No. 137
Seed biology	S M Pierce								▬	▬	▬	▬	
Commercial wildflower resources	G W Davis						▬	▬	▬	▬	▬	▬	Occasional Report No. 43
Lowlands conservation	M L Jarman and C J Burgers				▬	▬	▬	▬	▬				SANSPR No. 87
Fynbos systems ecology	J Miller and F J Kruger					▬	▬	▬		▬	▬	▬	SANSPR No. 105
Hydrology & hydrobiology	J Bosch						▬	▬					
Swartboskloof	B W van Wilgen							▬	▬	▬	▬	▬	SANSPR No. 104, Occasional Report No. 7
Nutrient cycling	D T Mitchell, A B Low and W D Stock						▬	▬	▬	▬	▬	▬	Occasional Report No. 18
Invasive biota	I A W Macdonald					▬	▬	▬					SANSPR Nos. 85 and 111. Occasional Report No. 19

[1] South African National Programmes Report.

beautiful biome. The dramatic mountain and coastal landscapes, the diversity and beauty of the flora, and the rich and mixed cultural heritage of the Cape result in an environment attractive to any active mind — let alone an ecologist. Yet the term 'fynbos' was seldom used, even by biologists, before the launch of the FBP. Today it is part of the vernacular — used in the press, radio, TV, and as a brand name in commodities.

The FBP was due to run from 1977 to 1986, but funding continued into 1989 to allow the completion of several studies. The project brought vigour and excitement to terrestrial ecology in South Africa. Much of the intellectual energy stimulated by the project remains — but the tradition will soon be lost if not provided with financial resources and leadership. The scientific results, reflected in this volume, indicate some of what was learnt. But much remains to be done — in-depth intercontinental comparisons between fynbos, kwongan, chaparral, etc.; the initiation of experiments to examine possible impacts of global change phenomena; the study of the below-ground component of fynbos ecosystems; the detailed study of interactions at the biome's boundary with karoo, forest, and grassland biomes; and many more topics.

Simultaneous to these ecological studies, the role of fynbos in a changing socio-economic environment must be assessed. Now, more than ever before, South Africa needs scientists and managers capable of addressing the environmental challenges of a post-apartheid society. The generation of young ecologists produced by the FBP has much to offer. But on them rests a responsibility to stimulate a new wave of research and action directed more sharply towards the needs of the 1990s.

REFERENCES

BOND P and GOLDBLATT P (1984). Plants of the Cape Flora. A descriptive catalogue. *Journal of South African Botany.* Supplementary Volume **13**, 1–455.

BOOYSEN P de V and TAINTON N M (eds) (1984). *Ecological effects of fire in South African ecosystems.* Springer, Berlin.

CAMPBELL B M, COWLING R M, BOND W J, KRUGER F J (1981). Structural characterization of vegetation in the fynbos biome. *South African National Scientific Programmes Report* **52**, CSIR, Pretoria 1–19.

DAY J A (1983). Mineral nutrients in mediterranean ecosystems. *South African National Scientific Programmes Report* **71**, CSIR, Pretoria 1–216.

DEACON H J, HENDEY Q B and LAMBRECHTS J J (1983). Fynbos palaeoecology: a preliminary synthesis. *South African National Scientific Programmes Report* **75**, CSIR, Pretoria 1–216.

HALL A V and VELDHUIS H A (1985). South African Red Book Data Book: Plants — fynbos and karoo biomes. *South African National Scientific Programmes Report* **117**, CSIR, Pretoria 1–144.

HUNTLEY B J (1987). Ten years of cooperative ecological research in South Africa. *South African Journal of Science* **83** 72–9.

HUNTLEY B J (ed) (1989). *Biotic diversity in Southern Africa: concepts and conservation* Oxford University Press, Cape Town.

JARMAN M L (1986). Conservation priorities in lowland fynbos. *South African National Scientific Programmes Report* **87**, CSIR, Pretoria 1–25.

KRUGER F J (1978). A description of the Fynbos Biome Project. *South African National Scientific Programmes Report* **28**, CSIR, Pretoria 1–55.

KRUGER F J and BIGALKE R C (1984). Fire in fynbos In *Ecological effects of fire in South African Ecosystems.* Springer, Berlin.

MACDONALD I A W and JARMAN N L (1984). Invasive alien organisms in the terrestrial ecosystems of the fynbos biome, South Africa. *South African National Scientific Programmes Report* **85**. CSIR, Pretoria 1–72.

MACDONALD I A W, KRUGER F J and FERRAR A A (eds) (1987). *The ecology and management of biological invasions in Southern Africa.* Oxford University Press, Cape Town.

MOLL E J, CAMPBELL B M, COWLING R M, BOSSI L, JARMAN M L and BOUCHER C (1984). A description of major vegetation categories in and adjacent to the fynbos biome. *South African National Scientific Programmes Report* **83**. CSIR, Pretoria 1–29.

SIEGFRIED W R and DAVIES B R (1982). Conservation of ecosystems: theory and practice. *South African National Scientific Programmes Report* **61**, CSIR, Pretoria 1–97.

YOUNES T (1991). Seventy years of IUBS: Assets, Constraints and Potential for International Cooperation. *Biology International* **22**, 2–11.

2

Selective regime and time

H J Deacon, M R Jury and F Ellis

INTRODUCTION

Ecosystems change by the biological processes of evolution, dispersal, and extinction. These processes operate within an environmental setting. The lithology or composition of the rocky substratum, the physiography or landforms, the climate, and the surface cover of soils are environmental factors that constrain changes in ecosystems. This is because they affect the fitness of organisms that make up the component communities to survive and reproduce. It is in this sense that the environment serves as the selective regime.

On geological time-scales the environmental setting, too, is subject to change. At a persistent if slow rate, erosion reduces relief and exposes different rock types, and conversely deposition produces new mantling sediments. Soils, the product of weathering, form in successive generations on materials that have been subjected to prior weathering. Climates are highly variable in the long and short term. The component parts of the selective regime have operated in a temporal framework and time itself becomes an important variable. The history of ecosystems is dynamic because of the changing relationship between organisms and their environment (Deacon 1988). It is all too evident that in any model of ecosystem change in modern times, the effects of human activities — the anthropogenic factor — have increased to the extent of becoming of overriding importance in selection.

In discussions of the evolution of the special characteristics of the fynbos, notably the range of growth forms, the numerous species in certain families and the patterns of diversity, the role of nutrient-poor sandy substrates, summer dry climates, and to a lesser extent topography have been emphasized (Kruger and Taylor 1979; Cowling and Holmes this volume; Cowling et al. this volume; Le Maitre and Midgley this volume; Stock et al. this volume). It is evident, however, that such environmental constraints have functioned in different ways at different times. The landscape and the substrates, for example, have existed in near modern form for many millions of years while summer dry climates are geologically much more recent. The flora and fauna include elements that have been described as bizarre relicts (Livingstone 1975) implying they are archaic taxa, and components of the biota are again older than the inception of summer dry climates. Fynbos ecosystems are the product of ongoing biological evolution under a selective regime and their present state is a point on a continuum of change. The purpose of this chapter is to provide a time-scale for the evolution of fynbos ecosystems and to discuss ways in which environmental or abiotic factors have been relevant in the course of their evolution.

ROCK SEQUENCES: STRATIGRAPHY, LITHOLOGY, AND TIME

There are rocks of different ages and types represented in the fynbos landscape and a fuller description of the geology of the region is given in Theron (1983) (see also Tankard et al. 1982; Deacon 1983a; Dingle et al. 1983). The oldest are sediments of the Malmesbury, Kango, Kaaimans, and Gamtoos Groups (Figures 2.1, 2.2). These are parts of the infilling of a basin-like area or geosyncline that developed during the final stages of the welding together of the Namaqualand section of the earth's crust almost 1 000 Myr ago. The sediments, ranging in lithology from clays to conglomerates, were folded and sheared in a Late Precambrian episode of mountain building (Saldanian orogeny) and intruded by granites (Cape Granite Suite). Subsequent erosion

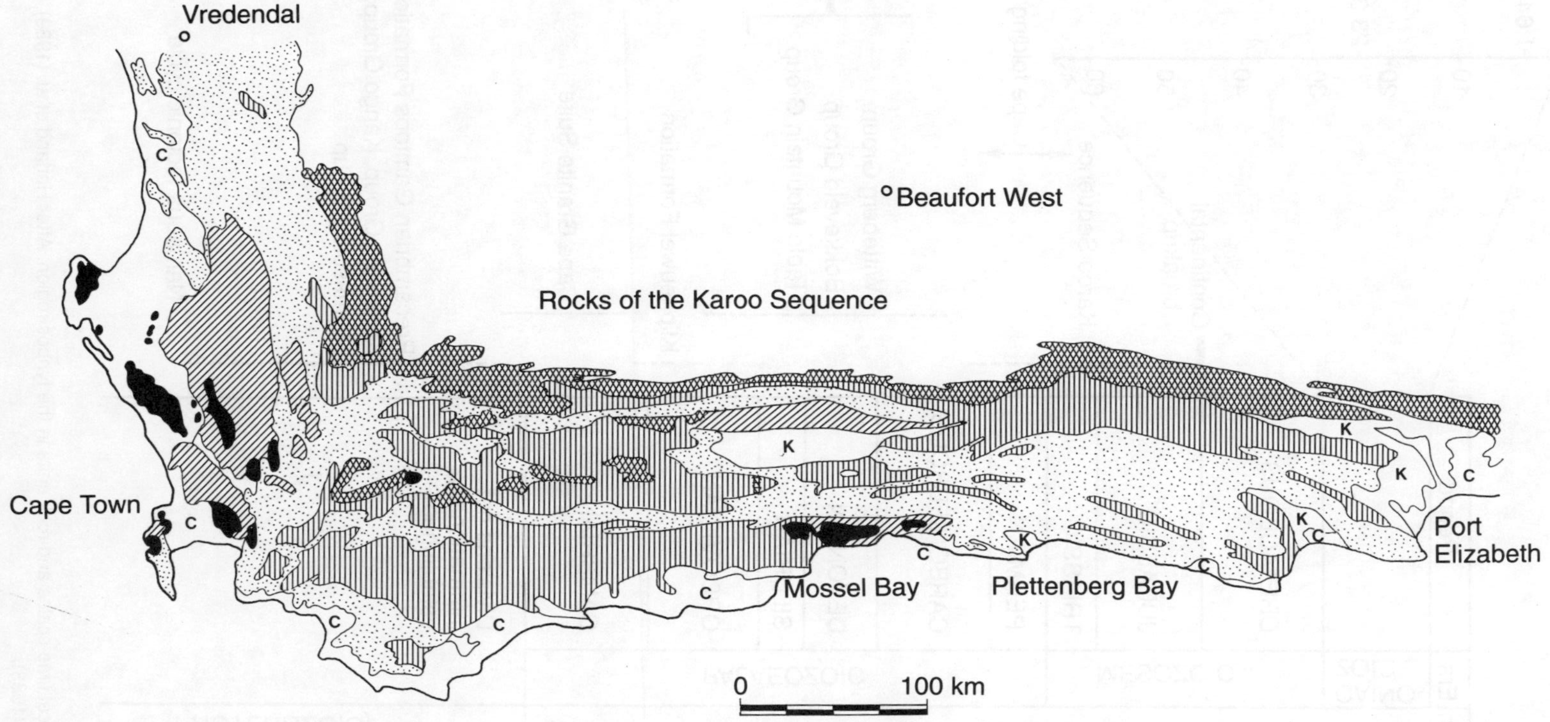

FIGURE 2.1 A geological map of the fynbos biome (based on Theron 1983) which shows the distribution of the main lithostratigraphic units. Rocks of the Granite Suite (solid colour), Malmesbury Group, and to a lesser extent the Cango Group to the east (diagonal hatching) and the Bokkeveld Group (vertical hatching), tend to be associated with relatively base-rich substrates. Rocks, particularly of the Table Mountain Group (stippled) and the Witteberg Group (cross hatching), tend to be associated with relatively base-poor substrates. The Cainozoic (C) and Cretaceous (K) Groups are also shown.

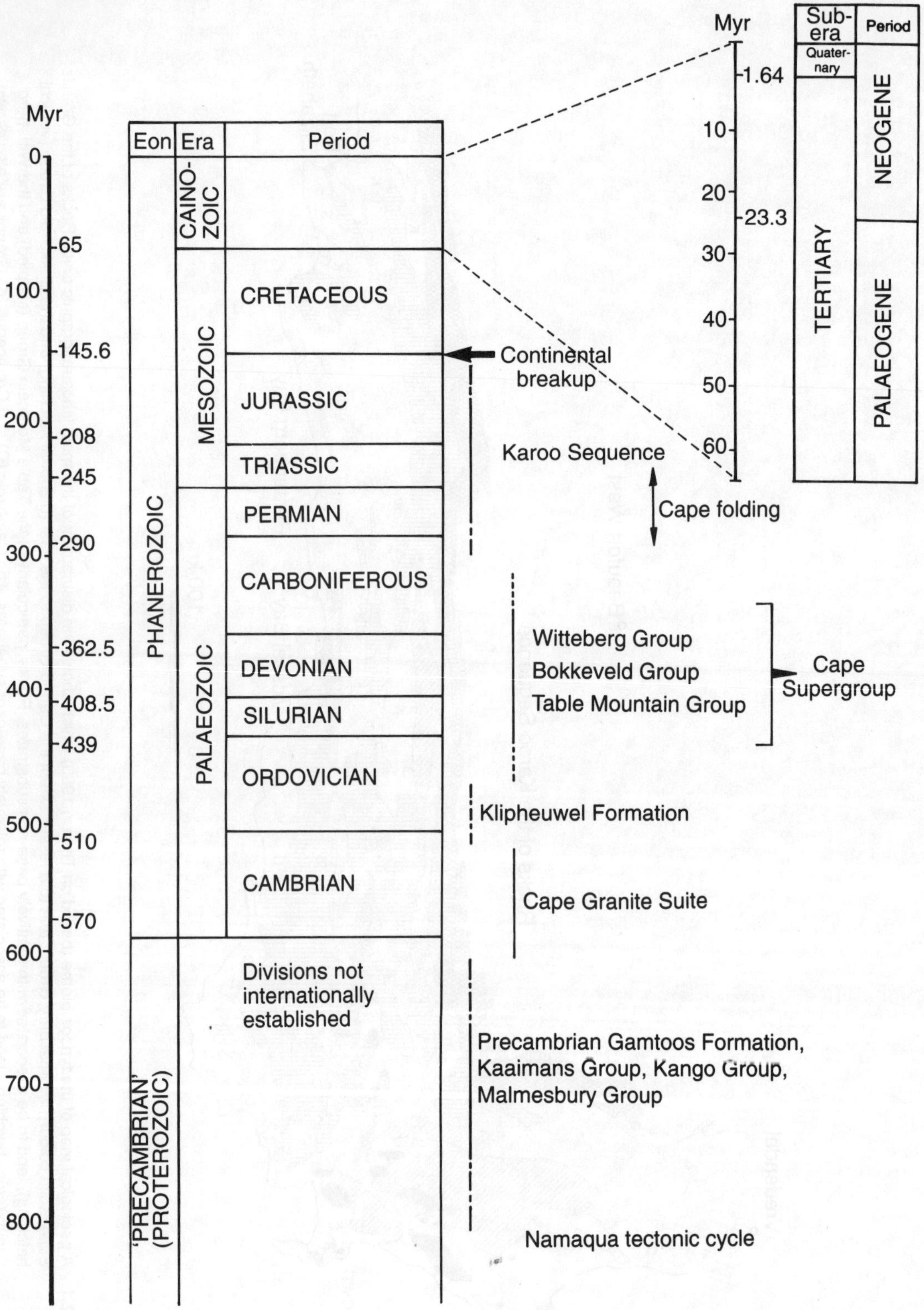

FIGURE 2.2 Geological time-scales and rock units in the fynbos region. After Harland et al. (1989), SACS (1980), and Theron (1983).

reduced these older Cape mountains to a level plain. Malmesbury Group rocks are exposed over an extended area in the south-western Cape but exposures of other groups associated with fault lines are more restricted. Together with the granites, the Precambrian rocks are important because they form inliers of substrates that are base rich. On weathering these predominantly clayey substrates release exchangeable cations like calcium, potassium, magnesium, and sodium that are important in soil formation and plant nutrient cycling (Stock and Allsopp this volume).

On the surface, planed over these older rocks, a new cycle of sedimentation was initiated by renewed subsidence and the accumulation of the great thickness of detrital materials that go to make up the rocks of the Cape Supergroup. Three groups are recognized. From the oldest to the youngest, they are the Table Mountain Group, the Bokkeveld Group, and the Witteberg Group. Within each group a number of formations — that is strata with distinctive lithologies — have been defined. These sediments that accumulated in what is known as the Cape basin are the near-shore or terrestrial, coarser, sandy (arenaceous) facies of the Table Mountain and Witteberg Groups, and the off-shore mudstone (argillaceous) facies of the Bokkeveld Group. It is the pile of medium to coarse (almost pure quartz) sands in this sequence (the Peninsula Formation of the Table Mountain Group alone has a thickness of more than 3 000 metres) that is typically associated with base-poor substrates and acid ground waters common in the region. The contrast is with the rocks of the Bokkeveld Group which like those of the Precambrian, form substrates of higher base status. The Cape Supergroup rocks are fossiliferous and include trace fossils such as trails and tracks and the remains of invertebrates and fish. These make it possible to place the accumulation of the Cape Supergroup within the range of the Late Ordovician to Carboniferous, or between 450–300 Myr ago.

The accumulation of sediments in the Cape basin ceased with uplift and the migration of the main centre of deposition northwards to the broad epicontinental Karoo basin. The initial sedimentation of the Karoo sequence — the Dwyka Formation — represents a continent-wide glaciation, and these sediments were folded together with those of the Cape basin during an episode of mountain building known as the Cape orogeny in the Permian and Triassic (278–215 Myr). As the Cape Fold Mountains were elevated, they became a southerly source of the supply of sediments to the Karoo basin. The Karoo deposition terminated in the Jurassic (208–145 Myr) and increased volcanic activity at that time marked the beginning of the break-up of Gondwanaland.

The coast was outlined as the African and South American plates moved apart. The separation was complete by 100 Myr when the Falkland Plateau drifted past the Cape, opening up the South Atlantic Ocean (Figure 2.3). Fluviatile and estuarine Cretaceous sediments are preserved on-shore in fault-controlled basins and are a record of the early stage of continental break-up. However, with the establishment of new base levels for erosion, the main depo-centres moved off-shore. The Early Cretaceous formations include fossils of invertebrates, small dinosaurs, plants, leaves, wood, and sporopollens, but no evidence of flowering plants — the angiosperms. The dispersal of the angiosperms took place in the mid-Cretaceous at a more advanced stage of continental separation and early angiosperm microfossils are recorded from this later age, some 100 Myr ago, in deep sea sediments accumulated in the Cape-Argentine basin — the proto-South Atlantic (McLachlan and Pieterse 1978).

There are colluvial screes, alluvial gravels, and pedocretes (indurated fossil soils), as well as veneers of coastal deposits that are broadly Cainozoic (0–65 Myr) in age (Figure 2.4). None of these has any great thickness but their importance is that they mantle older materials of different lithology. In areal distribution the most important are the shallow water limestones and coastal sands of the Mio-Pliocene (10–2 Myr), Bredasdorp, and Alexandria Formations (Le Roux 1989). These are a record of a number of transgressions and regressions of the sea in embayments along the coast like Saldanha, Agulhas, Still Bay, and Algoa, related to changes in the volumes of water locked up in the polar ice-caps. Large bodies of sands deposited on the outer margin of the coastal platform by these former high sea levels have been reworked in subsequent times and form

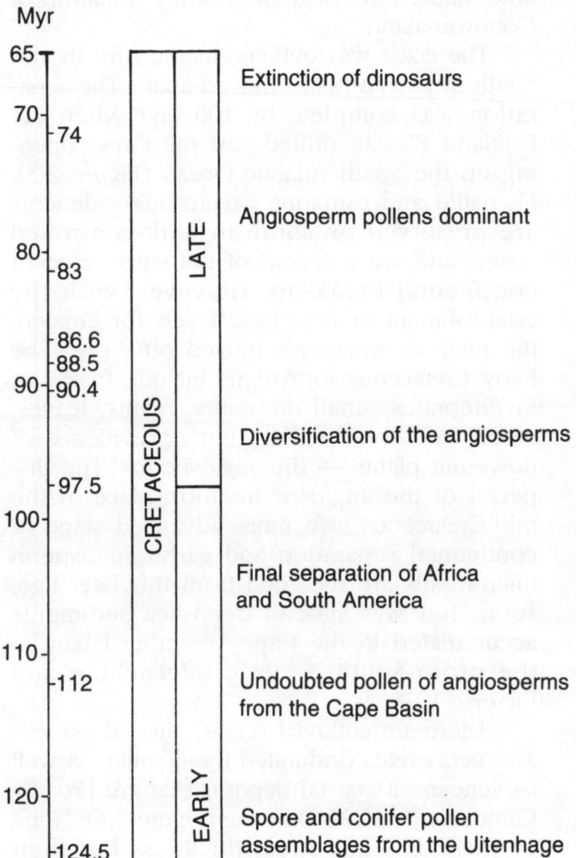

FIGURE 2.3 The chronostratigraphy of the Cretaceous Period and the palaeontological record (after Deacon 1983a).

extensive coastal dunefields. In the Pleistocene, high sea levels of the warmer interglacials did not rise more than a few metres above the present sea level on these stable coasts (Van Andel 1989). More significant have been the periods of Pleistocene low sea levels when the coastal forelands were more extensive (up to a 100 kilometres wider over the Agulhas bank) thus creating alternative routes for the dispersal and migration of plants and animals.

The geological history has determined the lithology, and the lithology in turn has determined the distribution of substrates of high and low base status. Precambrian and Bokkeveld rocks, together with argillaceous Cretaceous and calcareous coastal sands, form areas of substrates with higher base status, whereas the quartzose members of the Cape Supergroup and the older leached coastal sands contribute to the extent of substrates with an extremely low base status. Selection for substrate specialists has been important in the evolution of fynbos vegetation (Cowling et al. this volume).

PHYSIOGRAPHY

Mountain building in the Cape orogeny formed megafolds and monoclines striking in a west-north-west direction in the western zone and northward verging folds, with an east-west strike in the eastern zone (Söhnge 1983). The syntaxis — where the two trends of folding meet — is in the Caledon region of the south-western fynbos biome and this is the area of highest terrain diversity. Erosion has reduced the mountains to little more than high hills. Resistant sandstone lithologies and erodable fault lines, together with the linear structure imposed by folding, have tended to block out sections of the mountains along the strike. The mountains thus form strings of elevated island-like masses of quartzose substrates.

The physiography of the Cape Fold Mountains makes them important as a setting for the fynbos. This is most obvious in the correspondence between the syntaxis of the folding with its broken terrain and the main centre of species richness in flora. The linear sections of mountain have promoted diversity in the plant associations on a landscape rather than a local community scale. Regional richness is more significant than alpha diversity (Cowling et al.

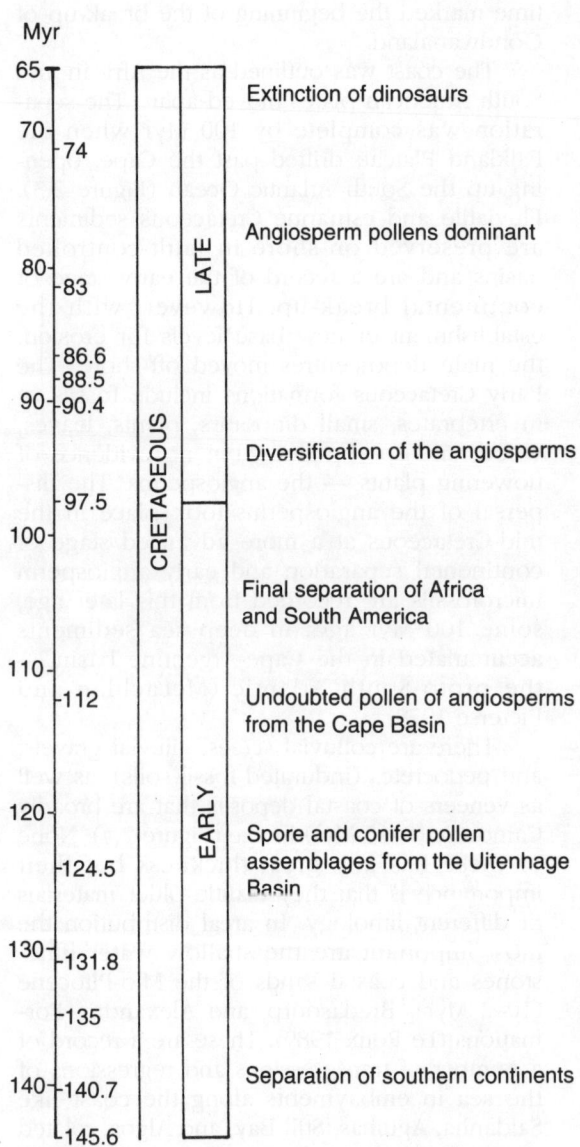

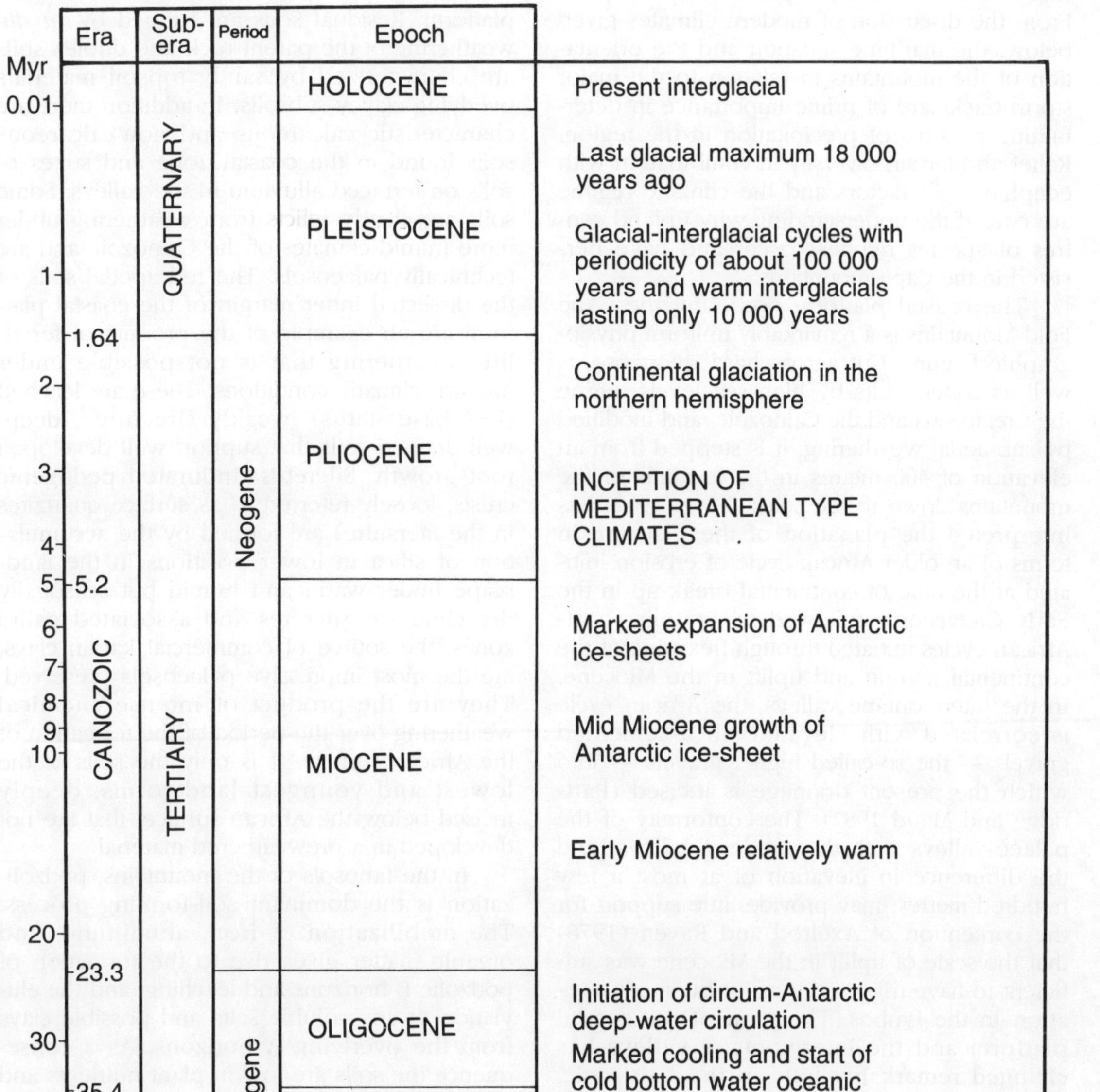
Era
Sub-era
Period
Epoch
Myr
0.01
1
1.64
2
3
4
5
5.2
6
7
8
9
10
20
23.3
30
35.4
40
50
56.5
60
65
CAINOZOIC
QUATERNARY
TERTIARY
Neogene
Palaeogene
HOLOCENE
PLEISTOCENE
PLIOCENE
MIOCENE
OLIGOCENE
EOCENE
PALAEOCENE
Present interglacial
Last glacial maximum 18 000 years ago
Glacial-interglacial cycles with periodicity of about 100 000 years and warm interglacials lasting only 10 000 years
Continental glaciation in the northern hemisphere
INCEPTION OF MEDITERRANEAN TYPE CLIMATES
Marked expansion of Antarctic ice-sheets
Mid Miocene growth of Antarctic ice-sheet
Early Miocene relatively warm
Initiation of circum-Antarctic deep-water circulation
Marked cooling and start of cold bottom water oceanic circulation
Warm humid climates, ice-free globe
Separation of Australia and Antarctica

FIGURE 2.4 The chronostratigraphy of the Cainozoic Era and general palaeoclimatic trends (after Deacon 1983a).

this volume). The moderate height of the mountains has meant that even under the reduced temperatures of the Pleistocene, extreme cold was not a limiting factor. However, elevation has been important in offsetting the effects of arid periods in the past. From the discussion of modern climates given below, the maritime situation and the orientation of the mountains in relation to the major storm tracks are of prime importance in determining patterns of precipitation in the region. Relief and terrain diversity in combination with edaphic (soil) factors and the climatic regime are part of the understanding why upland centres of species richness occur and have persisted in the Cape mountains.

The coastal platform bordering the Cape Fold Mountains is a remarkably uniform physiographic feature. Cut across hard quartzose as well as softer rocks by high sea levels during the Cretaceous and the Cainozoic, and modified by sub-aerial weathering, it is stepped from an elevation of 400 metres in the foothills of the mountains down to the coast. King (1962) has interpreted the planation of the platform in terms of an older African cycle of erosion initiated at the time of continental break-up in the Early Cretaceous, followed by younger post-African cycles initiated through flexuring of the continental margin and uplift in the Miocene. In the intermontane valleys, the African cycle is correlated with elevated silica-cemented gravels — the so-called higher gravels — into which the present drainage is incised (Partridge and Maud 1987). The conformity of the palaeo-valleys with the modern valleys, and the difference in elevation of at most a few hundred metres, may provide little support for the contention of Axelrod and Raven (1978) that the scale of uplift in the Miocene was sufficient to have triggered a major burst of speciation in the fynbos. The form of the coastal platform and the intermontane valleys has changed remarkably little in the Cainozoic, underscoring the antiquity of the landscape.

SOILS AND PEDOGENESIS

Soil units are defined on the basis of morphological characteristics including the horizons present, the occurrence of stone lines, and the textures which reflect the sizes of the aggregates of mineral grains in the soils. The units that can be recognized and mapped show a direct relationship to parent materials and terrain (Schloms et al. 1983; Figure 2.5). Lithosols — which are shallow, weakly developed soils — are associated with the mountains. There are residual and duplex soils associated with the more clayey rock substrates of the coastal platform. Residual soils are formed by *in situ* weathering of the parent rock and duplex soils are characterized by sandy topsoil materials overlying clayey subsoils. In addition there are characteristic calcareous and non-calcareous soils found in the coastal zone and suites of soils on terraced alluvium in the valleys. Some soils are clearly relicts from weathering under more humid climates of the Cainozoic and are technically palaeosols. The red apedal soils of the dissected inner margin of the coastal platform are an example of the product of ferrallitic weathering that is not possible under modern climatic conditions. These are leached (low base status), weakly structured, deep, well drained soils that support well developed root growth. Silcretes (indurated pedogenic crusts, loosely referred to as surface quartzites in the literature) are formed by the accumulation of silica in lower positions in the landscape under warm and humid but seasonally dry climates. Silcretes and associated pallid zones, the source of commercial kaolin clays, are the most impressive palaeosols preserved. They are the product of intense chemical weathering over the period of the formation of the African surface. It is only the soils of the lowest and youngest land forms, deeply incised below the African surface, that are not developed in a preweathered material.

In the lithosols of the mountains, podzolization is the dominant soil-forming process. The mobilization of iron, aluminium, and organic matter gives rise to the formation of podzolic B horizons and leaching, and the eluviation of any soluble salts and possible clays from the overlying A horizons. As a consequence the soils are low in plant nutrients and in particular they are deficient in phosphates and bases (Stock and Allsopp this volume). Joints and fissure systems feed localized wet sites where seepage peats and hydromorphic soils develop and lend to the range of sites for colonization by plants. Highly variable and localized conditions of soil depth, soil moisture, and aspect have selected for microhabitat specialists. The contrast is with the

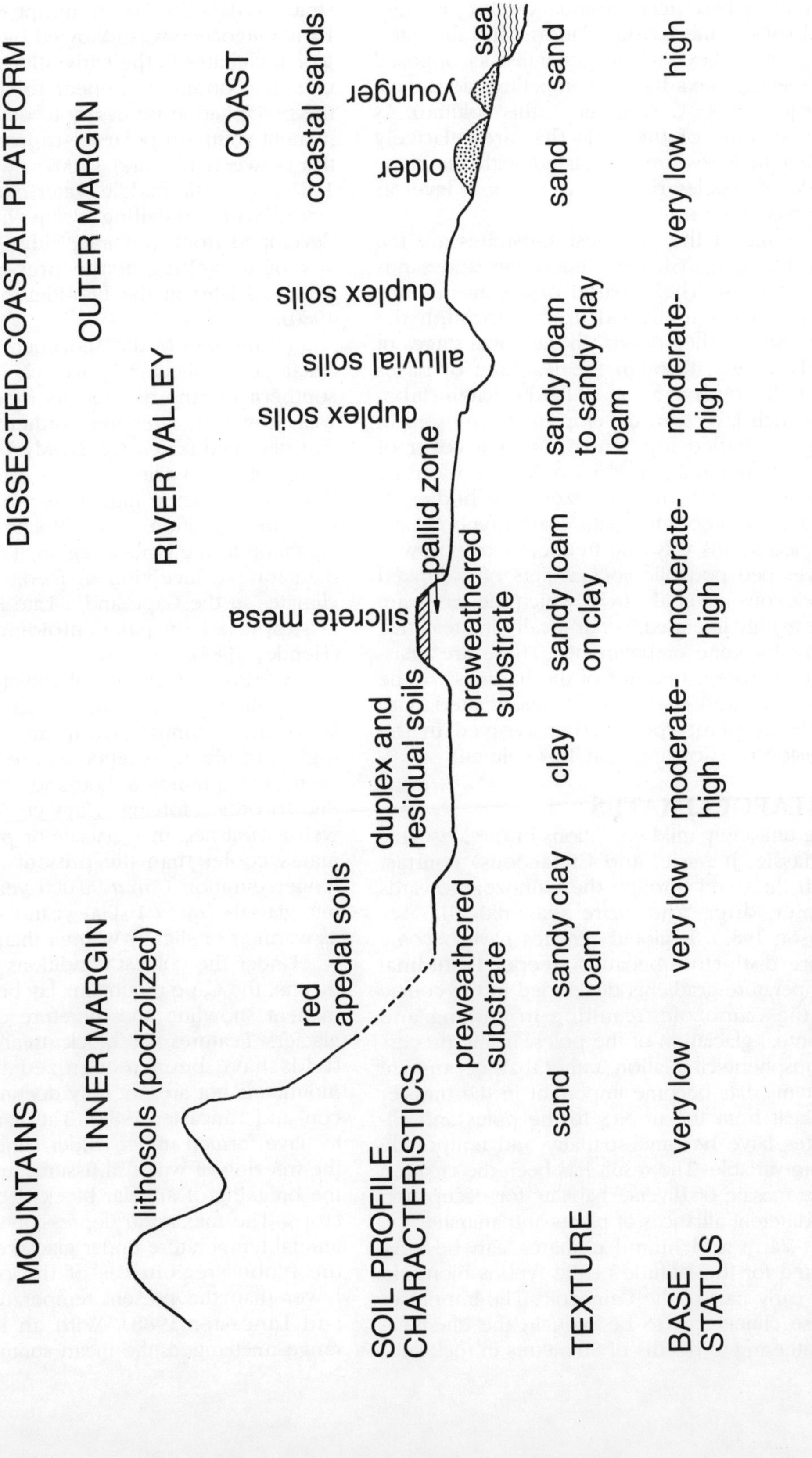

FIGURE 2.5 Topography and soils in the fynbos biome.

nominally base richer residual duplex and alluvial soils of the coastal platform and the intermontane valleys where generalist as opposed to specialist taxa have a competitive advantage (Deacon 1990; Cowling et al. this volume). As a rule none of the soils that are relatively richer in bases are associated with the same scale of species richness at regional level as the base-poor soils.

Some of the youngest substrates are the Mio-Pliocene to Recent limestones and sands of the coast. Their coastal origin means they were all originally calcareous through the inclusion of shell debris. In the initial stages of pedogenesis, through the leaching of bases from the top soil, a soft or hard calcium carbonate rich layer can develop in the subsoil. In the decalcified top sands, under a cover of vegetation and a pH of 5.5 or less, iron can be mobilized to form a podzol. The bodies of acid sands associated with coastal fynbos have formed in this way and frequently exhibit well developed podzolic soils. Inliers of indurated calcareous materials, from which the bases are not readily liberated, occur locally in areas that were Pliocene embayments. These are resistant to erosion because of the hardness of the materials and they may be associated with endemic plant species that evolved in the Pleistocene (Cowling et al. this volume).

PALAEOCLIMATES

The uniformly mild conditions of the Mesozoic (Triassic, Jurassic, and Cretaceous) contrast with the trend through the Cainozoic towards cooler, drier, and more seasonal climates (Tyson 1986). Regional climates have become more distinctive because steeper latitudinal temperature gradients developed in the course of the Cainozoic, resulting in cooling and eventual glaciation of the poles. In the process atmospheric circulation, rather than ocean gyre systems, has become important in the transfer of heat from the tropics to the poles and climates have become spatially and temporally more variable. The result has been the creation of a mosaic of diverse habitats for occupation by different alliances of plants and animals.

Warm and humid climates can be suggested for the latitude of the fynbos biome in the early part of the Cainozoic. The imprint of these climates is to be seen in the chemical weathering to depths of 50 metres in rock substrates (Glass 1977). An abrupt cooling in the Early Oligocene was followed by a recovery of global climates in the Early Miocene. The latter climatic conditions appear to have supplied the precipitation necessary to develop the permanent southern polar ice-cap on East Antarctica between 16.5 and 13 Myr (Woodruff et al. 1981). Antarctic middle water circulation associated with upwelling along the west coast developed from this time, although the intensity of upwelling at the present scale was achieved later in the Plio-Pleistocene (Siesser 1980).

By the end of the Miocene, with the West Antarctic ice-sheet in place, circulation in the southern oceans reached its modern form. It was, however, after the northern hemisphere was glaciated (*circa* 3.2–2.5 Myr) that the symmetry of zonal climates was established and the South Atlantic high pressure cell became fixed in a position which blocks summer precipitation to the fynbos region. This provides a date for the inception of mediterranean-type climates in the Cape and a Late Pliocene date is supported by palaeontological evidence (Hendey 1983).

A feature of the global climates of the last two million years — the Pleistocene — has been the rhythmic growth and retreat of the high latitude ice-sheets. These phenomena occur with a hundred thousand year periodicity due to orbital forcing (Hays et al. 1976). It is well established that glacials or periods of climates cooler than the present are of much longer duration (*circa* 90 000 years) than the interglacials (*circa* 10 000 years) with climates as warm as or slightly warmer than the present.

Under the coldest conditions of the Pleistocene, the Cape mountains lay below the permanent snowline and therefore did not carry glaciers. Features like block streams and block fields have been recognized in the Cape mountains but are not fully documented (Deacon and Lancaster 1988). They are presumed to have formed under colder climates through the freezing of water in fissures and joints and the breaking of angular blocks from rock outcrops. The maximum depression of the mean annual temperature under glacial conditions in the fynbos region was of the order of 5°C lower than the present temperature (Deacon and Lancaster 1988). With an interannular range unchanged, the mean summer temperatures at a glacial maximum were close to the present mean winter temperatures and were correspondingly lower.

Other parameters of climate have departed from the modern norm during the Pleistocene although they are less well documented in quantitative terms. Precipitation has been above and below that in present times and the area of winter rainfall at times has been more extensive. Pleistocene glacial climates were drier in general because evaporation from cooler oceans was lower. The fynbos region was not subjected to a Marion Island-type moist climate during Pleistocene glacials because the position of the westerlies did not move equatorwards more than a degree of latitude, south of southern Africa (Deacon 1983b). The CLIMAP study (CLIMAP Members 1976) of global conditions 18 000 years ago was able to map the position of the intertropical convergence zone in the southern ocean from deep-sea cores, and this ocean boundary which corresponds to the westerlies shows little displacement. Similarly the latitudinal position of the South Atlantic high would not have been markedly different from that of the present, although this high pressure cell may have been larger. The frontal systems that have their origin in the belt of the westerlies and bring winter precipitation to the region would have dominated the climates of the fynbos biome in the Pleistocene as in the present.

The influence of the season of precipitation on the distribution of vegetation is particularly apparent in the grassy fynbos that occurs in the eastern part of the biome (Cowling and Holmes this volume). Under conditions of biseasonal rainfall, subtropical pioneers — including grasses like *Themeda triandra* — have been able to invade the fynbos. Palaeontology and stable isotope analyses provide proxy evidence for the inception of biseasonal climates in the latter half of the Holocene — the last 5 000 years (Deacon and Lancaster 1988). Prior to this and for much of the Pleistocene, summer dry climates predominated over the whole region.

A factor contributing to the maintenance of bizarre relicts in the biota as well as neoendemic species has been the alleviation of Ice Age aridity by the reliability of winter rainfall systems. At the same time the persistence of summer dry conditions in the Pleistocene has limited recruitment of plant taxa from the summer rainfall margin (Linder et al. this volume). This has helped to maintain the endemic character of the biota. On a geological time-scale, the mediterranean-type climate in the Cape is recent in that it has been the dominant climate of the last few million years. It is under this regime that the alliances of plant taxa that characterize fynbos ecosystems developed. At generic and higher taxonomic levels, the flora has become impoverished relative to that of earlier geological times through extinction (Coetzee et al. 1983; Linder et al. this volume). However, impoverishment has not been as marked as that in the inland upland centres of species richness recognized by Nordenstam (1969). The climates since the Miocene, and in particular those of the Pleistocene, have been a major factor in the evolution of what is less a distinctive flora than a geologically recent specialized vegetation (Axelrod and Raven 1978; Deacon 1990; cf. Linder et al. this volume).

FIRE

There has been considerable interest in the role of fire in fynbos ecology. Fire history has been studied on interannular and decadal scales in field records and reports. Studies of fire history on longer time-scales based on measuring the flux of charcoal particles in geologically or chronometrically dated sedimentary sequences are at an initial stage. With the decreased dominance of humid forest cover in the Miocene, fire would have played a more significant role in the selective regime. The oldest direct evidence for fire in the region is in the Pliocene (Hendey 1983). It is assumed that the frequency of fires may have been higher with the advent of summer dry climates in the Pleistocene. Although no quantitative measures can be offered, fire stick farming was practised in the fynbos at least since the Late Pleistocene and is thought to have resulted in a quantum increase in the incidence of fires over that of the natural fire regime. The significance is that selection for alliances of taxa that have resilience to disturbance by fire began in the Miocene and has had increased importance in the Pleistocene (Le Maitre and Midgley this volume). The latter time would include the appearance of people in the landscape.

Fire is not an independent variable

because the substrates and soils, through the availability of nutrients, control the production of fuel in the form of litter and the climate the seasonal incidence of fires (Kruger 1983; Van Wilgen 1985; Le Maitre and Midgley this volume). The result of selection has been to produce what is more than a fire-prone vegetation; it is a 'pyrophylic' (fire loving) vegetation dominated by plants with life strategies tuned to the fire regime.

THE CURRENT CLIMATIC PATTERN

The scenario of modern climate in the fynbos biome has been reviewed by Fuggle (1981), Fuggle and Ashton (1979), and is best summarized by Campbell (1983). Other publications useful in describing the climate of the biome include Heydorn and Tinley (1980), Schulze (1984), Kamstra (1985), the Department of Agriculture (1986), Tyson (1986), and Preston-Whyte and Tyson (1988). Meteorological conditions within the fynbos biome are the product of a unique combination of geographic and climatic elements:

- The biome, located between 32–34°S and 18–24°E, is characterized by steep coastal mountains which align in a convex (exposed) fashion, and give rise to windward-leeward patterns such as föhn-like 'berg' winds.
- A ridge of high pressure extends from the South Atlantic anticyclone to the Cape coast; in summer the ridge lies near 37°S with attendant shallow and dry easterly winds, while in winter the ridge lies near 32°S and westerlies sweep the biome, providing orographic rainfall on west-facing slopes.
- The warm Agulhas Current is located to the south-east of the biome in contrast to the cold Benguela Current off the west coast, with implications for a moisture supply that is dependent on wind direction over the coastal plains.
- The persistent yet unstable westerly wind in the mid-latitudes gives rise to transient weather systems such as frontal depressions, ridging anticyclones, coastal lows, and cut-off lows which initiate sharp day-to-day variability in wind, moisture, and temperature.

Campbell (1983) draws attention to four climatic gradients within the biome: west-east, coast-interior, altitude, and north-south terrain aspect. The west-east gradient is significant in the following ways: solar radiation drops by 15% from west to east, seasonal and diurnal temperature ranges are greatest in the west, evaporation is similarly about 40% higher in the west particularly in summer, and a winter rainfall regime in the west contrasts with a bimodal regime (spring-autumn rainfall peaks) in the east. Such a gradient points to the windward-leeward pattern produced by south-east and easterly winds associated with the ridge of the South Atlantic anticyclone.

The coast-interior gradient is a climatic response to declining surface layer moisture with increasing distance from the warm Agulhas Current. Along the plains and slopes of the coastal platform there are lower annual temperatures, smaller seasonal and daily temperature ranges, greater cloudiness, higher rainfall, and lower evaporation than at interior stations with a more continental climate. Summer mists act to buffer the climate and minimize certain aspects of the day-to-day variability. Increased altitude causes decreased annual temperatures, decreased evaporation, and increased rainfall. Differences in terrain aspect allow north-facing slopes to receive considerably more solar radiation in winter and less orographic rainfall, particularly along the south coast.

To provide new insight to the mesoscale climate of the biome, Department of Agriculture data were taken from a vast and reliable network of stations and analysed for precipitation-evaporation balance. Seasonal trends were plotted for selected stations, many of which have been operating for up to 85 years. To indicate environmental stress, extremes of temperatures and wind were considered. Figure 2.6 shows the data grid and selected stations, together with the moisture balance for contrasting summer and winter months.

During January a tongue of extremely arid conditions sweeps southward from Namaqualand in the north-west to the Cape Flats in response to a leeward south-easterly wind descending from the range of mountains extending from the Hottentots Holland to the Cederberg. Moisture deficits over the Swartland-Piketberg district reach -400 mm month^{-1}. Over the windward coastal mountains near Cape Hangklip and Elgin, moisture deficits are reduced to -200 mm month^{-1} with orographic clouds limiting evaporation and enhancing

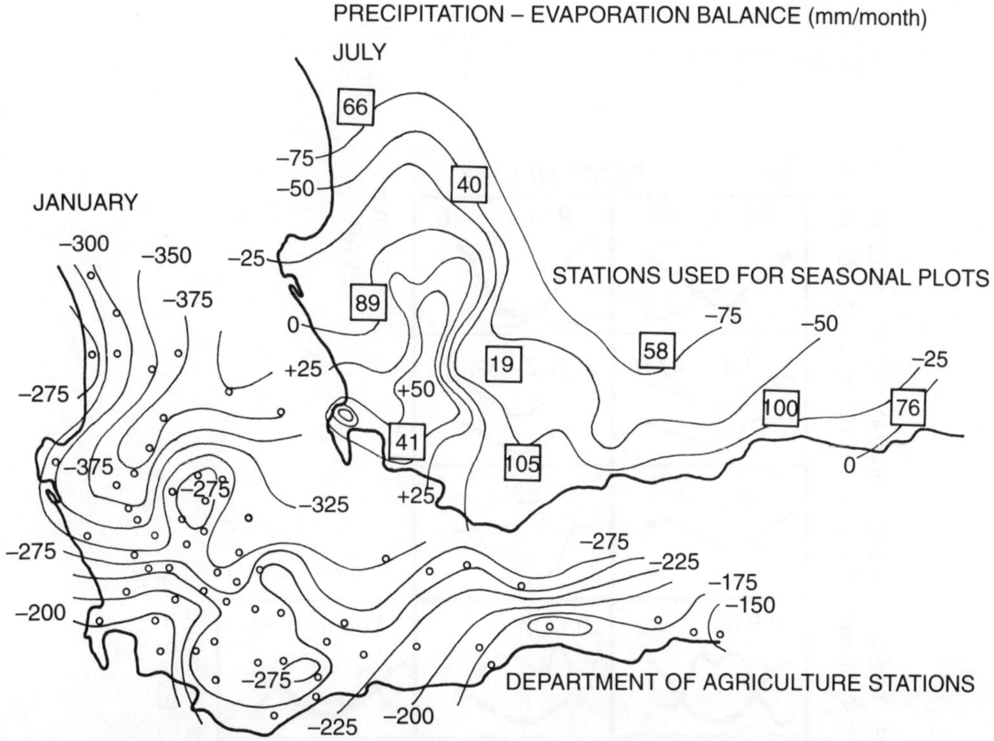

FIGURE 2.6 Precipitation/evaporation balance (mm month^{-1}) average for July (top) and January (bottom). Data used in the analyses are shown by dots (January); selected stations used for seasonal plots are shown by numbered squares (July).

rainfall. A sharp north-south gradient is obtained over the Cape Flats which reflects a similar north-south gradient in sea surface temperature near Cape Town. To the east of the Hottentots Holland Mountains moisture deficits remain in the -250 mm month^{-1} to -275 mm month^{-1} range. Over the south coast the moisture deficits decrease rapidly eastward to a minimum of -150 mm^{-1} where the warm Agulhas Current hugs the coast. A sharp north-south gradient is obtained from George to Oudtshoorn, indicating that winds bearing moisture from the east seldom rise over the coastal Outeniqua Mountains. Occasionally these easterly winds penetrate the Breede River Valley, where a 'moist' tongue is noted.

In July gradients of the moisture balance relax and become dominated by orographic rains over the west-facing Hottentots Holland-Cederberg mountain range. A sharp reduction in moisture is noted in the lee of these mountains in the vicinity of Worcester. A moisture surplus occurs south of 33°S and west of 20°E. Elsewhere a deficit remains during the mid-winter months. As mentioned previously, rains increase to the east of 20°E in the spring and autumn and would not be reflected in the July pattern.

Seasonal plots of monthly mean precipitation, daily evaporation, daily minimum and maximum temperatures, and daily wind run are shown in Figure 2.7, together with station number, elevation, and temperature and wind extremes. The layout of plots is in the geographic sense from north (top) to south (bottom) and west (left) to east (right). Comparing the different stations, the evaporation and temperature curves reflect the seasonal cycle of solar radiation. As much as 14 mm day^{-1} is evaporated from a class A pan at station 40 during the summer months. Precipitation there rarely exceeds 10 mm month^{-1}. Over the adjacent coastal plains (station 66) maximum temperatures have reached 47°C and wind speeds have peaked at 962 km day^{-1}. Such harsh environmental conditions in the southern rim of the Namib Desert may limit the northwards migration of fynbos elements. Winds are lowest in the mid-winter months and also follow the solar cycle. Looking further to the south

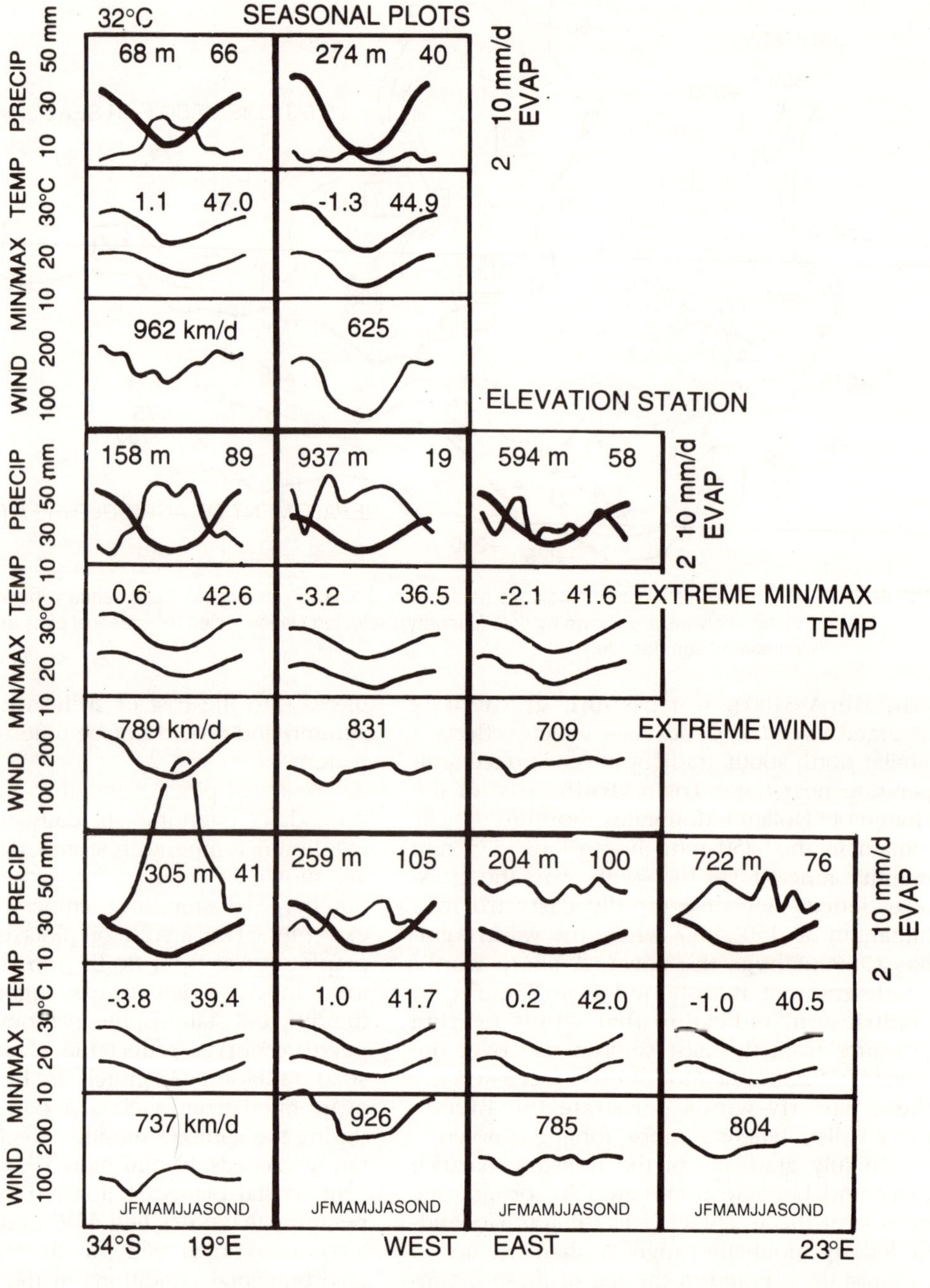

FIGURE 2.7 Seasonal plots for selected stations located in Figure 2.6. Graphs are presented as a column of three for each station for monthly mean precipitation and daily evaporation, with elevation and station number listed (top), daily minimum and maximum temperature with extremes listed (middle), and daily wind run with extremes listed (bottom).

and east, the obvious trend is for a more limited seasonal amplitude in all parameters. Station 19 at an elevation of 937 metres (the highest shown here) and station 41 at 305 metres near Elgin both show extreme minimum temperatures of -3.2 and -3.8°C respectively. Extreme wind conditions appear all along the coastal plains, with few stations reporting extreme wind runs of less than 700 km day^{-1}. The highest monthly averaged wind speed occurs at station 105 near Bredasdorp. Mean winds there exceed 300 km day^{-1} during the summer and possibly limit the distribution of forest except in the Elgin and George districts where a concave orientation of the coastal mountains reduces exposure to the wind. At mountain-top locations (stations 19, 41, and 100) winds are strongest during the winter, whereas the coastal plains — particularly in the west — exhibit a summer wind maximum.

Not shown in Figure 2.7 are the wind directions. However, the inspection of wind probabilities from other publications listed above indicates that winds tend to parallel the local orientation of the coast: being polarized in the SE-NW axis to the west of Cape Hangklip and in the E-W axis on the south coast. These strong coast-parallel winds are most evident at convex points such as Cape Point, Cape Agulhas, Cape Columbine, and Cape St Francis where mean annual speeds fall in the range 7 to 9 m s^{-1}. This is in excess of similar climatic regimes elsewhere in the world. At more concave coastal locations, such as Lamberts Bay and George, shallow sea breezes penetrate the low lying coastal plains and limit evaporation. Moving inland, winds often display a wider range of direction probabilities. Along the south coast some interesting changes in the direction probabilities have been noted. From Cape Agulhas to Plettenberg Bay easterly directions are most frequent during the summer, and these winds are driven by the ridge of the South Atlantic anticyclone. However Cape St Francis and Port Elizabeth summer wind directions are most often from the west to south-west. The Agulhas Current leaves the coast at Port Elizabeth and westerly storms in the mid-latitudes appear to track further northward there, essentially cutting off the influence of the South Atlantic ridge to the east of 25°E. Summer wind directions to the east of Port Elizabeth are north-easterly under the influence of the South Indian Ocean anticyclone. These winds carry air of a considerably more tropical nature.

An explanation for the bimodal nature of rainfall in the eastern portion of the fynbos region can be found in the higher incidence of ridging anticyclone and cut-off low pressure cells in the spring and autumn respectively. These transient weather systems rely on instabilities in the mid-latitude westerlies to provide southerly wind flow and consequent orographic rain against the south-facing mountains. Indeed, all the stations to the east of 20°E and south of 33°S (19, 105, 58, 100, and 76) exhibit bimodal rainfall distributions. The instabilities in the westerlies which produce southerly winds are most prominent in the spring and autumn when the westerlies shift across the latitudes. In mid-summer the westerlies have retreated poleward but are relatively stable, and similarly in mid-winter — the westerlies have advanced but are also stable. It should be pointed out that the weather systems associated with the westerlies — spring season ridging anticyclones, summer season coastal lows, autumn season cut-off lows, and winter season frontal depressions — all move rapidly eastward and blocking weather patterns persisting for months at a time are uncommon, indeed rare.

Interannual shifts in large-scale climatic patterns may have a significant impact on the moisture balance within the fynbos region, given the very close proximity of the South Atlantic Ridge. In some years, a slight encroachment of mid-latitude westerlies can displace the ridge such that summers are 'wet' as in 1982/3. Conversely the winter westerlies may fail to shift northwards, leaving the ridge over the fynbos region and reducing the large peak in winter rainfall at station 41, for example. Research is under way on feedback mechanisms between the atmosphere and ocean. Increased sea surface temperatures, which provide for an enriched supply of surface layer moisture, may lead to increased rainfall. However it is not clear whether the ocean leads or lags the atmosphere. With weather satellites and numerical models now providing a detailed description of marine weather systems, it is expected that great strides towards a better understanding of climatic processes will be made in the coming decade.

CONCLUDING DISCUSSION

It is fortuitous, an accident of geological history, that rock types with contrasting sandy and clayey lithologies occur on the southern margin of the African continent. It is these substrata, however, that in turn determine the base status of the soils and relief. It is also fortuitous that the south-western tip of the continent has become positioned in latitudes that are characterized by climates with a summer deficit in precipitation. It is a very restricted area of the south-western Cape that has a typical mediterranean-type climate at the present time, although the area was more extensive in the Pleistocene. Lithology, relief, soils, and climate together provide the unique setting for the fynbos.

Ecosystems are not special creations; they evolve through time. Although some of the families like the Proteaceae and the Restionaceae may have achieved a southern hemisphere or Gondwana-wide distribution in the Late Cretaceous (Linder et al. this volume) their prominence in present day fynbos communities does not imply a like antiquity for these communities. The alliances of taxa found in fynbos are a reflection of biogeographic remoteness and environmental selection in a range of mesic coastal and mountain habitats isolated by Pleistocene aridity.

A feature of fynbos communities is the unwieldy speciation in a few prominent genera (Cowling et al. this volume; Linder et al. this volume). The families to which these genera belong include some that achieved a wide geographical distribution early in the dispersion of the angiosperms and others that became widespread only in the Neogene (Miocene and later). The genus *Helichrysum* in the family Asteraceae is an example of the latter and *Erica* can be cited as an example of the former. Speciation has produced large numbers of vicariants with restricted distributions in the biome. Representatives of the same taxa with fewer derived characters may have a wide distribution outside the biome and this is the basis of the contention that speciation in the fynbos is secondary (Levyn 1964). Reconstruction of the palaeoecology of the biome (Coetzee et al. 1983) suggests the persistence of a widespread forest cover into the Miocene. This cover became attenuated with the increasing aridity of the Late Cainozoic. As a consequence there was the opportunity for gap filling by understorey taxa and by new invasives. The mix of genera that are rich in species has come about in this way.

In the context of the fynbos biome, characterized by a diverse terrain and the constraints of low nutrients, gap filling has favoured the evolution of micro-habitat specialists. Although the biome contains major and minor centres of species richness, these centres can be defined only in general terms (Nordenstam 1969; Deacon 1983a). The reason is that speciation was not an event but a continuing process involving opportunistic invasion, naturalization, and specialization; a process that operates at the level of the individual taxon. The degree to which any geographical area may show a marked pattern of species richness in a number of taxa that are not genetically closely related is some measure of the effectiveness of environmental selection. The main centre of species richness in the south-western Cape illustrates this well. It corresponds to the area of greatest physiographic and habitat diversity and the occurrence of purest quartzose substrates and therefore the lowest nutrient pools. It is positioned so that it has a relatively high and reliable precipitation to counter seasonal aridity. This combination of conditions has promoted and maintained species richness.

A result of more than a decade of research in the biome is the appreciation that the fynbos biota is not a bizarre relict. It is the product of dynamic biological changes in response to environmental selection. This appreciation has opened the way for the study of how different plants and animals, in their growth forms, reproduction, dispersal, fire response, and other strategies, cope with the variables of the abiotic and the biotic environment. Interest in the fynbos biota and fynbos ecology has not diminished because in the new wisdom the flora is considered less distinctive or because the vegetation is assessed as geologically recent and specialized. The fynbos biome retains a pre-eminent importance as a natural laboratory for ecological studies. Knowledge of the particular environmental conditions found in the biome contributes to this.

Plates

Plate 1 Restioid growth form, which is characteristic of fynbos. The species shown is *Cannomois taylori* from the Cederberg in the north-western Cape Floristic Region (Photo: J C Paterson-Jones).

Plate 2 Proteoid growth form, showing the serotinous cones of *Leucadendron coniferum*, growing on coastal sands at Betty's Bay in the south-west (Photo: J C Paterson-Jones).

Plate 3 Proteoid growth form, represented by the serotinous shrub *Protea punctata*, growing in the Cederberg (Photo: J C Paterson-Jones).

Plate 4 Proteoid growth form, represented by the locally endemic, myrmecochorous *Leucospermum fulgens*, growing at Potberg on the southern coastal forelands. This species is confined to colluvial soils derived in part from limestone (right) and sandstone (left) (Photo: J C Paterson-Jones).

Plate 5 Ericoid growth form, *Erica propinqua*, a limestone endemic growing at De Hoop on the southern Cape coastal forelands (Photo: J C Paterson-Jones).

Plate 6 Ericoid growth form, *Erica obliqua*, from the Hottentots Holland Mountains in the south-west. This species is one of the 526 ericas found in the Cape Floristic Region (Photo: J C Paterson-Jones).

Plate 7 One of the 1 340 lilioid geophytes in the Cape Floristic Region. Autumn flowering *Brunsvigia marginata* growing near Franschhoek in the south-west (Photo: J C Paterson-Jones).

Plate 8 Trees are a rare growth form in fynbos. *Widdringtonia cederbergensis* (Cupressaceae) is endemic to the Cederberg. Populations have declined in historical times as a result of logging and frequent fires (Photo: J C Paterson-Jones).

Plate 9 Proteoid fynbos on the lower slopes of the Kogelberg, near Cape Hangklip in the south-west. The proteoid shrub in the foreground is *Leucadendron sessile* (Photo: J C Paterson-Jones).

Plate 10 Proteoid fynbos on limestone at Hagelkraal on the Agulhas Plain on the south-western forelands. The prominent proteoid is *Leucadendron meridianum* (Photo: J C Paterson-Jones).

Plate 11 Ericaceous fynbos dominated by Ericaceae and the characteristic broad-leaved sedge, *Tetraria thermalis*, below Grootkop on the Cape Peninsula (Photo: J C Paterson-Jones).

Plate 12 Restioid fynbos on poorly drained coastal flats on the Cape Peninsula. *Elegia filacea* replaces the taller *E. cuspidata* along a drainage gradient (Photo: J C Paterson-Jones).

Plate 13 Asteraceous fynbos in the Cederberg. The fynbos is dominated by Asteraceae (yellow flowering *Euryops speciosissimus*) and ericoid shrubs (*Passerina*, *Phylica*, etc.). The large shrub on the rock outcrop is *Heeria argentea* (Anacardiaceae), a characteristic species of mountain thickets in the west (Photo: J C Paterson-Jones).

Plate 14 Asteraceous fynbos on coastal dunes near Still Bay on the southern Cape coastal forelands. The dominant species are ericoid shrubs (*Metalasia*, *Agathosma*, *Phylica*) and broad-leaved sub-tropical thicket elements (*Rhus*, *Olea*, *Euclea*) (Photo: R M Cowling).

Plate 15 A south-eastern landscape, represented by the planed summits and soft slopes of the Zuurberg Mountains. There is grassy shrubland (foreground), grassland on the north-facing slope (left), grassy fynbos on the south-facing slope (right), and thicket in the valleys. The proteoid shrub in the foreground is the non-sprouting *Leucadendron salignum* (Photo: J C Paterson-Jones).

Plate 16 A remnant of renoster shrublands on shale-derived soils on the southern Cape coastal forelands. The dominant shrub is *Elytropappus rhinocerotis* (Asteraceae) (Photo: J C Paterson-Jones).

Plate 17 Afromontane forest on the southern Cape coastal platform near Knysna. The emergent trees are *Podocarpus falcatus* (Photo: J C Paterson-Jones).

Plate 18 Thicket on nutrient-rich, bottomland soils in the Baviaanskloof Mountains in the south-east. The growth form mix is complex and dominant species include evergreen shrubs (*Euclea*, *Cassine*), succulent shrubs (*Euphorbia*, *Portulacaria*), and woody spinescent vines (*Capparis*, *Azima*) (Photo: R M Cowling).

Plate 19 Thicket on coastal dunes near Velorenvlei on the west coastal forelands. Dominant shrubs in background include species of *Euclea*, *Olea*, *Rhus*, etc. The spring aspect shows numerous annuals (*Osteospermum*, *Ursinea*, etc.) in openings between shrubs (Photo: J C Paterson-Jones).

Plate 20 The early spring aspect of succulent karroid shrublands at Botterkloof, east of the Cederberg. Vegetation is dominated by Mesembryanthemaceae (perennials and annuals) and Asteraceae (annuals) (Photo: E J Moll).

Plate 21 Karroid shrublands at Tierberg north of the Swartberg in the southern Cape. The yellow flowering shrub is *Pteronia pallens* (Asteraceae). Other shrubs are largely Mesembryanthemaceae (Photo: R M Cowling).

Plate 22 Fynbos fauna. The endemic bontebok (*Damaliscus dorcas dorcas*) on the Cape Peninsula (Photo: J C Paterson-Jones).

Plate 23 Fynbos fauna. An orange-breasted sunbird (*Nectarina violacea*) on *Brunia albiflora* at Kirstenbosch on the Cape Peninsula (Photo: J C Paterson-Jones).

Plate 24 *Tritoniopsis* (*Anapalina*) *triticeae* (Iridaceae), a member of the late-summer guild of red species with red flowers pollinated by the butterfly *Aeropetes tulbaghia* (Satyridae) (Photo: S Johnson).

Plate 25 People in the fynbos. Degraded renoster shrubland remnants on the shale-derived soils of the west coastal forelands (Photo: E J Moll).

Plate 26 People in the fynbos. Urbanization has destroyed endemic-rich fynbos communities on recent sands in the Cape Town metropolitan area. This region has the highest concentration of rare and endangered species in the Cape Floristic Region (Photo: E J Moll).

Plate 27 People in the fynbos. Alien shrubs and trees (*Acacia* spp., *Pinus* spp.) form impenetrable thickets on the Cape Peninsula (Photo: J C Paterson-Jones).

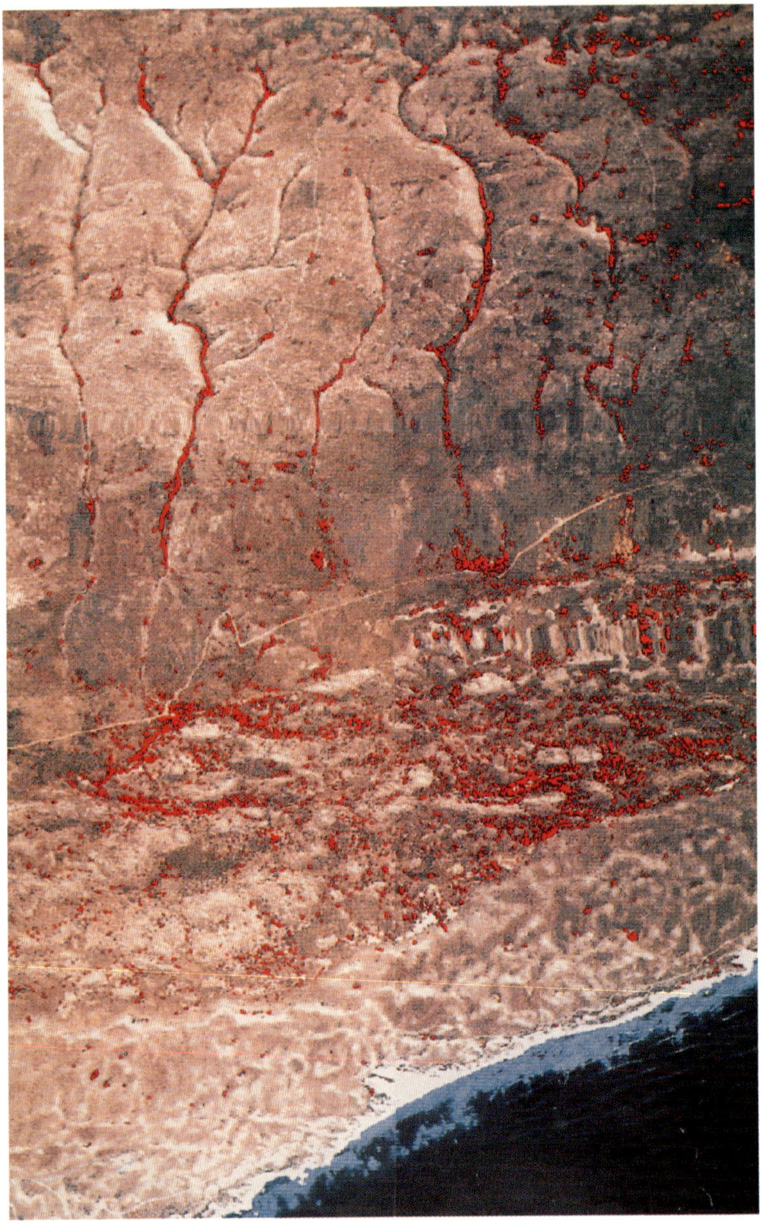

Plate 28 The distribution of *Acacia cyclops* (shown in red) at De Hoop on the southern Cape coast in 1989 as shown by thematic mapping using Daedelus (pixel size: 5–7.5 m^2). *Acacia cyclops* was not planted in the area, so the current distribution reflects the pattern of invasion from outside seed sources. Only a few patches of *A. cyclops* occur on the calcareous sands of the coastal dunefields at the bottom of the plate, and occurrence is not correlated with anthropogenic disturbance. *A. cyclops* is largely restricted to stream lines or pockets of deep soil in the limestone hills, inland of the dunefield. Drought is more intense on the shallow limestone soils and this may prevent the establishment of *A. cyclops* seedlings. Also, indigenous species with fleshy fruits (e.g. *Sideroxylon inerme* and *Rhus* spp.) occur mainly along stream lines; these provide foci of seed deposition and growth of *A. cyclops* clumps (Photo: Cape Department of Nature and Environmental Conservation).

ACKNOWLEDGEMENTS

Research reported in this chapter has been supported by the Foundation for Research Development, Pretoria, through the Special Programme for the Study of Climatic Change and the International Geosphere Biosphere Program (HJD, MJ). Support of the University of Stellenbosch (HJD, FE) and the University of Cape Town (MJ) is acknowledged. Some results reported follow from current research on the Late Pleistocene supported by the Centre for Science Development of the Human Sciences Research Council (HJD).

REFERENCES

AXELROD D I and RAVEN P H (1978). Late Cretaceous and Tertiary vegetation history of Africa. In *Biogeography and ecology of southern Africa.* (ed Werger M J A) Junk, The Hague, 77–130.

CAMPBELL B M (1983). Montane plant environments in the Fynbos Biome. *Bothalia*, **14**, 283–98.

CLIMAP PROJECT MEMBERS (1976). The surface of the Ice-Age earth. *Science* **191**, 1 131–7.

COETZEE J A, SCHOLTZ A and DEACON H J (1983). Palynological studies and the vegetation history of the fynbos. In *Fynbos palaeoecology: a preliminary synthesis.* (eds Deacon H J, Hendey Q B and Lambrechts J J N) *South African National Scientific Programmes Report* **75**, CSIR, Pretoria. 156–73.

DEPT. OF AGRICULTURE (1986). *Climate statistics for the Winter Rainfall Region up to 1986.* Agrometeorology Section, Elsenberg.

DEACON H J (1983a). An introduction to the fynbos region, time scales and palaeoenvironments. In *Fynbos palaeoecology: a preliminary synthesis.* (eds Deacon H J, Hendey Q B and Lambrechts J J N) *South African National Scientific Programmes Report* **75**, CSIR, Pretoria. 1–20.

DEACON H J (1983b). Comparative evolution of mediterranean-type ecosystems: a southern perspective. In *Mediterranean-type ecosystems: the role of nutrients.* (eds Kruger F J, Mitchell D T and Jarvis J U M) Springer-Verlag, Berlin. 3–40.

DEACON H J (1988). Palaeoecological perspectives on recent environmental change in southern Africa. In *Long-term data series relating to southern Africa's renewable natural resources.* (eds Macdonald I A W and Crawford R J M) *South African National Scientific Programmes Report* **157**, CSIR, Pretoria. 378–83.

DEACON H J (1990). Historical background of invasions in the mediterranean region of southern Africa. In *Biogeography of mediterranean invasions.* (eds Groves R H and Di Castri F) Cambridge University Press, Cambridge. 51–7.

DEACON J and LANCASTER N (1988). *Late Quaternary palaeoenvironments of southern Africa.* Clarendon Press, Oxford.

DINGLE R V, SIESSER W G and NEWTON A R (1983). *Mesozoic and Tertiary geology southern Africa.* Balkema, Rotterdam.

FUGGLE R F and ASHTON E R (1979). Climate. In *Fynbos Ecology: a preliminary synthesis* (eds Day J, Siegfried W R, Louw G N and Jarman M L) *South African National Scientific Programmes Report* **40**, CSIR, Pretoria. 7–15.

FUGGLE R F (1981). Macroclimatic patterns within the Fynbos Biome, Unpublished final report, Fynbos Biome Project, National Programme for Environmental Science, CSIR, Pretoria.

GLASS J G K (1977). Deep weathering of the southwestern Cape Granite and Malmesbury Group. *Technical Report of the Joint Geological Survey/University of Cape Town Marine Geoscience Group* **9**, 118–35.

HARLAND W B, ARMSTRONG R L, COX A V, CRAIG L E, SMITH A G and SMITH D G (1989). *A geologic time scale 1989.* Cambridge University Press, Cambridge.

HAYS J D, IMBRIE J and SHACKLETON N J (1976). Variations in the earth's orbit: pacemaker of the ice ages. *Science* **194**, 1 121–32.

HENDEY Q B (1982). *Langebaanweg: a record of past life.* South African Museum, Cape Town.

HENDEY Q B (1983). Palaeoenvironmental implications of the late Tertiary vertebrate fauna of the fynbos region. In *Fynbos palaeoecology: a preliminary synthesis* (eds Deacon H J, Hendey Q B and Lambrechts J J N) *South African National Scientific Programmes Report* **75**, CSIR, Pretoria. 100–1.

HEYDORN A E F and TINLEY K L (1980). Estuaries of the Cape, Part 1: synopsis of the Cape coast natural features, dynamics and utilization. *National Research Institute for Oceanology* **380**, CSIR, Stellenbosch.

KAMSTRA F (1985). Environmental features of the southern Benguela with special reference to the wind stress. In *South African ocean colour and upwelling experiment* (ed Shannon L V) Sea Fisheries Research Institute, Cape Town. 13–27.

KING L C (1962). *The morphology of the earth.* Oliver and Boyd, Edinburgh.

KRUGER F J (1983). Plant community diversity and dynamics in relation to fire. In *Mediterranean-type ecosystems: the role of nutrients.* (eds Kruger F J, Mitchell D T and Jarvis J U M) Springer-Verlag, Berlin. 446–72.

KRUGER F J and TAYLOR H C (1979). Plant species diversity in Cape fynbos. *Vegetatio* **41**, 85–93.

LE ROUX F G (1989). The lithostratigraphy of Cenozoic deposits along the south-east Cape coast as related to sea-level changes. Unpublished MSc thesis, University of Stellenbosch.

LEVYNS M R (1964). Migrations and origins of the Cape Flora. *Transactions of the Royal Society of South Africa* **37**, 85–105.

LIVINGSTONE D A (1975). Late Quaternary climatic change in Africa. *Annual Review of Ecology and Systematics* **6**, 249–80.

McLACHLAN I R and PIETERSE E (1978). Preliminary palynological results: site 361, leg 40, Deep Sea Drilling Project. *Initial Reports of the Deep Sea Drilling Project* **4**, 857–81.

NORDENSTAM B (1969). Phytogeography of the genus *Euryops* (Compositae): a contribution to the phytogeography of southern Africa. *Opera Botanica* **23**, 1–77.

PARTRIDGE T C and MAUD R R (1987). Geomorphic evolution of southern Africa since the Mesozoic. *South African Journal of Geology* **90**, 179–208.

PRESTON-WHYTE R A and TYSON P D (1988). *The atmosphere and weather of southern Africa.* Oxford University Press, Cape Town.

SACS, SOUTH AFRICAN COMMITTEE FOR STRATIGRAPHY (1980). Stratigraphy of South Africa, Part 1. Lithostratigraphy of the Republic of South Africa, South West Africa/Namibia, and the Republics of Bophuthatswana, Transkei and Venda. *Geological Survey of South Africa Handbook* **8**, Government Printer, Pretoria.

SCHLOMS B H A, ELLIS F and LAMBRECHTS J J N (1983). Soils of the Cape coastal platform. In *Fynbos palaeoecology: a preliminary synthesis* (eds Deacon H J, Hendey Q B and Lambrechts J J N) *South African National Scientific Programmes Report* **75**, CSIR, Pretoria. 70–86.

SCHULZE B R (1984). Climate of South Africa, Part 8, general survey, WB 28. *South African Weather Bureau* **60**, Government Printer, Pretoria.

SIESSER W G (1980). Late Miocene origin of the Benguela upwelling system off northern Namibia. *Science* **208**, 283–5.

SHNGE A P C (1983). The Cape Fold Belt perspective. *Geo-*

logical Society of South Africa Special Publication **12**, 1–6.
TANKARD A J, JACKSON M P A, ERIKSON K A, HOBDAY D K, HUNTER D R and MINTER W E L (1982). *Crustal evolution of southern Africa: 3.8 billion years of earth history.* Springer Verlag, New York.
THERON J N (1983). Geological setting of the fynbos. In *Fynbos palaeoecology: a preliminary synthesis* (eds Deacon H J, Hendey Q B and Lambrechts J J N) *South African National Scientific Programmes Report* **75**, CSIR, Pretoria. 21–34.
TYSON P D (1986). *Climatic change and variability in southern Africa.* Oxford University Press, Cape Town.
VAN ANDEL T H (1989). Late Pleistocene sea levels and the human exploitation of the shore and shelf of southern South Africa. *Journal of Field Archaeology* **16**, 133–55.
VAN WILGEN B W (1985). Derivation of fire hazard and burning prescriptions from climatic and plant ecological features of fynbos ecosystems. Unpublished PhD thesis, University of Cape Town.
WOODRUFF F, SAVIN S M and DOUGLAS R G (1981). Miocene stable isotope record: a detailed deep Pacific Ocean study and its paleoclimatic implications. *Science* **212**, 665–8.

3

Flora and vegetation

R M Cowling and P M Holmes

INTRODUCTION

In this chapter we attempt to provide an ecological biogeographic framework for this volume. We do this by asking the question: can distribution patterns of species and communities in the fynbos biome be explained by contemporary ecological factors? Such an analysis is potentially important since repeated and environmentally correlated patterns lead to a predictive understanding of ecological biogeography. Secondly, patterns which cannot be explained by contemporary ecological factors can provide the basis for historical hypotheses (Vuilleumier and Simberloff 1980; Endler 1982; see Linder et al. this volume). Thirdly, the analysis provides the ecological detail essential for understanding diversity, endemism, and speciation in the Cape Floristic Region (Linder 1985; Linder and Vlok 1991; Cowling et al. this volume).

The first part of the chapter deals with the flora of the Cape Floristic Region. The recent publication of a catalogue and analysis of the flora (Bond and Goldblatt 1984; see also Goldblatt 1978) is an important achievement. We do not provide a comprehensive review: readers are referred to Goldblatt's review of the literature and comparative analysis of the flora. Here we attempt to explain variation in the composition of regional floras from the south-western and south-eastern part of the Cape Floristic Region in terms of ecological gradients. This topic has not been adequately addressed in the literature and we report previously unpublished analyses.

The second part of the chapter provides an overview of vegetation-environment relationships in the fynbos biome. We largely confine our treatment to studies which appeared after the publication of two major reviews of fynbos ecology (Taylor 1978; Kruger 1979). Campbell's (1985a) structural classification of the vegetation of the fynbos biome mountains is the most significant publication to emerge since these reviews. Our analysis is based largely on Campbell's typology.

Note that the flora is discussed in relation to the Cape Floristic Region and the vegetation is discussed in relation to the fynbos biome (Figure 3.1). For details on the differences between these units, see the preface to this volume.

THE FLORA

A brief overview of the Cape flora

The Cape Floristic Region is a highly distinctive phytogeographical unit which is recognized as a floral kingdom of its own (Good 1974; Goldblatt 1978; Takhatajan 1986). Sixty-eight per cent of species, 19.5% of genera, and six families are endemic to the region. For equivalent-sized areas, the Cape Floristic Region has the highest recorded species density for any temperate or tropical region in the world (Bond and Goldblatt 1984; Cowling et al. 1989).

Most Cape taxa, including some of the largest genera (e.g. *Erica*, *Aspalathus*, *Agathosma*, *Cliffortia*, and *Muraltia*) (Table 3.1) show a pattern of high species concentration in the extreme south-west of the region (Levyns 1938, 1954; Weimarck 1941; Dahlgren 1963; Cowling 1983a; Oliver et al. 1983) (Figure 3.2). Other taxa well represented in the Cape Floristic Region such as *Euryops* (Nordenstam 1969), *Leucospermum* (A G Rebelo unpublished data), Mesembryanthemaceae (Jurgens 1986), and *Passerina* (Thoday 1925) have their greatest centres of diversity elsewhere in the region (Figure 3.2).

What are the biogeographical affinities of the non-endemic component of the flora? The Cape Floristic Region includes large areas of

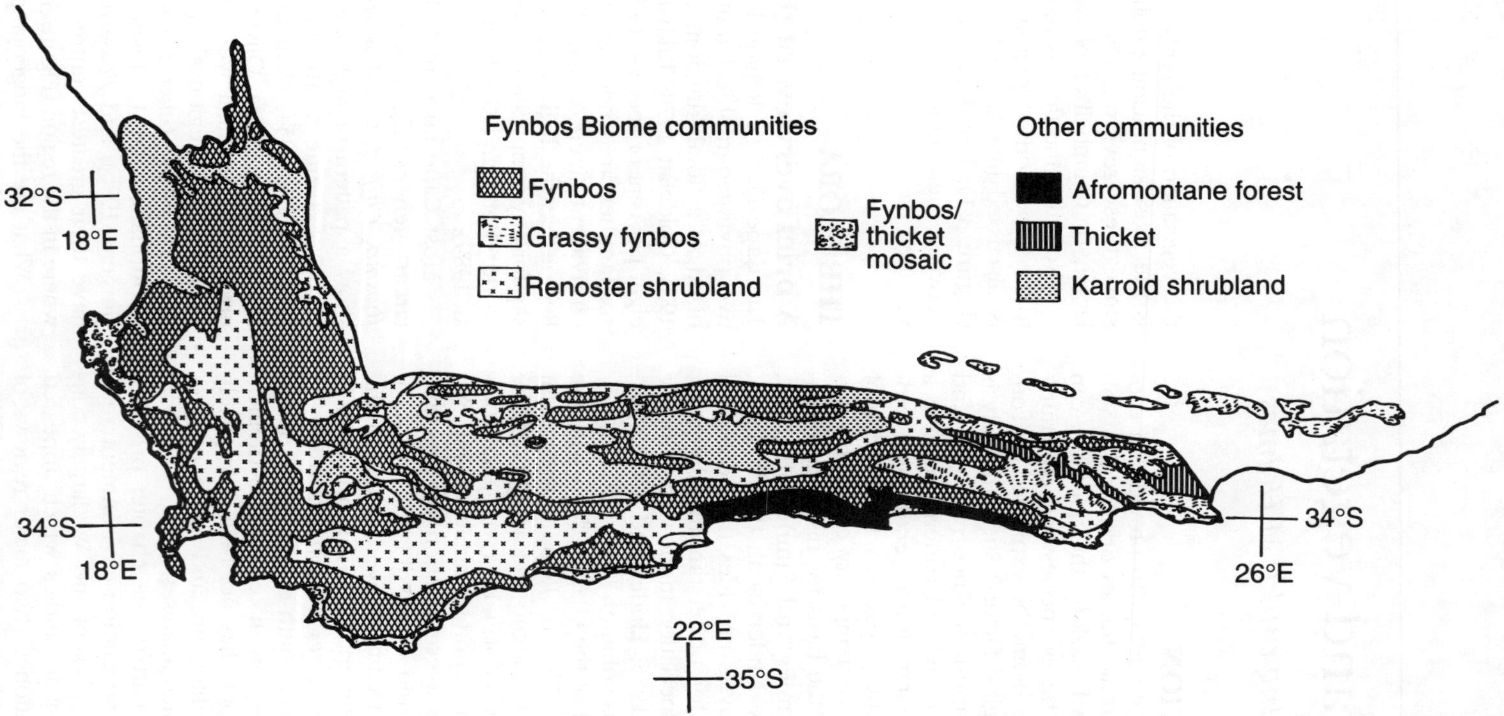

FIGURE 3.1 Major vegetation communities in the Cape Floristic Region (*sensu* Bond and Goldblatt 1984). The extension of grassy fynbos outside the region is also shown. Vegetation boundaries are based mainly on Moll and Bossi (1984a).

TABLE 3.1 Selected characteristics of floras (indigenous flowering plants only) in the south-western and south-eastern Cape Floristic Region (CFR). See Figure 3.4 for location of flora areas.

	South-western CFR			South-eastern CFR			
	Cape Peninsula	Cape Hangklip	Agulhas Plain	George	Knysna	Humansdorp	Total CFR
Area (km^2)	470	240	1 609	2 536	2 098	5 050	90 000
Genera	533	414	487	428	516	604	989
Species	2 256	1 383	1 751	1 204	1 416	2 240	8 504
Species/genus	4.2	3.3	3.6	2.8	2.7	3.7	8.6
Species/km^2	4.8	5.76	0.92	0.47	0.67	0.44	0.09
Ten largest families (%)	59.7	61.6	59.7	56.3	50.1	55.9	60.2
Ten largest genera (%)	18.1	21	19.8	21	18.3	14.2	20.4
Altitudinal range (m)	0–1 084	0–1 049	0–654	0–1 579	0–1 618	0–1 675	0–2 325
Rainfall range (mm yr^{-1})[1]	350–1 700	900–2 200	500–750	250–1 300	650–1 300	450–1 500	180–4 000
Cape shrublands (%)[2]	98	99	99	65	50	80	83
Karroid shrublands (%)	0	0	0	12	0	0	10
Afromontane forest (%)	2	< 0.5	< 0.5	18	45	10	4.5
Thicket (%)	< 0.5	< 0.5	< 0.5	5	5	10	2.4

[1] 1:250 000 Annual Rainfall Maps.
[2] Extent of vegetation types estimated from Figure 3.1. Cape shrublands include fynbos, grassy fynbos, renoster shrubland, and fynbos/thicket mosaic.

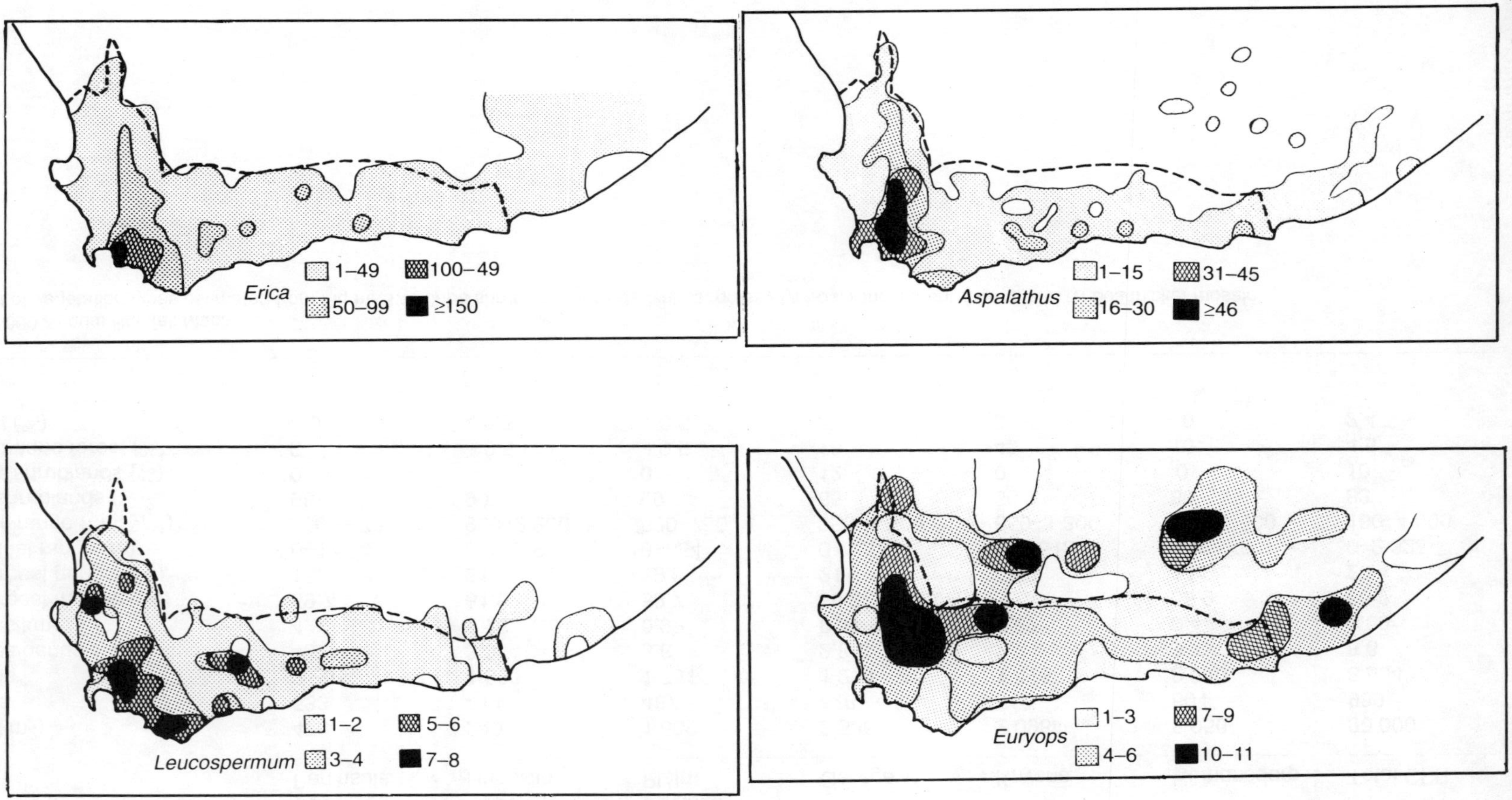

FIGURE 3.2 Isoflor maps for large genera in the Cape Floristic Region. All mapped by $^1/_4$° square except for *Euryops* ($^1/_2$°). Data from Oliver et al. (1983) (*Erica* and *Aspalathus*), A G Rebelo (unpublished data) (*Leucospermum*), and Nordenstam (1969) (*Euryops*). The thick dashed line represents the boundary of the CFR.

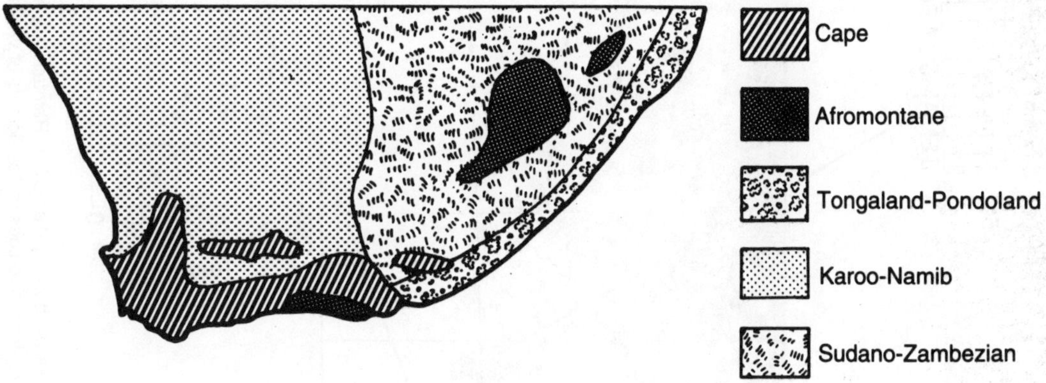

FIGURE 3.3 Phytochoria of southern Africa, after Werger (1978b). The Cape phytochorion is largely equivalent to the fynbos biome.

karroid shrubland (Figure 3.1): thus, the flora has a large component of species centred in the Karoo-Namib Region (Werger 1978a) (Figure 3.3). Many of these are succulents belonging mainly to the Mesembryanthemaceae, Crassulaceae, Asclepiadaceae, and Asteraceae. These species are richly represented in dry fynbos communities and renoster shrubland. The Afromontane flora (White 1978) is also prominent in the Cape Floristic Region. Typical Afromontane forest extends in isolated patches as far west as the Cape Peninsula. A large number of shrubs and herbs of Afromontane affinity occur in fynbos communities, especially in the east (Cowling 1983a). The affinities between the Afromontane and Cape Floristic Region floras have led both Tinley (1977) and Linder (1991) to suggest the recognition of a continent-wide Afrotemperate phytochorion. Shrubs and trees centred in the subtropical Tongaland-Pondoland Region (Moll and White 1978) penetrate into the fynbos biome primarily on lowland and relatively nutrient-rich sites. They are most richly represented in the south-east (Cowling 1983a).

Variation in the composition and characteristics of regional floras

Three similar-sized floras (1 200–2 200 species) from both the south-western and the south-eastern Cape Floristic Region were analysed in order to compare compositional characteristics. Data for the south-western Cape Floristic Region included two published lists — Cape Peninsula (Adamson and Salter 1950) and Cape Hangklip (Boucher 1977) — and an unpublished list for the Agulhas Plain (R M Cowling unpublished data) (Figure 3.4). The last-mentioned check-list was compiled from the PRECIS database (Gibbs-Russell 1985), recently published monographs, collections in three south-western Cape herbaria (BOL, COM, STE), and the collections and unpublished observations of R M Cowling. Nomenclature for Cape Hangklip and the Agulhas Plain was standardized in accordance with Bond and Goldblatt (1984) but this was only done for the larger genera in the Cape Peninsula flora. Floras from the south-eastern Cape Floristic Region were taken from check-lists for

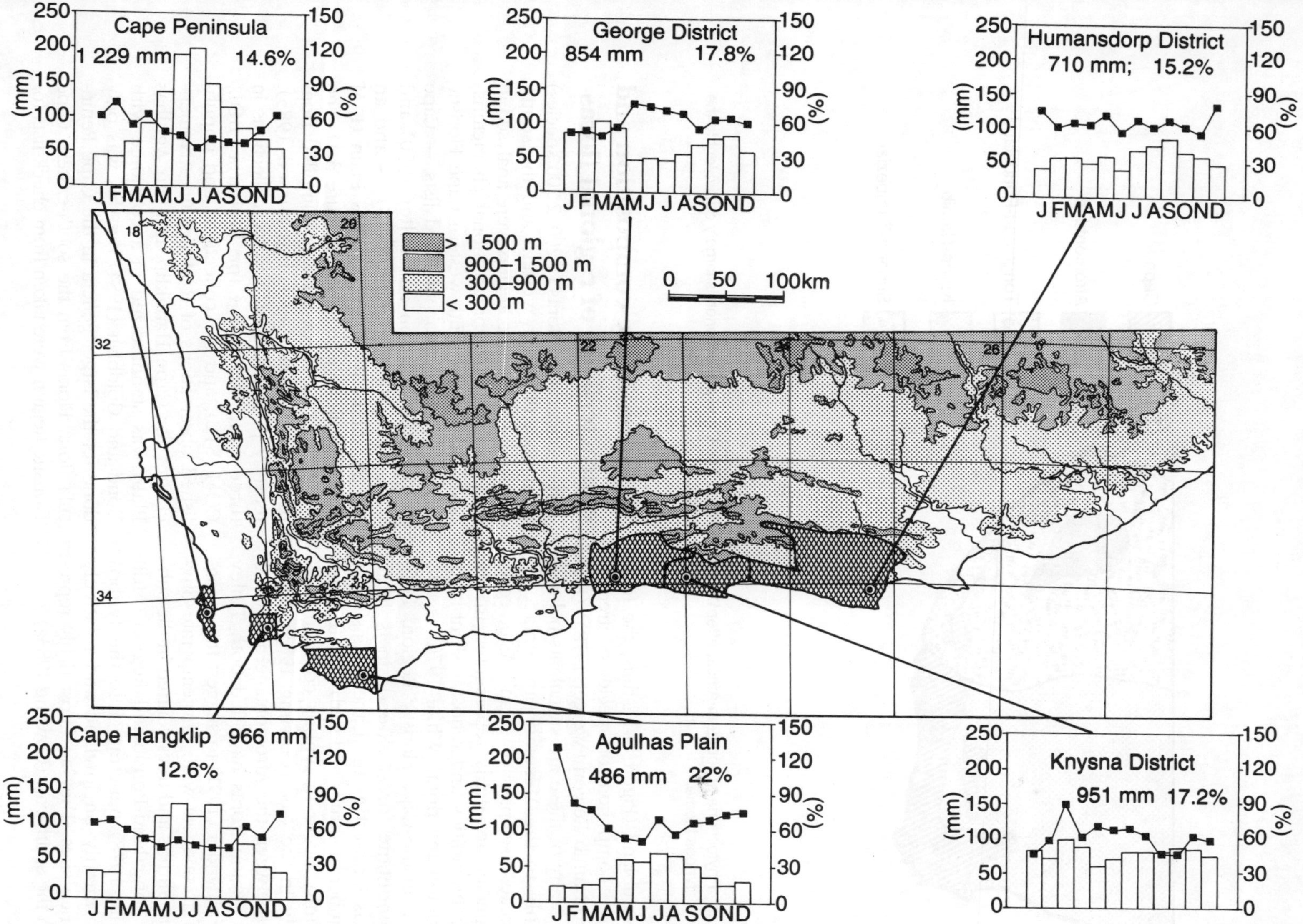

FIGURE 3.4 Location and representative rainfall data for the areas of the south-western and south-eastern floras shown in Table 3.1. Rainfall data are an artificial sequence of monthly totals generated for 30 years (Zucchini 1990). Bars refer to mean monthly rainfall and lines to coefficients of variation for monthly rainfall.

the George, Knysna, and Humansdorp magisterial districts (Fourcade 1941) (Figure 3.4). As these lists were compiled more than 50 years ago, and are certainly incomplete, they were supplemented from distribution data in Bond and Goldblatt (1984), and in the case of the Humansdorp district, from the collections and unpublished notes of R M Cowling. However, the sizes of these floras may well be underestimates, compared with the better collected areas in the south-west. Nomenclature of the larger genera was standardized in accordance with Bond and Goldblatt (1984).

We compiled the following characteristics of the floras: total number of species and genera, species-to-genus ratios, and number of species in the 10 largest families and genera. Relationships between floras were analysed using the last-mentioned data. A Kendall's rank correlation matrix (Siegel 1956) was generated on the basis of the rank order in each flora of the 25 genera which qualified as one of the largest 10 genera in any of the floras.

We also compared the frequency of species in different phytochorological groups (Cowling 1983a) in floras from coastal lowland regions of the south-western and south-eastern Cape Floristic Region. The floras were, for the Agulhas Plain (south-west) (R M Cowling unpublished data) and a sample flora from the Humansdorp district (south-east) (Cowling 1983b) (Figure 3.4). Both regions include a broadly similar range of habitats and plant communities, although part of the Humansdorp area has higher rainfall (> 600 mm yr^{-1}) than the Agulhas Plain (Cowling 1984; Cowling et al. 1988).

As there was no significant difference (U = 3, P = 0.35, Mann-Whitney U test for small samples, Siegel 1956) between the mean size of the three south-western ($\bar{x}$ = 1 796.7 ± (SD) 759.2) and the three south-eastern floras ($\bar{x}$ = 1 620 ± 948) analysed in Table 3.1, it is unlikely that comparisons of species-to-genus ratios and related compositional measures are a statistical artefact of variation in species number (Simberloff 1970; Jarvinen 1982). The south-western floras had a higher mean species-to-genus ratio ($\bar{x}$ = 3.7 ± 0.47) than the south-eastern floras ($\bar{x}$ = 3.1 ± 0.55) but this difference was not significant (U = 2, P = 0.2) (Table 3.1). However, there were significantly (U = 0, P < 0.05) higher proportions of species in the 10 largest families in the south-west ($\bar{x}$ = 60.3 ± 1.11) than the south-east ($\bar{x}$ = 54.1 ± 3.46). Results were not significant (U = 3, P = 0.35) for comparisons of the proportion of species in the 10 largest genera, although the mean value was higher for the south-west (19.6 ± 1.45) than the south-east (17.8 ± 3.41). Whereas the south-western areas were overwhelmingly dominated by Cape shrublands (largely fynbos), the south-eastern sites had a strong admixture of non-fynbos vegetation types (Table 3.1; Figure 3.1).

The family-level compositions of the south-western and south-eastern floras were very similar (Table 3.2). Considering the 10 largest families in floras, three (Campanulaceae, Proteaceae, and Restionaceae) present in the south-west were not recorded in the south-eastern floras; in the latter, only the Thymelaeaceae was not present in the south-west. Differences between the two regions were more pronounced at the generic level (Table 3.2). Of the pooled 10 largest genera recorded for the south-western floras, seven (*Elegia*, *Ficinia*, *Gladiolus*, *Lobelia*, *Phylica*, *Restio*, and *Tetraria*) were not recorded in the south-east. Eight genera (*Agathosma*, *Gnidia*, *Haworthia*, *Hermannia*, *Indigofera*, *Pelargonium*, *Polygala*, and *Satyrium*) recorded in the south-east were absent from the south-western lists. The correlation analysis of the generic data showed highly significant relationships among floras from the same areas of the Cape Floristic Region (Table 3.3). Of the nine south-western/south-eastern comparisons, only five were significant. All the George/south-western comparisons were non-significant, whereas all the Knysna ones were significant. Comparisons were also made between the regional floras and the (entire) Cape Floristic Region flora, although this was not strictly permissible since the regional floras do not, in this case, represent independent data sets (the task of removing each regional flora from the Cape Floristic Region flora for the purpose of this analysis proved too onerous). Notwithstanding this statistical problem, only the Cape Peninsula and Agulhas Plain floras were significantly correlated with the Cape Floristic Region flora.

There were major differences in the phytochorological spectra of the south-western flora (Agulhas Plain) and the south-eastern flora (Humansdorp) (Table 3.4). There was a higher

TABLE 3.2 Number of species in the 10 largest genera and 10 largest families in floras in the south-western and south-eastern Cape Floristic Region (CFR).

South-western CFR				South-eastern CFR								Total CFR	
Cape Peninsula		Cape Hangklip		Agulhas Plain		George		Knysna		Humansdorp			
Genera													
Erica	102	*Erica*	98	*Erica*	91	*Erica*	63	*Erica*	62	*Erica*	67	*Erica*	526
Aspalathus	47	*Aspalathus*	35	*Aspalathus*	60	*Aspalathus*	32	*Senecio*	38	*Senecio*	46	*Aspalathus*	245
Senecio	44	*Ficinia*	27	*Crassula*	29	*Pelargonium*	27	*Aspalathus*	26	*Pelargonium*	35	*Ruschia*	138
Ficinia	40	*Senecio*	22	*Gladiolus*	26	*Helichrysum*	25	*Helichrysum*	25	*Crassula*	31	*Phylica*	133
Crassula	33	*Restio*	22	*Phylica*	26	*Agathosma*	24	*Crassula*	23	*Helichrysum*	31	*Agathosma*	130
Oxalis	33	*Cliffortia*	19	*Senecio*	26	*Haworthia*	19	*Pelargonium*	22	*Aspalathus*	25	*Oxalis*	129
Helichrysum	28	*Elegia*	18	*Ficinia*	23	*Gnidia*	17	*Satyrium*	17	*Agathosma*	24	*Pelargonium*	125
Pelargonium	28	*Pelargonium*	17	*Muraltia*	23	*Muraltia*	16	*Indigofera*	16	*Thesium*	21	*Senecio*	113
Cliffortia	27	*Helichrysum*	16	*Thesium*	23	*Crassula*	15	*Cliffortia*	15	*Indigofera*	20	*Cliffortia*	106
Tetraria	26	*Lobelia*	16	*Helichrysum*	21	*Hermannia*	15	*Polygala*	15	*Oxalis*	18	*Muraltia*	106
Families													
Asteraceae	267	Asteraceae	152	Asteraceae	204	Asteraceae	148	Asteraceae	174	Asteraceae	306	Asteraceae	986
Fabaceae	155	Ericaceae	119	Iridaceae	149	Fabaceae	91	Fabaceae	105	Fabaceae	152	Ericaceae	672
Cyperaceae	139	Fabaceae	103	Fabaceae	142	Ericaceae	69	Orchidaceae	83	Liliaceae[1]	147	Mesembryanth[4]	660
Iridaceae	139	Restionaceae	100	Ericaceae	127	Liliaceae[1]	68	Ericaceae	69	Orchidaceae	119	Fabaceae	644
Poaceae	128	Iridaceae	98	Restionaceae	88	Orchidaceae	67	Poaceae	59	Cyperaceae	101	Iridaceae	612
Ericaceae	119	Cyperaceae	71	Poaceae	80	Iridaceae	62	Cyperaceae	55	Poaceae	100	Liliaceae[1]	347
Liliaceae[1]	119	Poaceae	64	Liliaceae[1]	76	Poaceae	55	Iridaceae	53	Scrophulariaceae[2]	99	Proteaceae	320
Orchidaceae	117	Proteaceae	62	Cyperaceae	63	Scrophulariaceae[2]	47	Scrophulariaceae[2]	48	Iridaceae	93	Restionaceae	310
Restionaceae	86	Orchidaceae	55	Proteaceae	59	Thymeleaceae	35	Liliaceae[1]	46	Ericaceae	81	Scrophulariaceae[2]	310
Scrophulariaceae[2]	77	Campanulaceae[3]	44	Mesembryanth[4]	58	Mesembryanth[4]	34	Restionaceae	30	Mesembryanth[4]	54	Rutaceae	259

[1] *Sensu lato.*
[2] Includes Selaginaceae.
[3] Includes Lobeliaceae.
[4] Mesembryanthemaceae excludes Aizoaceae.

TABLE 3.3 Correlation matrix (Kendall's *r*) showing relationships between south-western (italics) and south-eastern floras in the Cape Floristic Region (CFR). Twenty-five genera were used in the analysis (See Table 3.2 and text). CPE = Cape Peninsula, CHA = Cape Hangklip, AGP = Agulhas Plain, GEO = George, KNY = Knysna, HUM = Humansdorp. * = *P* < 0.001, ** = *P* < 0.01, * = *P* < 0.05.**

	CFR					
CPE	35*	*CHA*				
CHA	.13	.45**	*AGP*			
AGP	.44**	.61***	.38**	GEO		
GEO	.23	.21	.03	.26	KNY	
KNY	.17	.52***	.31*	.37**	.39**	HUM
HUM	.23	.51***	.25	.40**	.48***	.63***

TABLE 3.4 Phytochorological spectra for floras from the Agulhas Plain (south-western Cape Floristic Region) and the Humansdorp coastal plain (south-eastern Cape Floristic Region). Phytochoria as in Figure 3.3. CEN = Cape endemics, CAL = Cape-Afromontane linking species, CKL = Cape-Karoo-Namib linking species, CTL = Cape-Tongaland-Pondoland linking species, AEN = Afromontane endemics, ATL = Afromontane-Tongaland Pondoland linking species, WID = Phytochorological and ecological transgressor species (White 1978).

	Phytochorological groups								Total
	CEN	CAL	CKL	CTL	AEN	ATL	WID	Other	
Agulhas Plain									
No. species	1 338	63	142	38	12	15	132	11	1 751
%	76.4	3.6	8.1	2.2	0.7	0.9	7.5	0.6	100
Humansdorp									
No. species	342	63	51	39	84	68	94	138	879
%	38.9	7.2	5.8	4.4	9.6	7.7	10.7	15.7	100

frequency of Cape endemics, a lower frequency of both Cape-linking taxa (shrubs and herbs in the CAL, CKL, and CTL groups) (see Table 3.4 for explanation of categories) and of forest and thicket species (largely trees in the AEN and ATL categories), and lower frequency of other transgressor species (WID and other categories) in the Agulhas Plain than the Humansdorp flora (χ^2 = 1 714.7, df = 3, P < 0.001).

Explanations: ecology versus history

There were considerable differences in the composition of regional floras in the south-western and south-eastern Cape Floristic Region. The results of this analysis and other studies reveal the following trends:

• A concentration of species of Cape Floristic Region taxa (e.g. *Erica*, *Aspalathus*, Restionaceae etc.) in the extreme south-west with a marked decline eastwards, particularly east of the Breede River (Levyns 1938; Weimarck 1941; Dahlgren 1963; Cowling 1983a; Oliver et al. 1983) (Figure 3.2).

• Higher species-to-genus ratios and proportion of species in the 10 largest genera and families in the south-western than the south-eastern floras.

• Compositional differences associated with the presence — among the 10 largest genera in south-eastern floras — of taxa not centred in the Cape Floristic Region (e.g. *Gnidia, Hermannia, Indigofera, Polygala*, and *Satyrium*).

• Compositional differences reflected in the lower frequency in the south-eastern flora of species endemic to the Cape Floristic Region.

To what extent can these differences be explained by contemporary ecological factors or the vicissitudes of geological and climatic history and their effects on speciation, extinction, and migration (Vuilleumier and Simberloff 1980; Endler 1982)? Below we evaluate the extent to which ecological factors can explain these patterns.

What are the differences in ecological conditions between the south-western and south-eastern Cape Floristic Region sites studied here? Physiographically, all the sites fall within the Cape Folded Belt and share a largely similar array of landforms and soils (see Deacon et al. this volume). In both the south-west (Boucher 1978; Taylor 1984a; Cowling et al. 1988) and the south-east (Bond 1981; Cowling 1984) the landscape is dominated by quartzitic mountains and a coastal plain underlain largely by shales with a mantle of Cainozoic, alkaline, and marine deposits on the coastal margin (Hendey 1983; Deacon et al. this volume). In general, altitudinal ranges are greater for the south-eastern sites (Table 3.1) which also differ in having dry intermontane valleys. Overall, the range of parent materials and soils is similar for both regions, although the south-west lacks Cretaceous deposits of the Uitenhage Group, and limestones of the Bredasdorp Formation have very limited exposure in the south-east. Irrespective of parent material, soils are less acidic, more base-saturated, and have a higher clay and fine sand component in the south-east than the south-west (Campbell 1983).

The range in annual rainfall is larger for the south-eastern than the south-western sites. However, the highest values are recorded for the latter and the lowest values for the former (Table 3.1). There is a major difference in seasonality between the two regions. Summer drought is most pronounced in the extreme west where rainfall is almost entirely, and very predictably, associated with north-westerly frontal conditions in winter (Jackson and Tyson 1971; Deacon et al. this volume) (Figure 3.4). In the south-east a high proportion of annual rainfall is associated with post-frontal and tropical circulation when moist air is advected by a high pressure cell across the warm Indian Ocean. Conditions favourable for rain from this source occur throughout the year so that rainfall is non-seasonal or weakly bimodal (spring and/or autumn peaks) (Heydorn and Tinley 1980). Here monthly rainfall during the winter months is usually the least predictable (Figure 3.4). The boundary between winter and non-seasonal rainfall regions in the southern Cape Floristic Region is in the vicinity of the Breede River (Kruger 1979).

The high concentration of Cape Floristic Region taxa and associated high species-to-genus ratios in the south-west have been widely attributed to the strongly developed mediterranean climate there (Thoday 1925; Weimarck 1941; Dahlgren 1963; Levyns 1964). Why should this be so? The higher proportion of summer rain in the south-east together with the more fertile soils there (Campbell 1983) would favour non-Cape Floristic Region taxa and vegetation types other than Cape shrublands (e.g. forest and thicket) (Cowling 1983a; Campbell 1985a, 1986a; see below). Presumably this would limit the area occupied by Cape shrublands and hence the number of Cape Floristic Region species (Weimarck 1941; Dahlgren 1963). Furthermore, this penetration of tropical species into Cape shrublands would reduce the diversity of Cape Floristic Region taxa in the south-east (e.g. the replacement in the south-east of evergreen hemicryptophytes belonging to the Restionaceae and Cyperaceae by tropical C_4 grasses) (Vogel et al. 1978; Campbell 1985a; Campbell and Werger 1988). Increased grassiness could have two other important consequences for the membership of Cape Floristic Region taxa in south-eastern fynbos communities: the faster growing grasses could outcompete seedlings of the slower growing fynbos species in the early post-fire stages (Richardson and Bond 1991); and the more rapid post-fire recovery of grassy communities could result in fire cycles shorter than juvenile periods, leading to the extinction of non-sprouting fynbos shrubs (Cowling 1984).

Oliver et al. (1983) have argued that the higher concentration of Cape Floristic Region taxa in the south-west is a result of longer ecological gradients (greater variation in rainfall and altitude) and greater edaphic heterogeneity than the south-east. This is clearly untenable for the floras we studied where climatic gradients were longer and edaphic heterogeneity as great in the south-east (e.g. Bond 1981; Cowling 1984) as the south-west (e.g. Boucher 1978; Cowling et al. 1988; see also Table 3.1).

It is unlikely that the contemporary pattern of rainfall seasonality and edaphic factors can explain completely the high concentration of Cape Floristic Region taxa in the south-west. Although vegetation types other than Cape shrublands do occupy a greater relative area in the south-east (Table 3.1), the absolute area covered by Cape shrublands is comparable to the south-west (Figure 3.1). The only tropical growth forms to penetrate fynbos to any extent are C_4 grasses and subtropical thicket shrubs and trees. C_4 grasses as a major vegetation component are largely confined to grassy fynbos in the extreme south-east (Cowling 1984; Campbell and Werger 1988; see below). Thicket shrubs are an important component only in coastal dune fynbos and the coverage of these species differs little between the south-east (Cowling 1984) and south-west (Taylor 1972). The number of forest and thicket taxa shows a gradual decline from the extreme south-east to the south-west (Cowling 1983a; Geldenhuys and McDevette 1989); there is no evidence for an abrupt decline west of the Breede River as would be predicted from hypotheses based on contemporary rainfall seasonality patterns.

It appears that although contemporary ecological factors do play a substantive role in explaining differences in the composition of floras across the Cape Floristic Region (e.g. a higher number of phytochorological transgressors in the south-east), historical factors must be invoked to explain the enormous concentration of Cape Floristic Region species in the south-west. In this volume Linder et al. argue that the massive speciation within a limited number of genera in the Cape Floristic Region can be attributed to the fact that the region has comprised an edaphic (low nutrient soils) and climatic (winter rainfall) 'island' since the onset of modern circulation patterns at the end of the Tertiary (Hendey 1983: Deacon et al. this volume). Since then environmental deterioration has brought about the widespread extinction of many genera and families and provided the stimulus for the explosive speciation within a limited number of sclerophyllous genera, pre-adapted for nutrient-poor, summer-dry conditions (see also Cowling 1983a). Subsequently, non-Cape Floristic Region taxa have been unable to invade the region, except in marginal or zonal habitats. Contemporary ecological conditions indicate that these barriers to invasion are weaker in the south-east than the south-west, as in the latter soils are less fertile and summer drought is more pronounced. However, the relatively depauperate floras of extensive tracts of non-grassy fynbos in the south-east (see Cowling et al. this volume) suggest that this barrier was more pronounced in the past.

Since glacial conditions prevailed during 90% of the Pleistocene (Deacon et al. this volume) it is logical to infer that glacial climatic conditions were markedly more different between the south-west and the south-east than during interglacials (including the current Holocene). The implication is that fynbos would have been largely replaced by non-fynbos communities in the south-east during much of the Quaternary when Cape Floristic Region taxa were speciating.

The precise nature of these differences in palaeoclimatic conditions is unknown (Deacon and Lancaster 1988; Deacon et al. this volume). Evidence suggests that in the south-east, glacial periods were drier and colder than interglacials (Scholz 1986; Deacon and Lancaster 1988). Thus, in the south-east, fynbos, subtropical thicket, and Afromontane forest would have been confined during glacials to refugia; karroid communities would have been widespread (Cowling 1983a). In the south-west, however, glacial conditions would have been wetter owing to an increased frequency of frontal rains which would not have affected the south-east (Cockcroft et al. 1987). Fynbos would have persisted in most habitats or even expanded its range. Speciation would have occurred over a wide range of ecological conditions for a longer period in the south-west, resulting in a great number of habitat specialists and local endemics as well as high floristic turnover and regional richness (see Cowling et al. this volume). With the onset of climatic change at the start of the Holocene (12 000 BP), species would have expanded into the south-east from refugia in upper mountain sites or from the south-west (fynbos taxa), and along the subtropical east coast (subtropical thicket and Afromontane forest taxa) (Cowling 1983a). Insufficient time has elapsed in the south-east for the evolution of specialists endemic to new habitats colonized by fynbos or, in the south-west for the eastwards migration of tropical taxa into habitats that can potentially support them (Geldenhuys and McDevette 1989). This would explain the low levels of endemism, low turnover, and regional richness of south-eastern fynbos (Cowling et al. this volume) and the gradual westwards depauperization of the tropical flora (Cowling 1983a; Geldenhuys and McDevette 1989).

What are the most productive avenues for exploring the implications of this hypothesis? Palaeoclimatic modelling is clearly important but likely to be confounded by the complexity of climatic determinants and the steep climatic gradients in the Cape Floristic Region (Campbell 1983; Deacon et al. this volume). Palynological studies do have potential to indicate gross changes in vegetation (e.g. fynbos-karoo transition) but can provide no hint of the temporal changes in the number of *Erica* species in a site (Sugden 1990). Furthermore, these studies are limited in the Cape Floristic Region by the paucity of suitably long sequences in the mountains (Deacon 1983) and the invariant generic and family composition of fynbos along broad climatic gradients (i.e. from 250–3 000 mm yr^{-1}) (Kruger 1979). Comparative cladistic biogeographical analyses of monophyletic lineages in the south-west and the south-east hold much promise for explaining patterns of speciation in relation to ecological gradients (Linder 1985; Linder and Vlok 1991).

VEGETATION

Vegetation classifications, however natural or quantitative, are invented by humans (Mayr 1968; Campbell 1986b). As such they are working hypotheses, subject to further modification in an iterative process termed successive approximation (Poore 1962). Nowhere else in southern Africa has the successive refinement of vegetation concepts aroused so much controversy as in the fynbos biome.

Until the publication of two major reviews of fynbos ecology about a decade ago (Taylor 1978; Kruger 1979), very little progress had been made with vegetation classification in the fynbos biome. These reviews did make advances over Acocks' (1953) very general treatment of the biome's vegetation but did not provide comprehensive and workable vegetation concepts for the region. Taylor (1978) attributed the poor state of fynbos typology partly to a history of botanical effort directed largely towards solving the numerous taxonomic problems of the Cape flora, and partly to the complexity of the vegetation.

Recently Campbell (1985a) produced a classification of the mountain vegetation of the fynbos biome based largely on structural characters. His vegetation concepts have been applied successfully in the lowlands of the biome by Cowling et al. (1988) and Rebelo et al. (1991). We use a slightly modified form of Campbell's scheme as the basis for vegetation description in this chapter (Figure 3.5; Table 3.5).

Our approach in reviewing vegetation classification and characterization in the fynbos biome is as follows: firstly we review the major typological controversies which have arisen since the publication of Taylor (1978) and Kruger (1979). Although largely parochial, these controversies stimulated considerable debate. Secondly, we justify our choice of Campbell's scheme and provide an outline of that scheme, concentrating on the series level of his hierarchy. This section is summarized in Table 3.5 and Figures 3.5 and 3.6. Finally, we discuss the determinants of the boundaries of structurally defined vegetation types in the fynbos biome.

Controversies in vegetation typology

WHAT IS FYNBOS?

As a vegetation concept, fynbos has been defined in a number of different ways. Taylor (1978) defined fynbos physiognomically as having the characteristic presence of restioids (wiry, aphyllous, evergreen graminoids of the Restionaceae), ericoids (shrubs with small rolled leaves), and proteoids (shrubs with large isobilateral leaves belonging to the Proteaceae); and floristically by the lack of single species dominance and/or the conspicuous presence of the Restionaceae. This definition of fynbos was upheld by Kruger (1979) who emphasized that 'the only constant and differential element is the Restionaceae'. It very quickly emerged that this definition was problematic. The notion of a lack of single species dominance was rejected (Bond 1981; Cowling 1984) as was the differential role of any single structural feature (Campbell 1985a). Furthermore, Taylor's (1978) definition did not distinguish between Cape fynbos and fynbos occurring in the Afromontane Region (Killick 1979). Cowling (1984) overcame this problem by defining Cape fynbos shrubland in terms of biogeographic as well as structural and ecological criteria. His definition states that: 'Sample floras show a phytochorological spectrum in which approximately 50% of the species are restricted to the Cape phytochorion as delimited by Werger (1978b).' Moll and Jarman (1984a) also distinguish between Cape and Afromontane fynbos on phytochorological criteria. They define Cape fynbos as: 'Evergreen sclerophyllous shrublands, on oligotrophic soils, comprising essentially Cape Floristic Kingdom elements, consisting predominantly of either functionally isobilateral picophyllous and/or microphyllous to mesophyllous-leaved shrubs and usually associated with evergreen aphyllous and/or narrow-leaved sclerophyllous hemicryptophytes.' Campbell (1985a) provides a largely structural definition of fynbos which, he argues, would exclude most Afromontane fynbos because of the absence or low cover of restioids. Campbell's polythetic and somewhat cumbersome definition of Cape fynbos shrubland is certainly the most comprehensive available. It is presented in the form of a key but can be summarized as follows: 'The essential feature of [Cape] fynbos is the presence of restioids . . . and that other important features are Ericaceae and proteoids. In addition, the following attributes are more or less differential: sedges, particularly leafy sedges; non ericaceous ericoids (with > 5% cover); ericoid Asteraceae; stoeboids (shrubs with crowded minute leaves in fascicles); leaf spinescence; Penaeaceae; and Bruniaceae.' The importance of restioids in defining fynbos is evident from the fact that over 60% of the fynbos plots sampled by Campbell in the Cape mountains had > 30% cover restioids and over 85% had > 10% restioids.

WHAT IS FALSE FYNBOS?

Acocks' (1953) treatment of the vegetation of the fynbos biome created a great deal of confusion (Cowling 1984). He recognized three fynbos veld types — namely macchia (= fynbos), coastal macchia, and false macchia. Acocks' concept of a veld type is an agroecological unit of vegetation 'whose range of variation is small enough to permit it to have the same farming potentialities'. He mapped false macchia as incorporating all the fynbos of the Cape mountains from the central Langeberg and Swartberg to the extreme south-east near Grahamstown. Acocks (1953) suggested that false macchia was derived from grassland or forest as a result of selective grazing, clearing,

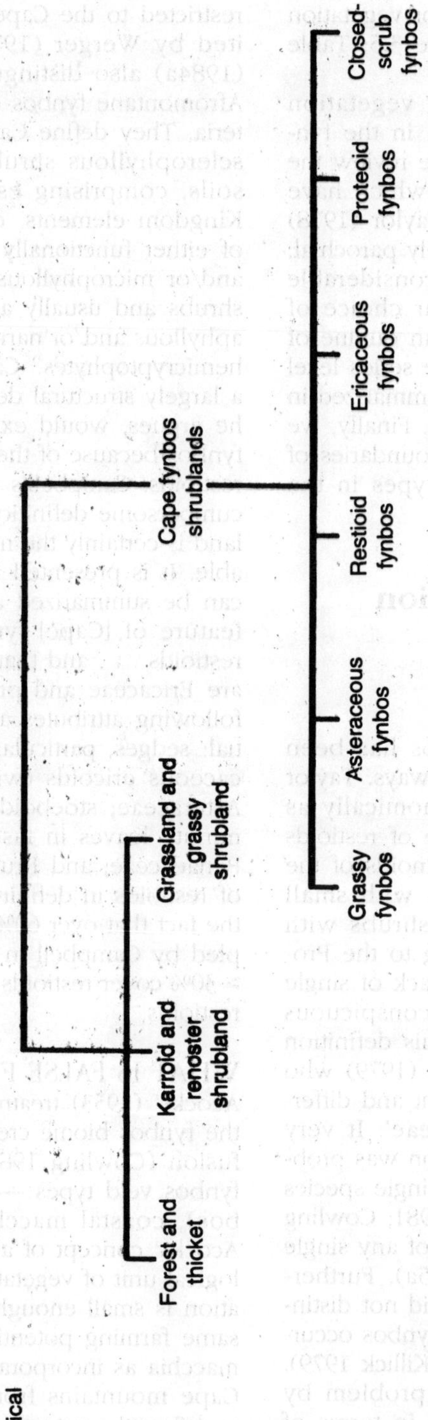

FIGURE 3.5 Hierarchical classification of fynbos biome plant communities, based on Campbell (1985a). Levels below the group and below the series are not shown for non-fynbos and fynbos respectively.

and characterized. See Figure 3.5 for the community hierarchy.

Community	Formation[1]	Differentiating features	Floristics[2]	Distribution[3]	Environment
Non-fynbos[4] Group: Forest and thicket					
Afromontane forest	Low or tall, microphyllous, evergreen forest	High cover of tall trees, microphylls and mesophylls; low cover of climbers and stems spinescence; well developed ground layer with broad-leaved sedges and ferns	*Halleria lucida, Podocarpus* spp., *Maytenus acuminata, Virgilia oroboides; Secamone alpinii, Chionanthus foveolata, Rapanea melanophloeos*	SW and SC mountains and forelands; generally restricted to small patches except in the central SC	Mesic, fire- and wind-protected sites with year-round moisture, mostly below 1 000 m; soils are mostly deep colluvium or alluvium and derived from a range of substrata; rainfall is usually above 800 mm yr^{-1}
Eastern forest and thicket	Low forest or tall, closed, microphyllous, evergreen shrubland	High cover of climbers and stem spinescence; well developed ground layer with leafy sedges and forbs	*Cassine aethiopica, Scutia myrtina, Pterocelastrus tricuspidatus, Sideroxylon inerme Schotia latifolia, Hippobromus pauciflorus, Rhus* spp.	E and SC forelands and E mountains; generally restricted to small patches	As above but often on more fine textured soils. Common in river valleys; rainfall is usually above 600 mm yr^{-1}
Western thicket	Mid-high to tall, open to closed, microphyllous, evergreen shrubland	High cover of small-leaved shrubs and a low cover of climbers and stem spinescence	*Heeria argentea, Hartogiella schinoides, Maytenus oleoides, Rhus angustifolia, Olea europaea* spp. *africana, Rhus lucida, Montinia caryophyllacea*	NW and SW, mostly on mountains, in clumps interspersed in fynbos; more rarely on NW forelands	Usually in sites too dry for Afromontane forest; on fire-protected sites especially rock outcrops and scree slopes
Dune thicket	Mid-high to tall, closed, microphyllous to nanophyllous, evergreen shrubland	High cover of nanophylls, climbers and stem spinescence; low cover of succulents	*Olea exasperata, Euclea racemosa, Rhus crenata, R. glauca, Sideroxylon inerme, Cassine maritima, Maytenus procumbens, Pterocelastrus tricuspidatus*	E, SC and SW forelands	On calcareous coastal dunes; rainfall is usually above 450 mm yr^{-1}.
Succulent thicket	Mid-high to tall, open to closed, nanophyllous, evergreen to semi-deciduous shrubland	High cover of nanophylls, climbers, stem spinescence and succulents; moderate cover of deciduous shrubs (in SW and NW)	*Euclea undulata, Rhus longilspina, Capparis sepiaria, Pappea capensis, Sideroxylon inerme, Euphorbia* spp., *Aloe* spp., *Crassula* spp. *Zygophyllum* spp., *Delosperma* spp.	E and SC forelands with a marginal intrusion into the SW and NW on coastal dunes	On deep, fertile and well-drained soils (including coastal dunes); rainfall usually less than 450 mm yr^{-1}

TABLE 3.5 continued.

Community	Formation[1]	Differentiating features	Floristics[2]	Distribution[3]	Environment
Group: Karroid and renoster shrubland					
Tall succulent shrubland	Mid-high to tall, open succulent and nanophyllous shrubland	High cover of succulents and nanophylls > 1.5 m	*Crassula rupestris* and other spp., *Portulacaria afra, Euclea undulata, Rhus glauca, Pappea capensis, Grewia robusta, Dodonaea viscosa*	SI mountains and valleys	On shallow to deep, usually fine textured soils on lower mountain slopes and hills; occasionally on steep, quartzite slopes; rainfall is usually less than 350 mm yr^{-1}
Succulent shrubland	Low, open to sparse, succulent and leptophyllous shrubland	Low cover of low (< 1.5 m) largely succulent shrubs. Few perennial grasses and ericoid-leaved shrubs	Very variable — *Pteronia* spp., *Relhania squarrosa, Tylecodon* spp., *Eberlanzia* spp., *Ruschia* spp. (and other Mesembryanthemaceae), *Euphorbia mauritanica* and other spp., *Crassula* spp.	I, NW and SI on inland lower slopes and valleys	Usually on non-quartzite-derived, deep to shallow, soils in the driest regions of the biome (rainfall < 250 mm yr^{-1}); occasionally on north-facing, dry quartzitic slopes and shale bands at higher altitudes (1 000–1 500 m)
Clanwilliam karroid shrubland	Low to mid-high, open, leptophyllous (ericoid or fleshy-leaved) shrubland	Low cover of fynbos characters (restioids, proteoids, non-ericaceous ericoids) and a high cover of elytropappoids[5]; high cover of grasses and fleshy-leaved shrubs	*Pteronia fasciculata, Relhania* spp., *Anthospermum tricostatum, Crassula rupestris, Ruschia spp., Rhus spp., Eriocephalus spp., Euryops speciosissimus, Passerina spp., Phylica villosa, Agathosma ovata*	I, NW and SI, restricted to lower mountain sites	Soils are deep (colluvial) or shallow lithosols always derived from quartzite; occurs on sites ecotonal between succulent shrubland and fynbos
Renoster shrubland	Low to mid-high, open to mid-dense leptophyllous shrubland, often with an open, grassy understorey	Differentiated from previous three series by a high cover of leptophylls and elytropappoids, and a high total cover	*Elytropappus rhinocerotis, Pteronia* spp., *Anthospermum aethiopicum, Dodonaea viscosa, Pentaschistis* spp., *Relhania* spp., *Helichrysum spp., Ruschia* spp., *Themeda triandra*	Throughout the fynbos biome on lower mountain, interior valley and coastal foreland sites	On fine textured soils usually on the ecotone between fynbos and succulent shrubland or tall succulent shrubland; very occasionally on quartzites

Community	Formation[1]	Differentiating features	Floristics[2]	Distribution[3]	Environment
Group: Grassland and grassy shrubland					
Grassy shrubland	Low to mid-high, open to mid-dense shrubland with an open grassy understorey	Herbaceous layer dominated by grasses with a high cover of non-elytropappoid shrubs	*Dodonaea viscosa*, *Phylica axillaris*, *Montinia caryophyllacea*, *Themeda triandra*, *Tristachya leucothrix*, *Pentaschistis eriostoma* and other spp., *Euryops* spp., *Rhus* spp., *Cymbopogon marginatus*	SI and E on lower mountain slopes and coastal forelands	Mostly confined to dissected valleys and Tertiary surfaces where soils are often deep, colluvial and relatively fertile; rainfall varies from 500–800 mm yr^{-1}
Grassland	Closed grassland often with a sparse shrub overstorey	Predominance of grasses and low cover of shrubs	*Themeda triandra*, *Tristachya leucothrix*, *Heteropogon contortus*, *Aristida junciformis*, *Elionurus muticus*, *Merxmuellera stricta*, *Pentaschistis* spp., *Restio triticeus*	E, as above	Generally on more mesic sites than above; soils are usually fine textured and deep, often on north-facing slopes where grassy fynbos is on south-facing slopes; rainfall is 600–800 mm yr^{-1}
Fynbos					
Group: Cape fynbos shrubland					
Grassy fynbos	Low, mid-dense to closed grassy letophyllous (ericaceous or ericoid), grassy shrubland	High grass cover and a relatively high cover of non-proteoid nanophylls and forbs	Variable: *Themeda triandra*, *Tristachya leucothrix*, *Restio triticeus*, *Leucadendron salignum*, *Phylica axillaris*, *Erica* spp., *Helichrysum* spp., *Rhus* spp., etc.	Mainly E but also SI and SC on lower mountain slopes and coastal forelands	Easternmost fynbos (high proportion of summer rain) on finer textured and more fertile soils than other fynbos types
Asteraceous fynbos	Low to mid-high, open to mid-dense, leptophyllous shrubland	Low total cover, often high grass and elytropappoid cover and a high cover of non-ericaceous ericoids	*Elytropappus*, *Pentaschistis*, *Phylica*, *Passerina*, *Agathosma* and other Rutaceae, *Aspalathus*, *Restio*, *Felicio*, *Hypodiscus*, *Cliffortia*, *Maytenus oleoides*, *Protea nitida*	Throughout the biome for mountains in E	Occupies the driest fynbos sites on a range of substrata, linking fynbos to karroid and renoster shrublands; widespread on coastal forelands on calcareous dunes and on shales and sil-ferricretes where rainfall is > 550 mm yr^{-1}; in the mountains, rainfall ranges from 450 to 950 mm yr^{-1} and soil depth averages less than 0.4 m

TABLE 3.5 continued.

Community	Formation[1]	Differentiating features	Floristics[2]	Distribution[3]	Environment
Restioid fynbos	Dwarf to tall, mid-dense to closed restioland with a sparse shrub stratum	High restioid cover (> 60%) and low shrub cover (< 30%); of all fynbos; it has the lowest constancy of tall (> 1.5 m) shrubs	Restionaceae, *Tetraria*, *Leucadendron*, *Pentaschistis*, *Phylica*, *Protea*, *Stoebe*, etc.	Throughout the biome except for the mountains in E; rare in SC	Occurs where conditions are limiting for shrub growth either owing to excessive waterlogging or drainage; it occupies more mesic sites than asteraceous fynbos; soils may be deep sands (e.g. on coastal forelands) or shallow and rocky; in the mountains restioid fynbos is a feature of dry, north slopes
Ericaceous fynbos	Low to mid-high, closed ericaceous shrubland	Like asteraceous fynbos a leptophyllous shrubland but has a high cover of restioids and the shrubs are mainly ericaceous ericoids. Shrub cover and total cover are also higher than Asteraceous fynbos	Ericaceae, Restionaceae, Bruniaceae, Peneaceae, Grubbiaceae, *Tetraria*, *Leucadendron*	Mostly SW and SC	Largely confined to south-facing and wet (rainfall average over 1 500 mm yr^{-1}) slopes of the coastal mountains; soils have a high carbon content, low pH and a high fine particle fraction
Proteoid fynbos	Low to tall, open to closed, proteoid[6] shrubland	Differs from other series by having > 10% cover of mid-high to tall non-sprouting, proteoid shrubs; included also are some communities on the coastal foreland where the canopy proteoids are low (< 1.5 m)	*Leucadendron conicum*, *L. coniferum*, *L. eucalyptifolium*, *L. gandogeri*, *L. laureolum*, *L. loeriense*, *L. meridianum*, *L. rubrum*, *L. uligonosum*, *L. xanthoconus*, *Protea aurea*, *P. compacta*, *P. eximia*, *P. laurifolia*, *P. lorifolia*, *P. mundii*, *P. neriifolia*, *P. punctata*, *P. obtusifolia*, *P. repens*, *P. susannae*	Throughout the biome except I; relatively rare in NW and E and most most extensive in the SC	Like grassy fynbos, proteoid fynbos occurs at lower altitudes and on soil than other fynbos series; occurs on a wide range of substrata including non-aeolian sand and limestone on the forelands; altitude ranges from sea-level to 950 m, and rainfall from 400 (coastal) to 1 100 mm yr^{-1}

Community	Formation[1]	Differentiating features	Floristics[2]	Distribution[3]	Environment
Closed-scrub fynbos	Mid-high to tall, open to closed, nano-microphyllous shrubland with an open cover of tall restioids	Similar to forest and thicket in the relatively high cover of meso-phyllous non-proteoid woody plants. Differs from forest and thicket by having a high cover (> 10%) of restioids and the frequent presence of Ericaceae	*Meterosideros angustifolia*, *Brachylaena neriifolia*, *Leptocarpus paniculatus*, *Salix mucronata*, *Cannomois virgata*, *Berzelia intermedia*, *Empleurum serrulatum*, *Leucadendron salicifolium*, *Elegia capensis*	Throughout the biome except in the E and I	Associated with well-drained riparian habitats in the mountains

1 According to Campbell et al. (1981).
2 Dominant species but in the case of fynbos series, only genera of families are listed.
3 In relation to regions shown in Figure 3.7, C = central, E = eastern, I = interior, NW = north-western, SC = southern coastal, SI = southern interior.
4 See text for distinction between fynbos and non-fynbos.
5 Shrubs with scale-like, pubescent leaves.
6 Shrubs with large isobilateral leaves belonging to the Proteaceae.

and fire. Cowling (1984) has argued that a false macchia concept is clearly untenable for the endemic-rich, non-grassy fynbos communities of the southern Cape. He also proposed that the notion of grassy fynbos (Table 3.5) as a recently derived vegetation type is also untenable.

Evidence refuting Acocks' (1953) hypothesis is the large number of regional and local endemics of fynbos affinity in grassy fynbos (Cowling 1983a). Secondly, while grassland, forest, and thicket communities are commonly found in the eastern mountains (Table 3.5, 3.6; Figure 3.6), their distribution can often be readily explained by site factors associated with soils, rainfall, and aspect: disturbance regime need not always be invoked (Cowling 1984; Campbell 1985a). Unfortunately, the notion that grassy fynbos is a false or derived vegetation type is entrenched in the agricultural community with often unfortunate consequences for resource management (Cowling 1984).

IS FYNBOS A HEATHLAND?

Specht (1979) proposed a global heathland concept comprising evergreen sclerophyllous communities restricted to low nutrient soils and having the presence, but not necessarily the dominance, of heath families belonging to the order Ericales. Specht's concept includes northern temperate heaths (Ericaceae-dominated, on nutrient-poor, podzolized, humic soils) as well as warm temperate to subtropical sclerophyll formations in Australia (e.g. kwongan) (Pate and Beard 1984) and South Africa (e.g. fynbos) (Kruger 1979). Heathlands are distinguished from shrublands which have an essentially seasonal herbaceous understorey (Specht and Moll 1983). In the fynbos biome, fynbos communities are regarded as heathlands whereas non-fynbos sclerophyllous shrublands (renoster shrubland, thicket) are classified as shrublands (Specht and Moll 1983). These latter communities are regarded as being analogous to other mediterranean-type shrublands (Boucher and Moll 1980) while fynbos is regarded as having stronger structural and functional similarities to other heathlands, irrespective of the climate in which they grow (Moll and Jarman 1984b; cf. Cowling and Campbell 1980; Milewski and Cowling 1985).

Bond (1981), Cowling (1984), and Campbell (1985b) have all rejected the concept of fynbos as a heathland. Their arguments are summarized below. Firstly, many fynbos communities lack heath families. Secondly, seasonal grasses are very common in some fynbos communities, particularly grassy fynbos. Thirdly, many non-fynbos communities in the Cape mountains (shrublands in Specht and Moll's (1983) terminology) also occur on nutrient-poor soils. As indicated by Campbell (1985b), a formulation of Specht's (1979) heathland concept based solely on soil nutrient status is problematic since many tropical savannas on nutrient-poor soils have no heathland floristic affinities.

Bond (1981) and Campbell (1985b) have proposed that in the fynbos biome the term heathland should apply to only ericaceous fynbos (Table 3.5) which best approximates the north temperate heathland concept. The similarity between north temperate heathlands and ericaceous fynbos probably relates to the occurrence of both vegetation types under relatively mild, moist oceanic climates on nutrient-poor soils. Moll and Jarman (1984b) have attempted to uphold Specht's (1979) heathland concept for fynbos by making *ad hoc* adjustments to accommodate the abovementioned criticisms. Although this typological controversy has now dissipated, the positive role it played in highlighting the role of nutrients as determinants of fynbos biome vegetation structure cannot be overlooked. A major limitation of the heathland concept was that it demanded the simplistic categorization of shrubland types in terms of soil nutrient status, a pitfall avoided by Campbell (1986a) in his treatment of the mountain vegetation of the fynbos biome.

COASTAL AND MOUNTAIN FYNBOS

Acocks (1953) was the first to distinguish between fynbos of the coastal lowlands (coastal macchia) and fynbos on the mountains (macchia and false macchia). This distinction was upheld by Taylor (1978), Kruger (1979), Moll et al. (1984), and Boucher (1987) (Table 3.7).

Taylor (1978) distinguished between coastal and mountain fynbos on entirely floristic grounds. Kruger (1979) stressed physiographic differences when he noted that

TABLE 3.6 Number of structurally defined communities in non-fynbos and seven fynbos series in each of the seven mountain regional community complexes recognized by Campbell (1985a). The communities are 'types' — the lowest recognized in Campbell's hierarchical scheme. See Table 3.5 for a description of the series and Figure 3.7 for a delineation of the regional complexes.

	Regional complex						
	North-western	Interior	Central	South-western	Southern-coastal	Southern interior	Eastern
No. transects	3	2	2	3	3	4	4
Non-fynbos	5	3	2	2	3	7	9
Grassy fynbos	0	0	0	1	2	5	5
Asteraceous fynbos	9	6	5	4	3	7	1
Restioid fynbos	7	1	11	8	2	6	0
Ericaceous fynbos	1	0	1	4	4	2	0
Proteoid fynbos	2	0	6	11	12	14	8
Closed-scrub fynbos	2	0	2	2	1	2	0
Total	26	10	27	31	26	41	23

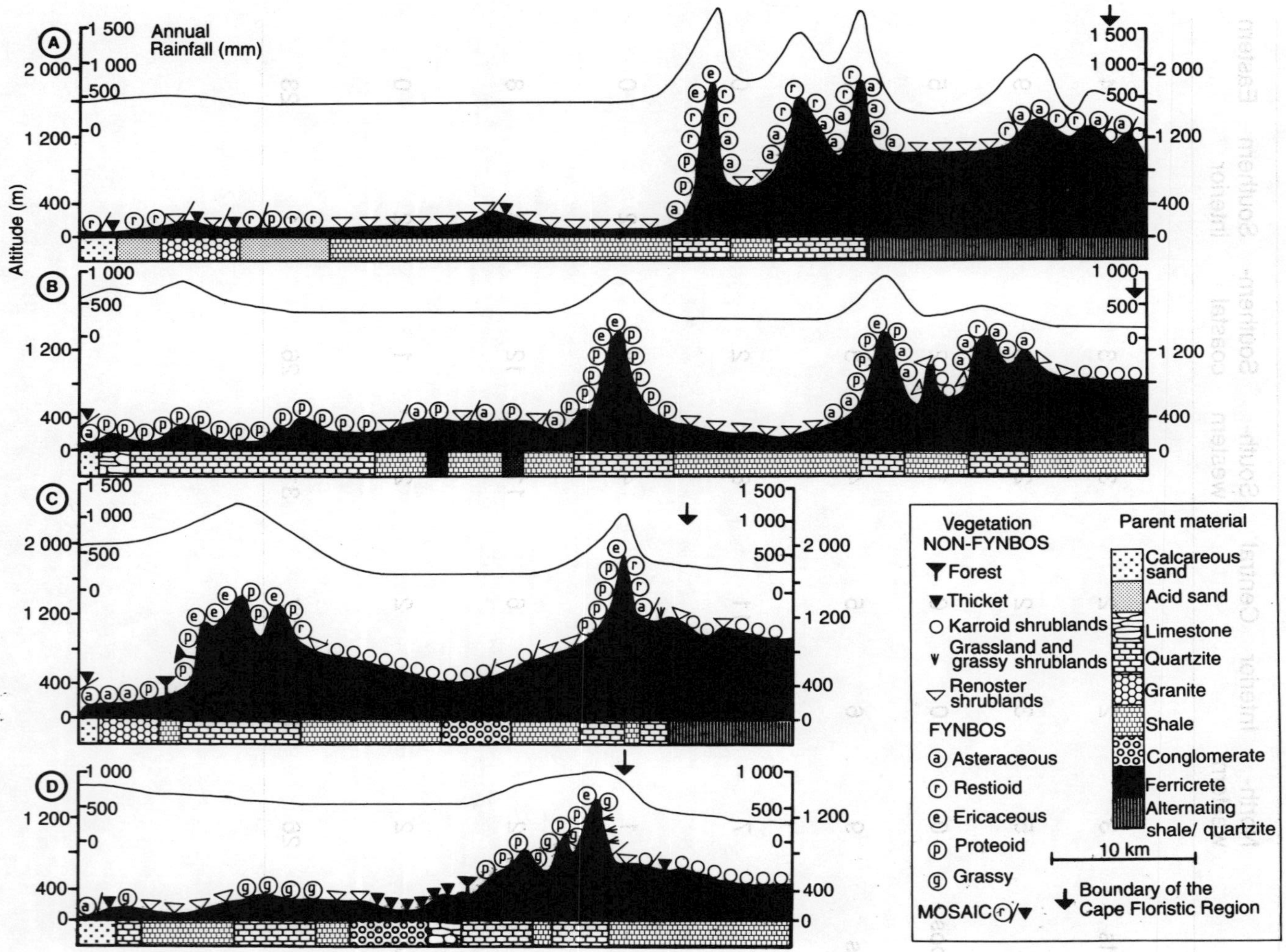

FIGURE 3.6 Distribution of vegetation types in relation to rainfall, topography, and geological substratum along four coast-to-interior transects in the fynbos biome. Locations of transects are shown in Figure 3.7.

This study (Based on Campbell 1985a)	Taylcr (1978)	Kruger (1979)	Moll et al. (1984)	Relevant phytosociological studies[1]
Non-fynbos				
Group: Forest and thicket				
Afromontane forest	–	–	Afromontane forest	Boucher (1978), Campbell and Moll (1977), Cowling (1983b), Geldenhuys (1989), Glyphis et al. (1978), McDonald (1988), McKenzie (1978), Werger et al. (1972)
Eastern forest and thicket	–	–	Kaffrarian thicket	Cowling (1983b)
Western thicket	Coast scrub	–	–	Boucher (1987), Boucher and Jarman (1977), McDonald (1988), Moffett and Deacon (1977), Van Wilgen and Kruger (1985), Werger et al. (1972)
Dune thicket	Strandveld, coast scrub	–	Kaffrarian thicket, south coastal strandveld	Boucher (1978, 1987), Boucher and Jarman (1977), Cowling (1983b), Taylor (1984a), Van der Merwe (1976, 1977)
Succulent thicket	Strandveld, coast scrub	–	Kaffrarian succulent thicket	Boucher (1987), Boucher and Jarman (1977), Cowling (1983b)
Group: Karroid and renoster shrubland				
Tall succulent shrubland	–	–	–	–
Succulent shrubland	–	–	–	Bond (1981)
Clanwilliam karroid shrubland	–	–	–	Bond (1981), Milton (1978)
Renoster shrubland	Coastal renoster-veld	–	South coastal renosterveld, south west coastal renoster-veld, west coast renosterveld	Boucher (1987), Cowling (1983b), Moffett and Deacon (1977)
Group: Grassland and grassy shrubland				
Grassy shrubland	–	Mountain fynbos – eastern moun-tains and hills	Grassy fynbos	Cowling (1983b), Moffett and Deacon (1977)
Grassland	–	Mountain fynbos – eastern moun-tains and hills	Grassy fynbos	Cowling (1983b)

Table 3.7 continued.

This study (Based on Campbell 1985a)	Taylor (1978)	Kruger (1979)	Moll et al. (1984)	Relevant phytosociological studies[1]
Fynbos				
Group: Cape fynbos shrubland				
Grassy fynbos	–	Mountain fynbos – eastern mountain and hills	Grassy fynbos	Cowling (1983b)
Asteraceous fynbos	Mountain fynbos – proteoid zone Arid fynbos Coastal fynbos Strandveld	Mountain fynbos tall broad sclerophyllous shrubland or open shrubland with heathland Arid fynbos Coastal fynbos – fynbos of coast sands	Dry mountain fynbos, Elim fynbos, dune fynbos, south coast strandveld, west coast strandveld	Bond (1981), Boucher (1978, 1987), Boucher and Jarman (1977), Cowling (1983b), McDonald (1988), McKenzie (1977), Moffett and Deacon (1977), Taylor 1984, Taylor and Van der Meulen (1981), Van der Merwe (1976, 1977), Van Wilgen and Kruger (1985), Werger et al. (1972).
Restioid fynbos	Mountain fynbos – hygrophilous fynbos – ericoid-restioid zone Coastal fynbos	Mountain fynbos – herbland – plateaux and depressions – restioid herb-lands Arid fynbos Coastal fynbos Fynbos on coastal sands	Dry mountain fynbos, mesic mountain fynbos, dune fynbos, south coast strandveld, west coast strandveld	Bond (1981), Boucher (1978, 1987), Boucher and Jarman (1977), Glyphis Laidler et al. (1978), Taylor (1985), Taylor and Van der Meulen (1981), Van Wilgen and Kruger (1985)
Ericaceous fynbos	Mountain fynbos – ericoid-restioid zone	Mountain fynbos – low ericoid open heath or open	Wet mountain fynbos	Bond (1981), Boucher (1978), Glyphis et al. (1978), Kruger (1974), Laidler et al. (1978), McDonald (1988), McKenzie et al. (1977)
Proteoid fynbos	Mountain fynbos – proteoid zone Coastal fynbos	Mountain fynbos – broad sclerophyll-ous scrub or open scrub – mixed sclero-phyllous scrub Coastal fynbos – fynbos on limestone	Mesic mountain fynbos, limestone fynbos, sand plain lowland fynbos	Bond (1981), Boucher (1978, 1987) Cowling (1983b), Kruger (1974), McDonald (1988), Taylor (1984a), Van der Merwe (1977)
Closed scrub fynbos	Mountain fynbos – hygrophilous fynbos	Mountain fynbos – broad sclero-phyllous closed scrub	Mesic mountain fynbos	Boucher (1978), Campbell and Moll (1977), McDonald (1988), Van Wilgen and Kruger (1985), Werger et al. (1972)

'Coastal fynbos is structurally like mountain fynbos and has the same plant families but occurs on the limestones and coastal forelands'. A problem with the floristic definition is that owing to high gamma diversity (geographic turnover) in the fynbos biome (Cowling et al. this volume), geographically distant mountain landscapes may have floras which show less similarity than juxtaposed mountain and coastal sites. Secondly, the physical and chemical characteristics of many lowland soils are no different from those in the mountains (Thwaites and Cowling 1988). Geographic location alone is a poor typological criterion.

Recently, Cowling et al. (1988) and Rebelo et al. (1991) have shown that vegetation of the fynbos biome lowlands can be easily accommodated by Campbell's (1985a) scheme. For example, the distinctive limestone communities mentioned by Taylor (1978) and Kruger (1979) were classified as proteoid fynbos (Table 3.7). Two new communities — dune asteraceous fynbos and dune thicket — were recognized to accommodate communities on calcareous coastal dunes. Elim fynbos, which Acocks (1953) and Moll et al. (1984) proposed as a distinctive type, was classified as asteraceous fynbos.

These studies have shown that the distinction between coastal and mountain fynbos on structural grounds is untenable. Campbell's (1985a) scheme has accordingly been adjusted by removing any reference to mountain fynbos (Figure 3.5).

FLORISTIC OR STRUCTURAL CLASSIFICATION?

The relative merits of floristic and structural approaches to classifying fynbos biome vegetation have been discussed by Bond (1981), Taylor and Van der Meulen (1981), Campbell (1986b), and Boucher (1987). There has been energetic support in the biome for the floristically based Braun-Blanquet method of sampling and synthesis as applied by the Zurich-Montpellier School of phytosociology (Werger 1974). Werger et al. (1972) concluded that this approach could be applied successfully to fynbos despite its floristic richness. There are several other examples where the Braun-Blanquet method has been successful. However, these have encompassed either small areas of fynbos (< 500 ha) (McKenzie et al. 1978; Glyphis et al. 1978; Laidler et al. 1978; Van Wilgen and Kruger 1985; Boucher and Shepherd 1988; McDonald 1988) or been located in relatively species-poor non-fynbos (Boucher and Jarman 1977; Campbell and Moll 1977; Cowling 1983b) and south-eastern fynbos (Van der Merwe 1976; Bond 1981; Cowling 1983b). In larger areas of south-western fynbos, both Boucher (1978) and Taylor (1984b) reported problems with the Braun-Blanquet approach which have subsequently been overcome (C Boucher personal communication). Boucher (1987), however, has successfully established a formal syntaxonomic hierarchy for fynbos and non-fynbos vegetation of the southern portion of the western Cape coastal forelands. Currently, phytosociological studies are being carried out in the north-western mountains (Cederberg) by H C Taylor and the south-central mountains (Langeberg) by D J McDonald. When these surveys are completed, there will be phytosociological surveys located in most of the phytogeographical regions in the fynbos biome. Both C Boucher and D J McDonald (personal communication) are confident that a formal taxonomic synthesis for the entire fynbos biome is feasible.

Campbell (1986b), on the other hand, argues against a formal floristic approach for classifying vegetation of the fynbos biome. His arguments centre on the floristic complexity of the fynbos. On practical grounds he maintains that the difficulty of identifying the many hundreds of species, both in the field and the herbarium, militate against this approach (see also Bond 1981). On theoretical grounds he contends that a floristic classification will answer biogeographic rather than ecological questions. This is due to the high gamma diversity of fynbos (Cowling et al. this volume) which results in environmentally and structurally similar communities having very different floristic composition (Bond 1981; Taylor 1984a; McDonald 1988). Werger et al. (1972) recognized this problem and suggested that it will be necessary to distinguish geographical races of an association, or regional associations with limited geographical extension.

The effects of fire on the composition of fynbos communities may also confound phytosociological studies. Cowling (1987) has suggested that the apparently random noise evident in phytosociological tables (e.g. Boucher

1978) could be explained by stochastic population processes associated with differential post-fire recruitment of fynbos species (Van Wilgen et al. this volume). Taylor (1984a) articulated this problem in the following statement: 'Precisely which species assumes local dominance appears to be largely determined by fire history — frequency, intensity and season of past burns.'

Campbell (1986b) advocates a physiognomic or structural-functional approach to overcome the taxonomic-floristic problems outlined above (see also Fosberg 1967; Beard 1973). In a flora with high levels of geographic turnover, a structural-functional classification is more likely to be effective in revealing ecological gradients than a floristic classification. He points out that all attempts at an overall fynbos classification have been structural (e.g. Marloth 1908; Adamson 1938; Taylor 1978; Kruger 1979).

An obvious problem with a structural classification of fire-prone shrublands is the effect of fire on physiognomy (Werger et al. 1972; Kruger 1979; Campbell 1986a; Boucher 1987). Structural classifications are inevitably based on the physiognomy of mature vegetation (e.g. Bond 1981; Campbell et al. 1981; Campbell 1985a) but at any given time vast areas of the mountains and lowlands are covered by immature vegetation. Furthermore, fire may dramatically alter population sizes of structurally diagnostic growth forms (e.g. non-sprouting proteoid shrubs) (Van Wilgen et al. this volume).

The success of Campbell's (1985a) structural approach is evident in the fact that he was able to provide a comprehensive typology for the mountains of the biome in a very short time. His scheme does, however, employ limited floristic (mainly family level) data which are diagnostically significant. At the current rate of research it is unlikely that a floristically based scheme would be available before the turn of the century. Nonetheless, mention must be made of the non-syntaxonomic importance of floristically based phytosociological studies. These time-consuming surveys generate a complex mass of data which is exceptionally valuable for testing a wide range of biogeographical and ecological hypotheses. In this volume, both Bond et al. and Cowling et al. have used phytosociological data for testing hypotheses on the determinants of community structure. Biologists and ecologists have a valuable source of inventory data for testing generalizations on the ecology and biogeography of the fynbos biome.

Vegetation types of the fynbos biome

Campbell's (1985a) structural scheme is the only comprehensive typology available, at this stage, for the entire fynbos biome. Therefore, we have adopted it here with minor modifications. His approach was a posteriori in that the classification was synthesized from empirical data. Campbell (1985a) sampled 507 plots on 22 transects located throughout the mountains of the fynbos biome (Figure 3.7). In each plot he assigned cover values to structural characters: over 400 different combinations of characters were recorded for the entire survey (see Campbell 1985a for details). Character choice was guided by hypotheses relating plant form to function (e.g. Campbell and Werger 1988; Stock et al. this volume).

The study also incorporated limited floristic data. Floristic information was used in the subdivision and characterization of structural characters (e.g. ericoid leaves were recorded by family) and dominant proteoid shrub species were used to differentiate subseries and types of proteoid fynbos. In addition, dominant species (> 10% cover) were noted in each plot.

The vegetation classification was produced using a table-sorting procedure based on visual inspection, similar to that used in the Braun-Blanquet approach (Campbell 1986b). The classification is hierarchical and communities at various levels are indicated by the syntaxonomically neutral terms group, series, subseries, and type (Figure 3.5). In this chapter we are concerned only with levels above the subseries. Communities below this level, particularly types, are very difficult to recognize in the field (Boucher 1978; personal observation). The subgroups eastern fynbos and mountain fynbos in Campbell's (1985) original classification have been rejected (see above and Cowling et al. 1988 for details). Other modifications which arise largely from the extension of Campbell's concepts to the lowlands of the fynbos biome (Cowling et al. 1988; Rebelo et

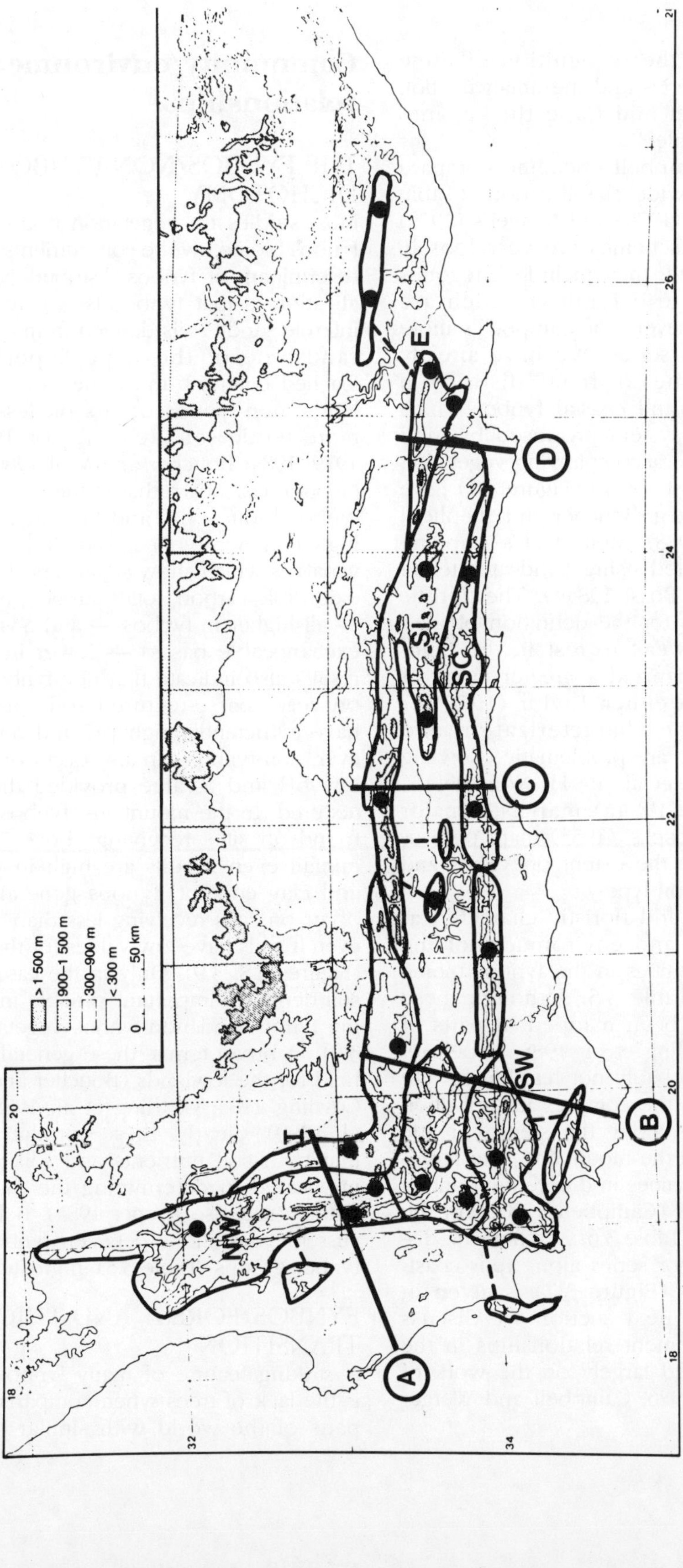

FIGURE 3.7 Location of Campbell's (1985a) transects (●) and the delimitation of his mountain regional community complexes. NW = north-western, I = interior, C = central, SW = south-western, SC = southern coastal, SI = southern interior, E = eastern. Also shown are the coast-to-interior transects (A–D) depicted in Figure 3.6.

al. 1991) include the recognition of dune thicket as a new series and the amalgamation of Mitchell thicket and Cape thicket into western thicket (Table 3.5).

How does Campbell's scheme compare with other biome-wide classifications (Table 3.7)? Both Taylor's (1978) and Kruger's (1979) structurally based schemes are very loosely and broadly defined: they include categories (e.g. strandveld, arid fynbos) which are accommodated in several of Campbell's more rigorously defined series. We have already commented on the artificial distinction between mountain and coastal fynbos which they uphold. The system proposed by Moll et al. (1984) and the accompanying vegetation map (Moll and Bossi 1984a) (Figure 3.1) have been widely used (e.g. Boucher and Shepherd 1988; McDonald 1988). Moll et al.'s mapping units were delineated using Landsat satellite imagery (Moll and Bossi 1984b). The floristic and structural basis for the definition of their categories is too weak to test in the field. Although they recognized a greater range of fynbos types than either Taylor (1978) or Kruger (1979), the characterization and delineation of these are problematic (Cowling et al. 1988; Rebelo et al. 1991). Nonetheless, Moll and Bossi's (1984a) map is a major improvement on Acocks' (1953) map, particularly with regard to the extent of fynbos and non-fynbos vegetation types.

The structural and floristic characterization, distribution, and environment of the major plant communities in the fynbos biome are presented in Table 3.5. Although these communities have been mapped in parts of the lowlands (Cowling et al. 1988; Rebelo et al. 1991), this is probably not feasible for the entire biome, given the complex juxtaposition of vegetation types along the steep climatic gradients present in the mountains. The representation of communities in different mountain regional complexes (Campbell 1985a) in the biome is shown in Table 3.6. A schematic distribution of the major series along four coast-to-inland transects (Figure 3.7) is given in Figure 3.6. In the next section we discuss community/environment relationships in the fynbos biome, based largely on the work of Campbell (1985a, 1986; Campbell and Werger 1988).

Community/environment relationships

THE FYNBOS/NON-FYNBOS DICHOTOMY

Most studies on vegetation boundaries in the fynbos biome invoke soil nutrients as the major determinant of fynbos distribution. The generalization is that fynbos is largely confined to infertile sandy soils derived from quartzites and sandstone of the Cape Supergroup, and leached colluvial and dune sands of the lowlands; non-fynbos occurs on less sandy and more fertile soils (e.g. Taylor 1978; Kruger 1979, 1984; Boucher and Moll 1980). Campbell (1986a) has shown that in the mountains of the fynbos biome, pH and clay in particular are higher in non-fynbos soils (Table 3.8). Other variables which show significant differences are oxidizable carbon, total nitrogen, medium sand — all higher in fynbos — and *S*-value (sum of exchangeable bases) — lower in fynbos. His results also indicate that non-fynbos can occur on very coarse-textured soils provided they have sufficiently high pH and *S*-value. However, non-fynbos can also occur on soils with a low pH and *S*-value provided they are fine-textured. In the mountains, fynbos is generally found on sites receiving above 700 mm yr^{-1} rainfall even if soils are high in pH, *S*-value, and clay content; fynbos generally does not occur on sites receiving less than 400 mm yr^{-1} even if soils have low values for these variables (Figures 3.8, 3.9). Only in the case of forest is soil depth an important variable in distinguishing fynbos and non-fynbos vegetation (Figure 3.9). In broad terms, these generalizations also hold for the lowlands (Boucher and Moll 1980; Cowling 1984; Cowling et al. 1988; Rebelo et al. 1990). Clearly, these correlations indicate that both soil nutrients and soil moisture are implicated in determining the distribution of fynbos (see also Kruger 1984). Below, we discuss the transitions between fynbos and non-fynbos groups (Table 3.5) in greater detail.

FYNBOS/FOREST AND THICKET TRANSITION

A striking feature of many fynbos landscapes is the lack of trees when compared with other parts of the world with similar climatic and

TABLE 3.8 Mean values for soil variables in fynbos and non-fynbos vegetation in the fynbos biome. Data for the mountains of the entire biome are from Campbell (1986b) and include a large range of geological substrata. Data from the south-eastern Cape (lowland and mountains) are from Cowling (1984) and are from T.M.G. quartzites. The same methods and laboratory were used for both groups of analyses.

	Mountains		S.E. Cape	
	Fynbos	Non-fynbos	Proteoid and grassy fynbos	Forest and thicket
pH	4 ***	4.5	4.4	4.6
Clay	7.1 **	11.3	1.7 ***	10.5
Oxidizable carbon	3.81 **	2.42	3.82 ***	11.84
S–value[1] (m/100g soil)	2.1 *	3.4	3.3 ***	13
Medium sand	34.1 *	26.3	–	–
Total N (%)	0.09 *	0.06	0.09 ***	0.44
Silt (%)	9.2	10.3	7.3 ***	18.7
Available P (ppm)	11.5	12.7	3.2 ***	10.1
n	52	20	16	8

*** = $P < 0.001$, ** = $P < 0.01$, * = $P < 0.05$

[1] *S*-value = sum of exchangeable cations

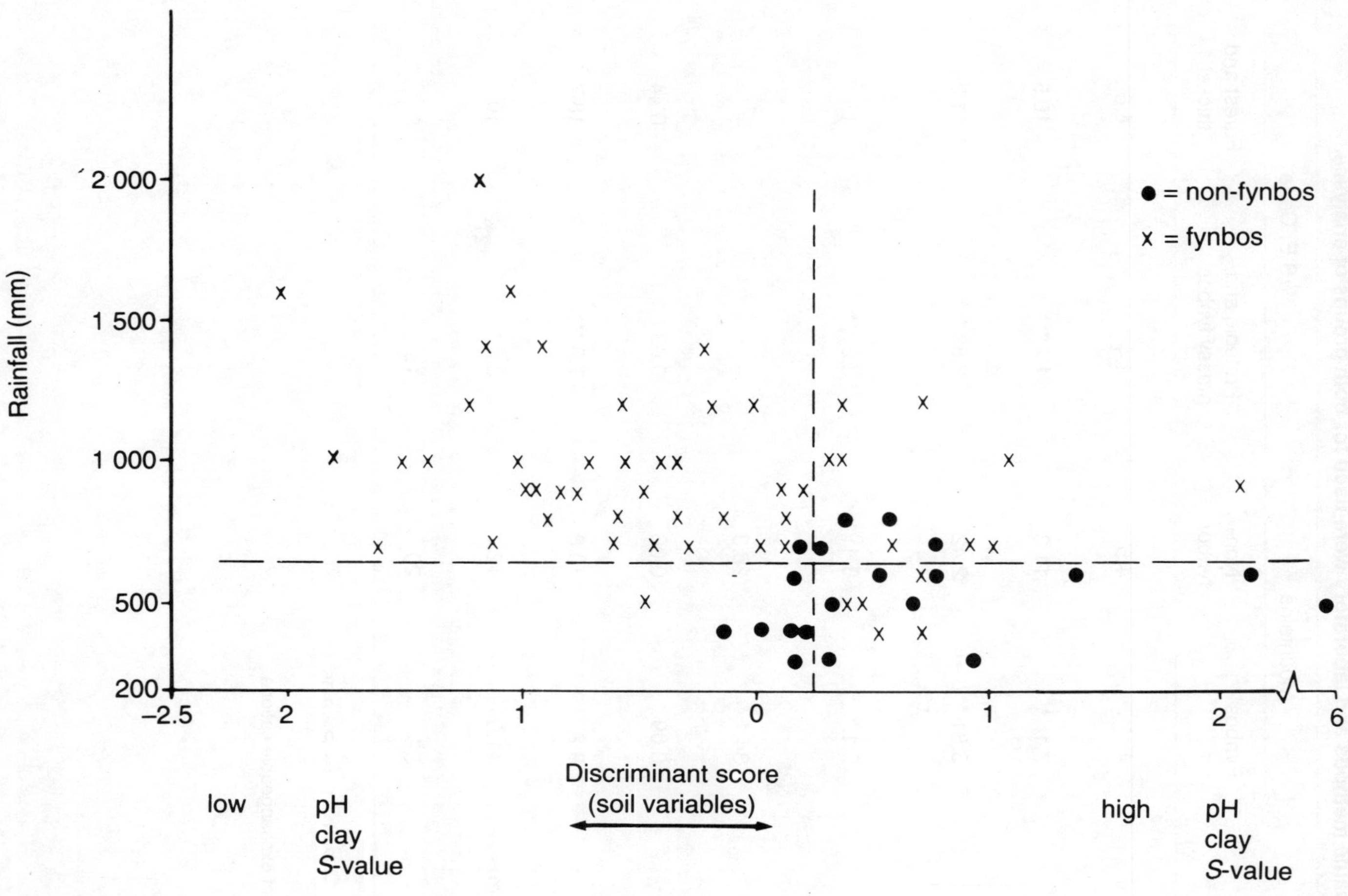

FIGURE 3.8 The relationship between annual rainfall and the discriminant scores from a discriminant function for fynbos and non-fynbos in the mountains of the fynbos biome. The plots used to construct the discriminant function are ordinated in the space defined by the axes. The dashed lines drawn in the two-dimensional space are the classification boundaries as given by the 600 mm isohyet and the discriminant function. Reprinted from Campbell (1986a) with the permission of Kluwer Academic Publishers.

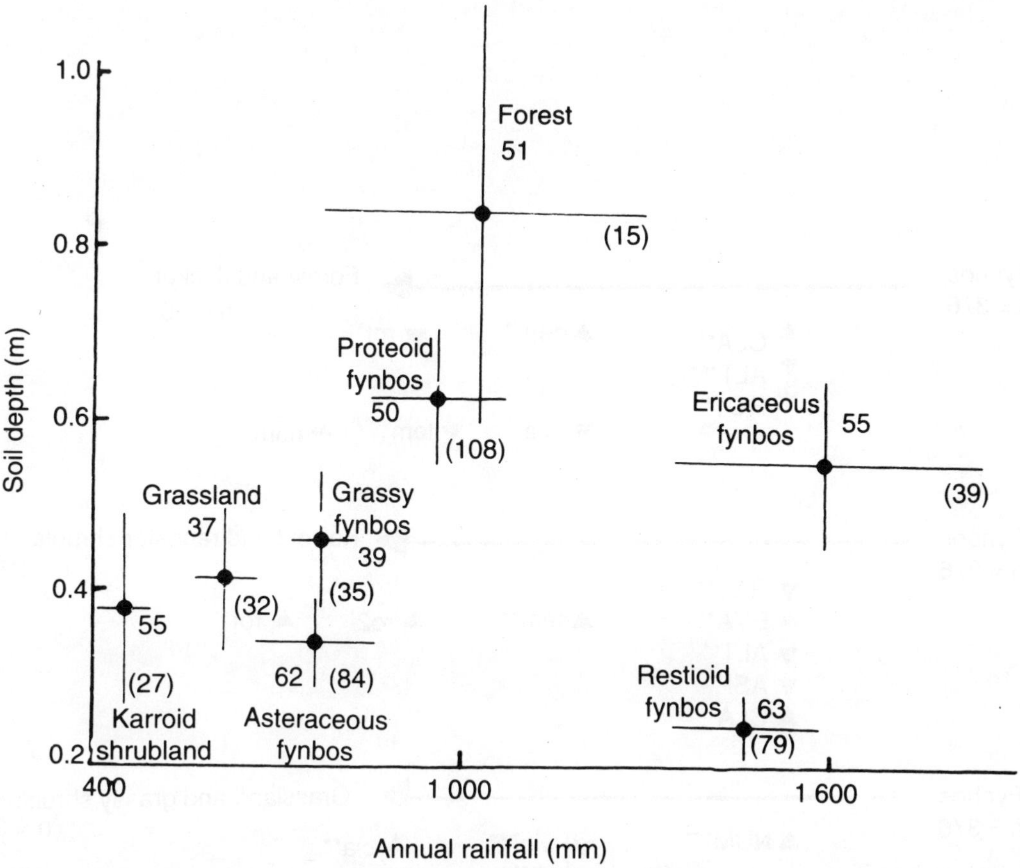

FIGURE 3.9 The major plant communities in the mountains of the fynbos biome ordinated within two-dimensional space defined by soil depth and annual rainfall. Ninety-five per cent confidence intervals for the means of these variables in each community are indicated. The numbers in the figure represent percentage winter rainfall for each community and, in brackets, number of plots per community. Reprinted from Campbell and Werger (1988) with the permission of Blackwell Scientific Publications.

edaphic conditions (Campbell et al. 1979; Richardson et al. this volume). For example, in south-western Australia, where winter rainfall exceeds 650 mm yr^{-1}, nutrient-poor soils support a woodland up to 10 metres tall; at higher rainfall (> 800 mm yr^{-1}) open forests up to 20 metres predominate (Beard 1984). Corresponding environments in the fynbos biome support fynbos, never taller than 5 metres (Richardson and Cowling 1991). Forest vegetation is of limited occurrence in the biome and is usually separated from fynbos by a distinct boundary. In this section we explore the ecological determinants of the fynbos forest and thicket boundary. Thicket, like forest, is a tropical African formation dominated by bird-dispersed species (Phillips 1931; Knight 1988) and is not fire-prone (Cowling 1984).

It is generally accepted that in the fynbos biome, Afromontane forest is restricted to fire-protected sites with deeper, moister, and more fertile soils than adjacent fynbos sites (Phillips 1931; Kruger 1979, 1984; Cowling 1984; Campbell 1986a; Manders 1990; Table 3.8). Thicket is strongly associated with deep, relatively fertile soils in fire-protected sites mainly in the lowlands (Cowling 1984; Cowling et al. 1988; Rebelo et al. 1991). In a discriminant analysis using environmental data from all of his plots in the mountains of the biome (n = 507, excludes soil chemical data), Campbell (1986a) showed that forest and thicket were largely associated with clay-rich soils (Figure 3.10). Other variables that differ between fynbos and forest and thicket reflect the latter's general restriction to low altitudes, mostly at

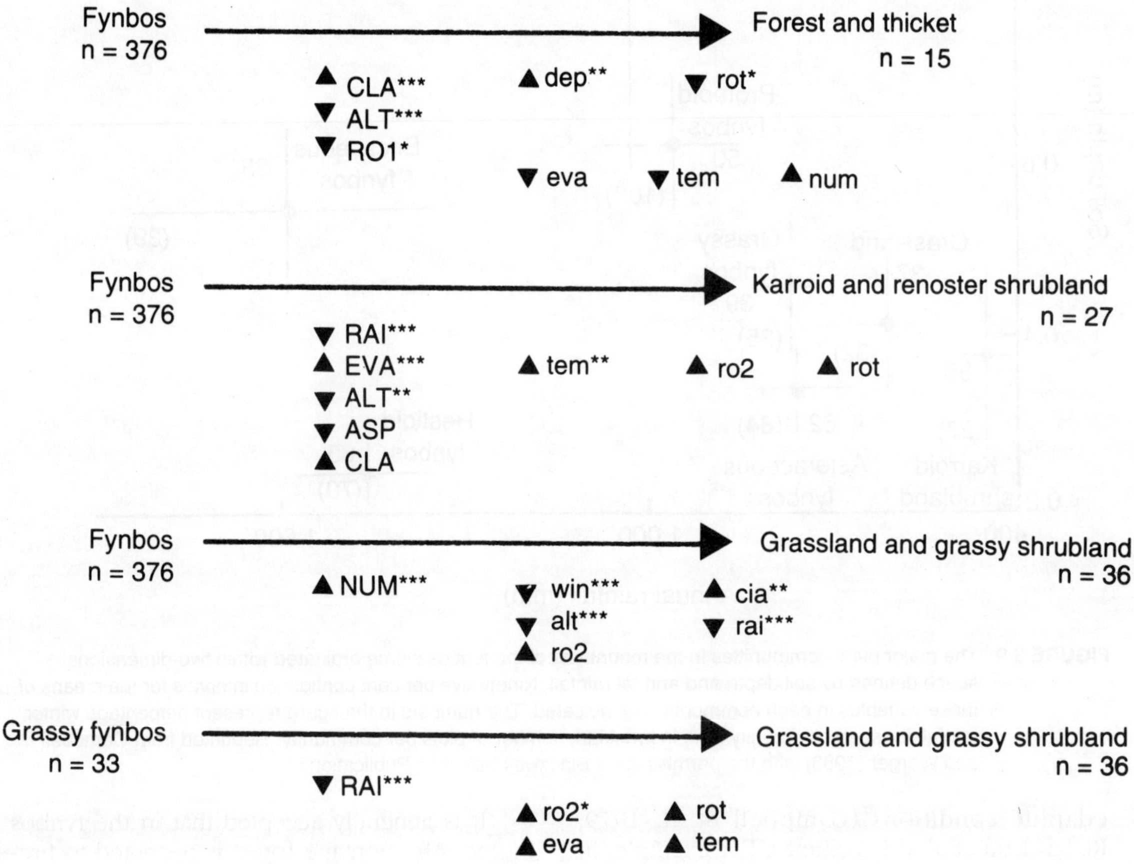

FIGURE 3.10 Step-wise discriminant analyses and significance test of environmental variables between fynbos and non-fynbos groups in the mountains of the fynbos biome. All the environmental variables listed are significantly different ($P < 0.05$) between groups (*** = $P < 0.0001$, ** = $P < 0.001$, * = $P < 0.01$). The variables in upper case are the variables in the discriminant function. The top variable in the list is the best discriminating variable, the second is the variable which accounts for most of the remaining difference between groups, and so on. The variables in lower case, not in the discriminant function, are listed next to the discriminating variables with which they are best correlated (results from Campbell 1983) and which, therefore, prevent their entry into the discriminant function (▲ = increasing, ▼ = decreasing). Abbreviations are as follows: alt = altitude, asp = aspect (hot to cool gradient), clay = % clay content, dep = soil depth, eva = annual pan evaporation, num = transect number (1–22 from north-west to south-east), rai = annual rainfall, rot = total rock cover, ro2 = rock cover 5–50 cm in diameter, tem = temperature range, win = % winter rainfall. From Campbell (1985b).

the base of southern and eastern coastal mountains. The longitudinal gradient (num in Figure 3.10) is associated with the increasing proportion of summer rain which would favour the essentially tropical, summer growing forest and thicket flora (Cowling 1983a; Pierce and Cowling 1984). The altitude restriction probably reflects the association of better developed soils with lower altitudes (Campbell 1983) although Bond (1981) suggests that a temperature-related tree-line exists in the southern Cape mountains.

In the Humansdorp region of the southeastern Cape, Cowling (1984) showed that on Table Mountain Group quartzites, forest and thicket soils have a significantly higher clay and silt content, *S*-value, oxidizable carbon, total nitrogen, and available phosphorus than fynbos soils (Table 3.8). Van Daalen (1981), however, has shown that in the Knysna region, Afromontane forest may develop on soils identical to those which support fynbos. Moreover, some soil chemical differences between fynbos and forest and thicket, particularly organic carbon and total nitrogen, may be plant-induced effects associated with different nutrient cycling processes of litter in the fire-free forest and thicket environment (Cowling 1984; Thwaites and Cowling 1988; Manders 1990; Stock and Allsopp this volume).

These correlative studies indicate that forest and thicket could potentially occupy a greater area of the fynbos biome than they do at present (Manders 1991). What then limits the spread of forest and thicket species into fynbos? Manders and Richardson (1991) have developed a general model for the establishment of nuclei of Afromontane forest in fynbos communities (see also McKenzie et al. 1978). Most forest species have bird-dispersed fruits and establish between fires, usually under tall shrubs which provide perch sites for frugivores (see also Knight 1988). Germination of forest species is enhanced by increased soil nutrient levels and litter cover associated with conditions beneath tall (usually proteoid) shrubs in fynbos (Bond and Stock 1989; Cowling and Gxaba 1990). In the absence of fire, long-lived forest species form nuclei which eventually develop into forest patches. Once established, these patches are highly fire-resistant (Manders 1990), largely due to the low flammability of fuel derived from forest species (Van Wilgen et al. 1990). Moreover, many forest species are capable of coppicing after fire (Phillips 1931). Manders and Richardson (1991) suggest that the exclusion of forest from fynbos is largely because the recruitment of forest species is not coupled with disturbance (fire). Conditions suitable for forest development may have more to do with the regeneration niche (Grubb 1977) of forest species than the maintenance of adult populations. An important but as yet unstudied phenomenon is the effect of site conditions on the growth rates, and the time taken to develop fire resistance of forest seedlings.

Pierce (1990) has developed a very similar model for the establishment of thicket in dune fynbos. Thicket seedlings establish from bird-dispersed propagules under emergent fynbos shrubs (perches) where conditions are also optimal for germination (R M Cowling unpublished data). It is possible that the widespread existence of thicket patches and species in dune fynbos relative to other fynbos communities (Taylor 1978; Cowling 1984) is a function of the generally higher soil nutrients in calcareous coastal dunes. This would facilitate the rapid growth of thicket seedlings enabling widespread recruitment between fires.

Manders (1991) has challenged the concept of fynbos and forest as distinct and mutually exclusive vegetation types. Bioclimatically, most of the fynbos biome could support some form of forest and thicket. Fire frequency in relation to the rate of forest and thicket development are the major determinants of their distribution in the biome (Manders 1991; Manders and Richardson 1991). These factors interact in complex and interesting ways to explain the invasion of fynbos by forest and thicket. Clearly, additional field experiments are required to unravel this problem.

FYNBOS/KARROID AND RENOSTER SHRUBLAND TRANSITION

In the mountains of the fynbos biome, Campbell (1986a) has shown that karroid and renoster shrubland is largely associated with interior, low rainfall areas (Figure 3.10). Karroid and renoster shrubland also occurs on soils with a higher pH than those of fynbos (Figure 3.11). Karroid shrubland often replaces fynbos on sandy soils at around 250–300 mm yr^{-1} and

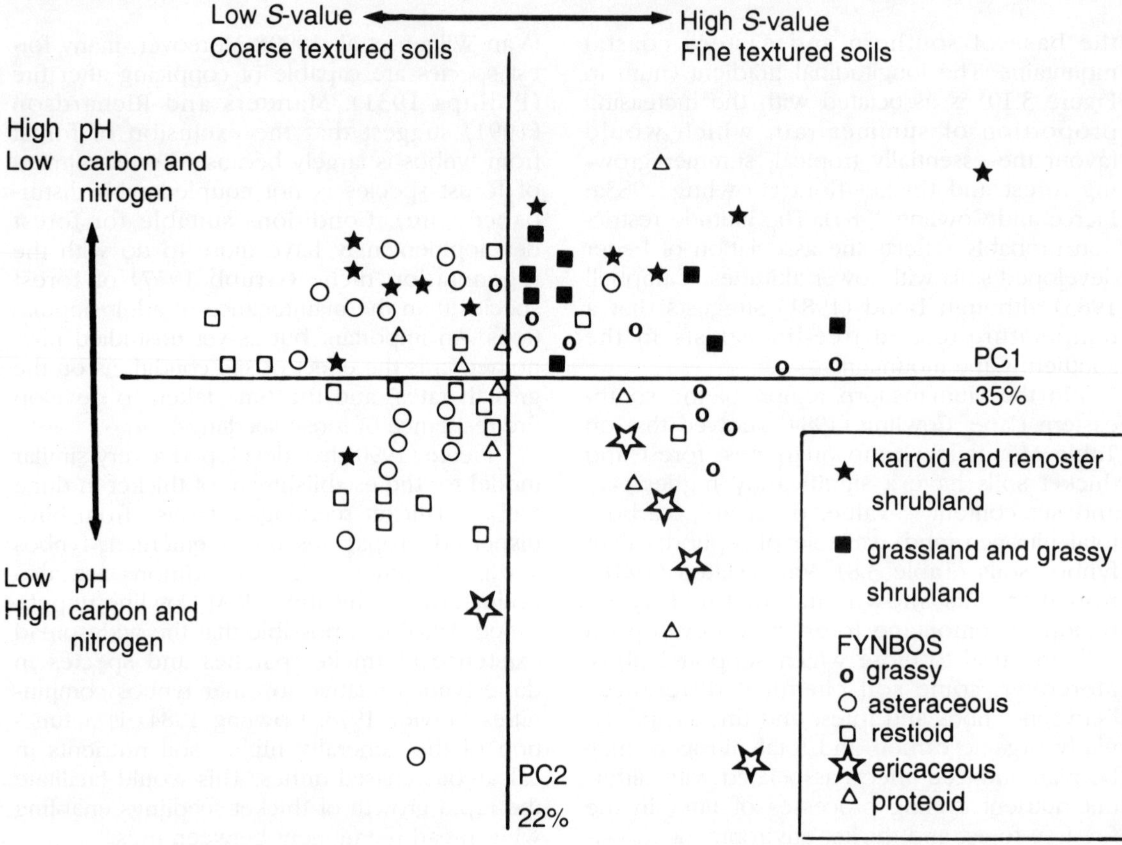

FIGURE 3.11 Soil community relationships as indicated by an ordination of 81 plots from the mountains of the fynbos biome. The ordination is a Principle Components Analysis of 10 soil variables. Reprinted from Campbell (1986a) with the permission of Kluwer Academic Publishers.

renoster shrubland replaces fynbos on clay-rich (shale and granite-derived) soils at around 600 mm yr^{-1} (Kruger 1979, 1984; Cowling 1984). There has been no experimental work on the physiological and biotic determinants of the fynbos/karroid and renoster shrubland boundary. Miller (1982) proposed that the transition between fynbos and karoo was a function of soil moisture and carbon economy: in karroid shrubland sites the carbon costs of evergreen leaves cannot be recovered and drought-deciduousness and succulent leaves are favoured (see also Bond 1981; Cowling and Campbell 1983; Campbell and Werger 1988). The transition from renoster shrubland to fynbos at higher rainfall on clay-rich substrata is usually attributed to a leaching-induced change in soil nutrient status (Kruger 1984). It is intriguing that these transitions are correlated with annual rainfall totals only: seasonality has no influence (Figure 3.10).

Disturbance regime has been widely implicated as a determinant of renoster shrubland boundaries. Cowling et al. (1988) and Boucher (1987) document the replacement, as a result of post-colonial burning and grazing, of fynbos by renoster shrubland on the south-western and western Cape coastal forelands respectively. Cowling et al. (1986) summarized evidence which suggests a grazing-induced replacement of grassland by renoster shrubland on the southern and south-eastern Cape forelands. Other studies suggest that renoster shrubland boundaries are edaphically determined and stable (e.g. Moffett and Deacon 1977). The extent to which *Elytroppapus*-dominated renoster shrubland is a product of post-colonial disturbance remains unresolved.

FYNBOS/GRASSLAND AND GRASSY SHRUBLAND TRANSITION

In the discriminant analysis of Campbell's (1986a) data from the mountains, the major discriminating variable for grassland and grassy shrubland was transect number (running from north-west to south-east) (Figure 3.10): grassy non-fynbos communities are associated with summer rainfall and/or the finer textured soils of the east (Campbell 1983; Figure 3.11). It is probably more pertinent to analyse the transition from grassy fynbos to grassland and grassy shrubland since both communities have eastern distributions (Table 3.5). Grassy fynbos is replaced by non-fynbos grassy communities on drier, more inland and north-facing sites under lower rainfall conditions (Figures 3.9, 3.10).

It is difficult to separate the role of soil nutrients and climate as determinants of grassiness in the eastern fynbos biome. This is because of the parallel west to east gradient of increasing proportion of summer rain and increasing soil clay content and *S*-value (Campbell 1983). Increased grassiness in the east is largely associated with a higher cover of seasonally green, C_4 grasses (Cowling 1984). The generally more fertile soils in the east, as well as on north slopes (Bond 1981) would favour grasses over evergreen hemicryptophytes such as restioids (Specht et al. 1977). However, higher growing season temperatures, resulting from a higher proportion of summer rain, would also favour C_4 grasses (Vogel et al. 1978) as would the higher radiation loads of north slopes (Cowling 1984). Thus, grassiness could be largely a result of climatic factors and their effects on growing season temperatures (Cowling 1984). Edaphic versus climatic controls of grassiness in the eastern fynbos biome will only be resolved by experimentation.

Disturbance, especially fire and grazing, has long been regarded as the major determinant of the fynbos/grassland boundary in the eastern fynbos biome (Acocks 1953; see below for the discussion on false fynbos). In the Afromontane Region of the eastern Cape, fynbos can be converted to grassland under certain fire regimes (Trollope 1973). However, in both the heavily grazed lowlands (Cowling 1984) and the undisturbed mountains (Campbell 1985a) of the eastern fynbos biome, there is no evidence that grazing effects determine the boundaries between grassland and fynbos.

ENVIRONMENTAL DISTINCTIONS BETWEEN FYNBOS SERIES

The determinants of the distribution of fynbos communities in the mountains of the fynbos biome are best described in terms of the three major environmental gradients described by Campbell (1983). These are: west-to-east gradient; coast-to-interior gradient; and an altitude-aspect gradient. On the coastal lowlands, where rainfall gradients are not so steep, edaphic factors alone are important determinants of community boundaries (e.g. Thwaites and Cowling 1988). We begin with the lowlands.

Rainfall on the lowlands is mainly less than 600 mm yr^{-1} and dry fynbos communities predominate. Calcareous coastal dunes support a mosaic of asteraceous fynbos and dune thicket (Figure 3.6). Restioid fynbos grows in drier (< 400 mm yr^{-1}) sites on inland and more leached dunes, shallow sands over bedrock, and in seasonally and permanently waterlogged sites. Asteraceous fynbos is found on remnant silcrete and ferricrete surfaces and on duplex, shale-derived soils where annual rainfall exceeds 600 mm yr^{-1}. Elsewhere, floristically distinct types of proteoid fynbos predominate on colluvial acid and neutral sands, quartzite, and limestone. More details on vegetation of the coastal lowlands can be found in Cowling (1984), Taylor (1984a), Boucher (1987), Cowling et al. (1988), and Rebelo et al. (1991).

Both rainfall and soils play a very important role in determining community boundaries along altitude and aspect gradients in the mountains. The deep and relatively fertile colluvial soils at the base of mountains support various types of proteoid fynbos (Figures 3.6, 3.9, 3.11). Ericaceous fynbos replaces proteoid fynbos at high altitude and high rainfall sites where soils may be permanently wet, relatively fine-grained, and rich in organic carbon. Soils on north slopes at these altitudes tend to be shallower and more droughty than south slopes. They support restioid fynbos, dominated by graminoids with shallow, intensive rooting systems capable of capturing most of the moisture which enters the soil profile (Stock et al. this volume). Because little water is available for deep percolation, shrubs are not a dominant feature of these communities (Bond 1981; Campbell 1986a). Asteraceous fynbos occurs at the dry end of the gradient (i.e. low altitude, north-facing slopes) where prolonged dry conditions favour shrubs capable of tapping deep underground water supplies. Both asteraceous and restioid fynbos occur on shallow soils. Nested within these broad-scale community changes are numerous boundaries between floristic communities often correlated with subtle environmental (usually edaphic) differences (e.g. Boucher 1978; Van Wilgen and Kruger 1985; McDonald 1988).

The coast-to-interior gradient is most strongly associated with decreasing rainfall (Campbell 1983). Thus, coastal mountains support largely proteoid and ericaceous fynbos, while in the interior mountains, restioid and asteraceous fynbos predominate (Figure 3.6).

The west-to-east gradient is largely one of decreasing summer-drought intensity and increasing soil fertility (Campbell 1983). The trend along this gradient is for grassy fynbos to replace proteoid fynbos, grassland to replace restioid fynbos, and grassy shrubland to replace asteraceous fynbos. The last-mentioned is most widespread in the strongly mediterranean climate region of the north-west, even at moderately high rainfall (Campbell 1986a).

SOIL NUTRIENTS IN PERSPECTIVE

A persistent problem in understanding the controls of community distribution within the fynbos biome is the difficulty of disentangling the effects of soil nutrients and soil moisture. Soil moisture plays a very important role in determining the distribution of structurally defined communities in the mountains (Bond 1981; Campbell 1986a). Taylor (1984a) found that soil moisture and drainage conditions could explain the distributions of floristically characterized communities all growing on infertile sands derived from quartzite. However, Cowling (1990) has reported almost complete floristic turnover between proteoid fynbos communities growing on physically similar but chemically different soils on the coastal lowlands.

When compared to non-fynbos vegetation, soil nutrients emerge as a strong determinant of fynbos community structure. Fynbos communities are relatively uniform in terms of growth-form composition (Marloth 1908; Kruger 1979). This was shown very clearly by Cowling and Campbell (1983) in a comparison of changes in growth-form mix and structural characters in fynbos and non-fynbos along parallel climatic gradients in the south-eastern fynbos biome. Non-fynbos soils were richer in all measured soil nutrients than corresponding fynbos soils. Non-fynbos showed pronounced structural change along the gradient, ranging from succulent thicket (1–3 metres tall) at the driest sites to Afromontane forest (> 15 m) at the wettest. By comparison, the fynbos gradient was very uniform in its growth-form composition: proteoid fynbos largely predominated along the entire transect. This and other studies (Cowling and Campbell 1980; Campbell and Werger 1988) lead unequivocally to the conclusion that nutrient-poor soils have an overriding effect on the structure of fynbos vegetation (see also Le Maitre and Midgley this volume).

CONCLUSIONS

There are strong relationships between contemporary ecological factors and the distribution of structurally defined plant communities in the fynbos biome. It is now possible to predict with reasonable accuracy changes in vegetation structure along environmental gradients. Much less is known about the functional significance of these growth form-environment correlations (see Stock et al. this volume). Furthermore, almost no research has been carried out on the relative importance of biotic (e.g. competition) and abiotic processes in explaining these patterns (see Bond et al. this volume). Nonetheless, Campbell's (1985b) study has gone a long way in providing a meaningful vegetation typology for the biome.

By contrast, flora-level patterns have been less studied and are not well understood. The higher proportion of phytochorological transgressor species in the south-east is easily understood in terms of contemporary climatic and edaphic conditions. The enormous concentration of species in a few genera in the south-west is less easy to explain. This remains one of the most intriguing enigmas of the Cape Floristic Region. This problem is addressed elsewhere in this volume by Cowling et al. and Linder et al.

ACKNOWLEDGEMENTS

R M Cowling thanks Bruce Campbell for creating light where there was darkness, and for sharing the experience. Eugene Moll provided enthusiasm and guidance throughout the research programme in describing and characterizing fynbos biome vegetation. John Beard, Charlie Boucher, Peter Linder, and Tony Rebelo made useful comments on earlier drafts of this chapter. The National Botanical Institute kindly allowed us the use of their base map. Many thanks Wendy Paisley and Alice Wiseman for expert typing services.

REFERENCES

ACOCKS J P H (1953). Veld types of South Africa. *Memoirs of the Botanical Survey of South Africa* **28**.

ADAMSON R S (1938). *The Vegetation of South Africa.* British Empire Vegetation Committee, London.

ADAMSON R S and SALTER T M (1950). *The flora of the Cape Peninsula.* Juta, Cape Town.

BEARD J S (1973). The physiognomic approach. In *Ordination and classification of communities.* (ed Whittaker R H) Junk, The Hague. 1–32.

BEARD J S (1984). Biogeography of the kwongan. In *Kwongan: plant life of the sandplain.* (ed Pate J S and Beard S J) University of Western Australia Press, Nedlands. 1–26.

BOND P and GOLDBLATT P (1984). Plants of the Cape Flora. A descriptive catalogue. *Journal of South African Botany Supplementary Volume* **13**, 1–455.

BOND W J (1981). Vegetation gradients in the southern Cape mountains. MSc thesis, University of Cape Town.

BOND W J, MIDGLEY J J and VLOK J (1988). When is an island not an island? Insular effects and their causes in fynbos shrublands. *Oecologia* **77**, 515–21.

BOND W J and STOCK W D (1989). The costs of leaving home: ants disperse seeds to low nutrient sites. *Oecologia* **81**, 412–17.

BOUCHER C (1977). A provisional check-list of the flowering plants and ferns in the Cape Hangklip area. *Journal of South African Botany* **43**, 57–80.

BOUCHER C (1978). Cape Hangklip area. II. The vegetation. *Bothalia* **12**, 455–97.

BOUCHER C (1987). A phytosociological study of transects through the Western Cape coastal foreland, South Africa. PhD thesis, University of Stellenbosch.

BOUCHER C and JARMAN M L (1977). The vegetation of the Langebaan area, South Africa. *Transactions of the Royal Society of South Africa* **42**, 241–72.

BOUCHER C and MOLL E J (1980). South African mediterranean shrublands. In *Mediterranean type shrublands.* (eds Di Castri F, Goodall D W and Specht R L) Elsevier, Amsterdam. 233–48.

BOUCHER C and SHEPHERD P A (1988). Plant communities of the Pella site. In *A description of the Fynbos Biome Project intensive study site at Pella.* (ed Jarman M L) Occasional Report 33. CSIR, Pretoria. 38–76.

CAMPBELL B M (1983). Montane plant environments in the fynbos biome. *Bothalia* **14**, 283–98.

CAMPBELL B M (1985a). A classification of the mountain vegetation of the fynbos biome. *Memoirs of the Botanical Survey of South Africa* **50**, 1–115.

CAMPBELL B M (1985b). Montane vegetation structure in the fynbos biome. PhD thesis, Rijksuniversiteit te Utrecht.

CAMPBELL B M (1986a). Montane plant communities of the fynbos biome. *Vegetatio* **66**, 3–16.

CAMPBELL B M (1986b). Vegetation classification in a floristically complex zone: the Cape Floristic Region. *South African Journal of Botany* **52**, 129–40.

CAMPBELL B M, COWLING R M, BOND W J and KRUGER F J (1981). Structural characterization of vegetation in the fynbos biome. *South African National Scientific Programmes Report* **52**, CSIR, Pretoria.

CAMPBELL B M, McKENZIE B and MOLL E J (1979). Should there be more tree vegetation in the mediterranean climate region of South Africa? *Journal of South African Botany* **45**, 453–7.

CAMPBELL B M and MOLL E J (1977). The forest communities of Table Mountain, South Africa. *Vegetatio* **43**, 43–7.

CAMPBELL B M and WERGER M J A (1988). Plant form in the mountains of the Cape, South Africa. *Journal of Ecology* **76**, 637–53.

COCKROFT M J, WILKINSON M J and TYSON P D (1987). The application of a present-day climatic model to the late-Quaternary in southern Africa. *Climatic Change* **10**, 161–81.

COWLING R M (1983a). Phytochorology and vegetation history in the south eastern Cape, South Africa. *Journal of Biogeography* **10**, 393–419.

COWLING R M (1983b). Vegetation studies in the Humansdorp region of the fynbos biome. PhD thesis, University of Cape Town.

COWLING R M (1984). A syntaxonomic and synecological study in the Humansdorp region of the fynbos biome. *Bothalia* **15**, 175–227.

COWLING R M (1987). Fire and its role in coexistence and speciation in Gondwana shrublands. *South African Journal of Science* 83, 106–11.

COWLING R M (1990). Diversity components in a species-rich area of the Cape Florisitic Region. *Journal of Vegetation Science* **1**, 699–710.

COWLING R M and CAMPBELL B M (1980). Convergence in vegetation structure in the mediterranean communities of California, Chile and South Africa. *Vegetatio* **43**, 191–7.

COWLING R M and CAMPBELL B M (1983). A comparison of fynbos and non-fynbos coenoclines in the lower Gamtoos River Valley, south eastern Cape, South Africa. *Vegetatio* **53**.

COWLING R M, CAMPBELL B M, MUSTART P, McDONALD D, JARMAN M L and MOLL E J (1988). Vegetation classification in a floristically complex area: the Agulhas Plain. *South African Journal of Botany* **54**, 290–300.

COWLING R M, GIBBS RUSSELL G E, HOFFMAN M T and HILTON-TAYLOR C (1989). Patterns of plant species diversity in southern Africa. In *Biotic diversity in southern Africa. Concepts and conservation.* (ed Huntley B J) Oxford University Press, Cape Town. 19–50.

COWLING R M and GXABA T (1990). Effects of a fynbos overstorey shrub on understorey community structure: implications for the maintenance community-wide species richness. *South African Journal of Ecology* **1**, 1–7.

COWLING R M, PIERCE S M and MOLL E J (1986). Conservation and utilization of South Coast Renosterveld, an endangered South African vegetation type. *Biological Conservation* **37**, 363–77.

DAHLGREN R (1963). Studies on Aspalathus: phytogeographical aspects. *Botanika Notiser* **116**, 431–72.

DEACON H J (1983). The comparative evolution of mediterranean-type ecosystems: a southern perspective. In *Mediterranean-type ecosystems: the role of nutrients.* (ed Kruger F J, Mitchell D T and Jarvis J U M) Springer, Berlin. 3–40.

DEACON J and LANCASTER N (1988). *Late Quaternary palaeoenvironments of southern Africa.* Clarendon Press, London.

ENDLER J A (1982). Problems in distinguishing historical from ecological factors in biogeography. *Systematic Zoology* **22**, 441–52.

FOSBERG F R (1967). A classification of vegetation for general

purposes. In *Guide to the check-list for IBP areas.* (ed Petersen G F) Blackwell, Oxford.
FOURCADE H G (1941). Check-list of the flowering plants of the division of George, Knysna, Humansdorp and Uniondale. *Memoirs of the Botanical Survey of South Africa* **20**, 1–127.
GELDENHUYS C J (1989). Environmental and biogeographic influences on the distribution and composition of the southern Cape forests (veld type 4). PhD thesis, University of Cape Town.
GELDENHUYS C J and McDEVETTE C R (1989). Conservation status of coastal and montane evergreen forests. In *Biotic diversity in southern Africa. Concepts and conservation.* (ed Huntley B J) Oxford University Press, Cape Town. 224–38.
GIBBS RUSSELL G E (1985). PRECIS, the National Herbarium's computerized information system. *South African Journal of Science* **81**, 662–5.
GLYPHIS J P, MOLL E J and CAMPBELL B M (1978). Phytosociological studies on Table Mountain, South Africa. I. The Back Table. *Journal of South African Botany* **44**, 281–9.
GOLDBLATT P (1978). An analysis of the flora of southern Africa: its characteristics, relationships and origins. *Annals of the Missouri Botanical Garden* **65**, 369–436.
GOOD R (1974). *The geography of flowering plants.* Longmans, London.
GRUBB P J (1977). The maintenance of species richness in plant communities: the importance of the regeneration niche. *Botanical Review* **52**, 107–45.
HENDEY Q B (1983). Cenozoic geology and palaeogeography of the fynbos region. In *Fynbos palaeoecology: a preliminary synthesis.* (ed Deacon H J, Hendey Q B and Lamprechts J J N) *South African National Programmes Report* **75**, CSIR, Pretoria.
HEYDORN A E F and TINLEY K L (1980). Estuaries of the Cape. Part 1. Synopsis of the Cape Coast. *CSIR Research Report* **380**, CSIR, Pretoria.
JACKSON S P and TYSON P D (1971). Aspects of weather and climate over southern Africa. *Environmental Studies Occasional Paper* **6**, 1–11.
JARVINEN O (1982). Species-to-genus ratios in biogeography: a historical note. *Journal Biogeography* **9**, 363–70.
JURGENS N (1986). Untersuchungen zur okologie sukkulenter Pflanzen des sudlichen Afrika. *Mitteilungen aus dem Institut fur Allgemeine Botanik Hamburg* **21**, 129–365.
KILLICK D J B (1979). African mountain heathlands. In *Heathlands and related shrublands of the world* (ed Specht R L). Elsevier, Amsterdam. 97–116.
KNIGHT R S (1988). Aspects of plant dispersal in the southwestern Cape with particular reference to the roles of birds as dispersal agents. PhD thesis, University of Cape Town.
KRUGER F J (1974). The physiography and plant communities of Jakkalsrivier catchment. MSc thesis, University of Stellenbosch.
KRUGER F J (1979). South African heathlands. *Heathlands of the world. A. Descriptive studies.* (ed Specht R L) Elsevier, Amsterdam. 19–80.
KRUGER F J (1984). Patterns of vegetation and climate in the mediterranean zone of South Africa. *Actualites Botaniques* **131**, 213–25.
KRUGER F J and TAYLOR H C (1979). Plant species diversity in Cape fynbos: gamma and delta diversity. *Vegetatio* **47**.
LAIDLER D, MOLL E J, CAMPBELL B M and GLYPHIS J P (1978). Phytosociological studies on Table Mountain, South Africa. 2. The Front Table. *Journal of South African Botany* **44**, 291–5.
LEVYNS M R (1938). Some evidence bearing on the past history of the Cape flora. *Transactions of the Royal Society of South Africa* **26**, 401–24.
LEVYNS M R (1954). The genus *Muraltia. Journal of South Africa Botany Supplementary Volume* **2**, 1–247.
LEVYNS M R (1964). Migrations and origins of the Cape flora. *Transactions of the Royal Society of South Africa* **37**, 85–107.
LINDER H P (1985). Gene flow, speciation and species diversity patterns in a species-rich area: the Cape Flora. In *Species and speciation.* (ed Vrba E S) Transvaal Museum, Pretoria.
LINDER H P (1991). A review of African Restionaceae. *Contributions from the Bolus Herbarium* **14**. (in press).
LINDER H P and VLOK J J (1991). The morphology, taxonomy and evolution of *Rhodocoma* (Restionaceae). *Plant Systematics and Evolution.* (in press).
MANDERS P T (1990). Fire and other variables as determinants of forest/fynbos boundaries in the Cape Province. *Journal of Vegetation Science* **1**, 483–90.
MANDERS P T (1991). The relationship between forest and mountain fynbos communities in the southwestern Cape Province of South Africa. PhD thesis, University of Cape Town.
MANDERS P T and RICHARDSON D M (1991). Colonization of Cape fynbos communities by forest communities. *Forest Ecology and Management* (in press).
MARLOTH R (1908). *Das Kapland.* Gustav Fischer, Jena.
MAYR E (1968). Theory of biological classification. *Nature* **220**, 545–8.
McDONALD D J (1988). A synopsis of the plant communities of Swartboschkloof, Jonkershoek, Cape Province. *Bothalia* **18**.
McKENZIE B (1978). A quantitative and qualitative study of the indigenous forests of the southwestern Cape. MSc thesis, University of Cape Town.
McKENZIE B, MOLL E J and CAMPBELL B M (1977). A phytosociological study of Orange Kloof, Table Mountain, South Africa. *Vegetatio* **34**, 41–53.
MILLER P C (1982). Some bioclimatic and pedologic influences on the vegetation in the mediterranean-type region of South Africa. *Ecologia Mediterranea* **8**, 141–56.
MILTON S J (1978). Plant communities of the Andriesgrond, Clanwilliam District. *South African Archeological Bulletin* **32**.
MILEWSKI A V and COWLING R M (1985). Anomalies in the plant and animal communities in similar environments at the Barrens, Western Australia and Caledon Coast, South Africa. *Proceedings of the Ecological Society of Australia* **14**, 199–212.
MOFFETT R O and DEACON H J (1977). The flora and vegetation of the surrounds of Boomplaas Cave: Cango Valley. *South African Archaelogy Bulletin* **32**, 127–45.
MOLL E J and WHITE F (1978). The Indian Ocean Coastal Belt. In *Biogeography and ecology of southern Africa.* (ed Werger M J A) Junk, The Hague. 561–98.
MOLL E J and BOSSI L (1984a). *Vegetation map of the fynbos biome.* Government Printer, Pretoria.
MOLL E J and BOSSI L (1984b). Assessment of the natural vegetation of the fynbos biome of South Africa. *South African Journal of Science* **80**, 355–8.
MOLL E J and JARMAN M L (1984a). Clarification of the term fynbos. *South African Journal of Science* **80**, 351–2.
MOLL E J and JARMAN M L (1984b). Is fynbos a heathland? *South African Journal of Science* **80**, 352–5.
MOLL E J, CAMPBELL B M, COWLING R M, BOSSI L, JARMAN M L and BOUCHER C (1984). A description of the major vegetation categories in and adjacent to the fynbos biome. *South African National Scientific Programmes Report* **83**, CSIR, Pretoria.
NORDENSTAM B (1969). Phytogeography of the genus *Euryops* (Compositae): a contribution to the phytogeography of southern Africa. *Opera Botanica* **23**, 1–77.
OLIVER E G H, LINDER H P and ROURKE J P H (1983). Geographical distributions of present-day Cape taxa and their phytogeographical significance. *Bothalia* **14**, 427–40.
PATE J S and BEARD J S (eds) (1984). *Kwongan: plant life of the sandplain.* University of Western Australia Press, Nedlands.
PHILLIPS J F V (1931). Forest succession and ecology in the

Knysna region. *Memoirs of the Botanical Survey of South Africa* **14**, 1–327.
PIERCE S M (1990). Patterns and processes in south coastal dune fynbos: population, community and landscape level studies. PhD thesis, University of Cape Town.
PIERCE S M and COWLING R M (1984). The phenology of fynbos, renosterveld amd subtropical thicket in the south eastern Cape, South Africa. *South African Journal of Botany* **4**.
POORE M E D (1962). The method of successive approximation in descriptive ecology. *Advances in Ecological Research* **1**, 35–68.
REBELO A G, COWLING R M, CAMPBELL B M and MEADOWS M E (1991). Plant communities of the Riversdale Plain. *South African Journal of Botany* **57**, 10–28.
RICHARDSON D M and BOND W J (1991). Determinants of plant distribution: evidence from pine invasions. *American Naturalist* (in press).
RICHARDSON D M and COWLING R M (1991). Why is mountain fynbos invasible and which species invade? In *Swartboschkloof — a mountain fynbos ecosystem.* (eds Van Wilgen B W, Richardson D M, Kruger F J and Van Hensbergen B J) Springer Verlag, Berlin (in press).
RUTHERFORD M C and WESTFALL R H (1986). Biomes of southern Africa. An objective characterization. *Memoirs of the Botanical Survey of South Africa* **54**, 1–98.
SCHOLTZ A (1986). Palynological and palaeobotanical studies in the southern Cape, MA thesis, University of Stellenbosch.
SIEGEL S (1956). *Non-parametric statistics for the behavioural sciences.* McGraw Hill, London.
SIMBERLOFF D S (1970). Taxonomic diversity of island biotas. *Evolution* **24**, 23–47.
SPECHT R L (1979). Heathlands and related shrublands of the world. In *Heathlands and related shrublands of the world.* (ed Specht R L) Elsevier, Amsterdam. 1–18.
SPECHT R L, CONNOR D J and CLIFFORD H T (1977). The heath-savannah problem: the effect of fertilizer on sand-heath vegetation of North Stradbroke Island, Queensland. *Australian Journal of Ecology* **2**, 179–86.
SPECHT R L and MOLL E J (1983). Heathlands and sclerophyllous shrublands — an overview. In *Mediterranean-type ecosystems. The role of nutrients.* (ed Kruger F J, Mitchell D T and Jarvis J U M) Springer-Verlag, Berlin. 41–65.
STORY R (1952). A botanical survey of the Keiskammahoek District. *Memoirs of the Botanical Survey of South Africa* **27**.
SUGDEN J (1990). Late Quaternary palaeoecology of the central and marginal uplands of the Karoo, South Africa. PhD thesis, University of Cape Town.
TAKHATAJAN A (1986). *Floristic regions of the world.* University of California Press, Berkeley.
TAYLOR H C (1972). Notes on the vegetation of the Cape Flats. *Bothalia* **10**, 637–76.
TAYLOR H C (1978). Capensis. In *Biogeography and ecology of southern Africa.* (ed Werger M J A) Junk, The Hague.
TAYLOR H C (1984a). Vegetation survey of the Cape of Good Hope Nature Reserve. II. Descriptive account. *Bothalia* **15**.
TAYLOR H C (1984b). Vegetation survey of the Cape of Good Hope Nature Reserve. I. The use of association-analysis and Braun-Blanquet methods. *Bothalia* **15**, 245–58.
TAYLOR H C and VAN DER MEULEN F (1981). Structural and floristic classification of Cape mountain fynbos on Rooiberg, Southern Cape. *Bothalia* **13**, 557–67.
TINLEY K L (1977). Framework for the Gorongosa ecosystem. DSc thesis, University of Pretoria.
THODAY D (1925). The geographical distribution and ecology of *Passerina. Annals of Botany* **39**, 175–208.
THWAITES R N and COWLING R M (1988). Soil-vegetation relationships on the Agulhas Plain, South Africa. *Catena* **15**.
TROLLOPE W S W (1973). Fire as a method of controlling macchia (fynbos) vegetation on the Amatole mountains of the eastern Cape. *Proceedings of the Grassland Society of Southern Africa* **8**, 35–41.
VAN DAALEN J C (1981). The dynamics of the indigenous forest-fynbos ecotone in the southern Cape. *South African Forestry Journal* **119**, 14–23.
VAN DER MERWE C V (1976). Die plantekologiese aspekte en bestuursprobleme van die Goukamma Natuurreservaat. MSc thesis, University of Stellenbosch.
VAN DER MERWE C V (1977). 'n Plantegroeiopname van die De Hoop Natuurreservaat. *Bontebok* **1**, 1–29.
VAN WILGEN B W, HIGGINS K B and BELLSTEDT D U (1990). The role of vegetation structure and fuel chemistry in excluding fire from forest patches in the fire-prone fynbos shrublands of South Africa. *Journal of Ecology* **78**, 210–22.
VAN WILGEN B W and KRUGER F W (1985). The physiography and fynbos vegetation communities of the Zacharianshoek catchments, southern Cape, South Africa. *South African Journal of Botany* **51**, 379–99.
VOGEL J C, FULS A and ELLIS R P (1978). The geographical distribution of Krantz grasses in South Africa. *South African Journal of Science* **74**, 9–15.
VUILLEUMIER F and SIMBERLOFF D (1980). Ecology versus history as determinants of patchy and insular distribution of high Andean birds. *Evolutionary Biology* **13**, 235–379.
WEIMARCK H (1941). Phytogeographical groups, centres and intervals within the Cape Flora. *Lunds Universitets Arsskrif.* NF Adv. Bd. **37**, 1–143.
WERGER M J A (1974). On concepts and techniques applied in the Zurich-Montpellier method of vegetation survey. *Bothalia* **11**, 309–23.
WERGER M J A (1978a). The Karoo-Namib Region. In *Biogeography and ecology of southern Africa.* (ed Werger M J A) Junk, The Hague. 233–99.
WERGER M J A (1978b). Biogeographical division of southern Africa. In *Biogeography and ecology of southern Africa.* (ed Werger M J A) Junk, The Hague. 301–462.
WERGER M J A, KRUGER F J and TAYLOR H C (1972). A phytosociological study of the fynbos and other vegetation at Jonkershoek, Stellenbosch. *Bothalia* **10**, 599–614.
WHITE F (1978). The Afromontane Region. In *Biogeography and ecology of southern Africa.* (ed Werger M J A) Junk, The Hague. 465–510.
ZUCCHINI W (1990). RAIN — a programme to generate artificial rainfall sequences. Applied Statistics and Decision Science Unit, University of Cape Town.

4

Plant diversity and endemism

R M Cowling, P M Holmes and A G Rebelo

INTRODUCTION

The widely documented increase in species richness for many taxa with decreasing latitude (e.g. Dobzhansky 1950; Fischer 1960; Pianka 1966; Brown 1988) is paralleled by a decline in geographical range size (Gentry 1986; Major 1988; Stevens 1989). These patterns suggest a relationship between species richness and narrow endemism (Cody 1986a; Gentry 1986) which is poorly understood (Myers and Giller 1988). Stevens (1989) argues that the richness of low latitude communities results from the transient establishment of habitat specialist species (stenotypes) with narrow distribution ranges out of their preferred habitat (the 'mass effect' of Shmida and Wilson 1985). The low range of climatic conditions experienced by organisms in low latitude environments promotes the evolution of habitat specialization (Stevens 1989).

Certain mediterranean climate regions provide a striking exception to these latitudinal trends (Cody 1986a; Gentry 1986). The most notable is the Cape Floristic Region of South Africa which has one of the highest species densities and levels of endemism for any temperate or tropical continental region (Goldblatt 1978; Linder 1985; Gentry 1986; Major 1988; Cowling et al. 1989). Following Stevens' (1989) argument, the great annual range of climatic conditions experienced in mediterranean regions (hot dry summers, cool wet winters) should favour the evolution of broad climatic tolerances and wide distribution ranges, resulting in lower species richness. What then is special about these species- and endemic-rich mediterranean climate regions?

Species diversity is a complex phenomenon which has eluded general explanations. The topic is confused because of the failure to consider the wide range of both ecological and historical factors which may determine richness (Whittaker 1977; Vuillemier and Simberloff 1980; Cowling 1983a; Rickleffs 1987; Brown 1988; Diamond 1988). Furthermore, many studies have been vague about spatial scales and few have considered the relationship between independent diversity components (richness and turnover) in determining regional richness (cf. Auerbach and Shmida 1985; Cody 1986a; Cowling 1990). Generally, the patterns and determinants of species turnover (beta and gamma diversity) have been poorly studied. Despite the unique richness of the Cape Floristic Region, its well known flora, and its accessibility for research, there have been relatively few studies on alpha diversity (Campbell and Van der Meulen 1980; Bond 1983; Cowling 1983a; Cowling and Gxaba 1990; Cowling 1990) and turnover (Kruger and Taylor 1979; Cowling and Campbell 1984; Cowling 1990).

Patterns and determinants of endemism have also been poorly studied in the Cape Floristic Region, especially when compared to the California Floristic Province (Stebbins and Major 1965; Raven and Axelrod 1978) and parts of Eurasia (see Major 1988 for a review). A major gap in the literature for all endemic-rich areas are studies which seek to characterize endemics in terms of taxonomic, biological, and habitat attributes.

In this chapter we analyse diversity and endemism in the Cape Floristic Region. Our emphasis is on fynbos (Cowling and Holmes this volume) although reference is made to other vegetation types. In analysing diversity, we have attempted to explain patterns of regional richness by studying the patterns, determinants, and interactions of different diversity components. In dealing with endemism we have concentrated on developing a biological profile of an endemic species. We also discuss the implications of our analysis for speciation. We conclude by discussing the relationship between diversity and endemism in the Cape Floristic Region.

DIVERSITY

Approach

In developing our approach we asked the following questions: How and why does species richness vary across the Cape Floristic Region? What controls the different diversity components, and how do these interact to explain the way species are packed into a landscape?

The size of the regional species pool can be attributed to three independent components of diversity, namely alpha, beta, and gamma diversity (Whittaker 1972; Cody 1975, 1983; Westoby 1985). Alpha diversity refers to the number of species in a homogeneous community. Beta diversity refers to species turnover along habitat or environmental gradients. Gamma diversity, as used here, refers to species turnover among equivalent habitats along geographical gradients (Cody 1975, 1983). In this sense, it is identical to Whittaker's (1972) delta diversity. Landscapes may be rich in species either because they have species-rich communities, or because turnover along environmental and geographical gradients is high: the richest landscapes have both high alpha diversity and high turnover (Figure 4.1).

We collated data on the three independent diversity components for sites located along a major environmental gradient in the Cape Floristic Region. The gradient, which ranges from the south-western to the south-eastern Cape Floristic Region (see Figure 4.2 for delimitation of the subregions), is characterized by an increasing proportion of summer rain and increasing soil nutrient status irrespective of geological substratum (Campbell 1983; Cowling and Holmes this volume; Deacon et al. this volume). The transect is at roughly the same latitude and equidistant from the coast (Figure 4.2): temperature variation across it is not significant (Campbell 1983). There is, however, a trend for maximum elevations and annual rainfall totals to be higher in the south-west than in the south-east. Our detailed comparisons are confined to broadly physiographically similar sites across the region. Relevant rainfall data are shown in Figure 4.3.

Wherever possible, we deduce predictions from the appropriate theory on the controls of the three independent diversity components. These predictions were tested using available data. An analysis of the variation in the values of the diversity components in the south-eastern and south-western Cape Floristic Region enabled us to predict tentatively regional richness in these two areas. We tested these predictions on regional richness by analysing species-area data from the two areas. Finally, we compared these patterns with those from other species-rich regions of the globe.

Alpha diversity

There are a number of theories and models dealing with the controls of alpha diversity in plant communities (Grubb 1977; Shmida and Ellner 1984; Cody 1986b; Fagerström 1988). The development of theory has been influenced profoundly by the notion that plant communities comprise assemblages of sessile organisms which all require the same basic resources of light, water, carbon dioxide, nitrogen, etc. It is difficult on the basis of these resources to invoke sufficient numbers of niches to explain coexistence in species-rich communities (Grubb 1977; Werner 1978; Silvertown and Law 1987). In communities of sessile organisms, competition for space is all-important: any plant that can hold a space succeeds in pre-empting resources required by other plants (Werner 1978; Yodzis 1986). Thus, much of classical competition theory is regarded as inappropriate for explaining coexistence and diversity in plant communities. Nonetheless, theories explaining coexistence in terms of Gaussian competition have been developed for plant communities (e.g. Tilman 1982; Cody 1986b).

DATABASE

Data on alpha diversity for vegetation types of the Cape Floristic Region are shown in Table 4.1. All plots are standard 1 m^2 (10 replicates) to 1 000 m^2 diversity plots (Shmida 1984). We discuss richness at the 1 m^2 and 1 000 m^2 levels. The former describes point diversity (Whittaker 1977) where species numbers are presumably controlled by biological interactions (Grime 1979; Bond 1983). We also regard richness at the 1 000 m^2 level as a measure of

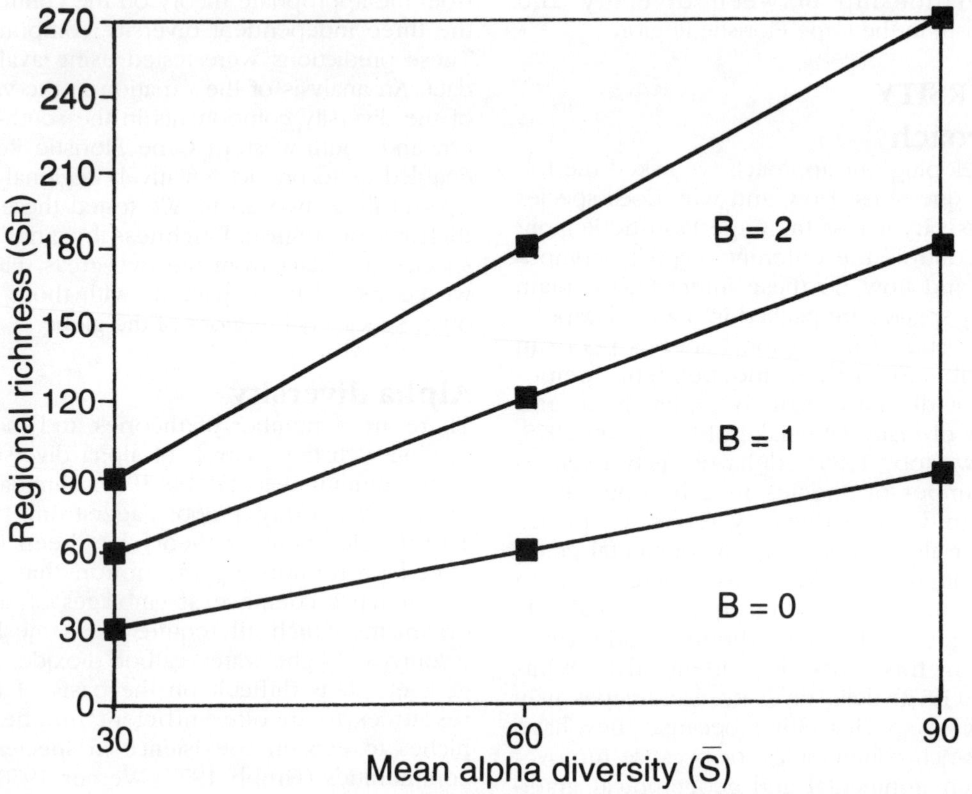

FIGURE 4.1 The effect on regional richness of the interactions between alpha and beta diversity. Alpha diversity ($\bar{S}$) is the mean number of species per community sampled at three sites along a hypothetical habitat gradient. Beta diversity (B) = $(g(H) + l(H))/2\bar{S}$ where $g(H)$ is the number of species gained along the gradient H and $l(H)$ represents the number of species lost along H (Wilson and Shmida 1984). B equals exactly the number of community changes or $B + 1$ distinct communities. Thus, regional richness (S_R) = $S(B + 1)$. Note that the incorporation of a measure of gamma diversity would also affect S_R. Then, $S_R = (B_n + 1)\ (B_m + 1)$ where B_n = beta diversity and B_m = gamma diversity (Cowling 1990).

alpha diversity although significant turnover or internal beta diversity may be associated with habitat heterogeneity at this scale (Whittaker 1977). We assume that plots are random samples and representative of the array of habitats within each vegetation type and region. For fire-prone vegetation (fynbos and renoster shrubland) only mature vegetation (15–30 years) was sampled. Succession in these communities is consistent with the inhibition model (Connell and Slayter 1977) with substantially higher richness in the young than the mature phase (Kruger and Bigalke 1984; Hoffman et al. 1987; Cowling and Pierce 1988). Other communities were not markedly disturbed.

NICHE DIFFERENTIATION

Niche differentiation is an equilibrium model which assumes that niches of coexisting plant species are dissimilar enough for competitive exclusion not to occur (Newman 1982; Crawley 1986). Thus, the number of species in a community is limited by the number of niches (MacArthur and Levins 1967). For coexisting species, niche differentiation will be manifest in growth form variation (Cody 1986b). We would expect, therefore, a positive relationship between growth form richness and species richness (Cody 1989). Furthermore, environmental conditions which favour the coexistence of many growth forms should also

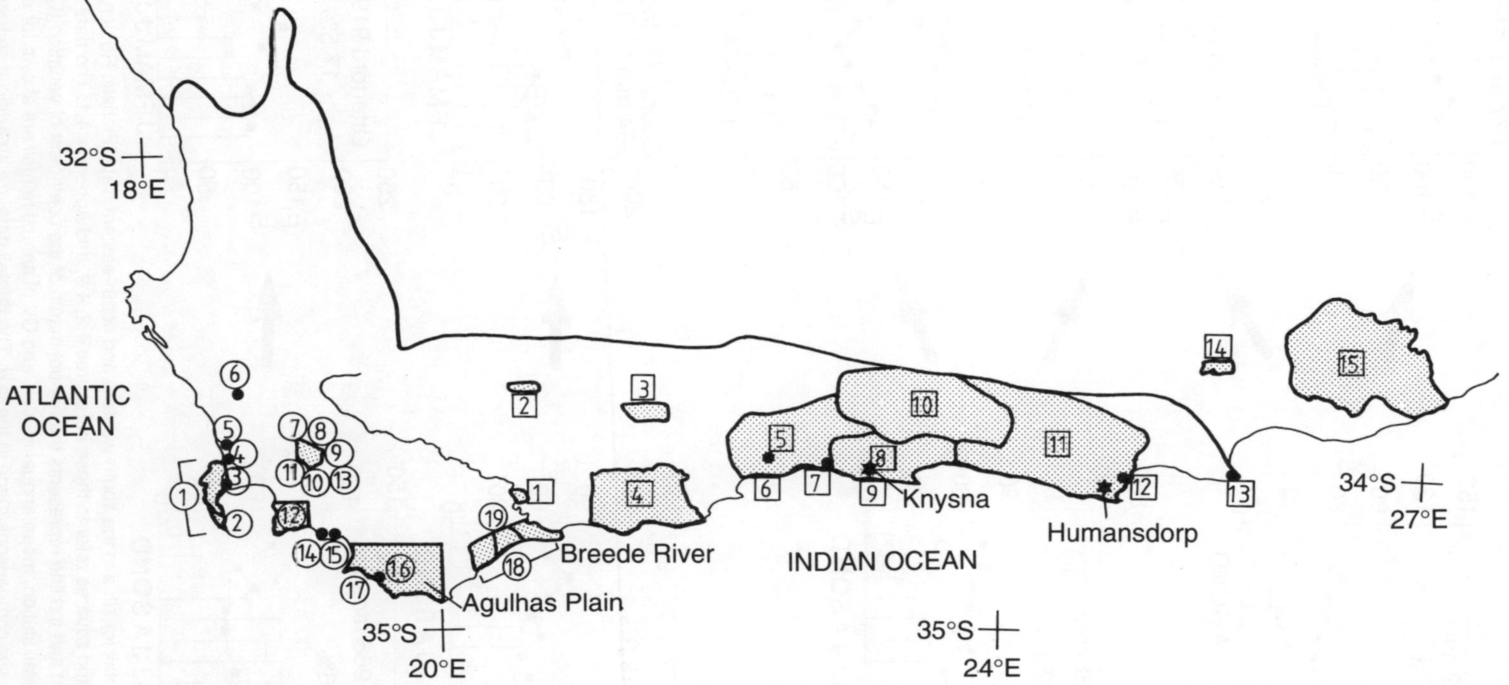

FIGURE 4.2 The sites used for diversity comparisons and species-area analysis (Figure 4.9). The numbers in circles refer to south-western sites (south-west of the Breede River) and the numbers in squares refer to south-eastern sites. Site numbers, names, and dominant vegetation types are given in Appendix 4.1.

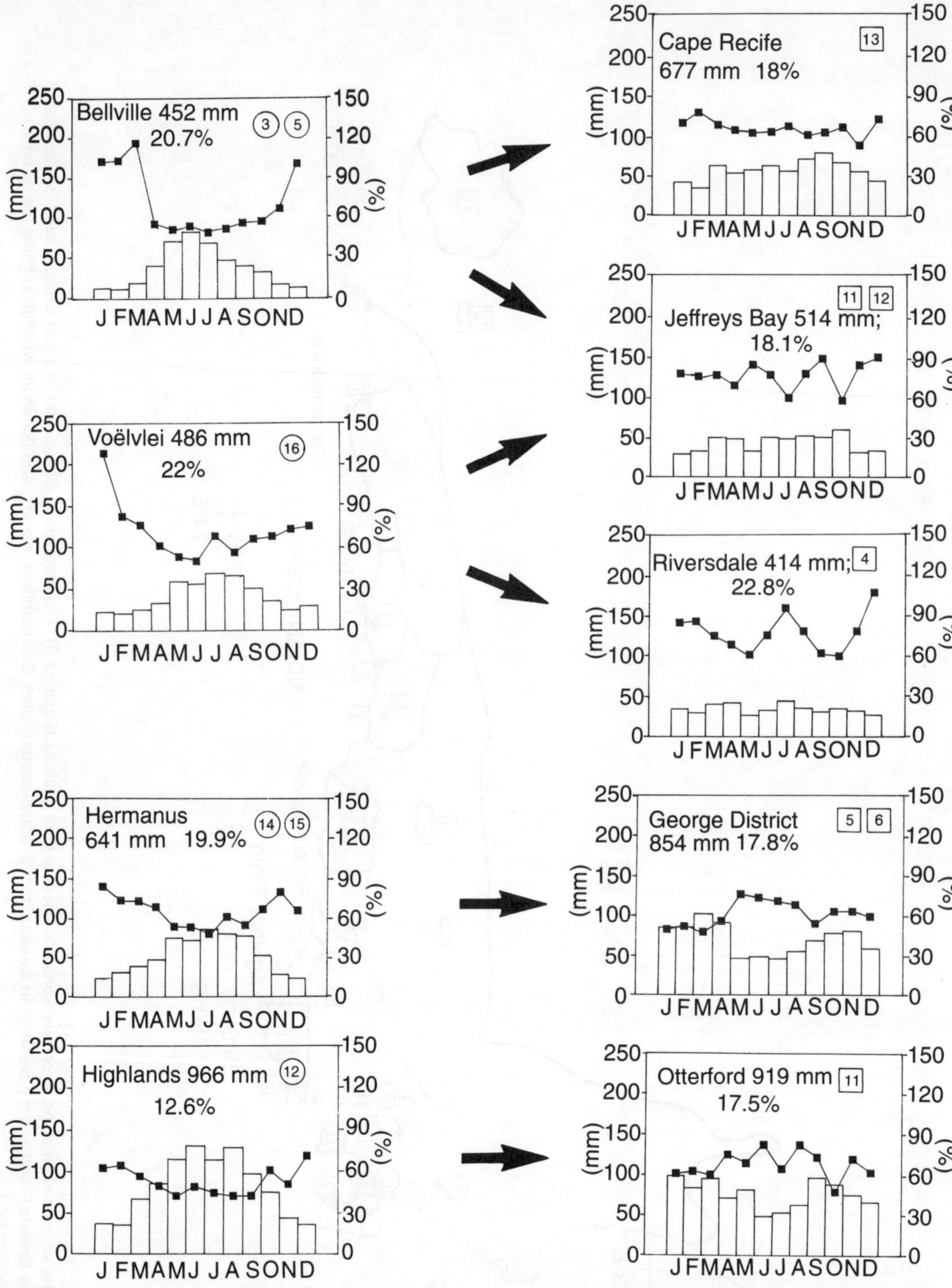

FIGURE 4.3 The rainfall for sites in the south-western and south-eastern Cape Floristic Region. The numbers in circles and squares refer to those in Figures 4.2, 4.9, and Appendix 4.1. The bars are mean monthly rainfall (mm) and the line represents the annual march of the coefficient of variation (CV) of monthly rainfall. The rainfall station, mean annual rainfall, and CV of annual rainfall are shown for each site. Arrows indicate site comparisons made in the text. The rainfall data are an artificial sequence generated for a 30 year period (Zuccini 1990).

TABLE 4.1 The patterns of alpha diversity in vegetation types in and adjacent to the Cape Floristic Region. The data are means ± S.D. (range). The significant differences ($P < 0.05$) between means, using Tukey's multiple range test (Zar 1984), are indicated by dissimilar superscripts in the columns for each scale. Forest data are excluded from the ANOVA and range test. Fynbos — west refers to the region west of the Breede River; south to the region between the Breede River and Knysna; and east to the region east of Knysna (see Figure 4.2). Karroid shrublands in the west are succulent shrublands characteristic of the succulent karoo biome; those in the east are grassy shrublands characteristic of the Nama karoo biome (Rutherford and Westfall 1986). Data are from Kruger (1979), Bond (1983), Cowling (1983a, 1990), Hoffman (1989), and the unpublished data of C Hilton-Taylor and F J Kruger.

		Scale			
Vegetation	*n*	$1 m^2$		$1\,000 m^2$	
Fynbos — west	33	16.7 ± 5.3^{a}	(7.8–25.0)	66.5 ± 20.6^{a}	(31–126)
Fynbos — south	14	14.5 ± 3.5^{a}	(9.8–10.2)	64.5 ± 16.8^{ab}	(41–104)
Fynbos — east	15	16.6 ± 5.0^{a}	(6.7–26.6)	66.2 ± 15^{ab}	(41– 93)
Renoster shrublands	11	17.3 ± 8.3^{ab}	(5.1–33.0)	84.2 ± 29.8^{ab}	(28–143)
Karroid shrublands — west	33	9.1 ± 4.4^{c}	(4.2–23.3)	71.0 ± 23.6^{ab}	(30–113)
Karroid shrublands — east	9	8.1 ± 1.5^{c}	(5.1– 9.8)	55.9 ± 8.7^{b}	(47– 74)
Subtropical thicket	11	10.7 ± 3.1^{bc}	(4.5–14.9)	58.7 ± 18.3^{ab}	(35– 98)
Forest	3	6.9 ± 3.7	(3.8–11.0)	43.7 ± 15.3	(26– 53)
F ratio		11.33		2.19	
P		0.00		0.05	

favour high species richness (Cody 1986b, 1989). Therefore, we predict that species richness would be higher in the south-eastern Cape Floristic Region where predictable amounts of year-round precipitation (Figure 4.3) would favour a large range of growth forms capable of exploiting a wide seasonal diversity of growing conditions (Cody 1986b, 1989). In the south-west, growth form diversity (and species richness) would be lower owing to the pronounced seasonality associated with summer drought conditions.

There was no significant difference in alpha diversity for fynbos vegetation across the Cape Floristic Region (1 m² data: $F = 0.053$, $P > 0.05$; 1 000 m² data: $F = 0.036$, $P > 0.05$; Table 4.1). At the 1 000 m² level renoster shrubland was significantly richer in the south-west ($\bar{x} = 107.5$, $n = 4$) than in the south-east ($\bar{x} = 70.9$, $n = 7$) ($t = 2.38$, $P < 0.05$) but not at the 1 m² scale ($\bar{x} = 20.9$ and 15.3 respectively, $t = 1.09$, $P > 0.05$). There were inadequate data for comparing east-west patterns for forest and thicket vegetation. However, Geldenhuys and MacDevette (1989) note a trend for declining species richness in 100 m² plots for Afromontane forest from the south-east ($\bar{x} = 27$) to the south-west ($\bar{x} = 17$).

Significant correlations between growth form richness and species richness exist for both non-fynbos and fynbos vegetation in both the south-eastern and south-western Cape Floristic Region (Figure 4.4). However, the explained variance (R^2) for the fynbos data sets was much lower than for non-fynbos. There was no significant difference ($t = 0.36$, $P > 0.5$) between growth form richness in south-eastern ($\bar{x} = 7.45$) and south-western fynbos ($\bar{x} = 7.6$). This is probably because low soil nutrients limit the array of growth forms in fynbos (Cowling and Holmes this volume). The data also show a high number of species per growth form in fynbos which raises the problem of how trophically similar species coexist (Shmida and Ellner 1984). Bond et al. (this volume) discuss Cody's (1986b) leaf niche differentiation hypothesis for coexistence in fynbos of physiognomically similar proteoid shrubs.

In non-fynbos there were significant differences between the south-east and the south-west for both growth form diversity ($\bar{x}_{SE}$ = 9.15, $\bar{x}_{SW} = 6.95$, $t = 2.79$, $P < 0.005$) and species richness ($\bar{x}_{SE} = 36.9$, $\bar{x}_{SW} = 18.1$, $t = 7.13$, $P < 0.001$). This suggests that rainfall seasonality may limit the mix of growth forms, and hence species richness of non-fynbos vegetation. Contradictory evidence comes from karroid shrublands which are richer in species at the 1 000 m² scale in the west than the east (Table 4.1). However, richness in the west is largely associated with the dwarf, leaf succulent shrub guild, comprising mainly Mesembryanthemaceae (R M Cowling unpublished data).

Tilman (1982, 1986) has developed an equilibrium model which predicts high alpha diversity on nutrient-poor soils because the ratios of limiting nutrients result in a wider range of nutritional trophic niches in these habitats. It is tempting to suggest that this model can explain patterns of species richness along soil nutrient gradients in the Cape Floristic Region, but there have been no empirical tests. Both Grime (1979) and Tilman (1982) predict that the relationship between species richness and soil fertility should be humpbacked with the highest richness occurring in relatively nutrient-poor habitats. The absolute levels of nutrients which favour high diversity are, however, unspecified. There have been a number of correlative studies relating soil nutrients to species richness in the Cape Floristic Region. Bond (1983) found a significant negative relationship between point diversity and soil fertility measured as the sum of exchangeable cations (*S*-value) for fynbos in the south-eastern Cape Floristic Region. However, *S*-values were not independent of the moisture regime in his study area. On the Agulhas Plain — a climatically uniform region of the south-west — Cowling (1990) found no significant relationship between point diversity and a wide range of soil chemical and physical variables. A similar lack of relationship was observed for fynbos, non-fynbos, and combined data sets (100 m² plots) in the Humansdorp region of the south-eastern Cape Floristic Region (Cowling 1983a). There is no evidence in the Cape Floristic Region for the strong inverse relationship between species richness and soil phosphorus, as is the case for some temperate Australian regions (Adams et al. 1989).

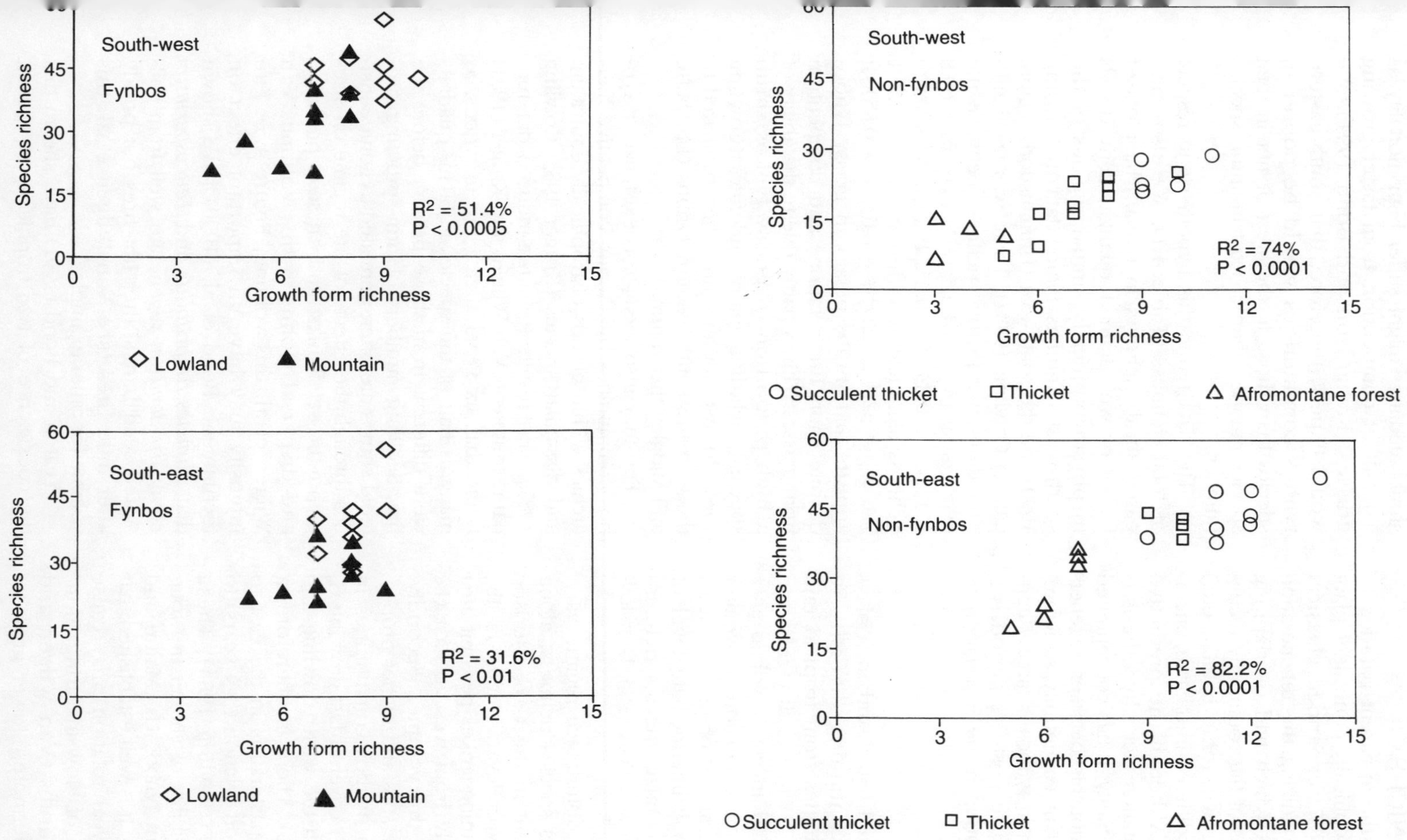

FIGURE 4.4 The relationship between growth form richness and species richness in fynbos and non-fynbos phytosociological plots from the south-western and the south-eastern Cape Floristic Region. Regressions fitted using the method of least squares. The south-western data are from Boucher (1978, 1987) and the south-eastern data from Cowling (1983b). The appropriate rainfall data in Figure 4.3 are: Bellville for south-western succulent thicket, thicket, and lowland fynbos; Highlands for south-western mountain fynbos and Afromontane forest; Jeffreys Bay for south-eastern succulent thicket, thicket, and lowland fynbos; Otterford for south-eastern mountain fynbos and Afromontane forest. The growth forms used in the analysis are: geophytes, non-geophytic forbs, evergreen graminoids, deciduous graminoids, small-leaved evergreen shrubs < 1 m and > 1 m tall, large-leaved (nanophyllous and larger) evergreen shrubs < 1 m and > 1 m, deciduous shrubs < 1 m and > 1 m, succulent shrubs < 1 m and > 1 m, evergreen trees, and vines.

REGENERATION NICHE

In order to explain the apparent paradox of the coexistence of trophically equivalent plant species, Grubb (1977) expanded classical niche theory by recognizing the regeneration niche. Thus, species which had overlapping requirements in terms of the commonly recognized niche axes (growth form, habitat, etc.) might coexist if they differed in their regeneration requirements. Although the predictive power of the regeneration niche hypothesis is weakened by its generality, it has concentrated attention on the relationship between regeneration biology and plant community structure. Because regeneration in fynbos is largely confined to the immediate post-fire period, we discuss the regeneration niche in relation to disturbance.

DISTURBANCE

It is generally accepted that disturbance plays a major role in maintaining diversity in species-rich communities from tropical rain forests (Connell 1978) to the abyssal zone (Grassle 1989). Disturbance is invoked as a process to prevent the exclusion of weaker competitors. Theoretical and empirical studies have shown that some intermediate level of disturbance will maximize species richness (Huston 1979; Walker and Peet 1983; Denslow 1985).

How would the disturbance regime vary, in a way that affects species richness, along the west-east gradient in the Cape Floristic Region? There are two relevant trends. Firstly, the incidence of lightning strikes per unit area increases significantly from west to east (Le Maitre and Midgley this volume). Secondly, owing to more fertile soils and higher proportion of summer rain, fynbos communities in the east have a higher cover of fast-growing C_4 grasses than those in the west (Cowling and Holmes this volume). There is, therefore, a strong incentive for pastoralists in the east to burn fynbos on a short rotation (3–4 years) to promote grassiness (Cowling 1984). These trends raise the possibility of shorter fire rotations in south-eastern fynbos (Le Maitre and Midgley this volume). It is well established that frequent fires (3–4 year rotation) may cause the local extinction of slow-maturing, non-sprouting fynbos shrubs (Van Wilgen and Kruger 1981). Furthermore, slower growing shrub seedlings might suffer high mortality as a result of competition from faster growing grasses (Richardson and Bond 1991). We would expect, therefore, that south-eastern fynbos communities would be poorer in species than those in the west; if not in total species number, then at least in shrub species number.

The data refute this hypothesis in respect of total richness (Table 4.1). We also compared shrub richness in five 0.1 ha plots of fynbos on well drained, quartzite-derived soils in physiographically similar regions of the south-west (Agulhas Plain, data in Cowling 1990) and the south-east (Humansdorp region, data in Cowling 1983a). There was no significant difference in total richness between the two regions ($\bar{x}_{SW}$ = 61.4 ± (SD) 14.6, $\bar{x}_{SE}$ = 69.8 ± 17.3, t = 0.83, $P > 0.05$) but nor was shrub richness significantly different ($\bar{x}_{SW}$ = 32.2 ± 7.4, $\bar{x}_{SE}$ = 32.4 ± 12.7, t = 0.03, $P > 0.05$). It appears that shrubs persist in grassy fynbos despite short fire rotations and competition from grasses. This is partly due to the ability of a high proportion of grassy fynbos shrub species, including many endemics (Cowling 1984), to resprout after fire when compared to those in the south-western fynbos (Le Maitre and Midgley this volume).

Fire may also mediate coexistence in fynbos communities by altering competitive hierarchies within groups of trophically equivalent and dissimilar species (Kruger 1983; Cowling 1987). This results from fire-induced differential recruitment. Van Wilgen and Kruger (1981) found that six years after a fire in fynbos, as many as half of the species in 50 m^2 quadrats were different from those present before the fire. Seedling numbers of non-sprouting proteoid shrubs are often unrelated to adult densities through the effects of fire regime on sizes of pre-fire seedbanks, and on seed predation and dispersal (Johnson this volume; Van Wilgen et al. this volume; Figure 4.5). Fire intensity may have an important effect on recruitment. Bond et al. (1990) have shown that *Mimetes fimbrifolius* and *Leucospermum conocarpodendron* regenerate well from soil-stored seed banks after hot fires but poorly after mild fires (see Van Wilgen et al. this volume for other examples).

Yeaton and Bond (1991) have shown that the coexistence of two trophically equivalent

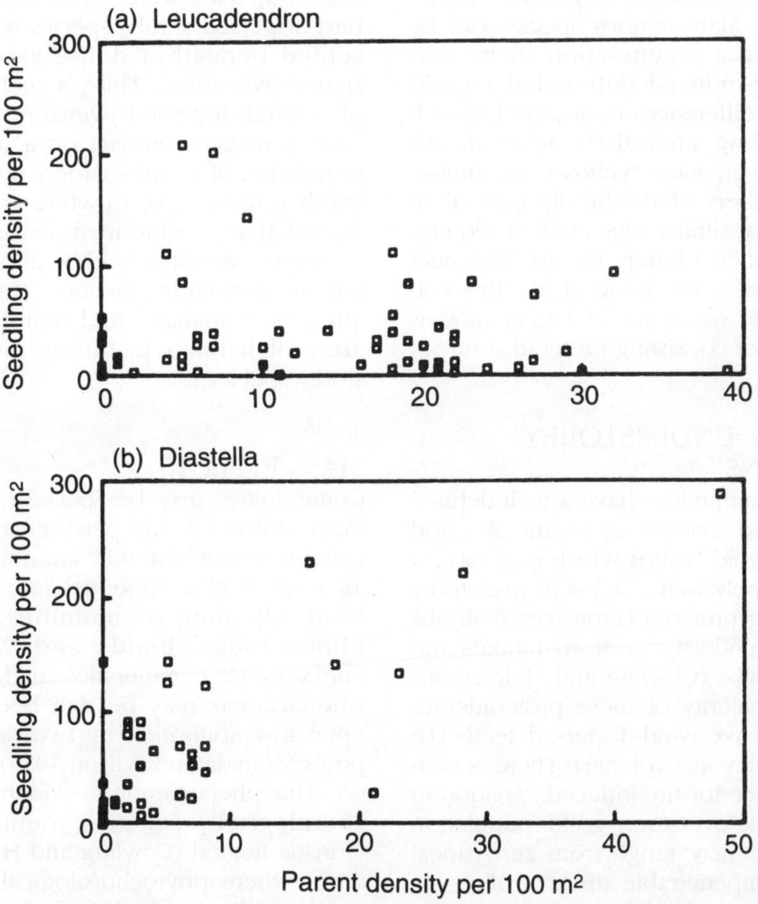

FIGURE 4.5 The relationship between parent and seedling density of a proteoid overstorey shrub, *Leucadendron laureolum*, with wind-dispersed seeds and an understorey shrub, *Diastella divaricata*, with myrmecochorous seeds, two years after a hot summer fire on the southern Cape Peninsula. $n = 66$ 10 x 10 m plots. Reprinted from Cowling and Gxaba (1990) with the permission of the South African Institute of Ecologists.

proteoid shrubs is maintained in the short-term by differences in their dispersal biology (see also Bond et al. this volume). However, long-term coexistence would require occasional fires which reduce the population sizes of *Protea lepidocarpodendron*, the superior competitor. The generalization is that variable fire regimes maintain coexistence by preventing the long-term dominance of any single species (Cowling 1987). Many fynbos species can be separated along a regeneration niche axis defined by fire-induced differential recruitment, owing to differences in dispersal, germination, or seedling survival (Bond et al. this volume). However, most fynbos communities have large numbers of trophically equivalent species with very similar regeneration requirements. This implies a lottery for establishment sites (Fagerstrom 1988). Bond et al. (this volume) discuss the extension of lottery models to communities of coexisting proteoid shrubs.

OVERSTOREY-UNDERSTOREY INTERACTIONS

Many fynbos communities have a well defined understorey and overstorey strata. A good example is proteoid fynbos which may have a sparse to extremely dense 1.5–4 m overstorey of non-sprouting proteoid (Proteaceae) shrubs and a 0.25–1 m understorey of graminoids and small-leaved shrubs (Cowling and Holmes this volume). The majority of these proteoids are serotinous and have wind-dispersed seeds (Le Maitre and Midgley this volume). There is considerable evidence for fire-induced variation in recruitment of the overstorey guild: population sizes after a fire may range from zero (local extinction) to impenetrable thickets of up to 84 000 stems ha^{-1} (Van Wilgen et al. this volume).

Specht and Morgan (1981) generalize that understorey species richness would be inversely related to overstorey cover in woodlands and shrublands on nutrient-poor soils. Studies in fynbos have shown that species richness is negatively related to the density of the proteoid shrub overstorey (Campbell and Van der Meulen 1980; Kruger and Bigalke 1984; Cowling and Gxaba 1990; Esler and Cowling 1990). Cowling and Gxaba (1990) found that stands with a sparse (*c.* 0–500 stems ha^{-1}) proteoid overstorey had double the number of understorey species in 1 m^2 plots than dense (*c.* 10 000–30 000 stems ha^{-1}) stands and that this pattern was evident in both mature and recently burnt fynbos. They also found compositional differences which were attributed to overstorey effects. For example, myrmecochorous and non-sprouting shrubs species were under-represented while bird-dispersed shrubs species were over-represented beneath a dense as opposed to a sparse overstorey. Thus, a mobile population of a wind-dispersed overstorey proteoid may have a negative impact on a more stationary population of an understorey myrmecochorous shrub (Figure 4.5). Cowling and Gxaba suggested that fire-induced spatial variation in overstorey density could be important in maintaining community richness through the creation of a spatially and temporally dynamic array of habitats favouring different understorey species.

MASS EFFECTS

Communities may be species-rich because of mass effects i.e. the persistence of species in communities where they are not self-sustaining as a result of a continual rain of propagules from adjoining communities (Shmida and Ellner 1984; Shmida and Wilson 1985). Furthermore, communities in floristically transitional areas may be rich because they are open to colonization by two or more species pools (Shmida and Wilson 1985).

This phenomenon is evident in the phytochorologically complex south-eastern Cape Floristic Region (Cowling and Holmes this volume) where phytochorological diversity is a good predictor of both fynbos and non-fynbos alpha diversity (Cowling 1983a) (Figure 4.6). In this area mesic, upland fynbos, and forest communities have lower phytochorological diversity and species richness than drier, lowland communities which comprise a rich admixture of Cape, karroid, and tropical species. Predictably, phytochorological diversity was unrelated to species richness on the Agulhas Plain in the south-west (Cowling 1990) since this area has a small pool of taxa not endemic to the Cape Floristic Region (Cowling and Holmes this volume).

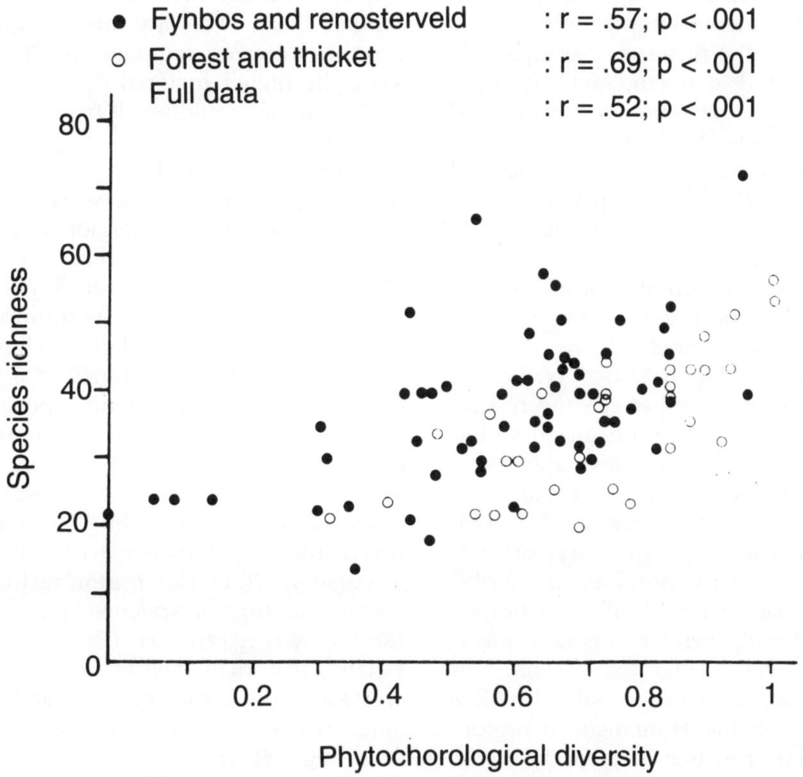

FIGURE 4.6 The relationship between species richness and phytochorological diversity in 100 m^2 plots from the Humansdorp region of the south-eastern Cape Floristic Region (from Cowling 1983a). Phytochorological diversity was computed using the Shannon-Wiener function in which p_i is the number of species in a phytochorological group (Cowling 1983c), expressed as a function of the total number of species in a sample. Reprinted from *Vegetatio* with the permission of Kluwer Academic Publishers.

HISTORY AND THE REGIONAL SPECIES POOL

Rickleffs (1987) suggests that there may be a strong link between alpha diversity and regional richness. He argues that since the size of a regional species pool is determined, in part, by historical processes, local species richness in environmentally similar sites need not converge. Rickleffs cites Shmida's (1981) data showing non-convergence in alpha richness (1 000 m^2 samples) between plant communities in the mediterranean climate regions of California and Israel. Westoby (1985, 1988), however, argues that for organisms such as plants, which do not have a widely dispersed life history phase, alpha diversity is uncoupled from the size of the regional species pool. He cites as support the convergent richness of 1 000 m^2 samples from shrublands on nutrient-poor soils in the mediterranean climate regions of the Cape and south-western Australia. If Rickleffs (1987) is correct, then we would expect uniformity in regional richness across the Cape Floristic Region in parallel with the uniformity in alpha richness. If Westoby (1988) is correct, then there is some strong overriding determinant of alpha richness in fynbos. We take up these points in the section on regional richness.

Contemporary ecological factors may also influence the size of the regional species pool. For example, the lower alpha richness of fynbos 'islands' surrounded by Afromontane forest in the south-eastern Cape Floristic Region was attributed to the lower species pool on these isolated fragments (Bond 1983). However, differences in the richness of comparable areas of 'island' and 'mainland' fynbos could be attributed to the absence of short-lived species on the islands which are seldom burnt (Bond et al. 1988). The very low richness of phytochorologically pure fynbos in the uplands of the Humansdorp region (Figure 4.6) may have an ecological or historical explanation, both of which assume a relationship between local and regional richness. The ecological hypothesis attributes lower regional richness of the small patches of fynbos to extinctions due to an island effect; the historical hypothesis argues that insufficient time has elapsed for the migration into these uplands of mesic fynbos species since climatic amelioration some 12 000 BP (Cowling 1983a). Evidence refuting the ecological hypothesis comes from a study by Bond et al. (1988) which showed that an island effect in floristically and structurally similar vegetation was evident only in areas smaller than 600 ha. This is considerably smaller than the upland fynbos patches sampled in the Humansdorp region. Nevertheless, more research is needed on testing independent predictions of historical and ecological hypotheses on species richness in the Cape Floristic Region.

SYNTHESIS

Despite differences in climate, disturbance regime and, to a lesser extent, soils, alpha diversity in fynbos does not vary across the Cape Floristic Region. On average, fynbos supports 15 species per square metre and 65 species per 1 000 m^2. Some of the variance in the data can be explained by fire effects, between-guild interactions, mass effects, and area and isolation. If this uniform species density indicates that fynbos communities are saturated, then what are the controls on species numbers? Rice and Westoby (1983) and Westoby (1988) have suggested low nutrient soils as an explanation for the convergent species richness of temperate Australian and Cape shrublands. Rainfall seasonality was not regarded as important. Whittaker (1977) invoked the greater age of these southern, Gondwanan landscapes as the explanation for the higher alpha diversity of Australian and Cape mediterranean communities relative to those in California, Chile, and the Mediterranean Basin. This argument implies a relationship between regional and local richness, since the higher regional richness of the Cape and Australia would result from a longer speciation history.

Forest and thicket communities show increasing richness from west to east in the Cape Floristic Region. Regional species pools for these communities show a similar trend (Cowling 1983b, c; Geldenhuys and MacDevette 1989). In these communities, local richness may be controlled by climatic factors which permit the coexistence of a wide range of growth forms, and hence species. It could also be determined by, as yet, unstudied area effects (forest and thicket patches are smaller and more isolated in the south-west) (Geldenhuys and MacDevette 1989). Furthermore, winter rainfall and lower soil nutrients may adversely affect the regeneration of many forest and thicket species in the south-west (see Cowling and Holmes this volume). Finally, historical factors associated with the westwards dispersion of forest and thicket taxa since the Holocene may also be important (Cowling 1983a).

Beta diversity

Despite the importance of beta diversity in determining regional species richness (Whittaker 1977), empirical and theoretical insights on its controls are very limited. Whittaker (1977) has explained the evolution of beta diversity as resulting from displacement and divergence after the arrival of additional species along habitat gradients. Shmida and Wilson (1985) suggest that habitat diversity is the ultimate determinant of beta diversity. However, areas with equivalent habitat diversities may have vastly divergent rates of species turnover (Cowling et al. 1989). Historical processes associated with the evolution of habitat specialists must also be considered.

There are very few studies on beta diversity in the Cape Floristic Region. Cowling (1990) observed almost complete turnover between sites (1 000 m^2 samples) on different soil types along three crude soil fertility gradients on the Agulhas Plain (Figure 4.7). He

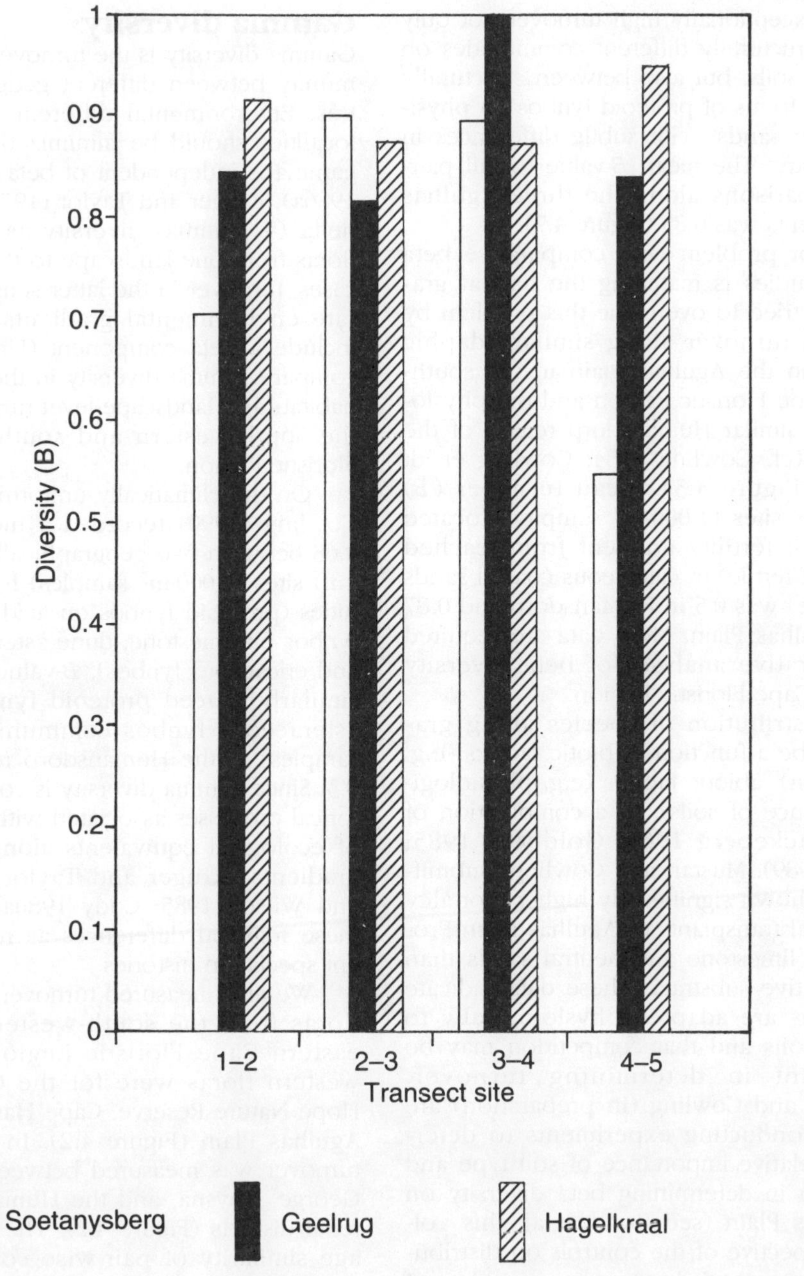

FIGURE 4.7 The species turnover between 1 000 m^2 plots along three gradients of increasing soil fertility on the Agulhas Plain in the south-western Cape Floristic Region. The turnover (*B*) is defined in Figure 4.1. Each transect spans about 5 km. *B*-values for the entire transects were 3.06, 3.28, and 3.82 for the Soetanysberg, Geelrug, and Hagelkraal transects respectively, indicating that each transect comprised between four and five distinct communities ($B + 1$). See Thwaites and Cowling (1988) and Cowling (1990) for details on soil and floristic data respectively for each of the sites. Reprinted from Cowling (1990) with the permission of the International Association for Vegetation Science (OPULUS Press Uppsala).

recorded exceptionally high turnover not only between structurally different communities on contrasting soils, but also between structurally convergent forms of proteoid fynbos on physically similar sands, with subtle differences in soil chemistry. The mean *B*-value for all pair-wise comparisons along the three Agulhas Plain gradients was 0.85 (Figure 4.7).

A major problem with comparative beta diversity studies is matching the habitat gradients. We tried to overcome this problem by comparing turnover along similar edaphic gradients on the Agulhas Plain in the south-western Cape Floristic Region and the physiographically similar Humansdorp region of the south-east (cf. Cowling 1984; Cowling et al. 1988; see Figure 4.3). Mean turnover (*B*) among four sites (1 000 m^2 samples) located along a soil fertility gradient from leached sands (least fertile) to calcareous coastal sands (most fertile) was 0.5 in Humansdorp and 0.87 on the Agulhas Plain. More data are required on comparative analyses of beta diversity across the Cape Floristic Region.

The distribution of species along gradients can be a function of biotic factors (e.g. com-petition), abiotic factors (e.g. physiological intolerance of soils), or a combination of these (Kruckeberg 1969; Goldberg 1985; Moloney 1989). Mustart and Cowling (submitted) have shown significantly higher mortality of reciprocal transplants of Agulhas Plain Proteaceae on limestone and neutral sands than on their native substrata. These data indicate that species are adapted physiologically to particular soils and that competition may be unimportant in determining turnover. Richardson and Cowling (in preparation) are currently conducting experiments to determine the relative importance of soil type and competition in determining beta diversity on the Agulhas Plain (see Bond et al. this volume). Irrespective of the controls on distribution, there appear to be a greater number of habitat or edaphic specialists on the Agulhas Plain than in the Humansdorp region. Since the array of habitats compared are broadly similar, this difference in turnover rates suggests divergence in the evolution of edaphic specialists between the two regions. We return to this subject in our discussion on endemism.

Gamma diversity

Gamma diversity is the turnover within a community between different geographical localities. Environmental differences between the localities should be minimized to ensure that gamma is independent of beta diversity (Cody 1986a). Kruger and Taylor (1977) have studied delta (= gamma) diversity as the change in floras from one landscape to the next. In most cases, turnover in the latter sense will incorporate environmental gradients and will thus include a beta component (Linder 1985). We compare gamma diversity in the sense of both habitat- and landscape-level turnover, between the south-western and south-eastern Cape Floristic Region.

On the climatically uniform Agulhas Plain, Cowling (1990) recorded a mean *B*-value of 0.68 between two geographically distant (*c.* 30 km) sites (1 000 m^2 samples) for four communities (proteoid fynbos on acid sand, proteoid fynbos on limestone, dune asteraceous fynbos, and ericaceous fynbos). *B*-values between two similarly spaced proteoid fynbos and dune asteraceous fynbos communities (1 000 m^2 samples) in the Humansdorp region averaged 0.3. Since gamma diversity is controlled by historical processes associated with the speciation of ecological equivalents along geographical gradients (Kruger and Taylor 1979; Shmida and Wilson 1985; Cody 1986a), we interpret these regional differences as reflecting different speciation histories.

We also measured turnover between three floras from the south-western and south-eastern Cape Floristic Region. The south-western floras were for the Cape of Good Hope Nature Reserve, Cape Hangklip, and the Agulhas Plain (Figure 4.2). In the south-east turnover was measured between the floras of George, Knysna, and the Humansdorp magisterial districts (Figure 4.2). The mean percentage similarity of pair-wise comparisons between south-western floras was significantly higher than south-eastern floras, indicating lower gamma diversity in the former region (Table 4.2). This trend was particularly pronounced for families largely comprising genera centred in the Cape Floristic Region (Ericaceae and Restionaceae). Species turnover of the Proteaceae in eighth degree squares is much higher in the south-western than the south-

TABLE 4.2 **The percentage similarity (Sörenson 1948) between total floras and selected families in the south-western and south-eastern Cape Floristic Region. Data for the Cape of Good Hope Nature Reserve are from Taylor (1985); Cape Hangklip from Boucher (1977); Agulhas Plain from R M Cowling (unpublished); and George, Knysna, and Humansdorp from Fourcade (1941). Floras within the two regions were standardized for nomenclature. All alien plants and pteridophytes were excluded.**

	South-west				South-east				
	Cape of Good Hope	Cape Hangklip	$X \pm$ S.D.		Knysna	Humansdorp	$X \pm$ S.D.	U[1]	Sig.
Total flora									
Agulhas Plain	45.8	42.5		George	51.7	48.3			
Cape of Good Hope		47.0	45.1 ± 2.3	Knysna		52.8	50.9 ± 2.3	0	*
Ericaceae									
Agulhas Plain	30.4	40.5		George	62.3	50.0			
Cape of Good Hope		35.0	35.3 ± 5.1	Knysna		58.3	56.9 ± 6.3	0	*
Fabaceae									
Agulhas Plain	41.2	43.8		George	51.6	36.2			
Cape of Good Hope		40.7	41.9 ± 1.7	Knysna		55.8	47.9 ± 10.3	3	NS
Proteaceae									
Agulhas Plain	32.1	45.6		George	50.0	32.6			
Cape of Good Hope		34.1	37.3 ± 7.3	Knysna		48.2	43.6 ± 9.7	2	NS
Restionaceae									
Agulhas Plain	49.3	44.7		George	51.9	59.3			
Cape of Good Hope		31.4	41.8 ± 9.3	Knysna		65.5	58.9 ± 6.8	0	*
Rutaceae									
Agulhas Plain	31.0	26.1		George	58.4	43.1			
Cape of Good Hope		51.4	36.1 ± 13.4	Knysna		40.0	47.2 ± 9.9	2	NS

[1] Mann-Whitney U test for very small samples. * = $P < 0.05$, NS = not significant

eastern Cape Floristic Region (Figure 4.8; see also Cody 1986a).

Since these geographical gradients incorporate some degree of environmental and thus habitat change, it is important to ask whether the ecological gradients between the three south-western sites are longer than those between the south-eastern sites. The south-western sites are all coastal areas with a broadly similar range of habitats and plant communities (Boucher 1978; Taylor 1984; Cowling et al. 1988). However, the proportion of summer rain increases marginally from the Cape of Good Hope to Cape Agulhas (Figure 4.3). Moreover, Cape Hangklip has many high altitude, high rainfall communities not found in the other sites, while Cape Agulhas includes the largest area of fynbos on limestone-derived soils and some unique communities on shale- and laterite-derived soils. The south-eastern sites are climatically and physiographically similar, comprising a mesic, narrow coastal plain arising abruptly to wet, coastal mountains (*c.* 1 500 m). On the landward side of the coastal mountains, there is a drier valley region. The range of vegetation types in the three areas is broadly similar (Phillips 1931; Bond 1981; Cowling 1984). However, the George district includes an interior basin dominated by karroid shrubland and the Humansdorp district includes extensive subhumid plains (often underlain by Cretaceous deposits) which support renoster shrubland and subtropical thicket: these habitats are unique to the two sites. The crude level of our analysis makes it impossible to apportion with any accuracy the degree to which turnover can be explained by historical or ecological factors. However, it appears that the magnitude of environmental change along the geographical gradients for the two regions is broadly similar so that historical processes — in particular greater speciation of ecological equivalents in the south-west — could play a substantial role in explaining differences in turnover between landscapes.

Regional richness

As defined earlier (see also Figure 4.1) regional richness is a product of the interactions between alpha, beta, and gamma diversity. We have shown that alpha diversity does not vary across the Cape Floristic Region but our very limited data indicate that turnover (both beta and gamma) is higher in the south-west than the south-east. Although the data on turnover are inadequate to provide a realistic generalization, we have, nonetheless, used these to predict the regional richness of a hypothetical landscape in the south-west and the south-east (Table 4.3). As expected, the predicted regional richness in the south-west is higher (1.6x) than in the south-east.

We tested this predicted difference in regional richness by analysing species-area data for 19 and 15 regional floras in the south-western and south-eastern Cape Floristic Region respectively (see Figure 4.2 and Appendix 4.1). All sites had an average rainfall in excess of 400 mm. Cape shrublands (fynbos or renoster shrubland) were the dominant vegetation type in all sites except for the Albany and Bathurst districts (Martin and Noel 1960) which are on the eastern periphery of the Cape Floristic Region. Many of the south-eastern sites had a stronger admixture of non-fynbos vegetation types than those in the south-west (cf. Figure 3.1 in Cowling and Holmes this volume). The check-lists vary in their completeness and coverage (all exclude aliens but only some include pteridophytes). However, error is unlikely to be biased with respect to region. For both data sets the best fit was the double logarithmic form of the power function (Arrhenius 1921; Williamson 1988) by the method of least squares. Thus,

$$\log S = k + z \log A$$

in which S and A are species number and area respectively, and k and z are constants. Some of the sites were nested within larger sites (Figure 4.2) and, thus, were not independent data. Removal of these did not result in significantly different regression statistics, so they were retained.

For both data sets species number was strongly and significantly ($P < 0.0001$) correlated with area (Figure 4.9; Table 4.4). Both regressions explained more than 90% of the variance, and parameter estimates were robust (cf. Boecklen and Gotelli 1984; Table 4.4). This is probably a result of the high k- and z-values for the regressions (Dunn and Loehle 1988). The slopes of the two curves were not significantly different (t = 0.796, df = 30) but the elevations were (t = 3.203, df = 31, $P < 0.005$) (Zar 1984). In the south-west, low-

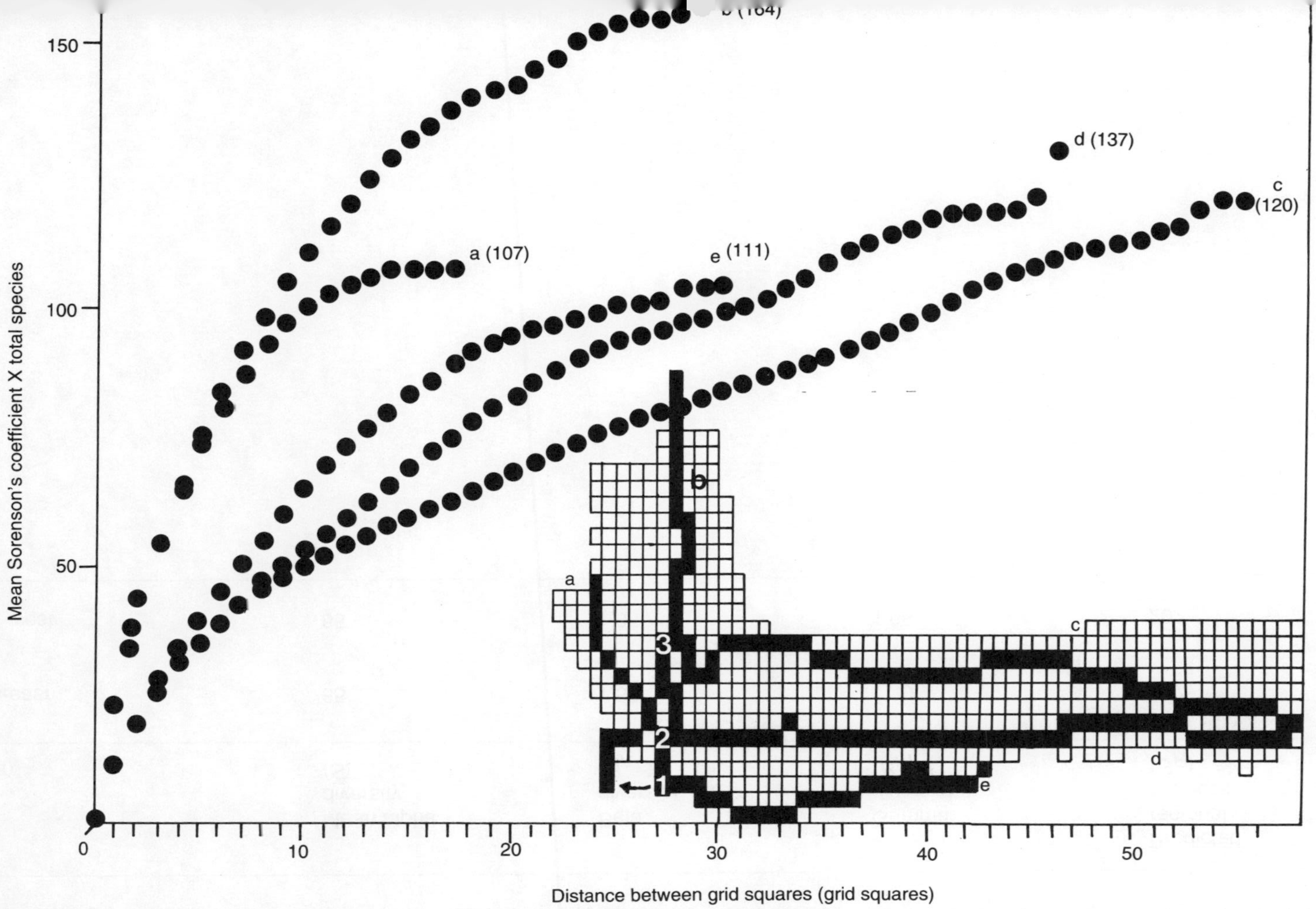

FIGURE 4.8 Species turnover for Proteaceae in eighth degree squares along transects a) from Cape Hangklip (1) to the Cape Peninsula and the western forelands; b) from Cape Hangklip (1) northwards along the western mountains; c) from Groot Winterhoek (3) eastwards along the southern, interior mountains; d) from Hottentots Holland (2) eastwards along the southern, coastal mountains; e) from Cape Hangklip (1) eastwards along the southern forelands. Turnover is calculated as the mean Sörenson's (1948) coefficient of grid squares, separated by different distances along the transects (Cody 1986a). These coefficients range from 0–1 and have been scaled to the total number of species (in parentheses) on the transects. (A G Rebelo unpublished data).

TABLE 4.3 **The regional richness predicted for two hypothetical, physiographically identical fynbos-dominated landscapes in the south-western and south-eastern Cape Floristic Region. In each hypothetical landscape three stations were sampled along analogous habitat gradients. Two 1 000 m^2 plots were sampled for each station at the extreme boundaries of each landscape. Predicted regional richness (S_R) = $\bar{S}(B_n+1)(B_m+1)$ where $\bar{S}$ is the mean alpha diversity of all plots in all habitats (n = 6 for each landscape) and B_n and B_m are turnover (Wilson and Shmida's (1984) measure) between and within habitats, respectively (Figure 4.1; Cowling 1990). The estimates of values for diversity components are from real data given in the text.**

Region	Mean alpha diversity ($\bar{S}$)	Beta diversity (B_n+1)	Gamma diversity (B_m+1)	Predicted regional richness (S_R)
South-east	65	1.50	1.30	127
South-west	65	1.85	1.68	202

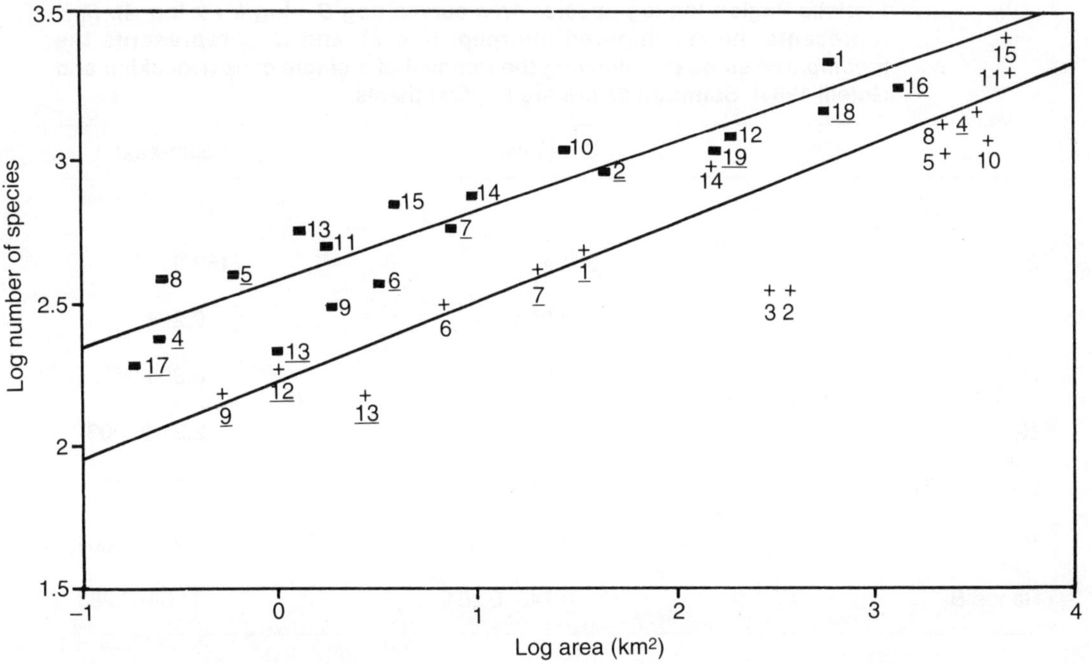

FIGURE 4.9 The species-area relationships for floras from the south-western (■) and south-eastern (+) Cape Floristic Region. The site numbers and localities are shown in Figure 4.2 and Appendix 4.1. Underlined numbers refer to lowland sites.

land sites fell largely below the regression line (Figure 4.9). When analysed separately, the *k*- and *z*-values for the lowland sites were 297.2 and 0.256 respectively (R^2 = 0.958, n = 10). Corresponding values for the montane sites were 415.9 and 0.257 (R^2 = 0.909, n = 9). However, neither the slopes (t = 0.007, df = 15) nor the elevations (t = 0.166, df = 16) of the montane and lowland site regressions were significantly different.

The mean area and richness for the south-western sites were 9.71 km² and 618 species respectively. Corresponding data for the south-eastern sites were 146.6 km² and 642.5 species. However, for both data sets, area ranges overlapped extensively. Since the slopes were homogeneous, it was possible to factor out area by comparing the ratio of *k* (intercept) values for the two regressions (Gould 1979). The intercept ratio was 2.35. Thus, a given area in the south-western fynbos biome has more than double the number of species than the same area in the south-east (cf. Table 4.3).

There is still a possibility that the higher regional richness in the south-west might simply be a result of longer ecological gradients than the south-east (Linder submitted). Campbell (1983) has shown that in the south-western mountains, upper altitudinal and rainfall ranges are greater than the south-east. We overcame this problem by comparing species density (spp. km^{-2}) of physiographically matched sites of approximately equal area in the two regions (see Figure 4.3 for rainfall data). Examples are given below, listing the south-western site first (see also Figures 4.2, 4.9; Appendix 4.1):

- Rondevlei (167.1); Cape Recife (51.5);
- Tygerberg (548.5); Seekoei (153.9);
- Vogelgat (115.8); Moordkuils (50.7);
- Agulhas Plain (1.1); Riversdale Plain (0.6).

Thus, the south-western sites had between 1.8 and 3.8x the species density of similar sized and physiographically equivalent south-eastern sites. Moreover, some of the larger sites in the south-east, which encompass a great diversity of climo-edaphic environments

TABLE 4.4 The regression coefficients for south-western and south-eastern Cape Floristic Region log-log species area curves ($\log S = \log k + z \log A$). $B_{0,-1}$ represents the recomputed intercept ($\log k$), and $B_{1,-1}$ represents the recomputed slope (z) following the removal of a single case (Boecklen and Gotelli 1984). Standard errors are in parenthesis.

	South-west	South-east
n	19	15
k	384.4	163.3
z	0.253	0.274
R^2	0.9	0.904
B_0	2.54 (0.002)	2.21 (0.003)
Range of $B_{0,-1}$	2.52–2.56	2.21–2.24
B_1	0.253 (0.001)	0.274 (0.001)
Range $B_{1,-1}$	0.243–0.264	0.264–0.283

and vegetation types — e.g. the George district (Bond 1981) and the Humansdorp district (Cowling 1984) — have as many species as south-western sites between 10 and 30 times smaller (the Cape of Good Hope Nature Reserve and the Cape Peninsula respectively). The presence in the south-east of several species pools associated with distinct vegetation types and phytochoria is clearly not a major determinant of regional richness within the Cape Floristic Region (cf. Danin 1978). Furthermore, these data show unequivocally that there is no relationship between alpha diversity and the size of regional species pools (at this scale) in the Cape Floristic Region (cf. Rickleffs 1987).

The lower regional richness of lowland relative to montane south-western sites recorded in this study is consistent with Linder's (submitted) data on species number in quarter degree grids. It would seem that fewer species are packed along the predominantly edaphic gradients of the lowlands than along the steep climatic gradients of the mountains (Linder submitted). This floristic complexity in the south-western mountains is amply revealed in the 'noisy' phytosociological tables compiled from this region (e.g. Boucher 1978; see also Cowling and Holmes this volume).

In summary, this analysis suggests that the higher regional richness in the south-western Cape Floristic Region is associated with higher turnover within and between habitats. As previously stated, studies on floristic turnover are very limited (Whittaker 1977; Shmida and Wilson 1985). However, we conclude that historical processes associated with different rates of evolution of habitat specialists (beta diversity) and ecological equivalents (gamma diversity) must be invoked to explain differences in regional richness between the south-western and south-eastern Cape Floristic Region. We return to the question of speciation in the section on endemism.

Comparisons with other species-rich areas

Many areas in the world other than the Cape Floristic Region have very high plant species richness at the regional level. These include some tropical rain forest areas (Ashton 1969; Gentry 1982), the South West Botanical Province of Australia (Beard 1981; Lamont et al. 1984), the California Floristic Province (Raven and Axelrod 1978), and parts of the Mediterranean Basin (see Davis et al. 1986; Table 4.5). The Cape Floristic Region and the three last-mentioned areas have predominantly mediterranean climates and support extensive areas of fire-prone shrublands and woodlands. The Australian and Cape regions have soils which are generally more nutrient-poor than those in California and the Mediterranean Basin. Tropical rain forests are structurally complex formations occurring on a wide range of habitats in high rainfall tropical climates across the globe (Richards 1952).

The richness of neotropical and Asian (but not African) tropical rain forest regions is comparable to the Cape Floristic Region; regional richness of the other mediterranean regions is lower (Table 4.5). What is the relative contribution of the three independent diversity components to regional richness in these species-rich regions?

Below we compare the diversity components and regional richness of the Cape with those of the California Floristic Province, South West Botanical Province, and tropical rain forest areas. The Mediterranean Basin was excluded from the analysis for lack of data. Even for some of the other regions data are meagre and, for some diversity components, non-existent.

Alpha diversity (species/1 000 m^2) is markedly different in the four regions (Table 4.6). Values for sclerophyllous shrublands and woodlands from the South West Botanical Province average about 69 species (George et al. 1979; Naveh and Whittaker 1979; Hnatiuk and Hopkins 1980) which is comparable to fynbos communities (Bond 1983; Cowling 1983a; Rice and Westoby 1983; Table 4.1). Chaparral of the California Floristic Province is poorer in species (an average of 30 species per 1 000 m^2) (Naveh and Whittaker 1979) than fynbos. Tropical rain forests have by far the highest alpha richness: mean values for lowland areas in the neotropics are 140 spp./ 1 000 m^2 (Gentry 1988); Australia 140 (Rice and Westoby 1983); Africa 129 (Gentry 1988); and Asia 193 (Gentry 1988). Thus, tropical rain forests the world over (Gentry 1982, 1988)

TABLE 4.5 The areas and numbers of species in some species-rich regions of the globe.

Region	Area (10^6 km^2)	Number of spp.	Species density (10^3 spp./10^6 km^2)
<u>Mediterranean climate regions</u>			
Cape Floristic Region[1]	0.09	8 550	94.4
California Floristic Province[2]	0.32	4 452	13.9
South west Botanical Province[3]	0.31	3 611	11.6
Greece[4]	0.13	*c.* 4 000	30.8
Portugal[4]	0.09	*c.* 2 500	27.8
<u>Tropical rain forest regions</u>			
Panama[5]	0.08	*c.* 8 300	103.3
Cuba[6]	0.11	*c.* 6 700	60.9
Guatamala[6]	0.11	*c.* 8 000	72.7
Peninsula Malaysia[4]	0.13	*c.* 8 000	61.5
Ivory Coast[4]	0.32	*c.* 4 700	14.7

[1] Bond and Goldblatt (1984)
[2] Raven and Axelrod (1978)
[3] Beard (1981)
[4] Davis et al. (1986)
[5] P Goldblatt (personal communication)
[6] Campbell and Hammond (1989)

have more than double the alpha richness of mediterranean communities on nutrient-poor soils, which in turn have about twice as many species as Californian chaparral.

Beta diversity along soil fertility gradients in south-western fynbos and south-western Australian kwongan is exceptionally high (Cowling et al. 1989; R M Cowling and E Witkowski unpublished data). However, relative to the Cape Floristic Region, the South West Botanical Province is topographically and climatically uniform. This results in shorter ecological gradients which lowers the contribution that beta diversity makes to regional richness (Kruger and Taylor 1979; Lamont et al. 1984). Linder (submitted) has shown that the long climatic gradients in the mountainous areas of the south-western Cape Floristic Region are very important determinants of regional richness. There are no quantitative data on beta diversity in the California Floristic Province but indications are that it is high. The Californian flora includes many locally en-demic, edaphic, and other habitat specialists (Stebbins and Major 1965; Raven and Axelrod 1978). Furthermore, ecological gradients in the topographical and climatically complex California Floristic Province are extremely long (Raven and Axelrod 1978). The importance of beta diversity in determining regional richness in the California Floristic Province is evident from a study by Richerson and Lum (1980). They found that regional richness in California could be largely explained by climatic and topographic variables: richness was highest in mountainous areas with high climatic and topographic diversity (and presumably high beta diversity) and lowest in dry, low-lying areas.

Beta diversity in tropical rain forest regions has not been studied. Both Ashton (1969) and Gentry (1986, 1988) mention high levels of edaphic specialization and turnover between different soil types in the neotropical lowland and Asian rain forests, respectively. This evidence for edaphically determined structure contrasts with Hubbell and Foster's (1986) random drift model for the maintenance of species richness in Barro Colorado — a neotropical forest region.

There are no data on within-habitat gamma diversity for any of the regions other than the Cape (this study). Species turnover along geographical gradients appears to be comparably high in the three mediterranean climate regions (Cody 1986a). Hopkins and Griffin (1984) mention an approximate turnover of 10 species per 25 km in kwongan areas of the South West Botanical Province. Lamont et al. (1984) report a 34% Sörenson's (1948) similarity in the floras between two kwongan sites 60 km apart. This is less pronounced than the 64% similarity between two fynbos sites on opposite sides of the same 2 km wide valley and the 40% similarity between sites less than 25 km apart (Kruger and Taylor 1979). The higher turnover in the Cape probably reflects, in part, the steep ecological gradients which parallel geographical gradients in the mountainous topography of the south-west (Linder 1985). Unfortunately, there are no data on floristic turnover for tropical rain forest regions.

Another way of analysing these patterns is to compare observed richness with the value predicted from the Cape species-area relationships in regions of different sizes in California, Australia, and tropical rain forests (Table 4.7). Equal sized areas of the California Floristic Province support about half as many species as the south-western Cape Floristic Region. This is probably due to the lower alpha diversity of Californian communities (Table 4.6). The fact that predicted regional richness for the south-eastern Cape Floristic Region was roughly comparable with California, even though the former has communities which are about twice as rich, indicates that turnover in the latter region is similar to the south-western Cape Floristic Region.

Regional richness in the smallest South West Botanical Province site was similar to the predicted value for the south-western Cape Floristic Region but, in larger sites, it was about half the value (Table 4.7). This is probably a result of the longer ecological gradients in the Cape (Table 4.6). On average, the Australian sites had 1.3x more species than predicted for the south-eastern Cape. This can be explained by the lower turnover in the latter area. However, this discrepancy is not evident in larger sized areas (> 1 000 km^2) when greater topographic heterogeneity is again encountered in the Cape.

Small areas (< 20 km^2) of neotropical rain forest have more than twice as many species

TABLE 4.6 A comparative qualitative analysis of diversity components and regional richness in four species-rich regions of the globe. See text for details.

	Alpha diversity	Beta diversity	Length of ecological gradients	Gamma diversity		Regional richness
				Within-habitat	Flora turnover	
Cape Floristic Region						
South-west	Moderate	Very high	Long	High	Very high	Very high
South-east	Moderate	Moderate	Long	Moderate	Moderate	Moderate
California Floristic Province	Low	High	Very long	?	?	Moderate
South West Botanical Province	Moderate	Very high	Short-moderate	?	Moderate	Moderate
Tropical rain forest regions	Very high	High (?)	Variable	?	?	Very high

TABLE 4.7 The regional richness in other species-rich parts of the globe. Corresponding richness for identical sized areas in the Cape Floristic Region (CFR) was predicted from species-area relationships separately for the south-west (SW) and the south-east (SE). Raw data in the species-area analysis for the CFR are given in Appendix 4.1. — indicates that area was not inclusive of the sample range (cf. Gould 1979).

Region	Area	No.	No. spp. predicted from CFR data	
	(km^2)	spp.	SW	SE
California Floristic Province[1]				
Tiburon Peninsula	15.1	370	637.6	344.0
San Francisco	115.2	640	1 158.8	600.9
Santa Barbara area	281.6	680	1 453.1	767.9
Santa Monica Mountains	819.2	640	1 904.1	1 029.4
Marin County	1 352.4	1 060	2 161.7	1 181.4
Santa Cruz Mountains	3 548.1	1 200	—	1 539.2
South West Botanical Province[2]				
Mt Lesuer	0.5	286	292.3	135.0
Mt Adams	13	290	667.0	330.1
South Eneabba	20	429	743.9	371.6
Tutanning	22.5	628	766.3	383.8
Two Peoples Bay	46	614	918.5	467.1
Stirling Range	1 156	874	77.8	1 131.6
Neotropical rain forests[3]				
Rio Palenque	1.7	1 033	398.0	188.9
Barro Colorado	15.6	1 318	698.4	347.1

1 Johnson et al. (1968)
2 Lamont et al. (1984)
3 Gentry (1982)

as predicted for the south-western Cape Floristic Region and more than four times as many as the south-east (Table 4.7). At this scale the difference would be largely due to the very much higher alpha diversity of the rain forests (Table 4.6). Since the richness of the Cape Floristic Region is similar to equal sized areas of tropical rain forest (Table 4.5), we predict that either beta or gamma turnover, or both of these, would be lower in the latter. Unfortunately there are insufficient data to test this.

By necessity this analysis of diversity components in species-rich regions has been very crude. In many cases, qualitative estimates of turnover were the best available data. Nonetheless, it suggests that the explanation for the high regional richness of the south-western Cape Floristic Region is unique: a combination of the very high turnover of moderately rich communities over long ecological gradients and the very high turnover of very rich floras along geographic gradients. The high within-habitat gamma diversity of south-western fynbos may also be unique. The importance of habitat specialization and geographic turnover in explaining the regional richness of the south-west suggests that the speciation history of the region may be unique.

ENDEMISM

Approach

How do levels of endemism vary across the Cape Floristic Region; are endemics a non-random sample in terms of taxonomic, habitat, and biological characteristics; and what are the relationships between endemism, speciation, and species richness? These are daunting questions (see Major 1988) and our attempts to provide answers are severely constrained by a lack of appropriate data relating, in particular, to the last two-mentioned questions.

Our approach has been to analyse the levels of endemism and taxonomic, edaphic, and biological aspects of endemism in two areas of the Cape Floristic Region. In order to explore the relationship between endemism and species richness in the region, we have chosen a flora from the species-rich south-west (Agulhas Plain) and the relatively species-poor south-east (Humansdorp) (Figure 4.10). The flora of the former comprises a full checklist although a subsample (species collected in 17 1 000 m^2 plots) (Cowling 1990) was used for analysing biological aspects of endemism. The Humansdorp sample flora comprises a list of species compiled from 97 100 m^2 plots and 20 1 000 m^2 plots. Both regions occupy a similar area and show strong physiographic similarities (Cowling 1984; Cowling et al. 1988). They are located on the coastal forelands and incorporate a wide diversity of edaphic habitats. The Humansdorp region has longer rainfall gradients (450–1 000 mm yr^{-1}) than the Agulhas Plain (450–600 mm yr^{-1}) and a much higher proportion of summer rainfall (Figure 4.3). Both regions have edaphic environments which are peculiar to them (e.g. Cretaceous conglomerate and sandstone in Humansdorp; Bredasdorp Formation limestone on the Agulhas Plain). However, in both regions the most widespread land forms are associated with Table Mountain Group sandstones (deep, colluvial, and shallow sands), Bokkeveld Group shales (duplex soils), and Recent calcareous sands (Cowling 1984; Thwaites and Cowling 1988). The Agulhas Plain is almost entirely dominated by fynbos and renoster shrubland, whereas about 15% of the Humansdorp region comprises forest and thicket.

The contemporary ecological conditions in both areas are conducive to the evolution of endemics. These include semi-arid to subhumid rainfall regimes and a high degree of edaphic heterogeneity (Stebbins and Major 1965; Raven and Axelrod 1978; Hopper 1979; Robinson and Gibbs Russell 1982; Papanicolaou et al. 1983; Kruckeberg and Rabinowitz 1985; Major 1988). The Humansdorp region, moreover, falls within a tension zone between four African phytochoria (Cowling 1983c; see Figure 3.3 in Cowling and Holmes this volume): ecotones between different biotic provinces are, under the motor of climatic change, considered areas of active speciation (Stebbins and Major 1965; Hopper 1979). However, before addressing the evolution of endemism in the Cape Floristic Region, we will clarify some concepts and analyse patterns of endemism.

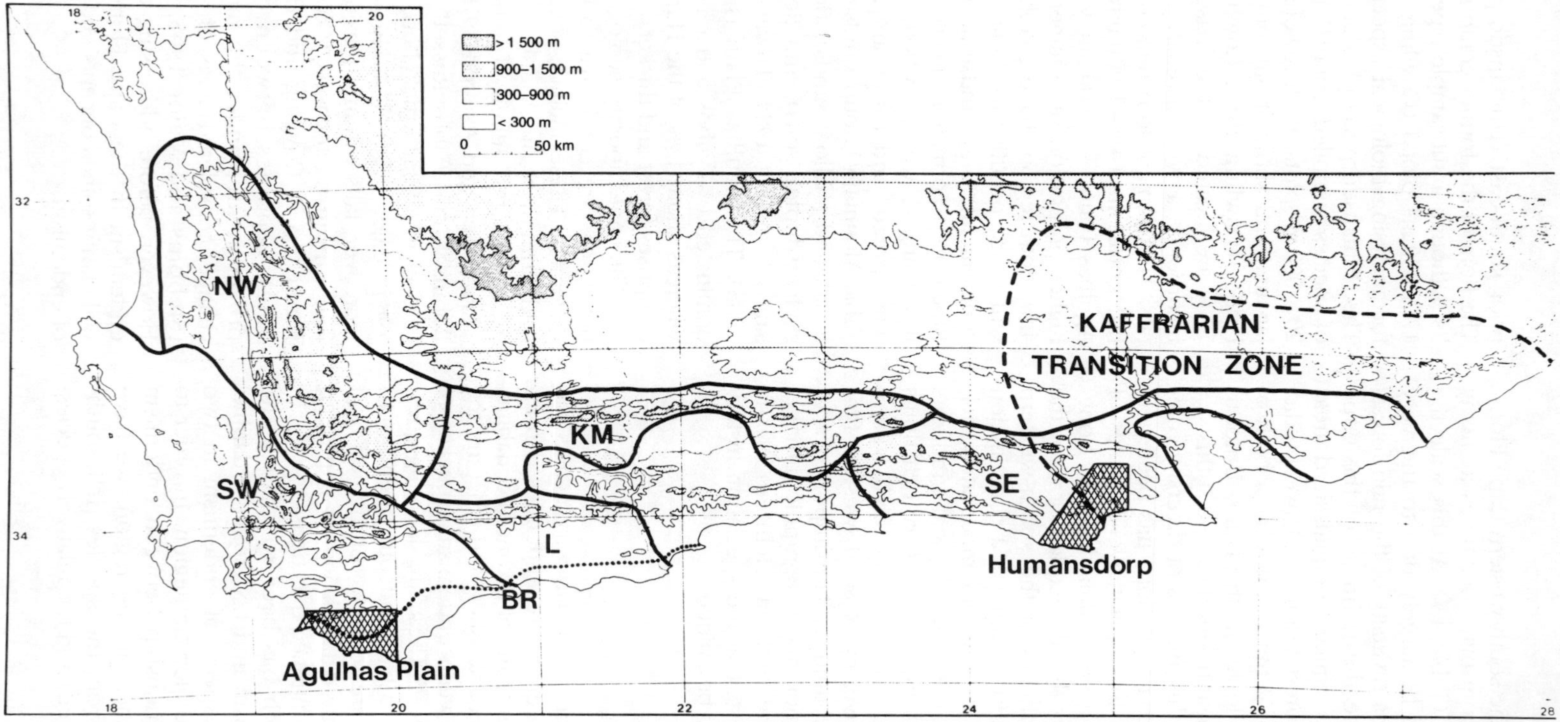

FIGURE 4.10 The location of Weimarck's (1941) major centres of endemism for the Cape Floristic Region. SW = South Western, NW = North Western, L = Langeberg, KM = Karoo Mountain, SE = South Eastern. Also shown is the Bredasdorp-Riversdale region (BR) and a non-Cape centre, the Kaffrarian Tension Zone.

Concepts and patterns of endemism in the Cape Floristic Region

CENTRES OF ENDEMISM

A centre or area of endemism is delimited by the more-or-less coincident distribution of taxa that occur nowhere else. As such, the Cape Floristic Region is a recognized centre (Goldblatt 1978; Bond and Goldblatt 1984; Linder et al. this volume). Weimarck's (1941) pioneering study was the first to recognize coherent centres of endemism within the Cape Floristic Region. He recognized five centres (Figure 4.10) on the basis of the distribution patterns of 462 taxa belonging to mainly Cape endemic genera in the Asteraceae, Boraginaceae, Cyperaceae, Iridaceae, Rosaceae, and Thymeleaceae. These centres have formed the basis for the analysis of distribution patterns for many subsequent monographs (e.g. Dahlgren 1963; Nordenstam 1969; Strid 1972). A recent analysis, using a clustering algorithm, of the distributions of an independent data set comprising 1 936 species belonging to characteristic or endemic Cape genera, largely corroborated Weimarck's centres (Oliver et al. 1983).

Criticisms of Weimark's system are that the centres are defined exclusively on the distribution of taxa largely endemic to the Cape Floristic Region and do not stress the importance of phylogenetic relationships (Nordenstam 1969). Linder et al. (this volume) suggest that Weimark's centres are not linked directly to historical explanations of distribution patterns. They cannot be regarded as areas in the classic biogeographic sense (Nelson and Platnick 1981) since studies have not been done to analyse the interrelationships of endemic species. The centres probably reflect differences in contemporary ecological conditions (i.e. they are phytochorological regions) but some of the montane centres probably do reflect different phylogenetic histories. Furthermore, Weimarck failed to emphasize the significance of some important ecological factors in delimiting his centres. For example, he did not recognize a well defined centre associated with the limestones of the south-western and south-eastern coastal region (Dahlgren 1963; Nordenstam 1969). We have called this the Bredasdorp-Riversdale Centre (Figure 4.10).

We define regional endemics as taxa confined to the centres of endemism shown in Figure 4.10. Thus, regional endemics in the Agulhas Plain flora include species restricted to the South Western and Bredasdorp-Riversdale Centres. In Humansdorp, regional endemics include taxa found only in Weimarck's South Eastern Centre or the Kaffrarian Transition Zone. The latter is a recognized centre of endemism in the south-eastern Cape for non-Cape taxa, especially karroid succulents (Croizat 1965; Nordenstam 1969; Cowling 1983c). Local endemics are species confined or nearly confined to the respective study areas. For the Agulhas Plain this area corresponds to the Bredasdorp Centre, a recognized subcentre within the South Western Centre (Weimarck 1941; Oliver et al. 1983; Midgley 1986; Rebelo this volume). The Humansdorp region falls on the boundary of two centres of local endemism recognized for the Proteaceae (and by correlation, many other Cape taxa) (Rebelo this volume). Local endemics were arbitrarily recognized as taxa confined or nearly confined to either of these centres.

Gentry (1986) refers to local endemics as species with distributions of 50 000 km^2 or less. All of our local endemics occupy ranges less than 2 000 km^2 (2.5% of the Cape Floristic Region); some less than 5 km^2. According to Gentry's terminology, our regional endemics would be classified as local endemics. Some local endemics may be locally abundant, whereas some widespread species may comprise only small populations at few or many localities (cf. Rabinowitz et al. 1986). Although categorization of endemics by rarity status would greatly improve the analysis, data were not available for most of the species included in the analysis.

LEVELS OF ENDEMISM

Levels of endemism in the Cape Floristic Region are among the highest in the world (Table 4.8). Comparable levels have been recorded for the mediterranean climate regions of California (Raven and Axelrod 1978) and south-western Australia (Beard 1981), for tropical rain forest areas (Gentry 1986), and certain oceanic islands (Bramwell 1979). All of the continental centres of endemism are also

TABLE 4.8 The percentage of endemic species in various continental areas and islands.

Region	Area (10^3 km^2)	No. spp.	% endemic
Continental			
Europe[1]	10 000	10 500	33
Australia[1]	7 716	*c.* 25 000	*c.* 85
E North America[1]	3 238	4 425	14
Southern Africa[1]	2 573	*c.* 21 000	*c.* 80
Guayana Highland[2]	1 000	*c.* 8 000	*c.* 75
California Floristic Province[1]	324	4 452	48
South West Botanical Province[1]	320	3 600	68
Greece[2]	129	*c.* 5 500	20
Cape Floristic Region[1]	90	8 578	68
Panama[3]	75	*c.* 6 800	*c.* 15
Santa Barbara region, CA[3]	7.3	1 390	10
Santa Rosa County, FL[3]	2.7	956	0.2
Agulhas Plain[4]	1.6	1 751	5.7
Marin County, CA[2]	1.4	1 313	0.8
Cape Peninsula[4]	0.5	2 256	8.1
Islands			
Japan[5]	377	4 022	34
British Isles[1]	308	1 443	1.2
New Zealand[1]	268	1 996	81
Hawaii[1]	16.6	1 897	92
Galapagos Islands[5]	7.9	543	43
Canary Islands[6]	7.2	*c.* 1 800	26

[1] Compiled in Bond and Goldblatt (1984)
[2] Compiled in Major (1988)
[3] Compiled in Gentry (1986)
[4] This study
[5] Porter (1979) includes subspecific taxa
[6] Humphries (1979)

regions of pronounced species richness (Major 1988; Table 4.5).

Regional endemism in the Cape Floristic Region, based on the distribution of 1 042 taxa (Figure 4.11), is very high — ranging from about 50% in the western winter rainfall centres (North Western and South Western) to between 18 and 28% in the centres of the non-seasonal rainfall zone. Similar patterns of regional endemism have been recorded, in a detailed and rigorous analysis, for the Proteaceae (Table 4.9).

Comparable data are available for the neotropics which indicate similar levels of endemism to the Cape Floristic Region (Gentry 1986; Prance 1987). Species endemism in the Choco Department of Colombia for 10 large genera (> 30 spp.) ranged from 5 to 72% ($\bar{x}$ = 38%) and for 11 large genera in Panama, the values ranged from 3 to 82% ($\bar{x}$ = 32%) (Gentry 1986). A broader analysis of the distribution of 8 117 recently monographed species in relation to nine neotropical phytogeographic regions (which are mostly orders of magnitude larger than the entire Cape Floristic Region) indicated an overall level of endemism of about 59% for canopy trees and lianas (range: 18–80%) and about 66% for epiphytes and palmettos (range: 42–86%) (see Figure 5 in Gentry 1986). Gentry has recorded extreme local endemism (distribution covering 5 to 10 km^2) of 10 and 24% in two isolated neotropical cloud forests. Highly localized endemics are common in the Cape Floristic Region (Goldblatt 1978; Bond and Goldblatt 1984) but these have not been quantified by either region or taxon, except for the Proteaceae where levels are very high (Table 4.9).

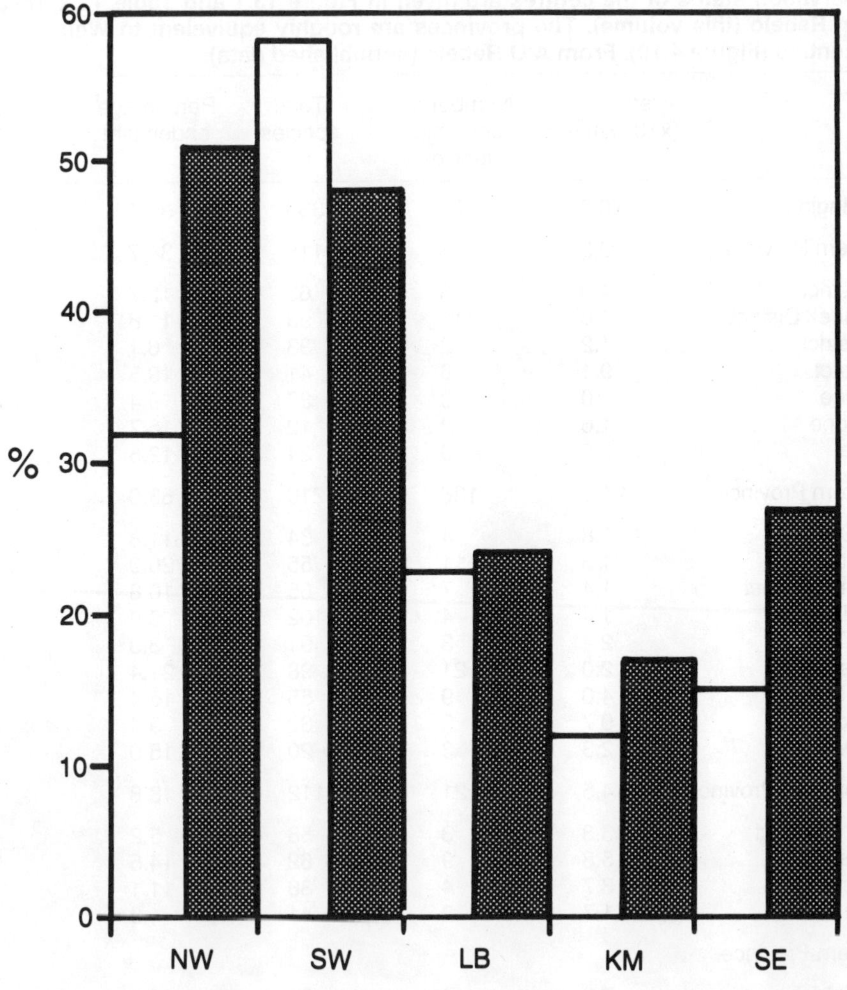

FIGURE 4.11 The distribution of taxa in relation to Weimarck's (1941) centres of endemism for the Cape Floristic Region (Figure 4.10). Included in the sample are 462 taxa from Weimarck (1941) and 680 additional taxa from nine genera (*Adenandra*, *Agathosma*, *Aspalathus*, *Leucadendron*, *Leucospermum*, *Muraltia*, *Paranomus*, *Sorocephalus*, and *Spatalla*). Reprinted from Cowling (1983c) with the permission of Blackwell Scientific Publications.

TABLE 4.9 **The endemism and richness of Proteaceae species (and distinct subspecies) within centres of endemism in the Cape Floristic Region. An endemic species is defined as having more than 95% of recorded populations within a centre. The area is in units (grid squares) of 12x12 km^2 and includes all vegetation types (including sea in coastal grid squares). Only grid squares containing Proteaceae species are included. Since Proteaceae species richness is significantly correlated ($P < 0.001$, df = 218) with the species richness of all major fynbos plant families (Rebelo and Siegfried 1990) and with the total species richness ($r^2 = 0.909$, $P < 0.001$) in the Cape Floristic Region at the quarter-degree grid (25x25 km^2) scale, the number of Proteaceae species per area is also given. The location of the centres, and the areas of fynbos and the conservation status of the centres are given in Figure 13.1 and Table 13.2, respectively, in Rebelo (this volume). The provinces are roughly equivalent to Weimarck's (1941) centres (Figure 4.10). From A G Rebelo (unpublished data).**

Region	Area (x10^3 km^2)	Number of endemic species	Total species	Percentage endemism	Spp. per 10^3 km^2
Cape Floristic Region	76.3	328	330	99.4	4.3
The North-western Province	19.2	41	118	34.7	6.1
Cederberg District	4.9	8	63	12.7	12.9
Great Winterhoek District	4.0	11	93	11.8	23.3
Picketberg District	1.2	2	33	6.1	27.5
Sandveld District	9.1	8	41	19.5	4.5
Sandveld Zone	4.0	3	32	9.4	8.0
Bokkeveld Zone	1.6	2	12	16.7	7.5
Gifberg Zone	3.5	3	24	12.5	6.9
The South-western Province	21.3	136	216	63.0	10.1
Malmesbury District	5.8	4	34	11.8	5.9
Peninsula District	1.3	11	55	20.0	42.3
Riviersonderend District	1.4	7	65	10.8	46.4
Franschhoek District	1.4	4	102	3.9	72.9
DuToitskloof District	2.4	3	91	3.3	37.9
Houwhoek District	2.0	21	98	21.4	49.0
Bredasdorp District	4.0	9	55	16.4	13.8
Potberg District	0.7	3	32	9.4	45.7
Mosselbay District	2.3	3	20	15.0	8.7
The Coastal Mountain Province	14.5	21	112	18.8	7.7
Koo Langeberg District	3.3	3	58	5.2	17.6
Langeberg District	5.8	9	62	14.5	10.7
Outeniqua District	3.7	4	36	11.1	9.7
Kouga District	1.7	2	41	4.9	69.7
The South-eastern Province					
Cockscomb District	7.6	4	21	19.0	2.8
The Inland Mountain Province	13.1	16	61	26.2	4.7
Swartberg District	8.2	13	57	22.8	7.0
Klein Swartberg Zone	0.9	4	25	16.0	27.8
Karoo Island Zone	2.0	4	43	9.3	21.5
Witteberg District	2.9	1	22	4.5	7.6
Karoo Outliers Gap	2.0	0	8	0.0	4.0
The Kamiesberg Outlier	0.6	2	2	100.0	3.3

Levels of endemism for 70 large genera in phytogeographical regions of the California Floristic Province are lower than values recorded for the Cape Floristic Region and the neotropics: values ranged from 5% (N. Sierra-Cascades) to 30% (S. California) ($\bar{x}$ = 18%) (Stebbins and Major 1965). Extreme local endemism is common in the Californian flora (Stebbins and Major 1965; Raven and Axelrod 1978) but there are no quantitative data.

Within the Cape Floristic Region, levels of both regional and local endemism are much higher in the south-west (Agulhas Plain) than the south-east (Humansdorp) (Table 4.10). Whereas all of the Agulhas Plain endemics belonged to genera centred in the Cape Floristic Region, this was the case for only 70% of the Humansdorp endemics. The remainder were of karroid (22%), subtropical (5%), and Afromontane (3%) affinity, largely endemic to the Kaffrarian Transition Zone.

PALAEOENDEMISM AND NEO-ENDEMISM

Following on Favarger and Constandriopoulos (1961), Stebbins and Major (1965) describe two major types of endemics — namely palaeoendemics and neoendemics. Palaeoendemics are taxonomically isolated taxa, usually monotypic sections, subgenera, or genera which may be regarded as evolutionary relicts. They rarely have close relatives in the same or adjacent areas. Neoendemics do have close relatives in the same or adjacent areas: they are usually subspecies, semispecies, or sibling species. The distinction between these two classes is often problematic (Hopper 1979; Kruckeberg and Rabinowitz 1985) but is greatly aided by cytological and cytogenetic data which enable the recognition of subclasses (schizoendemics, patroendemics, etc.) (Stebbins and Major 1965; Lewis 1972). Neoendemic species can occur in systematically isolated genera (Bramwell 1972) which are a feature of the Cape flora (Linder et al. this volume).

Although we lacked cytological and cytogenetic data, we argue that the endemic floras of both the Agulhas Plain and the Humansdorp region are overwhelmingly neoendemic. Firstly, nearly all of the endemics belong to species-rich genera and have sympatric or allopatric sister species. Examples of hypothetical sister species based on taxonomic judgement (Cronk 1987) among endemics of Cape affinity may be found in recent monographs (e.g. Rourke 1969, 1972; Williams 1972, 1982a, b). The systematic relationships of species within *Leucadendron* (Proteaceae) (Midgley 1986), *Leucospermum* (Proteaceae) (Rourke 1972), *Diosma* (Rutaceae) (Williams 1982b), and *Syndesmanthus* (Ericaceae) (E G H Oliver personal communication) suggest that recent speciation has occurred on the Agulhas Plain. There are only three monotypic genera among the Agulhas Plain endemics. These are *Caryotophora* (Mesembryanthemaceae), *Dymondia* (Asteraceae), and *Rhigiophyllum* (Campanulaceae). We are unable to judge whether these are palaeoendemics. However, both *Carytophora* and *Dymondia* are confined to geologically very young substrata (see below) which suggests neo-endemism. The Humansdorp flora includes two monotypic genera — *Smellophyllum* (Sapindaceae) and *Thamnus* (Ericaceae). *S. capensis* is a small tree confined to Afromontane forest and is probably a palaeoendemic (Cowling 1983c). Generic concepts within the Ericoideae are very fluid (Oliver 1988) so it is difficult to comment on the status of *Thamnus*.

A second line of evidence suggesting neoendemism is that many of the endemics in both regions are confined to sediments and soils which are geologically young. For example, the limestones, colluvial acid sands, duplex soils, and calcareous coastal sands of the coastal lowlands — which support most of the endemics in the south-west and the south-east — were only deposited or formed since the Middle to Late Pliocene, some 4 Myr (Hendey 1983). We assume that species confined to these surfaces are younger than this (Nordenstam 1969; Deacon 1983; Midgley 1987). Nearly all of the species of *Leucadendron* endemic to the Agulhas Plain have a high divergence index suggesting speciation some 3–4 Myr (Midgley 1987). We predict that phylogenetic analyses would show a preponderance of derived taxa in coastal lowland areas. It is important to note that our sample floras — both from areas on the geologically young coastal forelands — are biased with respect to the Cape Floristic Region as a

TABLE 4.10 The categories of endemism for floras from the Agulhas Plain (south-western Cape Floristic Region) and the Humansdorp region (south-eastern Cape Floristic Region). Local endemics are species confined to the Agulhas Plain (Thwaites and Cowling 1988) or the Humansdorp coastal plain (Cowling 1984) (Figure 4.10). Regional endemics for the Agulhas Plain are species confined to Weimarck's (1941) South Western Centre or Nordenstam's (1969) Bredasdorp-Riversdale centre; regional endemics for Humansdorp are species confined to Weimarck's (1941) South Eastern Centre or Cowling's (1983c) Kaffrarian Transition Zone (see Figure 4.10).

	Categories of endemism			Total
	Local	Regional	Non-endemic	
Agulhas Plain				
No. of species	100	413	1 238	1 751
%	5.7	23.6	70.7	
Humansdorp				
No. of species	4	105	770	879
%	0.5	11.9	87.7	

whole. The geologically ancient montane areas are likely to support a higher percentage of palaeoendemics (Rebelo this volume).

Taxonomic aspects of endemism

Are endemics a heterogeneous group taxonomically or do certain taxa have a higher than expected probability of being endemic? Other analyses have suggested that endemics are unequally divided among plant families (e.g. Tolmachev 1974a in Major 1988). We addressed this question by using chi-square analysis to test the hypothesis that the frequency of endemics in a family would not be significantly different from the frequency for the remaining flora (independent sample) from which it was drawn. We performed this analysis for three floras distributed across the Cape Floristic Region. Owing to problems with minimum expected cell count frequencies, we were only able to analyse the larger families (Table 4.11).

Families which tended to be significantly over-represented in the frequency of endemics were the Ericaceae, Proteaceae, Rutaceae, Thymelaeaceae, and Mesembryanthemaceae (Table 4.11). Under-represented families included the Poaceae, Asteraceae, and Orchidaceae. Interestingly, the Cyperaceae were under-represented in the Agulhas Plain and Humansdorp floras but over-represented in the Cape Peninsula flora. This analysis shows quite clearly that certain families are more strongly associated with endemism than others and that these taxonomic trends are broadly similar across the Cape Floristic Region.

There are few comparable data from other endemic-rich areas. Gentry (1986) has shown higher than average levels of endemism for *Amphitecna* (Bignoniaceae), *Anthurium* (Araceae), *Ardisia* (Myrsinaceae), and *Columnea* (Gesneriaceae) in neotropical rain forest areas. Raven and Axelrod (1978) list six families (Amaryllidaceae-tribe Allieae, Boraginaceae, Hydrophyllaceae, Onagraceae, Polemoniaceae, and Polygonaceae) which all contain a higher than average proportion of endemic species in the California Floristic Province. Tolmachev (1974a, b in Major 1988) has shown that the Fabaceae, Liliaceae, and Apiaceae have generally higher than average, and the Poaceae lower than average, levels of endemism in four floras from extra-tropical Eurasia.

Edaphic aspects of endemism

The restriction of endemic species to peculiar or isolated substrata (serpentine, limestone, laterite, quartzite, calcareous sands, etc.) is a widespread phenomenon in the endemic-rich

TABLE 4.11 **The frequencies of local, regional, and non-endemics in floras distributed across the Cape Floristic Region. The Cape Peninsula and Agulhas Plain are in the south-west and Humansdorp in the south-east. Data for regional endemics in the Cape Peninsula flora are unavailable. The Humansdorp flora contained only four local endemics which were therefore combined with regional endemics. The chi-square (χ^2) analysis tests the null hypothesis that the frequency of endemics in a family would not be different from the frequency in the total flora excluding that family. Only large families were used, in order to avoid excessively low predicted cell frequencies. NS = not significant; * = $P < 0.05$, ** = $P < 0.01$, *** = $P < 0.001$.**

Cape Peninsula				Agulhas Plain					Humansdorp			
	Endemics					Endemics				Endemics		
Family	Non	Local	χ^2	Family	Non	Regional	Local	χ^2	Family	Non	Regional	χ^2
All	2 074	182		All	1 238	413	100		All	770	109	
Apiaceae	41	1	1.66[1]NS	Asteraceae	151	50	2	9.53 **	Asteraceae	90	12	0.08[1] NS
Asteraceae	255	12	5.21 *	Campanulaceae	38	13	1	1.44 NS	Crassulaceae	19	1	0.45[1] NS
Campanulaceae	54	11	7.08 **	Cyperaceae	56	6	1	10.56 **	Cyperaceae	36	0	4.20[1] *
Cyperaceae	119	20	7.98 **	Ericaceae	44	58	22	82.16 ***	Ericaceae	14	11	20.75[1] ***
Ericaceae	89	30	49.78 ***	Fabaceae	89	41	8	3.20 NS	Euphorbiaceae	23	4	0.01[1] NS
Fabaceae	135	20	5.25 *	Iridaceae	111	31	7	1.16 NS	Fabaceae	51	10	0.96 NS
Iridaceae	132	7	1.83 NS	Liliaceae	63	11	1	7.23 *	Geraniaceae	15	1	0.14[1] NS
Liliaceae	113	6	1.55 NS	Mesembryanthem.	32	17	9	12.98 **	Iridaceae	33	3	0.25[1] NS
Mesembryanthem.	56	6	0.22 NS	Orchidaceae	36	2	0	10.92 **	Liliaceae	43	9	1.22 NS
Orchidaceae	114	3	5.04 *	Poaceae	62	3	0	19.97 ***	Mesembryanthem.	13	8	10.76[1] **
Poaceae	124	4	4.47 *	Polygalaceae	16	12	7	17.66 **	Poaceae	51	2	3.86 *
Proteaceae	32	9	11.08 **	Proteaceae	19	28	11	46.17 ***	Restionaceae	21	1	0.65[1] NS
Restionaceae	79	7	0.01 NS	Restionaceae	53	34	1	13.73 **	Rutaceae	9	8	16.05[1] ***
Scrophulariaceae	73	4	0.91 NS	Rutaceae	13	23	11	52.29 ***	Scrophulariaceae	16	1	0.20[1] NS
				Scrophulariaceae	48	8	0	7.27 *	Thymeleaceae	11	5	3.81[1] *
				Thymeleaceae	21	14	1	12.01 **				

[1] Chi-square with Yates correction

areas such as the Cape Floristic Region (Thoday 1925; Dahlgren 1968; Cowling 1983c), California Floristic Province (Raven and Axelrod 1978), South West Botanical Province (Hopper 1979), Greece (Papanicolaou et al. 1988), and neotropical rain forest regions (Gentry 1986; Brown and Prance 1987). Clearly, unusual substrata provide a strong selective force for the evolution of neoendemics (Raven 1964; Kruckeberg 1969; Kruckeberg and Rabinowitz 1985; MacNair 1987; Major 1988). Alternatively they may also provide a refuge from competition for palaeoendemics (Zedler et al. 1984; Major 1988).

What are the relationships between edaphic factors and levels of endemism on the Agulhas Plain and in the Humansdorp region? In both areas Cape endemics are significantly and negatively associated with measures of soil fertility (Table 4.12). These relationships are stronger in the Humansdorp region which includes many samples of non-fynbos vegetation on deep, relatively clay-rich and fertile soils. A broadly similar pattern was evident for regional endemics in Humansdorp which are concentrated in fynbos on nutrient-poor soils (Cowling 1983c). Soil variables were poorly correlated with both regional and local endemism on the Agulhas Plain where endemics occur on a wide range of substrata.

The frequency of both regional and local endemics in Agulhas Plain sample floras was similar (*c*. 22%) on all substrata except calcareous sand (χ^2 (excluding calcareous sand) = 6.5, P = 0.37) (Table 4.13). The inclusion of calcareous sand data resulted in a significant chi-square analysis ($\chi^2 = 21.9$, $P < 0.01$). Levels of regional endemism in Humansdorp were very similar on all substrata (*c*. 14%) except on shale (4%) (χ^2 excluding shale = 0.17, P = 0.98; χ^2 with shale = 9.01, P = 0.06). This reflects the generally lower levels of endemism in renoster shrubland on clay-rich soils compared with fynbos (Cowling 1983c).

Relative to regional endemics, local endemics on the Agulhas Plain were over-represented on limestone and shale/ferricrete (χ^2 = 41.5, $P < 0.0001$) (Table 4.14). Limestone endemics comprised 13% of the regional and 37% of the local endemic flora, although limestone only occupies about 5% of the Agulhas Plain. The extensive ferricrete surface remnants on the Agulhas Plain and the widespread presence of ferricrete as a stone-line in shale-derived duplex soils — an uncommon phenomenon in the Cape Floristic Region (Thwaites and Cowling 1988) — probably explain the high proportion of local endemics on this substratum. In Humansdorp the highest number of endemics were associated with quartzite, which is widespread in the study area, and calcareous sand, which occupies less than 15% of the region (Table 4.14). A significantly higher frequency of Agulhas Plain local endemics occurred on only one substratum (85%) than regional endemics (69%) (χ^2 = 10.5, $P < 0.005$). Furthermore, a significantly higher frequency of Agulhas Plain endemics (local and regional) occurred on a single substratum (72%) than Humansdorp endemics (59%) ($\chi^2 = 7.4$, $P < 0.01$).

These data suggest that edaphic specialization has been very important in the evolution of endemics on the Agulhas Plain and, to a lesser extent, in the Humansdorp region. The generally low nutrient status of most of the soils, and subtle differences in nutrient and moisture levels between soil types, represent a major selective force for plant speciation in the Cape Floristic Region (Thoday 1925; Rourke 1972; Linder 1985; Cowling 1987). At present, very little is known about edaphic specialization of fynbos species. There is a possibility that microsymbiont specificity may be implicated in the Ericaceae and certain legumes (e.g. *Aspalathus*) (Cowling et al. 1990; see also Tadros 1957). Soil chemistry appears to be important in preventing the invasion of limestone by calcifuge species (Mustart and Cowling submitted; Newton et al. 1991). Experiments are currently under way to determine the relative contributions of abiotic factors and competition in determining edaphically restricted distributions patterns on the Agulhas Plain (see Bond et al. this volume).

On the Agulhas Plain there are many locally and regionally endemic hypothetical sister taxa which occur on different but juxtaposed substrata. Examples are *Leucospermum cordifolium* (quartzite), *L. patersonii* (limestone) (Rourke 1972); *L. prostratum* (acid sand), *L. pedunculatum* (neutral sand) (Rourke 1972); *Protea obtusifolia* (limestone), *P. susannae* (neutral sand) (Rourke 1980); *Spatalla squamata* (acid sand); *S. ericoides* (neutral

TABLE 4.12 The significant correlations between % endemism and soil variables in 1 000 m^2 plots on the Agulhas Plain (*n* = 7) (R M Cowling unpublished data) and in 100 m^2 plots in the Humansdorp region (*n* = 97) (Cowling 1983b). Spearmans *r* for Agulhas data and Pearsons *r* for Humansdorp data.

	Agulhas Plain			Humansdorp	
Soil variable	Cape endemic[1]	Regional endemic	Local endemic	Cape endemic[1]	Regional endemic[2]
Soil depth		-	-	- - -	
% rock cover		+	+	+ + +	+ +
% sand				+ + +	+ +
pH				- - -	
% organic carbon	-			- - -	
Total nitrogen				- - -	
Available phosphorous	- - -			- - -	- - -
Available calcium	- - -			- - -	
S-value	- -			- - -	- -

\+ - $P < 0.05$
\+ + - - $P < 0.01$
\+ + + - - - $P < 0.001$

[1] Cape endemics are species endemic to the Cape phytochorion (*sensu* Werger 1978).
[2] Humansdorp regional endemics include four local endemics.

sand) (Rourke 1969); *Leucadendron meridianum* (limestone), and *L. coniferum* (neutral sand) (Williams 1972). Other species pairs may occur on the same substratum but have allopatric distributions, often separated by a few kilometres (e.g. on limestone: *Diosma haelkraalensis*, *D. guthriei* (Williams 1982b); on shale/ferricrete: *Leucadendron modestum*, *L. stelligerum* (Williams 1972). *Aulax umbellata* (Proteaceae) presents an interesting case. This species is widespread on acid sand and quartzite on the Agulhas Plain and adjacent lowland areas of the south-western Cape Floristic Region (Rourke 1987). A single small (< 50 individuals) population occurs on limestone on the Agulhas Plain (R M Cowling personal observation). Seedlings of this population have a significantly higher relative growth rate under laboratory conditions on limestone-derived sand than seedlings from adjacent populations growing on acid sand (Newton et al. 1991). Speciation, associated with selection for genotypes tolerant of the unusual substratum (cf. Raven 1964; Kruckeberg 1969), may have occurred. No morphological analysis of this population has been undertaken.

These patterns of edaphically differentiated sister taxa are less pronounced in Humansdorp. Groups worthy of investigation include *Euryops* (section *Psilosteum*) (Nordenstam 1969) and *Agathosma*. There is some evidence for calcareous sand races of *Muraltia*

TABLE 4.13 The levels of endemism in Cape shrublands (fynbos and renoster shrubland) on different substrata on the Agulhas Plain (south-western Cape Floristic Region) and in the Humansdorp region (south-east). Agulhas data are compiled from 1 000 m^2 plots (R M Cowling unpublished data) and Humansdorp data from 100 m^2 plots (Cowling 1983b). The four Humansdorp local endemics are included with the regional endemics. *n* = number of plots.

Agulhas Plain	Substratum				
	Quartzite *n* = 3	Acid sand *n* = 3	Shale/ ferricrete *n* = 2	Limestone *n* = 2	Calcareous sand *n* = 2
Non-endemic	111	66	95	59	84
Regional endemic	31	22	24	27	12
Local endemic	12	9	7	11	0
Humansdorp	Quartzite *n* = 9	Acid sand *n* = 5	Conglomerate *n* = 10	Shale *n* = 11	Calcareous sand *n* = 13
Non-endemic	109	59	139	114	125
Regional endemic	19	10	23	5	19

TABLE 4.14 The substratum specificity of endemics on the Agulhas Plain (south-western Cape Floristic Region) and the Humansdorp region (south-east). Agulhas flora = 1 751 spp., Humansdorp flora = 879 spp. The Humansdorp regional endemics include four local endemics.

	Substratum					
Agulhas Plain	Quartzite	Acid sand	Shale/ ferricrete	Limestone	Calcareous sand	> 1 substratum
Regional endemics	127	55	14	55	33	129
Local endemics	12	14	13	37	9	15
Humansdorp	Quartzite	Acid sand	Conglomerate	Shale	Calcareous sand	> 1 substratum
Regional endemics	26	4	9	4	21	45

squarrosa (Levyns 1954) and *Erica glumiflora* (R M Cowling personal observation).

Clearly, explanations for these distribution patterns should be based on phylogenetic analyses (Wanntorp et al. 1990). Is there a trend for derived species in a lineage to be associated with any particular substratum? Are these trends consistent with the time-scale of sediment deposition and soil formation? How do these patterns vary across the Cape Floristic Region? Unfortunately, phylogenetic data are lacking to test whether there are general speciational trends (Grant 1989) in association with different substrata.

In summary, edaphic endemism is pronounced on both the Agulhas Plain and Humansdorp. However, there is a notably lower proportion of species confined to single substratum in the latter area. Alkaline substrata (limestone on the Agulhas Plain and calcareous sand at Humansdorp), which occupy relatively small areas in each region, are associated with exceptionally high levels of endemism. Closely related endemics often occur on juxtaposed soil types indicating that ecological (i.e. edaphic) factors are a major force in speciation.

Biological aspects of endemism

Are endemics a random assemblage of species with respect to biological attributes such as growth form, dispersal mode, dispersal distance, etc? With the exception of a study by Harper (1979 in Kruckeberg and Rabinowitz 1985) we know of no other research which has rigorously addressed this question for plants.

We compared the association between endemism and biological attributes of species in the Agulhas Plain and Humansdorp floras (Table 4.15). The categorization of species with respect to attributes was based on data in Phillips (1951), Bond and Slingsby (1983), Bond and Goldblatt (1984), Rebelo (1987), and our own unpublished observations. In some cases the categories are too broad to be ecologically meaningful (e.g. soil-stored seed, insect pollination) but data were unavailable for finer subdivisions. Relationships were investigated separately for the two floras using two-way frequency tables (BMPD Program 4F, Dixon 1988). Chi-square was used to test for independence between the variables. Adjusted standardized deviates (Haberman 1973) exceeding three in absolute value were taken to indicate cells with unusually large deviations from the expected. Three-way frequency tables were computed to examine how some of these variables interact. Unfortunately log-linear modelling could not be used on the two data sets because the data were over-dispersed (i.e. there was insufficient data in many of the variable classes).

The chi-square tests were significant for all endemic class-attribute relationships with the exception of woody plant pollination (Table 4.15). However, patterns were often different for the two data sets. We list only the more salient trends. In both the Agulhas Plain and Humansdorp floras, shrubs were over-represented in the endemic classes. This pattern contrasts with the California Floristic Province where most neoendemics are annual dicots (Raven and Axelrod 1978). Widespread species tended to have higher than expected frequency of forbs and trees. There were more dwarf shrubs among local endemics in the Agulhas Plain flora than expected and more low shrubs than expected among Humansdorp regional endemics (Table 4.15). In both areas, taller woody growth forms were over-represented among widespread species. Ant-dispersed species were over-represented among local endemics in the Agulhas flora whereas vertebrate-dispersed species were more frequent than expected for widespread species. Similar trends were evident for Humansdorp, although the pattern for ant-dispersed species was strongest for Cape endemics. In both regions there was a trend for species with short distance dispersal to be over-represented in the narrow distribution classes. Gentry (1988) reported that mammal dispersal of fruits (short distance) is correlated with local endemism in the neotropics. The clearest trend with regard to seed storage was for species with non-storage (always vertebrate-dispersed fruits) to be over-represented among wides in both regions. An analysis of three-way frequency tables showed for the Agulhas Plain flora that 56% of locally endemic shrubs (comprising 84% of all local endemics) were ant-dispersed as were 41% of regional endemics. All of these species had soil-stored seed banks. Furthermore, 71% of locally endemic and 61% of regionally endemic shrubs had short disper-

TABLE 4.15 **The association between endemism and the biological attributes of species in the Agulhas Plain (n = 565) and the Humansdorp region (n = 879). Observed frequencies are listed with adjusted standardized deviates in parentheses (see text). LE = local endemics, RE = regional endemics, CE = Cape phytochorion (*sensu* Werger 1978) endemics, WD = widespread species. In the Humansdorp flora, the four local endemics were included with regional endemics to meet minimum expected cell count frequencies for chi-square (= χ^2) analyses.**

a) Growth form

	Agulhas Plain					Humansdorp region					
	Vine	Geophyte	Graminoid	Forb	Shrub and tree[1]	Vine	Geophyte	Graminoid	Forb	Shrub	Tree
LE	0(-1.0)	1(-2.6)	2(-2.7)	2(-1.7)	41(5.5)						
RE	0(-1.0)	10(-1.9)	28(2.1)	4(-2.9)	61(2.1)	2(-1.8)	17(1.3)	3(-3.3)	8(-2.7)	73(5.4)	5(-1.9)
CE	5(-0.6)	59(3.3)	58(0.4)	29(-1.4)	132(-1.6)	7(-2.7)	45(2.7)	49(3.1)	36(-1.8)	135(2.3)	2(-6.0)
WD	7(2.9)	19(-0.5)	24(-0.6)	34(5.4)	49(-3.5)	39(3.8)	40(-3.4)	56(-0.7)	96(3.5)	161(-5.8)	74(6.9)
	$\chi^2 = 80.3, P < 0.0001$					$\chi^2 = 119.7, P < 0.0001$					

b) Woody plant height

	Agulhas Plain					Humansdorp region					
	Dwarf shrub (< 0.25)	Low shrub (0.25–1 m)	Medium shrub (1–2 m)	Tall shrub (2–5 m)	Tree (>5)	Dwarf shrub (< 0.25 m)	Low shrub (0.25–1 m)	Medium shrub (1–2 m)	Tall shrub (2–5 m)	Low tree (5–10 m)	Tall tree (> 10 m)
LE	14(4.3)	21(-1.3)	5(-0.9)	1(-1.0)	0(-1.3)						
RE	7(-0.4)	41(1.2)	9(-0.6)	4(0.2)	0(-1.6)	12(1.7)	51(3.5)	7(-1.1)	3(-2.4)	5(-2.3)	0(-1.7)
CE	14(-1.2)	91(2.7)	24(0.4)	3(-2.5)	0(-2.9)	19(1.7)	84(3.9)	24(2.0)	8(-2.6)	2(-5.3)	0(-2.5)
WD	2(-2.1)	18(-3.7)	11(4.0)	9(4.0)	9(6.7)	15(-2.8)	78(-6.3)	26(-1.1)	42(4.2)	60(6.6)	14(3.6)
	$\chi^2 = 85.9, P < 0.0001$					$\chi^2 = 98.7, P < 0.0001$					

c) Dispersal mode

	Agulhas Plain					Humansdorp region				
	Wind	Vertebrate	Ant	Ballistic	Passive/ unknown	Wind	Vertebrate	Ant	Ballistic	Passive/ unknown
LE	13(-0.1)	1(-1.3)	25(4.6)	1(-0.4)	6(-3.3)					
RE	36(1.5)	0(-3.0)	37(2.6)	3(-0.2)	27(-2.1)	37(2.4)	9(-2.6)	26(2.3)	8(1.0)	28(2.4)
CE	86(0.7)	6(-4.3)	70(-0.6)	12(1.4)	109(1.5)	57(-1.8)	13(-6.6)	82(7.4)	17(0.7)	105(0.8)
WD	29(-2.1)	30(8.5)	14(-4.6)	2(-1.3)	58(2.3)	116(0.1)	123(7.9)	31(-8.5)	21(-1.3)	175(0.8)
	$\chi^2 = 117.5, P < 0.0001$					$\chi^2 = 125.5, P < 0.0001$				

d) Dispersal distance

	Short (< 10 m)	Medium (10–50 m)	Long (> 50 m)	Short (< 10 m)	Medium (10–50 m)	Long (> 50 m)
LE	32(0.8)	13(-0.1)	1(-1.3)			
RE	67(0.2)	36(1.4)	0(-3.0)	64(0.2)	37(2.4)	7(-3.0)
CE	190(1.4)	87(0.8)	6(-4.3)	205(6.7)	58(-1.7)	11(-6.8)
WD	74(-2.4)	29(-2.1)	30(8.5)	227(-6.4)	116(0.0)	123(8.4)
		$\chi^2 = 74.6$, $P < 0.0001$			$\chi^2 = 85.1$, $P < 0.0001$	

e) Seed storage

	Canopy-storage	Soil-storage	Non-storage	Canopy-storage	Soil-storage	Non-storage
LE	5(1.2)	41(0.5)	0(-1.9)			
RE	10(1.4)	93(1.2)	0(-3.1)	9(2.3)	90(1.4)	9(-2.7)
CE	20(0.5)	254(2.2)	9(-3.5)	14(1.0)	252(6.7)	8(-7.8)
WD	2(-2.7)	101(-4.1)	30(8.1)	12(-2.5)	321(-7.2)	133(9.1)
		χ^2 73.6, $P < 0.0001$			$\chi^2 = 90.0$, $P < 0.0001$	

f)Woody plant pollination

	Wind	Insect	Bird	Wind	Insect	Bird
LE	1(-1.5)	37(1.1)	3(0.0)			
RE	4(-0.5)	47(-1.2)	10(2.0)	7(0.8)	63(-1.9)	10(1.7)
CE	12(0.4)	111(-0.3)	13(0.0)	11(0.6)	116(-0.8)	12(0.4)
WD	7(1.4)	46(0.6)	1(-2.1)	14(-1.2)	226(2.1)	15(-1.7)
		$\chi^2 = 10.3$, $P > 0.10$			$\chi^2 = 5.73$, $P > 0.20$	

[1] Trees (9 spp.) in the Agulhas flora were included with shrubs.

sal distances. Sixty-eight per cent of regional endemics in the Humansdorp flora were shrubs. Of these, 32% were ant-dispersed and 64% had short distance dispersal.

There are other biological attributes not included in our analyses which are strongly associated with endemism. Wells (1969) has shown that speciation and local endemism in the Californian chaparral shrub genera, *Arctostaphylos* and *Ceanothus*, are related to the non-sprouting habitat. His argument is that non-sprouting species, with non-overlapping generations in a fire-prone environment, experience more sexual generations and stronger selection than the more genetically conservative sprouting species. We did not have adequate data to explore the association between mode of regeneration and endemism for our data sets. However, data for Proteaceae of the Cape Floristic Region (Table 4.16) show that local endemics are strongly over-represented in the non-sprouting category (χ^2 (local vs regional + wide) = 15.8, $P < 0.001$). Similar patterns are likely to exist for other Cape taxa. An example is the shrubby ant-dispersed genus *Acmadenia* (Rutaceae) where data on post-fire regeneration mode are available for 16 of the 32 species (William 1982a). All of these species were non-sprouting shrubs and only one was distributed in more than two local endemic centres (*sensu* Rebelo this volume; Figure 12.1).

Another attribute strongly associated with endemism is microsymbiont specificity. Both the fynbos Ericaceae and the Fabaceae have specific microbe-mediated nutrient-uptake (ericoid mycorrhizal and rhizobia respectively) (Lamont 1982). Cowling et al. (1990) have suggested that specificity for microbes could explain edaphic specialization and speciation in these groups. There was a trend for endemic species in both of these families to be over-represented in three Cape Floristic Region floras (Table 4.11). Furthermore, most species in the Ericaceae (672 spp.) and in *Aspalathus*, the largest fynbos legume genus (245 spp.), are non-sprouting shrubs (Le Maitre and Midgley this volume).

What generalizations emerge regarding the biological attributes of woody shrub endemism in the Cape Floristic Region? It appears that most local and regional endemics are dwarf to low, non-sprouting shrubs with soil-stored seeds which are ant-dispersed and/or form a symbiotic relationship with microbes. The differences in the endemism attributes between the Agulhas Plain and Humansdorp floras probably reflect the high proportion of endemics of non-Cape affinity in the latter. These species have a different array of biological attributes which have evolved in response to a selective regime fundamentally different from that in the Cape Floristic Region. Below we explore the implications of biological and other aspects of endemism for speciation of the Cape flora.

Implications for speciation

The dominant paradigm for speciation in the Cape Floristic Region is that it has been largely allopatric (Rourke 1972; Goldblatt 1978). The high floristic turnover between fynbos landscapes is suggestive of geographic speciation (Kruger and Taylor 1979). However, there is much evidence for patterns, such as occur on the Agulhas Plain, of geographic overlap in the distribution of closely related species. Linder (1985; Linder and Vlok 1991) suggests that sympatric speciation in response to steep ecological gradients is prevalent in the Cape flora. The following requirements must be met for sympatric speciation to occur:

- short gene flow distances;
- strong disruptive selection on sufficiently polymorphic traits in relation to spatially heterogeneous resources;
- the evolution of reproductive isolation either as a result of selection or pleiotropy;
- the extinction of intermediate individuals to minimize selection-recombination antagonisms under a wide range of ecological conditions; and
- the occasional and drastic reduction of marginal populations facilitating rapid speciation as a result of catastrophic selection or genetic drift (Lewis 1962; Grant 1981; Templeton 1981; Linder 1985; Kondrashov and Mina 1986; Rice and Salt 1988).

Much of the debate concerning allopatric or sympatric speciation is centred on the meaning of the word 'geographical' (Templeton 1981). Bearing in mind that most fynbos species have short gene dispersal distances (Linder 1985; Slingsby and Bond 1985; Table 4.15d), each distinctive habitat could represent a 'geographic' region without the usual physical barriers associated with allopatric specia-

TABLE 4.16 The relationship between endemism and fire response of Proteaceae in the Cape Floristic Region. Local endemics are confined to a single centre, regional endemics to more than two but fewer than 10 centres, and wides to more than 10 centres. See Figure 12.1 in Rebelo (this volume) for the location of centres.

	Local endemics	Regional endemics	Wide	Total
Sprouter	12	38	8	58
Non-sprouter	109	154	9	272

tion (Linder and Vlok 1991).

Stated thus, conditions are highly conducive to rapid 'sympatric speciation' in an area like the Agulhas Plain. The high level of edaphic specialization indicates that the complex mosaic of edaphic environments in the Cape Floristic Region represents an important selective regime. Populations of edaphically restricted species would be subject to drastic reduction as a result of fire in the short-term (Van Wilgen et al. this volume) and climatic change in the long-term (Goldblatt 1978). These processes would result in the extinction of intermediates (Cowling 1987) and a reduction in the population sizes of genetically divergent isolates in marginal habitats (e.g. limestone). The result could be rapid speciation. We now elaborate this model to explain why speciation has been so prolific in lineages which possess such traits as non-sprouting, ant-dispersed seed or microbe-mediated nutrient uptake. We begin with the non-sprouting, ant-dispersed combination.

There is a strong selective force to produce large protein-rich seeds in the fire-prone and nutrient-poor fynbos environment (Le Maitre and Midgley this volume). Large seeds enable rapid seedling growth and the avoidance of drought-induced mortality during the dry summer months. However, the commonly observed trade-off between seed size and number (Le Maitre and Midgley this volume) means that seed production is often low and highly variable. Ant dispersal has evolved to ensure that precious seeds are dispersed to sites safe from vertebrate predators (Johnson this volume). Low seed production results in small soil-stored seed banks which are often dependent on the current year's seed crop for maintenance (Pierce and Cowling 1991). Some fires may severely reduce the population sizes of non-sprouting, ant-dispersed species and thus promote population fragmentation and isolation, especially if the species are edaphically restricted to small habitat patches. In other words, these species are vulnerable to local extinction (Parker and Kelly 1989). Cowling and Bond (1991) have shown that shrubs with ant-dispersed seeds (and Ericaceae) were the species group most vulnerable to local extinction on small limestone habitat fragments on the Agulhas Plain. However, lineages that are extinction-prone are also prone to speciation. This certainly seems to be the case for *Clarkia*, a richly diversified genus of annuals in the California Floristic Province (Lewis 1962). Populations of *Clarkia* species, which have transient seed banks, are vulnerable to local extinction or severe population reduction as a result of environmentally extreme events. Speciation of populations at the margin of the species distribution or an unusual substrata (Raven 1964) occurs as a result of catastrophic selection whereby exceptionally adapted individuals survive extreme events (Lewis 1962, 1972). Similarly, the lack of buffering provided by persistent seed banks is thought to have contributed to diversity, endemism, and habitat specialization among non-sprouting species of the Californian chaparral genera, *Arctostaphylos* and *Ceanothus* (Wells 1969; Stebbins 1974; Parker and Kelly 1989).

Of relevance to our argument is Vrba's (1980, 1984) effect hypothesis which states that 'selection for proximal fitness may also, and incidentally, drive speciation' (Vrba and Gould 1986). Habitat specialists are more likely to be subject to directional selection in the event of environmental change leading to fragmentation, population divergence, vicariant speciation, and extinction. Thus, for fynbos shrubs, selection for non-sprouting and large, ant-dispersed seeds, which may be adaptive at the organism level, incidentally causes the multiplication of species. The potential of non-sprouting lineages for speciation probably has nothing to do with rapid generation turnover (cf. Wells 1969) which suggests a gradualist mode of speciation (Vrba 1980). Speciation occurs in punctuated events when populations of non-sprouting lineages undergoing strong selection are driven to very low numbers. Thus, traits prevail or even predominate in a flora because high speciation rates may overwhelm low survival rates. Microevolution at the organism level is uncoupled from macroevolution at the species level (Vrba and Eldredge 1984; Vrba and Gould 1986). The effect hypothesis is not an example of species selection since it does not require units of selection above the level of the species (Vrba 1980, 1984; Vrba and Gould 1986).

It could be argued that in a fire-prone environment, the sprouting ability is a more adaptive trait under a wider range of fire regimes than non-sprouting (James 1984; Le Maitre and Midgley this volume). However, selection for this trait will reduce the probability of population fragmentation and, thus, rates of lineage turnover. Clades which possess sprouting ability produce species that are resistant to extinction and hence, speciation: in Vrba's (1980) terminology these are survivors. The corollary is more interesting: the prevalence of non-sprouting in fynbos shrubs may have resulted not from any immediate adaptive advantage to the organisms, but rather from their greater potential for speciation. The preponderance of species with particular traits in communities and landscapes may be a stronger reflection of incidental speciation events than contemporary or historical ecological conditions (Fowler and MacMahon 1982; cf. Le Maitre and Midgley this volume).

This model similarly explains the tremendous diversification of lineages with microsymbiont-mediated nutrient uptake (e.g. Ericaceae and *Aspalathus*). Edaphic specialization would result from strong selection at the organism level to acquire new microsymbionts in different soil types (Cowling et al. 1990). It is appealing to suggest that the massive speciation of *Erica* is an effect of selection on seedlings for capturing more efficient microsymbionts in new edaphic environments.

CONCLUSIONS

Despite limited data, and the fact that analyses were often confined to coastal lowland sites, some general conclusions can be made regarding the patterns and evolution of species diversity and endemism in the Cape Floristic Region. Unquestionably, more data on these patterns, particularly turnover and endemism, are required from other parts of the region. Nonetheless, with regard to diversity, it appears that:

- Fynbos alpha diversity does not show much variation across the region, despite differences in environmental factors and disturbance regimes.
- Beta and gamma diversity are higher in the south-west than the south-east, probably because of different speciation histories in the two regions.
- Patterns of regional richness relate to species turnover (beta and gamma diversity) rather than alpha diversity: this explains the two-fold higher regional richness in the south-west than the south-east.
- The high regional richness of the Cape Floristic Region (particularly the south-west) relative to other species-rich areas of the globe is a function of the area's exceptionally high levels of turnover.

With regard to endemism, the following conclusions can be made:

- At a global scale the Cape Floristic Region has very high levels of local and regional endemism.
- Most of the endemics appear to be neoendemics, at least in the lowland floras studied here.
- Taxonomically, endemics are not a random assemblage — families such as the Ericaceae, Proteaceae, Rutaceae, etc. are significantly over-represented among endemics.
- Most endemics are edaphic specialists indicating the importance of soil type as a selective force — certain substrata (e.g. limestone) harbour disproportionally high numbers of endemics.
- Endemics are not random assemblages with respect to biological attributes. An endemic is most likely to be a dwarf to low, non-sprouting shrub with soil-stored, ant-dispersed seed and/or microsymbiont-mediated nutrient uptake.
- The possession of some of these traits,

though adaptive at the organism level, may incidentally cause lineage turnover and the multiplication of species.

We do not have explanations for all of these patterns. The differences in turnover (and levels of endemism) in physiographically similar landscapes in the south-western and south-eastern Cape Floristic Region suggest differences in speciation histories. The taxonomic and biological profiles of endemics in the two regions are broadly similar indicating similar speciation processes across the region. Clearly, the evolution of habitat specialists with narrow distribution ranges has been less pronounced in the south-east. This may be due to different Pleistocene climatic histories of the two regions (Cowling and Holmes this volume). Cowling and Holmes suggest that glacial conditions in the south-east were unsuitable for fynbos except in upland refugia. This hypothesis predicts that fynbos endemics should be concentrated in upland areas in the south-east. Corroborative data are available for the Humansdorp region (Cowling and Campbell 1983).

The abovementioned hypothesis assumes that differences in the ecological conditions between the south-western and south-eastern Cape Floristic Region were greater in the past than in the present. The possibility that contemporary ecological factors are responsible for the different patterns of diversity and endemism cannot be ruled out. For example, the pronounced summer drought of the south-west may promote speciation via catastrophic selection of non-sprouting lineages. While accepting differences in rainfall seasonality, sites used in the biome-wide comparisons were only crudely similar in terms of other ecological factors. A detailed comparison of diversity components in physiographically similar regions of the south-west and south-east would be of interest. The study area should include identical edaphic (in terms of parent material and soil form) and climatic (in terms of rainfall totals and temperature regimes) gradients. Physical factors, especially soil chemical data, should be rigorously quantified. It is possible that the generally more fertile soils of the south-east, irrespective of parent material (Campbell 1983), could result in less pronounced edaphic gradients and thus weaker selective forces.

Globally, the Cape Floristic Region, and fynbos in particular, has exceptionally high regional species richness and levels of endemism. These features are shared to a lesser extent with other mediterranean climate regions, such as the California Floristic Province (Raven and Axelrod 1978) and the South West Botanical Province of Australia (Hopper 1979). Together, these three areas comprise a striking deviation from the trend for plant species richness and endemism to increase with decreasing latitude (e.g. Pianka 1966; Gentry 1986; Brown 1988; Major 1988; Stevens 1989). In all these regions high regional richness is largely a function of turnover. Thus, regional richness has been determined mainly by the evolution of numerous habitat specialists many of which have narrow distribution ranges (Raven and Axelrod 1978; Hopper 1979; this study). In all cases, evolutionary divergence has been promoted by complex edaphic mosaics and the susceptibility of plant populations to fluctuations induced in the short-term by fire and summer drought, and in the long-term by climatic change (Raven 1973; Goldblatt 1978; Raven and Axelrod 1978; Hopper 1979; Cowling 1987). The Cape Floristic Region is peculiar in that it includes both the extensive array of infertile and peculiar soils typical of Australia and the topographical heterogeneity of California.

Finally, since speciation has played such an important role in the development of the Cape Floristic Region's exceptional species richness, more research is required on speciation processes in the region. If speciation is a largely ecological process (Linder 1985), then it should be possible to integrate ecological and phylogenetic studies (e.g. Linder and Vlok 1991). Speciation patterns in relation to the edaphic gradients, such as occur on the Agulhas Plain, hold particular promise for interesting research.

ACKNOWLEDGEMENTS

Some of the research in this chapter was funded by grants to R M C from the South African Nature Foundation and the Foundation for Research Development's Core Programme. A G R acknowledges funding from the University of Cape Town. We thank Peter Goldblatt, Peter Linder, and David Bell for commenting on the manuscript. Wendy Paisley and Alice Wiseman typed many drafts with grace and skill.

REFERENCES

ADAMS P, STRICKER P and ANDERSON D J (1989). Species-richness and soil-phosphorus in plant communities in coastal New South Wales. *Australian Journal of Ecology* **14**, 189–98.

ADAMSON R S and SALTER T M (1950). *The flora of the Cape Peninsula.* Juta, Cape Town.

ARRHENIUS O (1921). Species and area. *Journal of Ecology* **9**, 95–9.

ASHTON P S (1969). Speciation among tropical forest trees: some deductions in the light of recent evidence. *Biological Journal of the Linnean Society* **1**, 155–96.

AUERBACH M and SHMIDA A (1987). Spatial scale and the determinants of plant species richness. *Trends in Ecology and Evolution* **2**, 238–42.

BEARD J (1981). *Vegetation of Western Australia. 1:1 000 000 Series. Swan.* University of Western Australian Press, Nedlands.

BOECKLEN W J and GOTELLI N J (1984). Island biogeographic theory and conservation practice: species-area or specious area relationships. *Biological Conservation* **29**, 63–80.

BOND P and GOLDBLATT P (1984). Plants of the Cape Flora. A descriptive catalogue. *Journal of South African Botany Supplementary Volume* **13**, 1–455.

BOND W J (1981). Vegetation gradients in the southern Cape mountains. MSc thesis, University of Cape Town.

BOND W J (1983). On alpha diversity and the richness of the Cape flora: a study in southern Cape fynbos. In *Mediterranean-type ecosystems: the role of nutrients.* (eds Kruger F J, Mitchell D T and Jarvis J U M) Springer, Berlin. 225–43.

BOND W J and SLINSGBY P (1983). Seed dispersal by ants in shrublands of the Cape Province and its evolutionary implications. *South African Journal of Science* **79**, 231–3.

BOND W J, MIDGLEY J J and VLOK J (1988). When is an island not an island? Insular effects and their causes in fynbos shrublands. *Oecologia* **77**, 512–21.

BOND W J, LE MAITRE ROUX D and ERNSTEIN R (1990). Fire intensity and regeneration of myrmecochorous Proteaceae. *South African Journal of Botany* **56**, 326–30.

BOUCHER C (1977). A provisional check-list of the flowering plants and ferns in the Cape Hangklip area. *Journal of South African Botany* **43**, 57–80.

BOUCHER C (1978). Cape Hangklip area. 2. The vegetation. *Bothalia* **12**, 455–97.

BOUCHER C (1987). A phytosociological study of transects through the Western Cape coastal foreland, South Africa. PhD thesis, University of Cape Town.

BOUCHER C and SHEPHERD P (1987). Plant communities of the Pella site. In *A description of the Fynbos Biome Project intensive study site at Pella.* (ed Jarman M L) *Ecosystem Programmes Occasional Report* **33**, CSIR, Pretoria. 38–76.

BRAMWELL D (1972). Endemism in the flora of the Canary Islands. In *Taxonomy, phytogeography and evolution.* (ed Valentine D H) Academic Press, London, 141–60.

BRAMWELL D (ed) (1979). *Plants and islands.* Academic Press, London.

BROWN J H (1988). Species diversity. In *Analytical biogeography. An integrated approach to the study of animal and plant distributions.* (eds Myers A A and Giller P S) Chapman and Hall, London and New York, 57–89.

BROWN K S Jr and PRANCE G T (1987). Soils and vegetation. In *Biogeography and Quarternary history in tropical America.* (eds Whitmore T C and Prance G T) Clarendon Press, Oxford, 19–45.

CAMPBELL B M (1983). Montane plant environments in the fynbos biome. *Bothalia* **14**, 283–98.

CAMPBELL B M and VAN DER MEULEN F (1980). Patterns of plant species diversity in fynbos. *Vegetatio* **43**, 43–7.

CAMPBELL B M and WERGER M J A (1988). Plant form in the mountains of the Cape, South Africa. *Journal of Ecology* **76**, 637–53.

CAMPBELL D G and HAMMOND D (eds) (1989). *Floristic inventory of tropical countries: the status of plant systematics, collections, and vegetation, plus recommendations for the future.* New York Botanical Garden, New York.

CODY M L (1975). Towards a theory of continental diversities: bird distribution over mediterranean habitat gradients. In *Ecology and evolution of communities.* (eds Cody M L and Diamond J) Harvard University Press, Cambridge, Mass. 214–57.

CODY M L (1983). Continental diversity patterns and convergent evolution in bird communities. In *Mediterranean type ecosystems. The role of nutrients.* (eds Kruger F J, Mitchell D T and Jarvis J U M) Springer, Berlin. 357–91.

CODY M L (1986a). Diversity, rarity and conservation in mediterranean-climate regions. In *Conservation biology. The science of scarcity and diversity.* (ed Soule M E) Sinauer, Sunderland, Mass. 122–52.

CODY M L (1986b). Structural niches in plant communities. In *Community ecology.* (eds Diamond J and Case T J) Harper and Row, New York. 381–405.

CODY M L (1989). Growth-form diversity and community structure in desert plants. *Journal of Arid Environments* **17**, 199–209.

CONNELL J H (1978). Diversity in tropical rain forests and coral reefs. *Science* **199**, 1 302–9.

CONNELL J H and SLAYTER R D (1977). Mechanisms of succession in natural communities and their role in community stability and organization. *American Naturalist* **111**, 1 119–44.

COWLING R M (1983a). Diversity relations in Cape shrublands and other vegetation in the south eastern Cape, South Africa. *Vegetatio* **45**, 103–27.

COWLING R M (1983b). Vegetation studies in the Humansdorp region of the fynbos biome. PhD thesis, University of Cape Town.

COWLING R M (1983c). Phytochorology and vegetation history in the south eastern Cape, South Africa. *Journal of Biogeography* **10**, 393–419.

COWLING R M (1984). A syntaxonomic and synecological study in the Humansdorp region of the fynbos biome. *Bothalia* **15**, 175–227.

COWLING R M (1987). Fire and its role in coexistence and speciation in Gondwanan shrublands. *South African Journal of Science* **83**, 106–12.

COWLING R M (1990). Diversity components in a species-rich area of the Cape Floristic Region. *Journal of Vegetation Science* **1**, 699–710.

COWLING R M AND BOND W J (1991). How small can reserves be? An empirical approach in Cape fynbos. *Biological Conservation* **55** (in press).

COWLING R M and CAMPBELL B M (1983). A comparison of fynbos and non-fynbos coenoclines in the lower Gamtoos River Valley, south-eastern Cape, South Africa. *Vegetatio* **53**, 161–78.

COWLING R M and CAMPBELL B M (1984). Beta diversity along fynbos and non-fynbos coenoclines in the lower Gamtoos River Valley, south-eastern Cape. *South African Journal of Botany* **50**, 187–9.

COWLING R M and GXABA T (1990). Effects of a fynbos overstorey shrub on understorey community structure: implications for the maintenance of community-wide species richness. *South African Journal of Ecology* **1**, 1–7.

COWLING R M and PIERCE S M (1988). Secondary succession in coastal dune fynbos: variation due to site and disturbance. *Vegetatio* **76**, 131–9.

COWLING R M, CAMPBELL B M, MUSTART P, McDONALD D,

JARMAN M L and MOLL E J (1988). Vegetation classification in a floristically complex area: the Agulhas Plain. *South African Journal of Botany* **54**, 290–300.

COWLING R M, GIBBS RUSSELL G E, HOFFMAN M T and HILTON-TAYLOR C (1989). Patterns of plant species diversity in southern Africa. In *Biotic diversity in southern Africa. Concepts and conservation.* (ed Huntley B J) Oxford University Press, Cape Town, 19–50.

COWLING R M, STRAKER C J and DEIGNAN M T (1990). Does microsymbiont specificity determine plant species turnover and speciation in Gondwanan shrublands? A hypothesis. *South African Journal of Science* **86**, 118–20.

CRAWLEY M J (1986). Introduction. In *Plant ecology.* (ed Crawley M J) Blackwell, Oxford. 1–50.

CROIZAT L (1965). An introduction to the subgeneric classification of *Euphorbia* L. with a stress on the South African and Malagasy species. *Webbia* **20**, 573–706.

CRONK Q C B (1985). The history of endemic flora of St Helena: a relictual series. *New Phytologist* **105**, 509–20.

DAHLGREN R (1963). Studies on *Aspalathus*: phytogeographical aspects. *Botaniska Notiser* **116**, 431–72.

DAHLGREN R (1968). Distribution and substrate in the South African genus, *Aspalathus. Botaniska Notiser* **121**, 504–34.

DANIN A (1978). Plant species diversity and ecological districts of the Sinai Desert. *Vegetatio* **36**, 83–93.

DAVIS S D, DROOP S J M, GREGERSON P, HENSON L, LEON C J, VILLA-LOBO J L, SYNGE H and ZANTOVSKA J (1986). *Plants in danger. What do we know?* International Union for Conservation of Nature and Natural Resources, Gland, Switzerland and Cambridge, United Kingdom.

DEACON H J (1983). The comparative evolution of mediterranean-type ecosystems: a southern perspective. In *Mediterranean-type ecosystems: the role of nutrients.* (eds Kruger F J, Mitchell D T and Jarvis J U M) Springer, Berlin. 3–40.

DEL MORAL R (1982). Control of vegetation on contrasting substrates: herb patterns of serpentine and sandstone. *American Journal of Botany* **69**, 227–38.

DENSLOW J S (1985). Disturbance mediated coexistence of species. In *The ecology of natural disturbance and patch dynamics.* (eds Picket S T A and White P S) Academics Press, New York. 261–84.

DIAMOND J M (1988). Factors controlling species diversity: overview and synthesis. *Annals of the Missouri Botanical Garden* **75**, 117–29.

DIXON W J (1988). *BMDP Statistical software manual.* University of California Press, Berkeley.

DOBZHANKSY T (1950). Evolution in the tropics. *American Scientist* **38**, 9–21.

DUNN C P and LOEHLE C (1988). Species-area parameter estimation: testing the null model of lack of relationship. *Journal of Biogeography* **15**, 721–8.

DUTHIE A V (1929). Vegetation and flora of the Stellenbosch Flats. *Annals of the University of Stellenbosch* **7**, 1–59.

ESLER K J and COWLING R M (1990). The effects of density on the reproductive output of *Protea lepidocarpodendron. South African Journal of Botany* **56**, 29–33.

FAGERSTRÖM T (1988). Lotteries in communities of sessile organisms. *Trends in Ecology and Evolution* **3**, 303–6.

FAVARGER R C and CONSTANDRIOPOULOS J (1961). Essai sur l'endemissme. *Bulletin Society Bot. Suisse* **71**, 384–408.

FISCHER A G (1960). Latitudinal variations in organic diversity. *Evolution* **14**, 64–81.

FOURCADE H G (1941). Checklist of the flowering plants of the division of George, Knysna, Humansdorp and Uniondale. *Memoirs of Botanical Survey of Southern Africa* **20**, 1–127.

FOWLER C W and MACMAHON J A (1982). Selective extinction and speciation: their influence on the structure and functioning of communities and ecosystems. *American Naturalist* **119**, 480–98.

GELDENHUYS C J and MACDEVETTE D R (1989). Conservation status of coastal and montane evergreen forest. In *Biotic diversity in southern Africa. Concepts and conservation.* (ed Huntley B J) Oxford University Press, Cape Town. 224–38.

GENTRY A H (1982). Patterns of neotropical plant species diversity. *Evolutionary Biology* **15**, 1–84.

GENTRY A H (1986). Endemism in tropical vs temperate plant communities. In *Conservation biology.* (ed Soule M) Sinauer Press, Sunderland. 153–81.

GENTRY A H (1988). Change in plant community diversity and floristic composition on environmental and geographic gradients. *Annals of the Missouri Botanic Gardens* **75**, 1–34.

GEORGE A S, HOPKINS A J M and MARCHANT N G (1979). Western Australian heathlands. In *Heathlands and related shrublands. A. Descriptive studies.* (ed Specht R L) Elsevier, Amsterdam. 211–30.

GIBBS RUSSELL G E (1985). PRECIS, the National Herbarium's computerized information system. *South African Journal of Science* **81**, 662–5.

GOLDBERG D E (1985). Effects of soil pH, competition, and seed predation on the distribution of two tree species. *Ecology* **66**, 503–11.

GOLDBLATT P (1978). An analysis of the flora of southern Africa: its characteristics, relationships and origins. *Annals of the Missouri Botanical Garden* **65**, 369–436.

GOULD S J (1979). An allometric interpretation of species-area curves: the meaning of the coefficient. *American Naturalist* **114**, 335–43.

GRANT V (1981). *Plant speciation.* 2nd ed. Columbia University Press, New York.

GRANT V (1989). The theory of speciational trends. *American Naturalist* **133**, 604–12.

GRASSLE J F (1989). Species diversity in deep-sea communities. *Trends in Ecology and Evolution* **4**, 12–15.

GRIME J P (1979). *Plant strategies and vegetation processes.* Wiley, Chichester.

GRUBB P J (1977). The maintenance of species richness in plant communities: the importance of the regeneration niche. *Botanical Review* **52**, 107–45.

GRUBB P J and HOPKINS A J M (1986). Resilience at the level of the plant community. In *Resilience in mediterranean-type ecosystems.* (ed Dell B, Hopkins A J M and Lamont B B) Junk, Dordrecht. 21–38.

HABERMAN S J (1973). The analysis of residuals in cross-classified tables. *Biometrics* **29**, 3–29.

HENDEY Q B (1983). Palaeontology and palaeoecology of the fynbos region: an introduction. In *Palaeoecology of the fynbos landscape: a preliminary synthesis.* (eds Deacon H J, Hendey Q B and Lamprechts J J N) *South African National Scientific Programmes Report* **75**. CSIR, Pretoria. 100–15.

HNATIUK R J and HOPKINS A J M (1980). Western Australian species-rich kwongan (sclerophyllous shrublands) affected by drought. *Australian Journal of Botany* **28**, 573–85.

HOFFMAN M T (1989). Vegetation studies and the impact of grazing in the semi-arid Eastern Cape. PhD thesis, University of Cape Town.

HOFFMAN M T, MOLL E J and BOUCHER C (1987). Post-fire succession at Pella, a South African lowland fynbos site. *South African Journal of Botany* **53**, 370–4.

HOPKINS A J M and GRIFFIN E A (1984). Floristic patterns. In *Kwongan-plant life of the sandplain.* (eds Pate J S and Beard J S) University of Western Australia Press, Nedlands. 69–83.

HOPPER S D (1979). Biogeographical aspects of speciation in the south-west Australian flora. *Annual Review of Ecology and Systematics* **10**, 399–422.

HUBBEL S P and FOSTER R B (1986). Biology, chance and history, and the structure of tropical tree communities. In

Community ecology. (eds Diamond J M and Case T J) Harper and Row, New York. 314–24.
HUMPHRIES C J (1979). Endemism and evolution in Macaronesia. In *Plants and islands.* (ed Bramwell D) Academic Press, London. 171–99.
HUSTON M (1979). A general hypothesis of species diversity. *American Naturalist* **133**, 81–101.
JAMES S (1984). Lignotubers and burls - their structure, function and ecological significance in mediterranean ecosystems. *Botanical Review* **50**, 225–66.
JOHNSON M P, MASON L G and RAVEN P H (1968). Ecological parameters and plant species diversity. *American Naturalist* **102**, 297–306.
KONDRASHOV A S and MINA M V (1986). Sympatric speciation: when is it possible? *Biological Journal of the Linnean Society* **27**, 1–23.
KRUCKEBERG A R (1969). Soil diversity and the distribution of plants with examples from western North America. *Madrono* **20**, 129–54.
KRUCKEBERG A R and RABINOWITZ D (1985). Biological aspects of endemism in higher plants. *Annual Review of Ecology and Systematics* **16**, 449–79.
KRUGER F J (1974). The physiography and plant communities of Jakkalsrivier catchment. MSc thesis, University of Stellenbosch.
KRUGER F J (1979). South African heathlands. In *Heathlands of the world. A. Descriptive studies.* (ed Specht R L) Elsevier, Amsterdam. 19–80.
KRUGER F J (1983). Plant community diversity and dynamics in relation to fire. In *Mediterranean-type ecosystems. The role of nutrients.* (eds Kruger F J, Mitchell D T and Jarvis J U M) Springer, Berlin. 446–72.
KRUGER F J and BIGALKE R C (1984). Fire in fynbos. In *Ecological effects of fire in southern African ecosystems.* (eds Booysen P de V and Tainton N M) Springer, Berlin. 67–114.
KRUGER and TAYLOR H C (1979). Plant species diversity in Cape fynbos: gamma and delta diversity. *Vegetatio* **47**, 85–93.
LAMONT B B (1982). Mechanisms for enhancing nutrient-uptake in plants with particular reference to mediterranean South Africa and Western Australia. *Botanical Review* **48**, 597–689.
LAMONT B B, HOPKINS A J M and HNATIUK R J (1984). The flora - composition, diversity and origins. In *Kwongan - plant life of the sandplain.* (eds Pate J S, Beard J S) University of Western Australia Press, Nedlands. 27–50.
LEVYNS M R (1954). The genus *Muraltia. Journal of South African Botany Supplementary Volume* **2**, 1–247.
LEWIS H (1962). Catastrophic selection as a factor in speciation. *Evolution* **16**, 257–71.
LEWIS H (1972). The origin of endemics in the California flora. In *Taxonomy, phytogeography and evolution.* (ed Valentine D H) Academic Press, London. 179–89.
LINDER H P (1985). Gene flow, speciation and species diversity patterns in a species-rich area: the Cape Flora. In *Species and speciation.* (ed Vrba E S) Transvaal Museum, Pretoria. 53–7.
LINDER H P (submitted). Environmental correlates of patterns of species-richness in the south-western Cape Province of South Africa. *Journal of Biogeography.*
LINDER H P and VLOK J (1991). The morphology, taxonomy and evolution of *Rhodocoma* (Restionaceae). *Plant Systematics and Evolution.* (in press).
MACARTHUR R H and LEVINS R (1967). The limiting similarity, convergence and divergence of coexisting species. *American Naturalist* **101**, 377–85.
MACNAIR M R (1987). Heavy metal tolerance in plants: a model evolutionary system. *Trends in Ecology and Evolution* **2**, 354–9.
MAJOR J (1988). Endemism: a botanical perspective. In *Analytical biogeography. An integrated approach to the study of animal and plant distributions.* (eds Myers A A and Giller P S) Chapman and Hall, New York. 117–46.
MARTIN A R H and NOEL A R A (1960). *The flora of Albany and Bathurst.* Rhodes University, Grahamstown.
MIDGLEY J J (1986). Aspects of phylogeny, evolution and biogeography of the genus *Leucadendron* (Proteaceae). *Palaeoecology of Africa* **17**, 193–200.
MIDGLEY J J (1987). The derivation, utility and implications of divergence index for the fynbos genus *Leucadendron* (Proteaceae). *Botanical Journal of the Linnean Society* **95**, 137–52.
MOLONEY K A (1989). The local distribution of perennial bunchgrass: biotic or abiotic control. *Vegetatio* **80**, 47–61.
MUSTART P M and COWLING R M (submitted). Role of the regeneration niche in determining the distribution of edaphically restricted fynbos *Leucadendron* and *Protea* species pairs. *Journal of Ecology.*
MYERS A A and GILLER P S (1988). Biogeographic patterns. Introduction. In *Analytical biogeography. An integrated approach to the study of plant and animal distributions.* (eds Myers A A and Giller P S) Chapman and Hall, London and New York. 15–21.
NAVEH Z and WHITTAKER R H (1979). Structural and floristic diversity of shrublands and woodlands in northern Israel and other mediterranean areas. *Vegetatio* **41**, 171–90.
NELSON G and PLATNICK N (1981). *Systematics and Biogeography.* Columbia University Press, New York.
NEWMAN I E (1982). Niche separation and species diversity in terrestrial vegetation. In *The plant community as a working mechanism.* (ed Newman I E) Blackwell Scientific Publications, Oxford. 61–75.
NEWTON I P, COWLING R M and LEWIS O A M. Growth of calcicole and calcifuge Agulhas Plain Proteaceae under glasshouse conditions. *South African Journal of Botany.*
NORDENSTAM B (1969). Phytogeography of the genus *Euryops* (Compositae): a contribution to the phytogeography of southern Africa. *Opera Botanica* **23**, 1–77.
OLIVER E G H (1988). Studies in the Ericoideae. VI. The generic relationship between *Erica* and *Phillipia* in southern Africa. *Bothalia* **18**, 1–10.
OLIVER E G H, LINDER H P and ROURKE J P (1983). Geographical distribution of present day Cape taxa and their phytogeographical significance. *Bothalia* **14**, 427–40.
OLIVIER M C (1983). An annotated check-list of the Angiospermae of the Cape Recife Nature Reserve. *Journal of South African Botany* **47**, 813–28.
PAPANICOLAOU K, BABALONAS D and KOKKINII S (1983). Distribution patterns of some Greek mountain endemic plants in relation to geological substrate. *Flora* **174**, 405–37.
PARKER T V and KELLY V R (1989). Seed banks in California chaparral and other mediterranean climate shrubs. In *Ecology of soil seed banks.* (eds Leck M A, Parker T V and Simpson R L) Academic Press, New York. 232–55.
PATE J S and BEARD J S (eds) (1984). *Kwongan - plant life of the sandplain.* University of Western Australian Press, Nedlands.
PHILLIPS J F V (1931). Forest succession and ecology in the Knysna region. *Memoirs of the Botanical Survey of South Africa* **14**, 1–327.
PHILLIPS E P (1951). *Genera of South African flowering plants.* Government Printer, Pretoria.
PIANKA E R (1966). Latitudinal gradients in species diversity: a review of concepts. *American Naturalist* **100**, 33–46.
PIERCE S M and COWLING R M (1991). Dynamics of soil-stored seed banks of six shrubs in fire-prone fynbos. *Journal of Ecology* (in press).
PORTER D M (1979). Endemism and evolution in Galapagos

Islands vascular plants. In *Plant and islands.* (ed Bramwell D) Academic Press, New York. 48–63.
PRANCE G T (1987). Biogeography of neotropical plants. In *Biogeography and Quarternary history in tropical America.* (eds Whitmore T C and Prance G T) Clarendon Press, Oxford. 46–65.
RABINOWITZ D, CAIRNS S and DILLON T (1986). Seven forms of rarity and their frequency in the flora of the British Isles. In *Conservation biology: the science of scarcity and diversity.* (ed Soule M E) Sinauer, Sunderland. 182–204.
RAVEN P H (1964). Catastrophic selection and edaphic endemism. *Evolution* **18**, 336–8.
RAVEN P H (1973). The evolution of Mediterranean floras. In *Mediterranean type ecosystems, origin and structure.* (eds Mooney H A and Di Castri F) Springer-Verlag, Heidelberg. 213–34.
RAVEN P H and AXELROD D I (1978). Origin and relationships of the California flora. *University of Californian Publications in Botany* **72**, 1–134.
REBELO A G (ed) (1987). A preliminary synthesis of pollination biology in the Cape flora. *South African National Scientific Programmes Report* **141**. CSIR, Pretoria.
REBELO A G and SIEGFRIED W R (1990). Protection of fynbos vegetation: ideal and real-world options. *Biological Conservation* **54**, 15–31.
REBELO A G, COWLING R M, CAMPBELL B M and MEADOWS M E (1991). Plant communities of the Riversdale Plain. *South African Journal of Botany* **57**, 10–18.
RICE W R and SALT G W (1988). Speciation via disruptive selection on habitat preference: experimental evidence. *American Naturalist* **131**, 911–17.
RICE B and WESTOBY M (1983). Plant species richness at the 0.1 ha scale in Australian vegetation compared to other continents. *Vegetatio* **52**, 129–40.
RICHARDS P W (1952). *The tropical rain forest.* Cambridge University Press, Cambridge.
RICHARDSON D M and BOND W J (1991). Determinants of plant distribution: evidence from pine invasions. *American Naturalist* (in press).
RICHERSON P J and LUM K L (1980). Patterns of plant species diversity in California: relation to weather and topography. *American Naturalist* **116**, 5 043–536.
RICKLEFFS R E (1987). Community diversity: relative roles of local and regional processes. *Science* **235**, 167–71.
ROBINSON E R and GIBBS RUSSELL G E (1982). Speciation environments and centres of diversity in southern Africa. 1. Conceptual framework. *Bothalia* **14**, 83–8.
ROURKE J P (1969). Taxonomic studies on *Sorocephalus* R.Br. and *Spatella* Salisb. *Journal of South African Botany Supplementary Volume* **7**, 1–124.
ROURKE J P (1972). Taxonomic studies on *Leucospermum* R.Br. *Journal of South African Botany Supplementary Volume* **8**, 1–194.
ROURKE J P (1980). *The Proteas of Southern Africa.* Purnell, Johannesburg.
ROURKE J P (1987). The inflorescence morphology and systematics of *Aulax* (Proteaceae). *South African Journal of Botany* **53**, 464–80.
RUTHERFORD M C and WESTFALL R H (1986). Biomes of southern Africa: an objective characterization. *Memoirs of the Botanical Survey of South Africa* **54**, 1–98.
RYCROFT H B (1953). A quantitative ecological study of the vegetation at Biesiesvlei, Jonkershoek. PhD thesis, University of Cape Town.
SHMIDA A (1981). Mediterranean vegetation in California and Israel: similarities and differences. *Israel Journal of Botany* **30**, 105–23.
SHMIDA A (1984). Whittaker's plant diversity sampling method. *Israel Journal of Botany* **33**, 41–6.
SHMIDA A and ELLNER S (1984). Coexistence of plant species with similar niches. *Vegetatio* **58**, 29–55.
SHMIDA A and WILSON M V (1985). Biological determinants of species diversity. *Journal of Biogeography* **12**, 1–20.
SILVERTOWN and LAW (1987). Do plants need niches? *Trends in Ecology and Evolution* **2**, 24–6.
SLINGSBY P and BOND W J (1985). The influence of ants on the dispersal distance and seedling recruitment of *Leucospermum conocarpodendron* (L.) Buek (Proteaceae). *South African Journal of Botany* **51**, 30–5.
SPECHT R L and MORGAN D G (1981). The balance between foliage projective covers of overstorey and understorey strata in Australian vegetation. *Australian Journal of Ecology* **6**, 193–202.
SÖRENSON T (1948). A method of establishing groups of equal amplitude in plant sociology based on similarity in species content. *Biological Skr. Danske Selsk* **5**, 1–34.
STEBBINS G L (1974). *Flowering plants: evolution above the species level.* Harvard University Press, Cambridge, Mass.
STEBBINS G L and MAJOR J (1965). Endemism and speciation in the California flora. *Ecological Monographs* **35**, 1–31.
STEVENS G C (1989). The latitudinal gradient in geographical range: how so many species coexist in the tropics. *American Naturalist* **133**, 240–56.
STRID A K (1972). A revision of genus *Adenandra* (Rutaceae). *Opera Botanica* **32**, 1–112.
TADROS T M (1957). Evidence of the presence of an edaphobiotic factor in the problem of serpentine tolerance. *Ecology* **38**, 14–23.
TAYLOR H C (1984). Vegetation survey of the Cape of Good Hope Nature Reserve. II. Descriptive account. *Bothalia* **15**, 259–91.
TAYLOR H C (1985). An analysis of the flowering plants and ferns of the Cape of Good Hope Nature Reserve. *South African Journal of Botany* **51**, 1–13.
TEMPLETON A R (1981). Mechanisms of speciation - a population genetic approach. *Annual Review of Ecology and Systematics* **12**, 23–48.
THODAY D (1925). The geographical distribution and ecology of Passerina. *Annals of Botany* **39**, 175–208.
THWAITES R M (1987). Preliminary investigation into the geomorphological history of the Agulhas Plain. *South African Geographical Journal* **69**, 165–8.
THWAITES R M and COWLING R M (1988). Soil-vegetation relationships on the Agulhas Plain, South Africa. *Catena* **15**, 333–45.
TILMAN D (1982). *Resource competition and community structure.* Princeton University Press, Princeton.
TILMAN D (1986). Resources, competition and the dynamics of plant communities. In *Plant ecology.* (ed Crawley M T) Blackwell, Oxford. 51–76.
VAN DER MERWE C V (1976). Die plantekologiesae aspek en bestuursprobleme van die Goukamma-Natuurreservaat. MSc thesis, University of Stellenbosch.
VAN DER MERWE P (1966). Die flora van die Swartboschkloof, Stellenbosch en die herstel van die soorte na 'n brand. *Annals of the University of Stellenbosch* **41**, 691–736.
VAN WILGEN B W and KRUGER F J (1981). Observations on the effects of fire in mountain fynbos at Zachariashoek, Paarl. *Journal of South African Botany* **47**, 195–212.
VAN WYK B E, VAN WYK C M and NOUVELLIE P E (1988). Flora of the Zuurberg National Park. 2. An annotated checklist of ferns and seed plants. *Bothalia* **18**, 221–32.
VUILLEUMIER F and SIMBERLOFF D (1980). Ecology versus history as determinants of patchy and insular distribution of high Andean birds. *Evolutionary Biology* **13**, 235–379.
VRBA E S (1980). Evolution, species and fossils: how does life

evolve. *South African Journal of Science* **76**, 61–84.
VRBA E S (1984). Evolutionary pattern and process in the sister group. Alcelaphini-Aepycertini (Mammalia: Bovidae). In *Living fossils.* (eds Eldredge N and Stanley S M) Springer, New York. 62–79.
VRBA E S and ELDREDGE N (1984). Individuals, hierarchies and processes: towards a more complete evolutionary theory. *Paleobiology* **10**, 146–71.
VRBA E S and GOULD S J C (1986). The hierarchical expansion of sorting and selection: sorting and ecological selection cannot be equated. *Paleobiology* **12**, 217–28.
WALKER J and PEET R K (1983). Composition and species diversity of pine-wiregrass savannas of the Great Swamp, North Carolina. *Vegetatio* **55**, 163–79.
WANNTORP H E, BROOKS D R, NILSSON T, NYLIN S, RONQUIST F, and STEARNS S C (1990). Phylogenetic approaches in ecology. *Oikos* **57**, 119–32.
WEIMARCK H (1941). Phytogeographical groups, centres and intervals within the Cape Flora. *Acta University Lund* **37**, 5–143.
WELLS P V (1969). The relation between mode of reproduction and extent of speciation in woody genera of the California chaparral. *Evolution* **23**, 264–7.
WERGER M J A (1978). Biogeographical division of southern Africa. In *Biogeography and ecology of southern Africa.* (ed Werger M J A) Junk, The Hague. 301–462.
WERNER P A (1978). Competition and coexistence of similar species. In *Topics in plant population biology.* (eds Solbrig O T, Jain S, Johnson G B and Raven P H) Mac-Millan, London. 287–310.
WESTOBY M (1985). Two main relationships among the components of species richness. *Proceedings of the Ecological Society of Australia* **14**, 103–7.
WESTOBY M (1988). Comparing Australian ecosystems to those elsewhere. *Bioscience* **38**, 549–56.
WHITTAKER R H (1972). Evolution and measurement of species diversity. *Taxon* **21**, 213–51.
WHITTAKER R H (1977). Evolution of species diversity in land communities. *Evolutionary Biology* **10**, 1–67.
WILLIAMS I J M (1972). A revision of the genus *Leucadendron. Contributions to the Bolus Herbarium* **3**, 1–425.
WILLIAMS I J M (1982a). Studies in the genera of the Diosmeae (Rutaceae): 13. A revision of the genus *Acmadenia* Bartl. Wendl. *Journal of South African Botany* **48**, 169–240.
WILLIAMS I J M (1982b). Studies in the genera of the Diosmeae (Rutaceae): 14. A review of the genus *Diosma* L. *Journal of South African Botany* **48**, 329–407.
WILLIAMSON M (1988). Relationship of species number to area, distance and other variables. In *Analytical biogeography. An integrated approach to the study of animal and plant distributions.* (eds Myers A A and Giller P S) Chapman and Hall, London and New York. 91–115.
WILSON M V and SHMIDA A (1984). Measuring beta diversity with presence-absence data. *Journal of Ecology* **72**, 1 055–64.
YEATON R I and BOND W J (1991). Competition between two shrub species: dispersal differences and fire promote coexistence. *American Naturalist* (in press).
YODZIS P (1986). Competition, mortality and community structure. In *Community ecology.* (ed Diamond J and Case T J) Harper and Row, Cambridge. 480–91.
ZAR J H (1984). *Biostatistical analysis* (2nd ed). Prentice-Hall, New Jersey.
ZEDLER P H, GAUTIER C R and JACKS P (1984). Edaphic restriction of Cupressus forbesii (tecate cypress) in southern California, U.S.A. - a hypothesis. In *Being alive on land.* (ed Margaris N S, Arianoustou-Farragitaki M and Oechel W C) Junk, The Hague. 237–43.
ZUCCINI W (1990). RAIN - a programme to generate artificial rainfall sequences. Applied Statistics and Decision Science Unit. University of Cape Town.

APPENDIX 4.1 The location and sources for sites (Figure 4.2) used in determining species-area relationships (Figure 4.8).

Site and number		Number of spp.	Area (km^2)	Vegetation	Source
South-western Cape Floristic Region					
1	Cape Peninsula	2 256	471.00	Fynbos	Adamson and Salter (1950)
2	Cape of Good Hope N.R.[1]	1 036	77.50	Fynbos	Taylor (1985)
3	Rondevlei N.R.	229	1.37	Fynbos, subtropical thicket	I A W MacDonald (personal communication)
4	Cape Flats N.R.	210	0.20	Fynbos, subtropical thicket	B Low (personal communication)
5	Tygerberg N.R.	373	0.68	Renoster shrubland	I A W MacDonald (personal communication)
6	Pella	379	2.70	Fynbos	Boucher and Shepherd (1987)
7	Stellenbosch Flats	585	10.40	Fynbos, renoster shrubland	Duthie (1929)
8	Biesiesvlei	364	0.27	Fynbos	Rycroft (1953)
9	Assegaaibos N.R.	306	1.68	Fynbos	CDNEC[2] (personal communication)
10	Jonkershoek N.R.	1 142	45.30	Fynbos	Kruger and Taylor (1979)
11	Swartboschkloof	483	1.82	Fynbos	Van der Merwe (1966)
12	Cape Hangklip	1 383	240.00	Fynbos	Boucher (1977)
13	Jakkalsrivier	533	1.58	Fynbos	Kruger (1974)
14	Fernkloof N.R.	773	14.46	Fynbos	I A W MacDonald (personal communication)
15	Vogelgat N.R.	697	6.02	Fynbos	I A W MacDonald (personal communication)
16	Agulhas Plain	1 751	1 609.25	Fynbos, renoster shrubland	R M Cowling (unpublished)
17	Hagelkraal	157	0.15	Fynbos	R M Cowling (unpublished)
18	Arniston-Breede	1 509	600.00	Fynbos	CDNEC (personal communication)
19	De Hoop	1 179	180.00	Fynbos	CDNEC (personal communication)

APPENDIX 4.1 continued.

South-eastern Cape Floristic Region					
1	Bontebok N.R.	446	27.86	Renoster shrubland, fynbos	I A W MacDonald (personal communication)
2	Anysberg	473	340.00	Fynbos, renoster shrubland, karroid shrubland	CDNEC (personal communication)
3	Rooiberg	481	250.00	Fynbos, renoster shrubland, karroid shrubland	CDNEC (personal communication)
4	Riversdale Plain	1 580	2 860.00	Fynbos, renoster shrubland, subtropical thicket	Rebelo et al. (1991)
5	George District	1 204	2 536.00	Fynbos, renoster shrubland, Afromontane forest, karroid shrubland	Fourcade (1941)
6	Moordkuils	313	6.17	Fynbos	J Vlok (personal communication)
7	Goukamma N.R.	380	20.55	Fynbos, subtropical thicket	Van der Merwe (1976)
8	Knysna District	1 416	2 098.00	Fynbos, Afromontane forest	Fourcade (1941)
9	Groot Eiland	150	0.67	Fynbos	Bond et al. (1988)
10	Uniondale	1 202	4 378.00	Fynbos, renoster shrubland, karroid shrubland	Fourcade (1941)
11	Humansdorp	2 240	5 050.00	Fynbos, renoster shrubland, Afromontane forest, subtropical thicket	Fourcade (1941)
12	Seekoei N.R.	217	1.41	Renoster shrubland, subtropical thicket	CDNEC (personal communication)
13	Cape Recife N.R.	173	3.36	Fynbos subtropical thicket	Olivier (1983)
14	Zuurberg N.R.	1 100	207.8	Fynbos, subtropical thicket, grassland	Van Wyk et al. (1988)
15	Albany and Bathurst	2 389	4 800.00	Grassland, subtropical thicket, fynbos, karroid shrubland	Martin and Noel (1960)

[1] N.R.= Nature reserve or national park.

5

History of the Cape flora

H P Linder, M E Meadows and R M Cowling

INTRODUCTION

Studies in the history of a biota are multifaceted (Myers and Giller 1988). The most common is the historical geography of the biota which investigates the way in which patterns of distributions may have changed in the past. Historical biogeography ranges from narratives fitted to sets of observations (narrative approaches), to rigorous hypothesis testing (i.e. cladistic approaches) (Ball 1975; Patterson 1981). While the narrative approach may produce a story that is rich in testable hypotheses, the relationship between it and the physical geographical history is not always clear. With the advent of vicariance biogeography this relationship has been made rigorous. Consequently, the relationship between extant patterns of distributions and historical changes in the geography can be tested (Rosen 1976, 1978; Humphries and Parenti 1986). A less common approach is historical ecology which studies evidence of extinct plant communities (Wanntorp et al. 1990). This approach has almost always been concerned with the reconstruction of extinct plant communities from fossil assemblages. Similarly, the composition of past floras has been determined from fossil studies. It should, however, also be possible to infer past floristic composition through a critical study of both the evolutionary histories of component taxa and the present floristic composition of floras.

In this chapter we attempt to reconstruct the history of the flora of the Cape Floristic Region (Goldblatt 1978). We start by tracing the origins or sources of the flora, using track analyses to establish initial hypotheses and cladistic biogeographic methods (Humphries et al. 1988) to provide historical explanations. We then discuss the unusual features of the contemporary flora and vegetation of the region and consider historical changes in floristic composition as inferred from palaeopalynology. We attempt to explain these phenomena by developing a model that invokes extensive climatically driven extinction towards the end of the Tertiary. In addition, we argue that recruitment from neighbouring floras was limited owing to the peculiar nature of the soils in the Cape Floristic Region ('edaphic island hypothesis').

We consider mainly the fynbos component of the flora of the Cape Floristic Region. The Afromontane forest component may have invaded in azonal habitats (Cowling and Holmes this volume) during the Holocene (Cowling 1983; Meadows and Linder 1989). Consequently, it is largely ignored. We consider the Karoo flora as being marginal to fynbos.

Fynbos is defined by structural, floristic, and phytogeographical criteria. Cowling and Holmes (this volume) provide details of the first two of these criteria. Phytogeographical criteria include the concentration of species in the south-western Cape Province, usually in the Caledon District (Levyns 1954, 1964), and a high level of endemism to the region (Goldblatt 1978). Although these criteria were developed to define the Cape flora, Linder (1990) has shown that fynbos has much in common with the flora of the Afromontane archipelago-like regional centre of endemism (White 1983) and may be considered part of a continent-wide Afrotemperate flora.

ORIGINS OF THE CAPE FLORA

The origins of the flora are best traced by phytogeographical analysis. The extra-regional distributions of the taxa, especially the species, are analysed to indicate the areas which share a floristic component with the study area. This generic component of phytogeography (Stott 1981) indicates with which floras the study flora previously had contact. The history of these contacts can then be inferred by comparison with known palaeoclimates and palaeogeography.

Track analysis

The first level of approach to the problem of locating areas and floras with possible previous contact with the Cape flora is by identifying the general tracks which link them. A track is the distribution area of a taxon. If several tracks coincide, they define a general track. General tracks can be built up of the tracks of species, genera, families, or any other taxonomic rank, or combinations of different ranks. Usually general tracks link several centres of endemism, or even different continents. This has been a favourite field for phytogeographers (Adamson 1958; Levyns 1964; Goldblatt 1978) as the data required are a by-product of taxonomic research. Goldblatt (1978) provides detailed listings of the families and genera showing various distribution patterns. These analyses have resulted in the recognition of a large endemic component in the Cape flora, but in addition there are several non-endemic components which provide the necessary information on the tracks. Three tracks can be recognized: a Gondwanan track, an African track, and a Boreal track.

Tracks are statistical constructs and do not directly imply any causal factors. There are two explanations available for the establishment of a track. The first suggests that the track is established by long-distance dispersal across the barriers in the track. This is the centre-of-origin approach and has long been the only generally accepted explanation (Humphries and Parenti 1986). Dispersalism does not lead to general explanations: idiosyncratic explanations usually relate to the dispersability of the taxa under study. The second type of explanation (vicariance) is that uninterrupted distributions were broken by the establishment of barriers, leading to geographical isolation and subsequent speciation. This vicariance approach (Croizat et al. 1974) has the advantage that it leads to general explanations. In addition to the taxon under study, the establishment of a barrier is likely to affect a wide range of taxonomically unrelated taxa with similar tracks. Furthermore, there should be evidence of geographical barriers which can often be dated, thus allowing the dating of events which shaped the evolutionary history of the group.

There are two tests that can be applied to establish the likelihood of dispersalism or vicariance. If the series of speciation events for a group spanning a track are consistent with the known geological events establishing the intervals in the track, then the hypothesis that the track has a vicariance explanation is not falsified. The second criterion is that several taxonomic groups show the same track and the same series of speciation events. It is more likely that this pattern would arise from vicariance than dispersal since the latter is essentially random in both space and time.

The Gondwanan track

The Gondwanan (Antarctic) track (Levyns 1962) (Figure 5.1) includes all southern hemisphere continents, and various islands. Antarctica would also be part of this track, but it has lost its trachaeophyte flora. The Gondwanan track includes both genera and families. *Gunnera* (Gunneraceae), *Ehrharta* (Poaceae) (Gibbs Russell and Ellis 1987), *Podocarpus*, *Dietes* (Iridaceae) (Goldblatt 1981), and *Bulbinella* (Asphodelaceae) (Dahlgren et al. 1985) are examples of links at generic level. There are only some 16 supra-generic taxa from Africa in this track e.g. Proteaceae, Haemodoraceae, Cunoniaceae, and Arundineae (Poaceae) (Linder 1989).

The only group that has been critically tested is the Restionaceae (Linder 1987). Cutler (1969) showed that the genera are mostly endemic to the southern continents, and later argued that some of the pattern may be due to the 'invasion' (presumably dispersal) of *Leptocarpus* into South America (Cutler 1972). However, the cladistic biogeographic pattern (Figure 5.2) shows that the Gondwanan breakup sequence matches the reduced area cladogram: South America and Australia are sister areas and, in combination, they form the sister area to Africa. This agrees with the presumed breakup sequence of Gondwana, where Africa was the first to break out of the super-continent, while Australasia, Antarctica, and South America remained in contact much longer (De Wit et al. 1988). This implies, at least for the Restionaceae, a vicariance explanation for the track. By implication, the Pro-

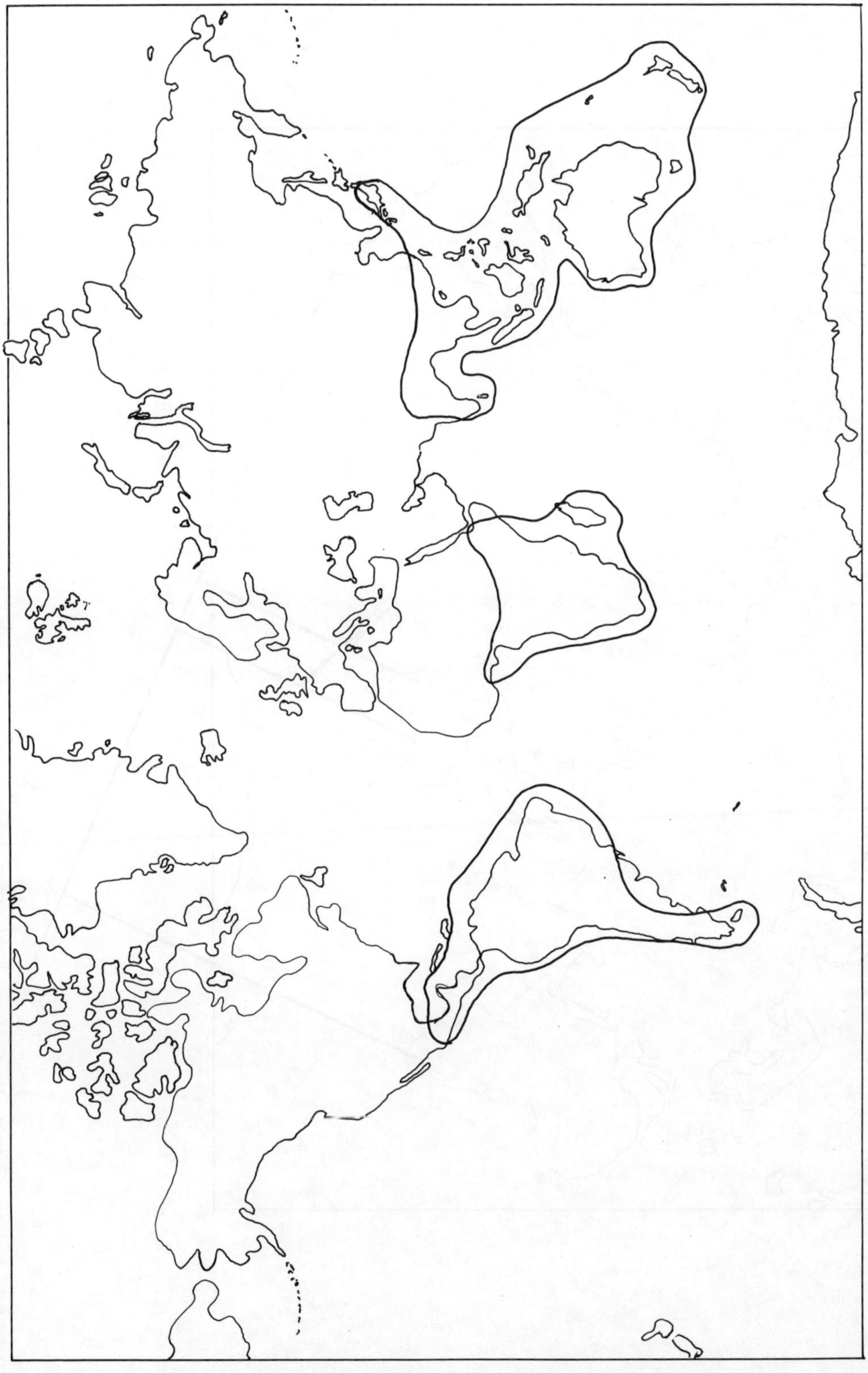

FIGURE 5.1 The Gondwanan track, as illustrated by the distribution of the family Proteaceae (after Good 1974). For the number of taxa in each region see Table 5.2.

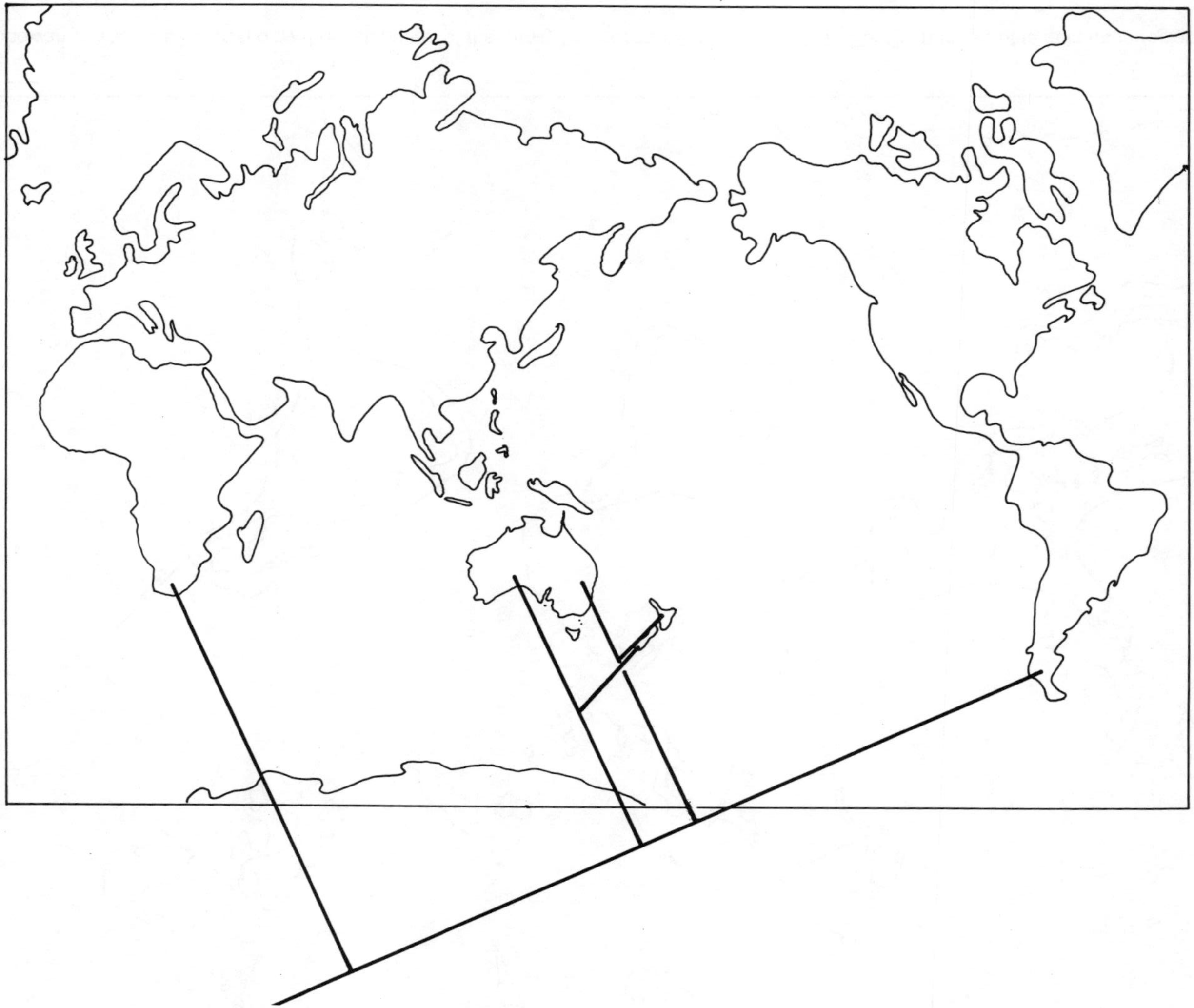

FIGURE 5.2 A cladogram for the Restionaceae (Linder 1987, 1991) superimposed over the distribution of the family, showing the relationships among the Gondwanan fragments. This sequence is in agreement with the breakup sequence of the supercontinent.

teaceae and the other taxa should also have a vicariance explanation, but critical tests are lacking. It has generally been suggested that the Proteaceae reached the African continent by dispersal from west Gondwana across the proto-Indian Ocean (Johnson and Briggs 1975; Axelrod and Raven 1978; Takhtajan 1986). These authors cite as evidence the fact that the breakup of Gondwana preceded the origin of the Angiosperms, and certainly the evolution of some of the genera. However, the time of origin of the Angiosperms has not been dated with confidence.

The African track

The strongest links in the Cape flora are with tropical Africa. This has been argued by Adamson (1958) who showed that the Cape flora was largely derived from the African, rather than the Antarctic flora. However, the flora of tropical Africa is phytogeographically complex, and several phytochoria can be recognized (White 1983). The only detailed study on the phytochorological links of the Cape flora is by Cowling (1983) which shows strong links between the Cape and Afromontane phytochoria in the south-eastern Cape Floristic Region (see also Cowling and Holmes this volume). The analyses of individual taxa by Weimarck (1933, 1936, 1941) and later authors all indicate links to the Afromontane phytochorion. This pattern represents the Afrotemperate track of Linder (1983, 1990) (see Figure 5.3). Typical taxa of this track are *Protea* (Rourke 1980), the Diseae (Orchidaceae) (Linder 1982), *Aristea* (Iridaceae) (Weimarck 1940), *Cliffortia* (Rosaceae) (Weimarck 1934), the Anthospermeae (Rubiaceae) (Puff 1986), the Mesembryanthema (Hartmann 1991), and *Pentaschistis* (Poaceae) (Linder and Ellis 1990). This track has not yet been critically analysed and the complex interrelationship of the tropical and temperate components is not yet understood.

The only test with a group that spans the length of the track has been on *Lotononis* (Fabaceae) (H P Linder and B-E van Wyk unpublished data). This shows a general relationship of (Cape Floristic Region — Namaqualand) ((Drakensberg — tropical Africa) Mediterranean Basin) (Figure 5.4). These results imply that the most closely related areas are the Cape Floristic Region and Namaqualand, and the various parts of Africa north of the Port Elizabeth-Grahamstown interval (the Kaffrarian interval of Weimarck 1941). These area relationships are consistent with the more localized analyses on the Coryciinae (Orchidaceae) (Kurzweil and Linder 1991), *Disperis* (Orchidaceae) (Manning and Linder 1991), and *Microloma* (Asclepiadaceae) (Bruyns and Linder 1991). Linder and Van Wyk (unpublished data) suggest that this pattern may reflect events in the Oligocene to Pliocene, during the desiccation of Africa, the establishment of the rift valleys, and the final uplift of the eastern margins of the Great Escarpment in southern Africa (Partridge and Maud 1987). This implies that the pan-African distribution patterns existed during the early part of the Tertiary, at least up to the Oligocene. The modern pattern developed with the establishment of strong polar climatic gradients, the formation of the tropical deserts, and the initiation of the winter rainfall climates at the southern and northern extremes of the continent. These factors led to the fragmentation of this African flora with the winter rainfall areas being the most floristically distinct.

The boreal track

The boreal track (Figure 5.5) includes taxa such as *Anemone*, *Ranunculus*, Dipsaceae, *Galium*, *Festuca*, *Alchemilla*, *Hieracium*, *Viola*, and *Scabiosa*. These taxa are mostly common in the northern hemisphere, including Europe, Asia, and North America. Some, like *Alchemilla*, *Anemone*, *Galium*, *Festuca*, and *Ranunculus* also occur on the East African high mountains (Hedberg 1965). Others show a southern African-northern hemisphere disjunction.

There have been no cladistic analyses of these taxa. Their tracks suggest migration from Europe to southern Africa along the East African uplands. As the East African uplands were formed between 30 and 20 million years ago (Grove 1983), the earliest occurrence of these taxa in southern Africa would have been in the Neogene. However, for some taxa (e.g. Ericaceae) the problem is clearly more complex. The Ericaceae is cosmopolitan in temperate areas, but the greatest diversity of subfamilies is in the boreal region. Africa and Australia have what appear to be vicariant taxa, with the Ericoideae in Africa and the Epacridaceae

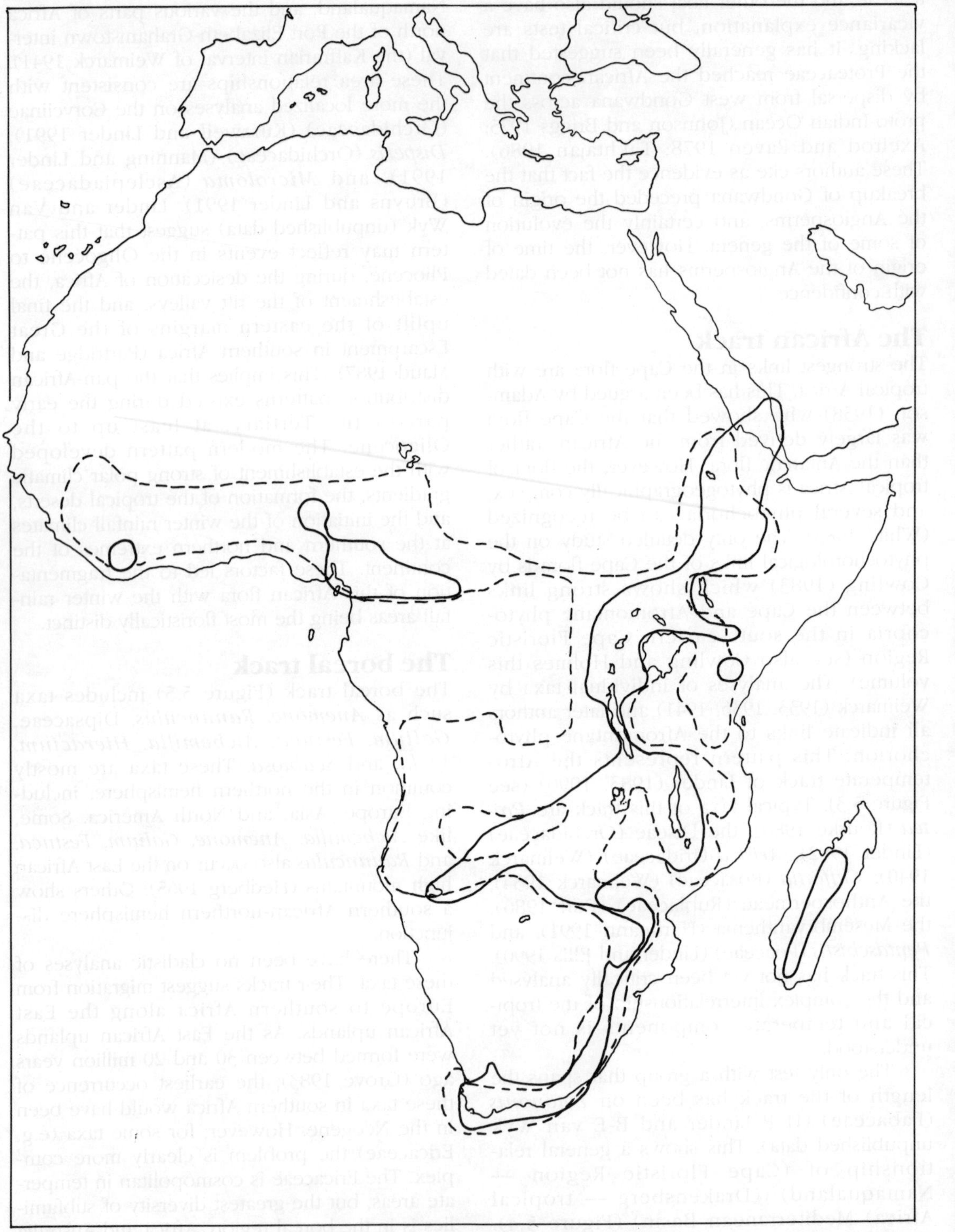

FIGURE 5.3 The African track, as illustrated by the distributions of the genera *Protea* (solid line, after Rourke 1980) and *Disa* (broken line, after Linder 1983).

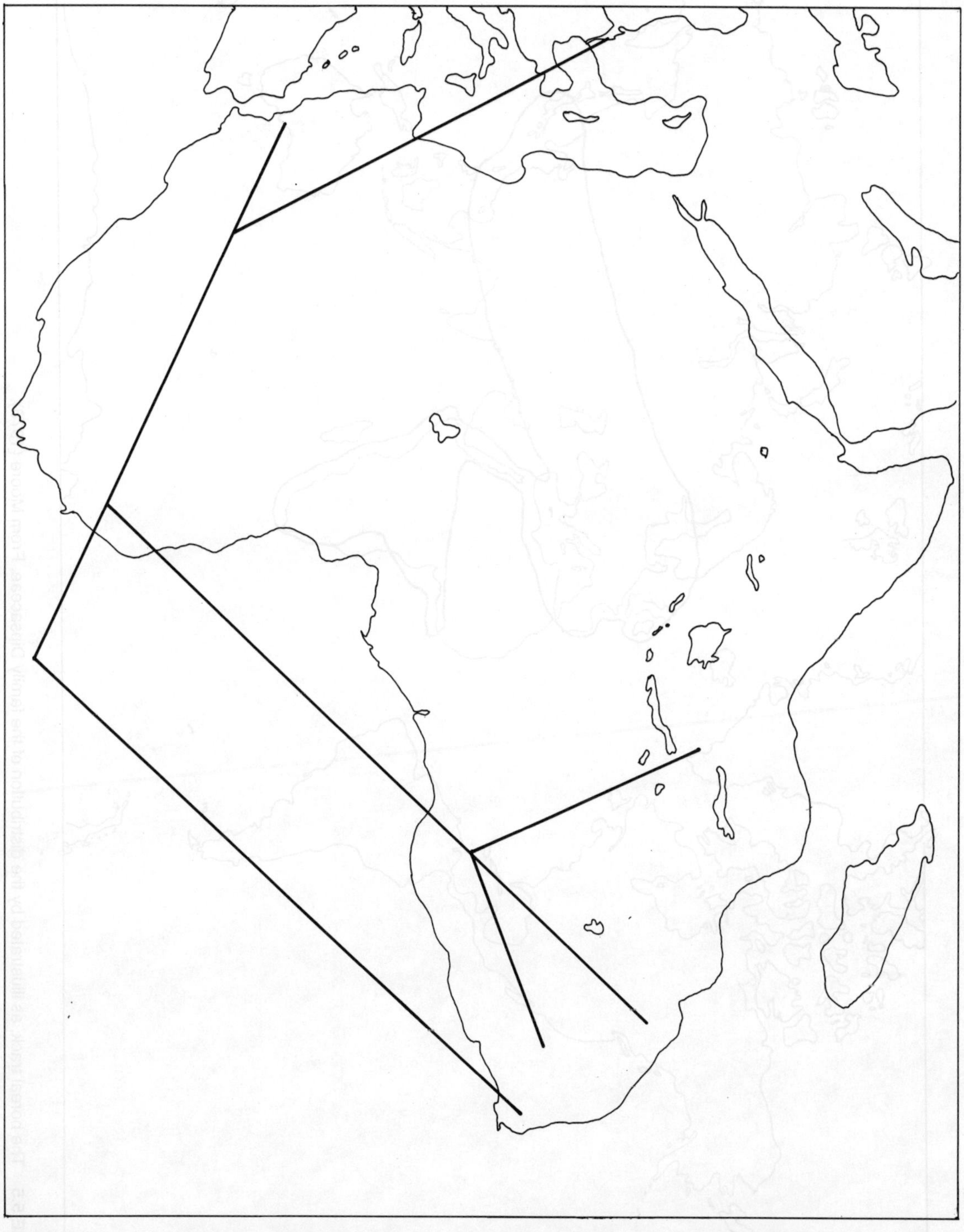

FIGURE 5.4 The cladogram of the genus *Lotononis* (H P Linder and B-E van Wyk unpublished data), superimposed over the distribution area of the genus, showing the relationships among the various parts of the African track.

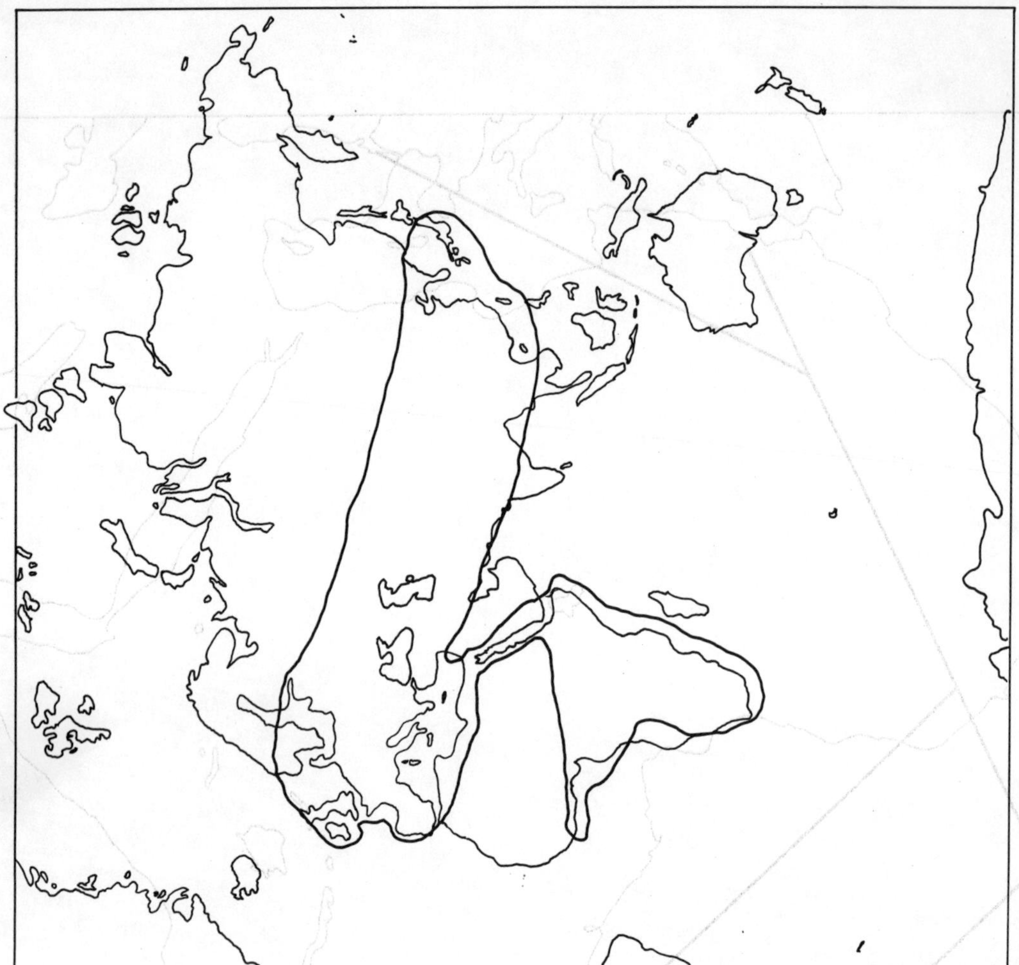

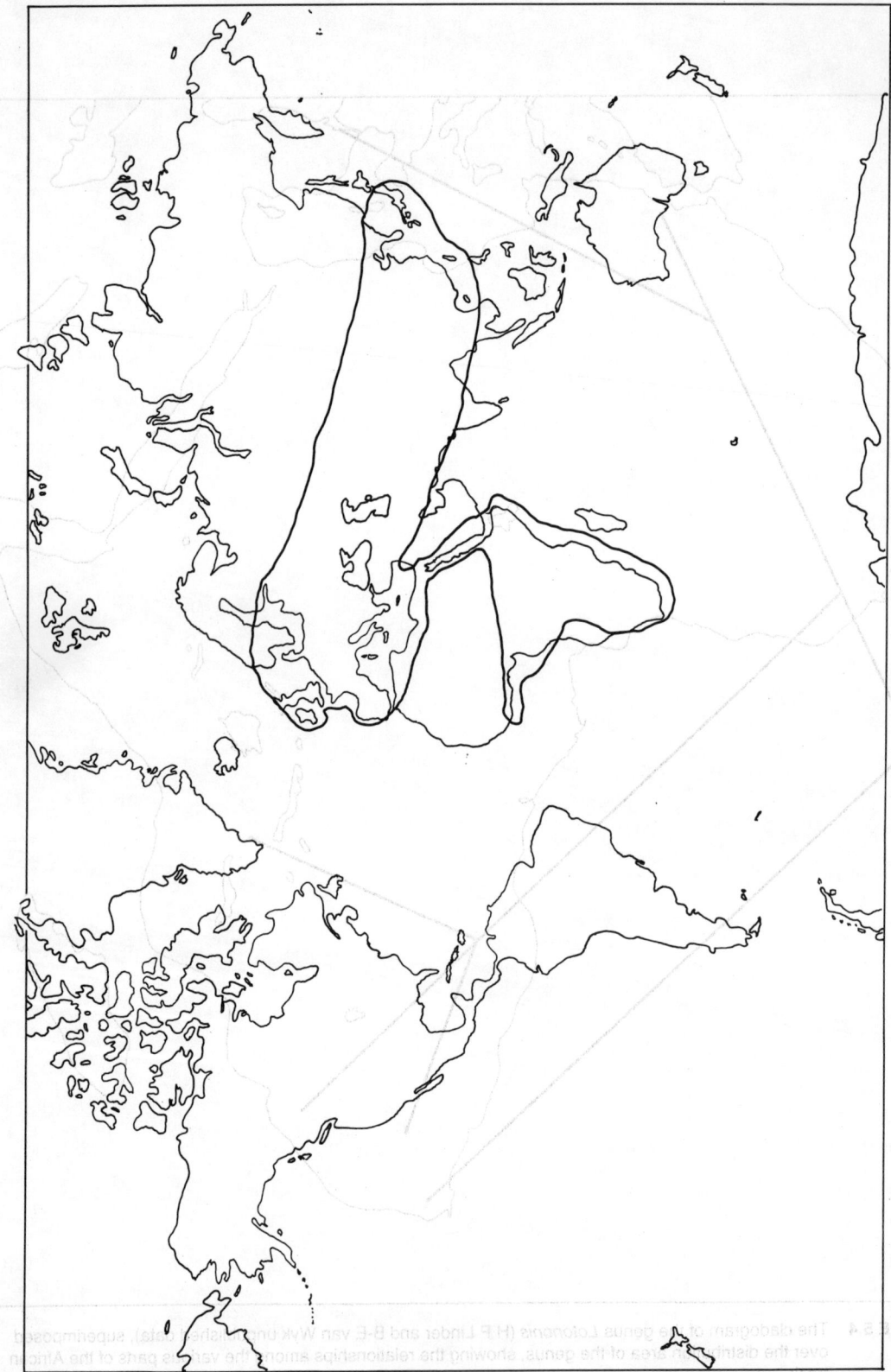

FIGURE 5.5 The boreal track, as illustrated by the distribution of the family Dipsaceae. From Moore (1978).

in Australia. But the Epacridaceae is not the sister taxon of the Ericoideae, which implies that Europe, rather than Australia, is the sister area. However, Ericaceae pollen has been recorded in the Cape since the Early Tertiary (Scholz 1985), indicating a trans-Tethys range for this family. Other intriguing boreal links include the close relationship between the African and Eurasian ground orchids (Diseae and Orchideae) (Dressler 1981; Burns-Balogh and Funk 1986). The Australian ground orchids (Diurideae) are clearly more distantly related. There is no indication of the age of the Orchidoideae, consequently the hypothesis that they may have migrated down the Afrotemperate track cannot be tested.

INTERNAL PHYTOGEOGRAPHY OF THE CAPE FLORA

Distribution patterns within the Cape flora were recognized from very early on. Drege (1844) published a map showing subdivisions within the Cape Floristic Region, based on his own extensive collections. With a highly improved distributional and taxonomic database, Weimarck (1941) delimited 'centres' and 'intervals' in the Cape flora. These were based on visual inspection of distribution maps. The analysis was repeated by Oliver et al. (1983) using multivariate statistics on a much larger data set. This essentially confirmed the findings of Weimarck. The centres are discussed in much more detail by Cowling et al. (this volume).

The historical implications of the centres are not clear. Although several monographs have indicated how the species are distributed among the centres (Dahlgren 1963; Nordenstam 1969; Strid 1972), these patterns have not been linked directly to historical explanations. Nor has any cladistic biogeographic study been performed to analyse the relationships among the centres. For some groups, at least, the centres appear to be more in the nature of statistical constructs, than as areas of strict endemism (Kurzweil and Linder 1991; H P Linder and B-E van Wyk unpublished data). In these groups there are no taxa strictly endemic to the centres, but owing to the overlapping patterns of species distributions, these centres can be retrieved by multivariate analysis. However, in other genera there are distinct sets of endemic taxa in each centre. A critical analysis of one of these genera could provide great insight into the relationships among the centres. Hypotheses on the vegetational history of various parts of the Cape Floristic Region may be tested phytogeographically, either by locating areas with only derived species or by showing that some areas have a high proportion of 'basal' species.

DIVERSITY PATTERNS IN THE CAPE FLORA

The remarkable richness of the Cape flora is a major factor in its recognition as a floral kingdom (Good 1974; Goldblatt 1978; Takhtajan 1986). However, 'richness' or 'diversity' are complex concepts (Bond 1989; Cowling et al. this volume). Richness can be applied to any measurable component that allows a comparison to other areas. Here we will explore three components: taxic diversity, by which we mean the number of taxa (species, genera, and families); phylogenetic diversity, in which we attempt to establish the number of 'ancient' or 'ancestral' taxa; and structural diversity, which refers to the morphological divergence within the flora.

Taxic diversity

Taxic diversity refers to the number of taxa in an area. These data can be assembled either as absolute figures (numbers of families, genera, or species) or as ratios of the numbers of genera per family or the number of species per family.

In order to establish a norm, floras from several areas were analysed: the Cape Floristic Region (Bond and Goldblatt 1984), Western Australia (Beard 1968), the Sydney area (Beadle et al. 1986), Southern California (Munz 1974), Natal (Ross 1972), and Swaziland (Compton 1976). The floras analysed are all from areas which were not glaciated during the Pleistocene, which avoids the effects of relatively 'young' floras, such as are found in north and central Europe. In addition, all samples are from mediterranean or subtropical climatic regions. We analysed the relationships between the number of species, genera and families, and area for all the floras. We wished to determine whether the flora of the Cape Floristic Region was in any way unusual in terms of these.

The Cape Floristic Region has a much higher species density than the other areas (Figure 5.6a). The genus to family ratio for the region is average (Figure 5.6b) whereas the number of species per family is considerably higher than the other floras (Figure 5.6c). This indicates that the richness at species level, rather than at generic or family level, in the Cape flora needs an explanation. This poses a clear question: why are there so many species in the Cape flora?

Goldblatt (1978) notes that the flora of the Cape Floristic Region has a remarkably large percentage of species (22%) found in the 10 largest genera — a figure exceeded only by such insular floras as Hawaii (42.1%) and New Zealand (26.3%). However, more interesting is the species per genus ratio which is 8.9 in the Cape Floristic Region — the highest recorded for the areas considered by Goldblatt (1978). These data are consistent with the data presented above, and reinforce the impression that the Cape flora is unusual only in its enormous richness at the species level.

Structural diversity

Overall structural and growth form diversity in fynbos and allied shrublands — which include the bulk of the Cape flora's species — is remarkably poor, given the long environmental gradients in the Cape Floristic Region (Cowling and Holmes this volume; Cowling et al. this volume). Adamson (1958) commented that 'there is a marked uniformity of life-form from sea level to the highest summits. There is an altitudinal zonation in species but not in life-forms'. Although it is possible to find representatives of almost all growth forms in fynbos (except epiphytic plants), the majority of these growth forms are rare.

It is generally accepted that the growth form uniformity in fynbos results from the overriding effects of low soil nutrients on vegetation structure (Cowling and Holmes this volume). However, morphological diversity in fynbos plants is low even when compared with plants from the South West Botanical Province of Australia which also has low nutrient soils and a broadly similar mediterranean-type climate (Milewski and Cowling 1985). For example 86% of the 70 Cape Floristic Region *Protea* species (Proteaceae) are shrubs less than two metres tall and only 4% are trees (data from Rourke 1980). For south-west Australian *Banksia* species (Proteaceae) which, like proteas in fynbos, are the dominant overstorey component, 44% of 54 taxa are shrubs more than two metres tall and 22% are trees (data from George 1984). The lack of taller growth forms in fynbos when compared to south-western Australian vegetation on edaphically and climatically similar sites is also discussed by Cowling and Holmes (this volume) and Richardson et al. (this volume). Variation in leaf shape and size of woody plants is also much more pronouced in Australian kwongan than fynbos (R M Cowling and E Witkowski unpublished data).

Phylogenetic diversity

Phylogenetic diversity is an indication of the number of ancient (Tertiary or older) lineages in a flora. A low proportion of these lineages may indicate a high degree of extinction; a high proportion shows that these ancient lineages have persisted in the flora. Note that this does not imply that the ancient lineages need to be speciose: merely that they should be present. Ideally, these data should be retrieved from the fossil record which, unfortunately, is poorly known in the Cape Floristic Region (see below). We have attempted to estimate the phylogenetic diversity from the extant flora. This is done by investigating the occurrence of palaeoendemics (Stebbins and Major 1965), as well as comparing the number of phyletic lines of common taxa surviving in Australia and southern Africa.

Generic and family endemism does give some indication of the degree of phylogenetic diversity and hence palaeoendemism. Goldblatt (1978) believes that there are many relictual species in the flora of southern Africa. One of the characteristics of relictual or palaeoendemic species is that they are taxonomically isolated. The Cape flora has 198 endemic genera (Goldblatt 1978) and eight (including Bruniaceae which has one extralimital species) endemic families (Dahlgren and Van Wyk 1988). Six of the eight families are isolated: they are either placed in their own orders, or their placement is controversial (Table 5.1). Several families have their closest

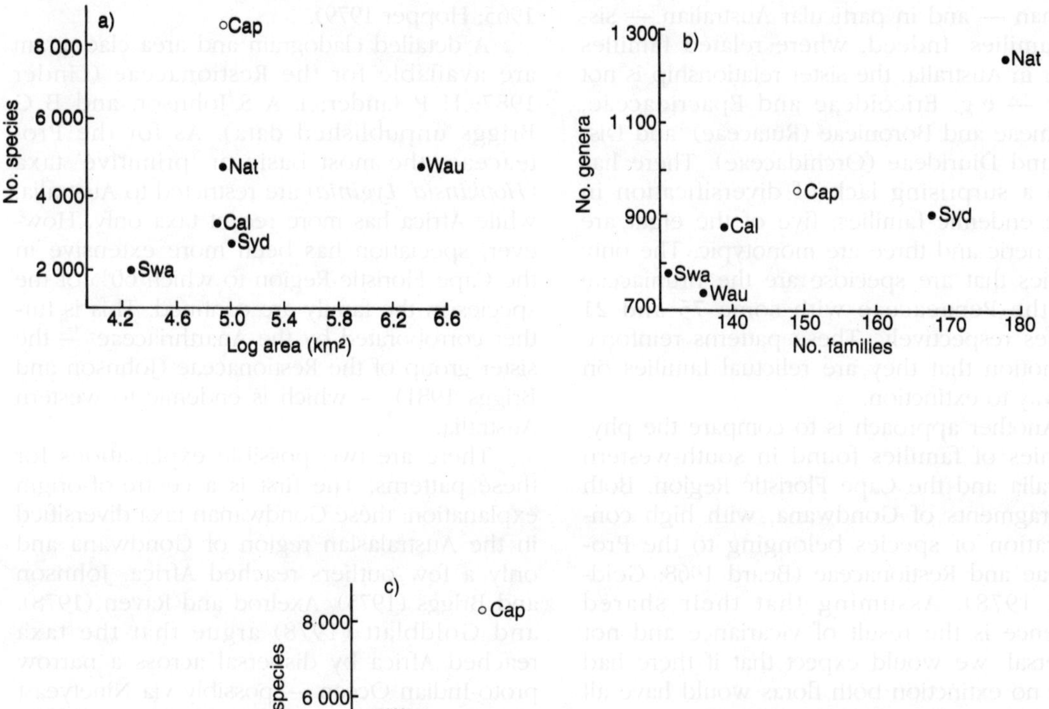

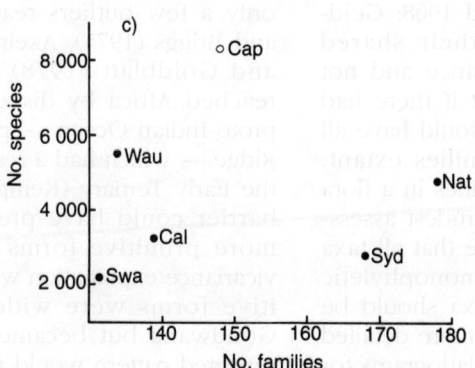

FIGURE 5.6 A comparison of the taxic diversity of the Cape Floristic Region (Cap) with that of Swaziland (Swa), southern California (Cal), Western Australia (Wau), Sydney (Syd), and Natal (Nat). In (a) it is shown that the Cape Floristic Region is exceptionally species-rich, in (b) it is clear that the number of genera per family is average in the region, while in (c) it is demonstrated that there are an unusually high number of species per family in this area.

relatives in southern Africa (e.g. Bruniaceae and Grubbiaceae, Penaeaceae and Rhynchocalycaceae). There is a curious lack of Gondwanan — and in particular Australian — sister families. Indeed, where related families occur in Australia, the sister relationship is not clear — e.g. Ericoideae and Epacridaceae, Diosmeae and Boronieae (Rutaceae), and Diseae and Diurideae (Orchidaceae). There has been a surprising lack of diversification in these endemic families: five of the eight are unigeneric and three are monotypic. The only families that are speciose are the Bruniaceae and the Penaeaceae with some 75 and 21 species respectively. These patterns reinforce the notion that they are relictual families on the way to extinction.

Another approach is to compare the phylogenies of families found in south-western Australia and the Cape Floristic Region. Both are fragments of Gondwana, with high concentration of species belonging to the Proteaceae and Restionaceae (Beard 1968; Goldblatt 1978). Assuming that their shared presence is the result of vicariance and not dispersal, we would expect that if there had been no extinction both floras would have all the phyletic lines of these families extant. Absence of any of the phyletic lines in a flora would indicate extinction. The crudest assessment of this would be to assume that all taxa in the existing taxonomy are monophyletic and, consequently, all higher taxa should be on all Gondwanan fragments. A more detailed approach is to analyse the area cladograms for the taxa. This would provide the basic information on the monophyletic groups rather than working through the classification.

Johnson and Briggs (1975) present a detailed supra-generic classification for the Proteaceae. From the distribution of the subfamilies, tribes, and subtribes (Table 5.2) it is clear that Africa has a very poor representation of higher taxa compared to Australia. The African genera — with the exception of *Brabejum stellatifolium* (Macadamieae: Macadamiinae) which is restricted to stream banks in the south-western Cape Floristic Region — belong to the Proteae. Similarly in the Myrtaceae: Leptospermoideae, *Metrosideros* is the only genus in the Cape Floristic Region. As in the case of the floras of California and south-western Australia, relictual or palaeoendemic taxa are largely confined to mesic sites which are relatively buffered against climatic change (Stebbins and Major 1965; Hopper 1979).

A detailed cladogram and area cladogram are available for the Restionaceae (Linder 1987; H P Linder; L A S Johnson and B G Briggs unpublished data). As for the Proteaceae, the most basic or 'primitive' taxa (*Hopkinsia*, *Lyginia*) are restricted to Australia, while Africa has more recent taxa only. However, speciation has been more extensive in the Cape Floristic Region to which 60% of the species in the family are confined. This is further corroborated by the Anarthriaceae — the sister group of the Restionaceae (Johnson and Briggs 1981) — which is endemic to western Australia.

There are two possible explanations for these patterns. The first is a centre-of-origin explanation: these Gondwanan taxa diversified in the Australasian region of Gondwana and only a few outliers reached Africa. Johnson and Briggs (1975), Axelrod and Raven (1978), and Goldblatt (1978) argue that the taxa reached Africa by dispersal across a narrow proto-Indian Ocean — possibly via Ninetyeast Ridge — which had a restionalean flora during the Early Tertiary (Kemp 1978). The dispersal barrier could have prevented some of the more primitive forms from crossing. The vicariance explanation would be that the primitive forms were widespread throughout Gondwana but became extinct in Africa. A repeated pattern would indicate the vicariance explanation to be the more parsimonious one. In the case of the Proteaceae, the cladogram is not sufficiently resolved to suggest confidently a vicariance-extinction explanation. The Australian subfamilies (Table 5.2) could all belong to one clade, implying that the Proteae are paraphyletic, which is highly unlikely. In the Restionaceae it is clear that if the vicariance assumption is valid, the phyletic lines that led to the Anarthriaceae, *Hopkinsia*, and *Lyginia* must have become extinct in Africa. As both the Restionaceae and the Proteaceae present the same general pattern, this might apply to all other Gondwanan taxa for which cladograms are not yet available. The vicariance hypothesis is, therefore, falsifiable.

The absence of 'expected' phyletic lines of the Proteaceae and Restionaceae in the Cape

TABLE 5.1 The number of genera, species, and closest relatives of the families endemic to the Cape Floristic Region. The data are from Dahlgren and Van Wyk (1988).

Family	No. of genera	No. of species	Relatives
Geissolomataceae	1	1	Very isolated, in its own order
Bruniaceae[1]	12	*c.* 75	Near Grubbiaceae, isolated
Grubbiaceae	1	3	Near Bruniaceae, isolated
Penaeaceae	7	21	Myrtales, probably near Rhynchocalycaceae
Roridulaceae	1	2	Cornales, distinct but not isolated
Retziaceae	1	1	Very isolated, possibly near Scrophulariales
Stilbaceae	5	12	Scrophulariales, possibly near Retziaceae
Lanariaceae	1	1	Near Tecophileaceae, isolated

[1] One species in southern Natal and Pondoland.

flora is in principle no different from the absence of other 'ancient' Gondwanan taxa e.g. *Nothofagus* (Fagaceae). This genus is common in the temperate forests of South America and Australasia, with a pollen record that dates to the Upper Cretaceous (Humphries 1983). The pollen is very distinctive and is not likely to be missed in the pollen record. Not only is *Nothofagus* absent throughout Africa, but its pollen has never been found there. This suggests a primitive absence rather than an extinction, and so appears to support the 'filter dispersal' notion of the Gondwanan connection across the proto-Indian Ocean.

FOSSIL DATA

The need to establish a fossil record

Data on the contemporary distribution patterns exhibited by the Cape flora can be effectively used in the reconstruction of its origins and development through time. However, a more complete understanding of the dynamic evolution of fynbos may be achieved only through an analysis of the fossil record. The fragmented nature of the fossil record for the last 60 million years in this region is a hindrance to solving the evolutionary puzzle presented by the Cape flora. However, sufficient palaeoenvironmental and palaeoecological information 'is available now to outline some of the important features of the history of the major vegetation types found in the (fynbos) biome' (Coetzee et al. 1983).

Problems with the fossil evidence

There has been much scepticism concerning how representative are the accumulations of plant microfossils, in particular pollen. To begin with, the chances of preservation over time are slim, particularly the time scales involved in reconstructing events of evolutionary significance. Furthermore, not all microfossils have identical levels of resistance to environmental conditions at the site of accumulation. The source of the fossil pollen, whether local or regional, is often imperfectly

TABLE 5.2 The distribution and number of supra-generic taxa of the Proteaceae. The data are from Johnson and Briggs (1975).

Subfamilies Tribes	Cape Floristic Region	Africa and Madagascar	Australasia	America
Persoonioideae	0	0	7	6
Persoonieae	0	0	6	6
Bellendeneae	0	0	1	0
Proteoideae	13	14	12	0
Conospermeae	0	1	10	0
Franklandieae	0	0	2	0
Proteeae	13	13	0	0
Sphalmioideae	0	0	1	0
Carnarvonioideae	0	0	1	0
Grevilleoideae	1	2	34	8
Oriteae	0	0	2	1
Knightieae	0	0	4	0
Embothrieae	0	0	7	3
Grevilleeae	0	0	3	0
Helicieae	0	0	4	0
Macadamieae	1	2	10	4
Banksieae	0	0	4	0

known. The differential rate of pollen and spore production might present difficulties for the identification of the relative abundance of particular elements in a fossil assemblage. Such problems relate to fossil analysis in general, but for the Cape Floristic Region in particular, there are some additional factors interfering with the interpretation of the fossil record. The record itself is fragmentary. The relevant time period is represented by the last 60 million years and yet there are only four or five fossil assemblages which are located within the Cainozoic (Figure 5.7). There are thus very few 'windows' into the past and many gaps over millions of years of evolutionary history about which we know nothing. The fossil assemblages are also geographically remote from each other. There is, therefore, no guarantee that the differences observed between, for example, the fossil flora at Arnot (dated as Palaeocene) and the fossil flora at Noordhoek (dated as Miocene) are definitively a function of evolutionary change over time. A further problem is that all fossil sites are in the lowlands, thus giving little insight into the flora of the quartzitic mountains which harbour, in fynbos vegetation, the vast majority of the species of the Cape flora (Goldblatt 1978). Until the fossil record is improved through the investigation of more assemblages, the interpretation of the history of the Cape Floristic Region will remain problematic.

The fossil record from the Late Cretaceous to the Pliocene

The fossil floras of the region in relation to the evolution of the Cape flora have been reviewed by Coetzee et al. (1983) and much attention has been focused on the Tertiary. The largely gymnospermous flora of the mid-Cretaceous in southern Africa was already infiltrated by angiosperm elements by the time the Late Cretaceous deposits sampled by the Deep Sea Drilling Project 361 were laid down (McLachlan and Pieterse 1978). This trend appears to have been maintained into the Palaeocene by which time there had already

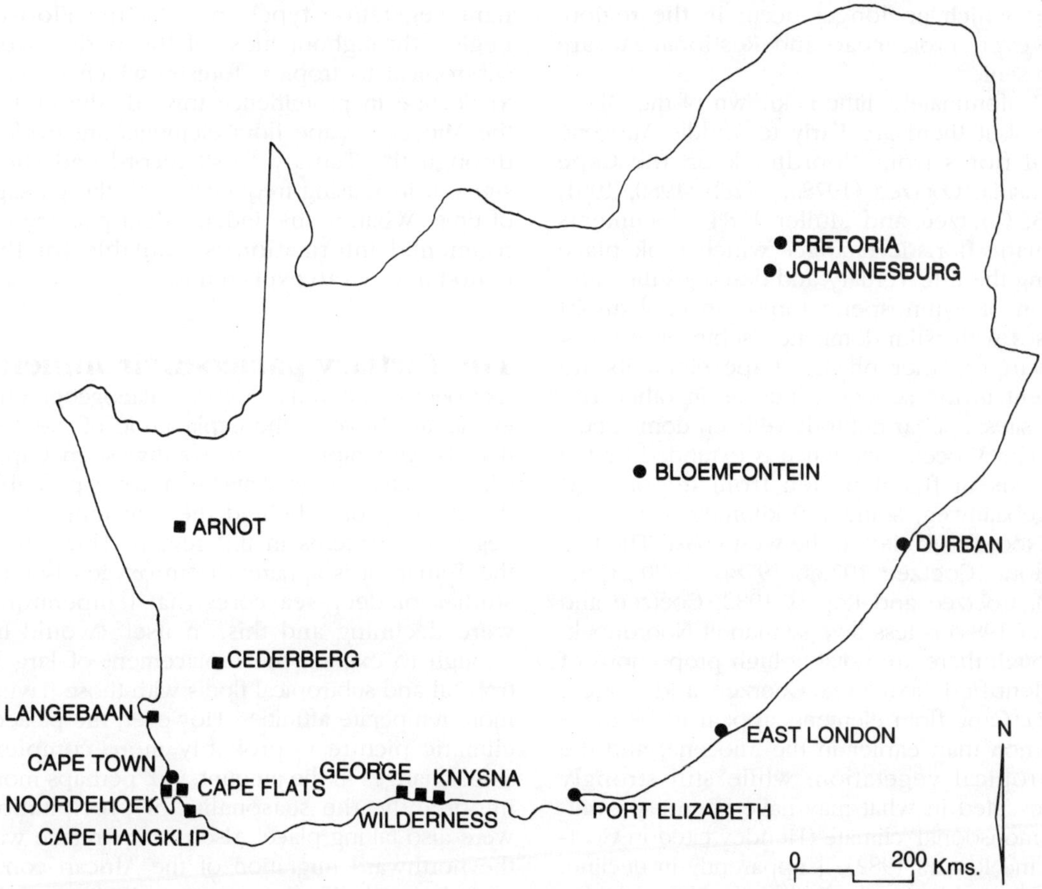

FIGURE 5.7 A map showing the plant microfossil localities (squares) mentioned in the text.

been widespread extinction of the mid-Cretaceous gymnospermous flora (Coetzee et al. 1983; Scholtz 1985). The event or series of events at the Cretaceous-Tertiary boundary which determined the pattern of global evolutionary history for the next 60 million years is not evident directly in the southern African fossil record. However, the Palaeocene flora from Arnot (Scholtz 1985) was already substantially different from that which prevailed a few million years earlier at the end of the Cretaceous. The palaeovegetation at the Arnot site, which currently lies on the ecotone between fynbos and karroid shrublands, was probably a dry subtropical forest. Although many pollen forms are present which are today extinct in the Cape, it is interesting that the Arnot flora contains several Cape flora elements, in particular the Proteaceae, Ericaceae, and Restionaceae (Scholtz 1985). Most of the affinities of this microflora, however, lie with tropical Africa.

The Eocene record is not well known from southern Africa, and the only appropriate fossil flora which has been tentatively assigned to this age is from the lignite deposits known as the Knysna Beds (Phillips 1927; Thiergart et al. 1963; Coetzee et al. 1983). A re-examination of the micro- and macro-fossils (Coetzee et al. 1983) has revealed that these beds represent a riverine facies. Again, the overall fossil assemblage appears to have been produced by a subtropical forest, in which the dominants included Casuarinaceae (now extinct in Africa) and possibly some varieties of palms (*Nipa*-type) which no longer occur in the region. However, Proteaceae and Restionaceae are represented.

Unfortunately, little is known of the Oligocene, but there are Early to Middle Miocene fossil floras from Noordhoek on the Cape Peninsula. Coetzee (1978a, 1978b, 1980, 1981, 1983; Coetzee and Muller 1984) documents dramatic floristic changes which took place during the Late Tertiary and describes the alternation of gymnosperm forests in cool, moist phases with palm-dominated subtropical forest in warmer, drier phases. Cape elements are evident in the sequence but, as in other Tertiary sites, appear not to have been dominant.

The Miocene sequence is extended by the analysis of fossil pollen from deposits at Langebaanweg, some 150 kilometres north of the Noordhoek site on the west coast. The fossil flora (Coetzee 1978a, 1978b, 1980, 1981, 1983; Coetzee and Rogers 1982; Coetzee and Muller 1984) is less diverse than at Noordhoek, although there are notably high proportions of unidentified taxa (see Coetzee and Rogers 1982). Cape flora elements appear to be more common than earlier in the Miocene, and the subtropical vegetation, while still strongly represented in what may have been a prevailing 'monsoonal' climate (Hendey cited in Coetzee and Rogers 1982), is apparently in decline, as indicated by markedly lower concentrations of palm pollen.

It is apparent that many of the tropical and subtropical rain forest elements, including such taxa as *Microcachrys* (Podocarpaceae), *Cupaniopsis* (Sapindaceae), *Casuarina* (Casuarinaceae), *Ascarinopsis* (Chloranthaceae), and Winteraceae (Coetzee and Muller 1984; Coetzee and Praglowski 1984), were also common to the other austral land masses, especially Australia (Stover and Partridge 1973; Kemp 1976), Ninetyeast Ridge (Kemp and Harris 1977), and New Zealand (Couper 1960; Mildenhall and Crosbie 1979; Mildenhall 1980). Coetzee et al. (1983) argue that this indicates a 'uniformly favourable' Miocene environment. Also, they propose a modern analogue in the form of the contemporary forests of Madagascar, where some of the taxa now extinct on the African mainland (e.g. Winteraceae) are still extant.

In summarizing the fossil pollen evidence for the Tertiary, it would appear that the dominant vegetation types in the Cape Floristic Region throughout most of the period were subtropical to tropical forests, which seemed to decline in prominence towards the end of the Miocene. Cape flora elements are evident through the Tertiary fossil record and show signs of increasing frequency with the passage of time. What, if any, independent palaeoenvironmental information is available for this period in order to explain this?

The Tertiary palaeoenvironment

As Coetzee (1983) has noted: 'palaeogeographic events are basic to the explanation of the Tertiary biogeography of the southwestern Cape'. Climatic changes in Antarctica were probably the driving force behind the development of vegetation patterns in this region. Throughout the Tertiary it is apparent from oxygen isotope studies of deep-sea cores that temperatures were declining and this, in itself, would be enough to explain the replacement of largely tropical and subtropical floras with those having more temperate affinities. However, the palaeoclimatic picture is probably more complex, since changes in the amount and, perhaps more importantly, the seasonality of precipitation were also taking place. Also of importance was the northward migration of the African continent during the Tertiary (Coetzee 1980).

Several palaeoclimatic features are especially noteworthy. Kennett (1980) describes what is essentially Tertiary climatic deterioration as a consequence of the breakdown of low latitude land connections after the fragmentation of Gondwana. Once Australia and Antarctica became separated, Antarctica was thermally isolated and located around the pole; a 5°C drop in the temperature of ocean bottom waters appears to have taken place around the Eocene/Oligocene boundary in response to this. Important palaeoclimatic events took place in the Late Tertiary. The East Antarctica ice-sheet became established around 14–11 Myr and there was a further marked increase in its size during the terminal Miocene, perhaps 5 Myr, at which stage the ice-sheet may have been up to 50% larger than today, with sea levels lowered by some 200 metres (Coetzee 1983). The glacio-eustatic drop in sea level around 6.6 to 5.2 Myr, experienced globally (Siesser and Dingle 1981), was undoubtedly a response to the increased magnitude of the Antarctic ice-sheet. This event coincided with more consistent upwelling of cold (Benguela) water off the coast of southern Africa and the establishment of west coast aridity (Siesser 1980). Coetzee (1980) points out that this end-Tertiary change in conditions would have produced an increasingly arid summer climate in the south-western Cape, a circumstance which would have favoured the Cape shrubland vegetation over the subtropical rain forest elements reliant on predictable summer rains.

Rates of extinction during the Tertiary

The literature on the fossil record indicates strongly that the end of the Tertiary was critical for the demise of the extensive subtropical forests in contemporary temperate zone regions. This is perhaps best summed up by Coetzee's (1978a) comment that the Pliocene saw 'the almost complete extermination of the Tertiary vegetation'. Although the fossil record is somewhat fragmentary, it is possible to examine rates of extinction with respect to the contemporary Cape flora by analysing the fossil occurrence of taxa. At the family level, this can be done using the list of taxa recorded for Late Cretaceous and Tertiary fossil localities published by Coetzee et al. (1983). Although this is not a complete list of all taxa recorded, it may be regarded as sufficiently representative as an indication of relative extinction rates.

Only 20% of the Late Cretaceous flora from Deep Sea Drilling Project Site 361 is still found in the contemporary Cape flora. In the Palaeocene (Arnot flora), this value had risen to 52% so that by that time more than half the flora was essentially modern. A progressive modernization is indicated through the Tertiary; the Eocene (Knysna) flora has a 62% Cape element, and the Early (Noordhoek) and Late (Noordhoek and Langebaanweg) Miocene floras 67% and 80% respectively. This is strongly suggestive of major developments in the flora of the region at the end of the Cretaceous. The differences between the Late Miocene flora and that of the present day Cape flora are, however, not as large as expected, since only one-fifth of the Late Miocene taxa had actually become extinct in the Cape by the Quaternary. The representativeness of the data set is, however, limited. For example, if the situation is viewed differently i.e. if the families present today are compared with those present during the Late Miocene, then the two floras appear to be very different.

It is also possible to examine the similarities between fossil floras from different geological epochs (Table 5.3). The analysis suggests that there was considerable turnover between the Late Cretaceous and Palaeocene and a rather more gradual turnover in taxa present during the Tertiary. The turnover between the Late Miocene and the present day cannot be computed in this way because the contemporary flora constitutes a more complete data set than any of the fossil floras. However, the absence of many contemporary families in the Late Miocene, albeit through preservation anomalies, certainly suggests substantial turnover at the end of the Tertiary.

Although these fossil data sets are somewhat incomplete, it is apparent that the flora of the south-western Cape was substantially altered on more than one occasion during the last 70 million years. The Late Cretaceous saw the demise of many tropical and subtropical taxa, while the Pliocene witnessed the introduction of many 'new' taxa which we now assign to the Cape flora.

Later events

It is unfortunate that the Pliocene and Quaternary geological times are not well represented by the fossil record, because it must have been at that time that the embryonic Cape flora, already in place by the Late Miocene, diversified prolifically. We do have some fossil clues from the terminal Quaternary in this region. Work in the Wilderness area (Martin 1968), George (Scholtz 1986), Cape Hangklip and Cape Flats (Shalke 1973), and Cederberg (Meadows and Sugden 1991) (Figure 5.7) has thrown light on the very recent dynamics of Cape vegetation. The Cederberg study reveals a continuous record of vegetation history spanning the last 14 500 years. Although this is shown to be a period of subtle shifts in plant community patterns in response to variations in environmental conditions (Deacon et al. this volume), fynbos vegetation persisted throughout the sequence. While it would be foolish to

TABLE 5.3 The rates of turnover in plant taxa between geological epochs as measured by Sörenson's (1948) coefficient of similarity (CC). The data are from Coetzee et al. (1983).

Epoch	CC[1]
Late Cretaceous to Palaeocene	0.30
Palaeocene to Eocene	0.41
Eocene to Early Miocene	0.68
Early Miocene to Late Miocene	0.87
Late Miocene to contemporary[2]	'low'

[1] CC = 2C/A + B where A is the number of taxa in epoch 1, B is the number of taxa in epoch 2, and C is the number of taxa in common.
[2] The turnover over this period was not calculated as the modern flora is not a fossil assemblage.

extrapolate these conclusions too far back into the Quaternary, the possibility that fynbos was not subjected to drastic climatic changes during the Late Quaternary may partly explain the massive speciation in the Cape Floristic Region. Meadows and Sugden (1991) tentatively suggest that greater speciation rates would have been favoured by moderate environmental fluctuations, while extinction would have dominated where the oscillations of climate were more severe. Until the fossil record is improved and has fewer gaps, speculation as to the effect of Pliocene and Quaternary environments on the Cape flora will continue (see also Cowling and Holmes this volume).

NARRATIVE HISTORY OF THE CAPE FLORA

In order to combine all the different patterns documented and discussed above, we present a narrative on the history of the Cape flora. Although narrative explanations are in dispute (Ball 1975; Humphries 1983), they may contain specific, testable hypotheses, and it is the final combination which goes beyond the data. As such, narratives are similar to evolutionary scenarios (Eldredge and Cracraft 1980).

The flora of the Cape Floristic Region dates back to the end of the Cretaceous when it was a Gondwanan flora, shared with the other southern continents. During the Tertiary, various taxa from Laurasia were added to the flora (elements such as the Rosaceae, Ericaceae, and Boraginaceae). In the first half of the Tertiary, this evolving African flora was probably not greatly diversified across the continent, as the polar climatic gradients were still poorly developed. The genera, and possibly the same species, extended from the Mediterranean Basin to the Cape. During the Mio-Pliocene the poles cooled, the climate became generally colder, and steeper climatic gradients were established from the equator to the poles. Simultaneously, vast areas in Africa were uplifted, resulting in the large domes and uplands characteristic of east, central, and southern Africa. The doming was associated with the development of the African rift valleys and the development of block uplift mountains such as the Ruwenzori Mountains and volcanic massifs such as Mt Elgon, Mt Kenya, and Mt Kilimanjaro. The final result is a temperate habitat that extended from the Cape to Ethiopia as a series of isolated islands in an otherwise tropical continent. The contemporary Cape and Afromontane floras evolved in these habitats.

At the end of the Tertiary the climate at the Cape deteriorated rapidly, changing from a tropical to a warm temperate forest climate and eventually to a summer dry mediterranean-type climate. The deterioration presented a 'bottleneck' which eliminated many taxa, leaving only a few families of mainly sclerophyllous plants which now dominate in fynbos. These sclerophylls were able to survive the sudden climatic deterioration (Cowling 1983), and extended their distribution ranges into the areas vacated by the retreating rain forests. This was accompanied by extensive speciation, leading to large numbers of very closely related species. This burst of speciation would have been especially extensive due to the complex physiography of the Cape Floristic Region, where virtually every mountain peak has a distinct climate (Linder 1985), and would represent a punctuation event in the equilibrium, as argued in the punctuated equilibrium model (Gould and Eldredge 1977). Under the drier, cooler, and more seasonal Pleistocene climatic conditions (Deacon et al. this volume), recolonization from the adjacent tropical floras was limited, especially on nutrient-poor substrates. Consequently, the products of the initial radiation of the sclerophylls persisted, as a speciose flora with a low higher-taxon diversity, lack of structural diversity, and lack of ancient phyla. Moderate climatic change and fire may subsequently have contributed to the speciation (Goldblatt 1978; Cowling et al. this volume), but cannot account for the diversity of sclerophylls relative to 'tropical' taxa, and indeed would not be necessary under this model.

This narrative scenario yields several testable predictions. Firstly, fossil floras from the Cape Floristic Region younger than the Tertiary-Pleistocene boundary would be overwhelmingly dominated by Cape flora taxa (Ericaceae, Proteaceae, Restionaceae, gnaphaloid Asteraceae, *Phylica* (Rhammnaceae), Iridaceae, arundinoid Poaceae etc.). This should be in stark contrast to the situation preceding the Tertiary-Pleistocene boundary, when tropical taxa should dominate. Although the existing fossil data are consistent with this, there are as yet no truly montane fynbos sites which sample this interface.

Secondly, nodes of diversification of fynbos genera outside the Cape Floristic Region would be in sites inimical to forest growth and extensive grasslands. The latter can support annual and biennial fires which prevent recruitment of fynbos species (Cowling and Holmes this volume; Le Maitre and Midgley this volume). This may be due to peculiar edaphic conditions, such as the quartzites of the Chimanimani Mountains in Zimbabwe (Phipps and Goodier 1962), or to harsh climatic conditions at the montane-alpine interface in the East African mountains, where fires may be limited due to the high and frequent rainfall (Hedberg 1964).

Thirdly, if the majority of the species originated from an event at the Tertiary-Pleistocene boundary, it could be expected that the species in the flora of the Cape Floristic Region would be largely well defined, with relatively few species complexes. This would be a difficult concept to quantify, but personal experience with the taxonomy of genera common to the Cape Floristic Region and tropical Africa would substantiate such an impression.

Finally, this narrative predicts that taxa of subtropical and temperate forest lineages endemic to the Cape Floristic Region should occupy marginal or azonal habitats which are moist and fire-protected, reflecting habitat conditions prior to the climatic change at the Tertiary-Pleistocene boundary (Hopper 1979). There are 43 such taxa endemic to the Cape Floristic Region. Of these nine are presumably palaeoendemic (monotypic at least in Africa), including *Brabejum* (Proteaceae), *Hartogiella* and *Maurocenia* (Celastraceae), *Heeria* (Anacardiaceae), *Hyaenanche* (Euphorbiaceae), *Laurophyllus* (Anacardiaceae), *Metrosideros* (Myrtaceae), *Platylophus* (Cunoniaceae), and *Smellophyllum* (Sapindaceae). All of these species are confined to forested and riparian sites or fire-free rocky outcrops; none occurs in fynbos (Bond and Goldblatt 1984; R M Cowling and H P Linder personal observation). Of the remaining endemics, 17 species belonging to genera such as *Cassine* and *Maytenus* (Celastraceae), *Diospyros* and *Euclea* (Ebena-ceae), *Brachylaena* (Asteraceae), *Cussonia* (Araliaceae), *Cryptocarya* (Lauraceae), *Olea* (Oleaceae), *Colpoon* (Santalaceae), *Rapanea* (Myrsinaceae), *Sterculia* (Sterculiaceae), *Podocarpus* (Podocarpaceae), and *Atalaya* (Sapindaceae), have their sister taxa in tropical Africa (H P Linder and R M Cowling unpublished data). The remaining few endemics belong to two ditypic genera — *Virgilia* (Fabaceae) and *Lachnostylis* (Euphorbiaceae) — endemic to the Cape Floristic Region. All of these endemics occur either in forests, rocky outcrops, and other marginal habitats such as coastal dunes (Cowling 1983; Bond and Goldblatt 1984). Only *Myrica* (Myricaceae, four endemics) and *Rhus* (Anacardiaceae, nine endemics) have sister species groups within the Cape Floristic Region, indicating *in situ* speciation. These two genera also have species which are components of fynbos vegetation (e.g. *Myrica humilis*, *M. kraussiana* (Killick 1969), *Rhus rimosa*, *R. scytophylla*, and *R. rosmarinifolia* (Bond and Goldblatt 1984; R M Cowling and H P Linder personal observation).

In terms of the regional patterns of richness in the Cape flora, it might be expected that in areas with richer soils, widespread tropical taxa would be able to invade, fill vacant niches, and block the Early Quaternary speciation. This is evident on the more nutrient-rich shale- and granite-derived soils of the south-western Cape lowlands, and the more nutrient-rich sandstone-derived soils of the south-eastern Cape Floristic Region (Cowling and Holmes this volume).

CONCLUSIONS

The history of the biosphere may be understood as a series of 'steady-states' in which the Darwinian uniformatism of Lyell operates, developing diversity at species level, followed by a 'crisis' which eliminates much of this diversity and allows relatively insignificant elements to radiate (Raup 1981). This is one possible way how the history of the Cape flora can be interpreted. There was a long period of diversification during the Tertiary, when elements from an ancient Gondwanan flora were mixed with elements that may have arrived during the Middle Tertiary from Eurasia. These taxa differentiated throughout Africa. With the deterioration of the climates over the whole continent and the establishment of the winter rainfall regime in the Cape Floristic Region, this ancient flora became regionalized. The flora of the Cape Floristic Region experienced a bottleneck which resulted in the loss of much taxic diversity associated with predominantly forest flora. The contraction of the forest provided new habitats which were occupied by range extensions associated with rampant speciation of the few taxa that survived the crisis. However, contrary to the steady-state/crisis model, in which taxic diversity is accumulated gradually, we would argue that a substantial proportion was generated during a massive burst of speciation associated with the range extensions of the sclerophyllous vegetation that survived the drastic climatic changes at the Tertiary-Pleistocene boundary.

ACKNOWLEDGEMENTS

A V Hall and A E van Wyk provided useful criticism on a draft of this chapter.

REFERENCES

ADAMSON R S (1958). The Cape as an ancient African flora. *The Advancement of Science* **58**, 1–10.

AXELROD D I and RAVEN P H (1978). Late cretaceous and tertiary vegetation history of Africa. In *Biogeography and Ecology of southern Africa.* (ed Werger M J A) Junk, The Hague. 79–130.

BAKER H A and OLIVER E G H (1967). *Ericas in Southern Africa.* Purnell Cape Town.

BALL I R (1975). Nature and formulation of biogeographical hypotheses. *Systematic Zoology* **24**, 407–30.

BEADLE N C W, EVANS O D and CAROLIN R C (1986). *Flora of the Sydney Region, New South Wales.* Reed Books Pty Ltd, Sydney.

BEARD J S (1968). *West Australian plants.* Surrey Beatty and Sons, New South Wales.

BOND P and GOLDBLATT P (1984). Plants of the Cape Flora. *Journal of South African Botany, Supplementary Volume* **13**.

BOND W J (1989). Describing and conserving biotic diversity. In *Biotic Diversity in Southern Africa.* (ed Huntley B J) Oxford University Press, Cape Town. 2–18.

BRUYNS P V and LINDER H P (1991). A revision of *Microloma* R.Br. (Asclepiadaceae — Asclepiadeae). *Botanische Jahrbuecher* **112**, 453–527..

BURNS-BALOGH P and FUNK V A (1986). A phylogenetic analysis of the Orchidaceae. *Smithsonian Contributions to Botany* **61**, 1–79.

CHISUMPA S M and BRUMMIT R K (1987). Taxonomic notes on tropical African species of Protea. *Kew Bulletin* **42**, 815–53.

COETZEE J A (1978a). Late Cainozoic palaeoenvironments of southern Africa. In *Antarctic Glacial History and World Palaeoenvironments.* (ed Van Zinderen Bakker E M) Balkema, Rotterdam. 115–27.

COETZEE J A (1978b). Climatic and biological changes in southwestern Africa during the late Cainozoic. *Palaeoecology of Africa* **10**, 13–29.

COETZEE J A (1980). Tertiary environmental changes along the south-western African coast. *Palaeontologia Africana* **23**, 197–203.

COETZEE J A (1981). A palynological record of very primitive angiosperms in Tertiary deposits of the south-western Cape Province, South Africa. *South African Journal of Science* **77**, 341–3.

COETZEE J A (1983). Intimations on the Tertiary vegetation of southern Africa. *Bothalia* **14**, 345–54.

COETZEE J A and MULLER J (1984). The phytogeographic significance of some extinct Gondwana pollen types from the Tertiary of the southwestern Cape (South Africa). *Annals of the Missouri Botanical Garden* **71**, 1 088–99.

COETZEE J A and PRAGLOWSKI J A (1984). Pollen evidence for the occurrence of *Casuarina* and *Myrica* in the Tertiary of South Africa. *Grana* **23**, 23–41.

COETZEE J A and ROGERS J (1982). Palynological and lithological evidence for the Miocene palaeoenvironment in the Saldanha region (South Africa). *Palaeogeography, Palaeoclimatology, Palaeoecology* **39**, 71–85.

COETZEE J A, SCHOLTZ A and DEACON H J (1983). Palynological studies and vegetation history of the fynbos. In *Fynbos palaeoecology: a preliminary synthesis* (eds Deacon H J, Hendey Q B and Lambrechts J J N) *South African National Scientific Programme Report* **75**, 156–73.

COMPTON R H (1976). The Flora of Swaziland. *Journal of South African Botany Supplementary Volume* **11**.

COUPER R A (1960). Southern hemisphere Mesozoic and Tertiary Podocarpaceae and Fagaceae and their palaeogeographic significance. *Proceedings of the Royal Society of London, Series B* **152**, 491–9.

COWLING R M (1983). Phytochorology and vegetation history in the south-eastern Cape, South Africa. *Journal of Biogeography* **10**, 393–419.

CROIZAT L, NELSON G AND ROSEN D E (1974). Centres of origin and related concepts. *Systematic Zoology* **23**, 265–87.

CUTLER D F (1969). Juncales. In *Anatomy of the Monocotyledons IV.* (ed Metcalfe C R) Clarendon Press, Oxford.

CUTLER D F (1972). Vicarious species of Restionaceae in Africa, Australia and South America. In *Taxonomy, phytogeography and evolution* (ed Valentine D H) Academic Press, London. 73–83.

DAHLGREN R M (1963). Studies on *Aspalathus*. Phytogeographical aspects. *Botaniska Notiser* **116**, 431–72.

DAHLGREN R M T, CLIFFORD H T and YEO P F (1985). *The families of the Monocotyledons.* Springer Verlag, Berlin.

DAHLGREN R and VAN WYK A E (1988). Structures and relationships of families endemic to or centered in southern Africa. *Monographs in Systematic Botany from the Missouri Botanical Garden* **25**, 1–94.

DE WIT M, JEFFREY M, BERGH H and NICHOLAYSEN L (1988). *Geological map of sectors of Gondwana reconstructed to their position ca 150 My.* Bernard Price Institute, University of Witwatersrand, Johannesburg.

DREGE J F (1844). *Zwei Pflanzengeographische Documente.* Flora 1 843 pt. 2.

DRESSLER R L (1981). *The Orchids. Natural history and classification.* Harvard University Press, Cambridge.

ELDREDGE N and CRACRAFT J (1980). *Phylogenetic patterns and the evolutionary process.* Columbia University Press New York.

GEORGE A S (1984). *The Banksia book.* Sydney, Kangaroo Press.

GIBBS RUSSELL G E and ELLIS R P (1987). Species groups in the genus *Ehrharta* (Poaceae) in southern Africa. *Bothalia* **17**, 51–65.

GOLDBLATT P (1978). An analysis of the flora of southern Africa: its characteristics, relationships, and origins. *Annals of the Missouri Botanical Garden* **65**, 369–436.

GOLDBLATT P (1981). Systematics, phylogeny and evolution of *Dietes* (Iridaceae). *Annals of the Missouri Botanical Garden* **68**, 132–53.

GOOD R (1974). *The geography of the flowering plants.* 4th edition. Longmans, London.

GOULD S J and ELDREDGE N (1977). Punctuated equilibria: the tempo and mode of evolution reconsidered. *Palaeobiology* **3**, 115–51.

GROVE A T (1983). Evolution of the physical geography of the East African Rift Valley region. In *Evolution, time and space* (eds Sims R W, Price J H and Whalley P E S) Academic Press, London. 115–55.

HARTMANN H (1991). Mesembryanthema. *Contributions from the Bolus Herbarium* **13**, 75–157.

HEDBERG O (1964). Features of Afroalpine plant ecology. *Acta Phytogeogr. Suecica* **49**, 1–144.

HEDBERG O (1965). Afroalpine flora elements. *Webbia* **19**, 519–29.

HOPPER S D (1979). Biogeographical aspects of speciation in the South-west Australian Flora. *Annual Review of Ecology and Systematics* **10**, 399–442.

HUMPHRIES C J (1983). Biogeographical explanations and the southern beeches. In *Evolution, time and space.* (eds Sims R W, Price J H and Whalley P E S) Academic Press, London. 335–65.

HUMPHRIES C J, LADIGES P Y, ROOS M and ZANDEE M (1988). Cladistic biogeography. In *Analytical biogeography. An integrated approach to the study of animal and plant distributions.* (eds Myers A A and Giller P S) Chapman and Hall, London. 371–404.

HUMPHRIES C J and PARENTI L R (1986). *Cladistic Biogeography.* Oxford Monographs on Biogeography, Clarendon Press, Oxford.

JOHNSON L A S and BRIGGS B G (1975). On the Proteaceae — the evolution and classification of a southern family. *Botanical Journal of the Linnean Society* **70**, 83–182.

JOHNSON L A S and BRIGGS B G (1981). Three old southern families — Myrtaceae, Proteaceae and Restionaceae. In *Ecological biogeography of Australia.* (ed Keast A) Junk, Utrecht. 429–69.

KEMP E M (1976). Early Tertiary pollen from Napperby, central Australia. Bureau of Mineral Resources. *Journal of Australian Geology and Geophysics* **1**, 109–14.

KEMP E M (1978). Tertiary climatic evolution and vegetation history in the Southeast Indian Ocean region. *Palaeogeography, Palaeoclimatology, Palaeoecology* **24**, 169–208.

KEMP E M and HARRIS W K (1977). *The palynology of early Tertiary sediments, Ninetyeast Ridge, Indian Ocean.* Special Papers in Palaeontology 19. Palaeontological Association, London.

KENNETT J P (1980). Palaeoceanographic and biogeographic evolution of the Southern Ocean during the Cenozoic, and Cenozoic microfossil datums. *Palaeogeography, Palaeoclimatology, Palaeoecology* **31**, 123–52.

KILLICK D J B (1969). The South African species of *Myrica. Bothalia* **10**, 5–17.

KURZWEIL H, LINDER H P and CHESSELET P (1991). The phylogeny and evolution of the *Pterygodium-Corycium* complex (Coryciinae, Orchidaceae). *Plant Systematics and Evolution* **175**, 161–223.

LEVYNS M R (1954). The genus *Muraltia. Journal of South African Botany, Supplementary volume* **2**.

LEVYNS M R (1962). Possible Antarctic elements in the South African flora. *South African Journal of Science* **58**, 237–41.

LEVYNS M R (1964). Migrations and origin of the Cape Flora. *Transactions of the Royal Society of South Africa* **37**, 85–107.

LINDER H P (1982). The orchids of the Cape Floral Region. In *Proceedings of the 10th World Orchid Conference.* (eds Stewart J and Van der Merwe C N) South African Orchid Council, Johannesburg. 59–63.

LINDER H P (1983). The historical phytogeography of the Disinae (Orchidaceae). *Bothalia* **14**, 565–70.

LINDER H P (1985). Gene flow, speciation, and species diversity patterns in a species-rich area: the Cape Flora. In *Species and speciation.* (ed Vrba E S) Transvaal Museum Monograph 4. 53–7.

LINDER H P (1987). The evolutionary history of the Poales/Restionales – a hypothesis. *Kew Bulletin* **42**, 297–318.

LINDER H P (1989). Grasses in the Cape Floristic Region: phytogeographical implications. *South African Journal of Science* **85**, 502–5.

LINDER H P (1990). On the relationship between the vegetation and floras of the Afromontane and the Cape regions of Africa. *Mitt. Bot. Inst. Hamb* **23b**. 777–90.

LINDER H P (1991). A review of the African Restionaceae. *Contributions from the Bolus Herbarium* **13**, 209–64.

LINDER H P and ELLIS R P (1990). Vegetative morphology and interfire survival strategies in the Cape Fynbos grasses. *Bothalia* **20**, 91–103.
MANNING J and LINDER H P (1991). Pollinators and evolution in *Disperis* (Orchidaceae), or why are there so many species? *South African Journal of Science* (in press).
MARTIN A R H (1968). Pollen analysis of Groenvlei Lake sediments, Knysna (South Africa). *Review of Palaeobotany and Palynology* **7**, 107–44.
McLACHLAN I R and PIETERSE E (1978). Preliminary palynological results: site 361, leg 40, Deep Sea Drilling Project. *Initial Reports of the Deep Sea Drilling Project* **4**, 857–81.
MEADOWS M E and LINDER H P (1989). A reassessment of the biogeography and vegetation history of the southern Afromontane Region. In *Biogeography of the mixed evergreen forests of southern Africa.* (ed Geldenhuys C J) Foundation for Research Development, Pretoria. 15–29.
MEADOWS M E and SUGDEN J M (1991). A vegetation history of the last 14 500 years on the Cederberg, SW Cape. *South African Journal of Science* **87**, 34–43.
MILDENHALL D C (1980). New Zealand late Cretaceous and Cenozoic plant biogeography: a contribution. *Palaeogeography, Palaeoclimatology, Palaeoecology* **31**, 197–233.
MILDENHALL D C and CROSBIE Y M (1979). Some porate pollen from the upper Tertiary of New Zealand. *New Zealand Journal of Geology and Geophysics* **22**, 499–508.
MILEWSKI A V and COWLING R M (1985). Anomalies in the plant and animal communities in similar environments at the Barrens, Western Australia, and the Caledon Coast, South Africa. *Proceedings of the Ecological Society of Australia* **14**, 199–212.
MOORE D M (1978). Dipsacaceae. In *Flowering plants of the World.* (ed Heywood V H) Oxford University Press, Oxford. 261–2.
MYERS A A and GILLER P S (eds) (1988). *Analytical Biogeography. An integrated approach to the study of animal and plant distributions.* Chapman and Hall, London.
MUNZ P A (1974). *A Flora of Southern California.* University of California Press.
NORDENSTAM B (1969). Phytogeography of the genus *Euryops* (Compositae). A contribution to the phytogeography of southern Africa. *Opera Botanica* **23**, 1–72.
OLIVER E G H, LINDER H P and ROURKE J P (1983). Geographical distribution of present-day Cape taxa and their phytogeographical significance. *Bothalia* **14**, 427–40.
PARTRIDGE T C and MAUD R R (1987). Geomorphic evolution of southern Africa since the Mesozoic. *South African Journal of Geology* **90**, 179–208.
PATTERSON C (1981). Methods of paleobiogeography. In *Vicariance biogeography, a critique.* (eds Nelson G and Rosen D E) Columbia University Press, New York. 446–500.
PHILLIPS J F V (1927). Fossil *Widdringtonia* in lignite of the Knysna Series, with a note on fossil leaves of several other species. *South African Journal of Science* **24**, 188–97.
PHIPPS J B and GOODIER R (1962). A preliminary account of the plants and ecology of the Chimanimani Mountains. *Journal of Ecology* **50**, 291–319.
PUFF C (1986). A biosystematic study of the African and Madagascan Rubiaceae-Anthospermeae. *Plant Systematics and Evolution, Supplementum* **3**.
RAUP D M (1981). What is a crisis? In *Biotic crises in ecological and evolutionary time.* (ed Nitecki M H) Academic Press, New York. 1–12.
ROSEN D E (1976). A vicariance model of Carribean biogeography. *Systematic Zoology* **24**, 431–64.
ROSEN D E (1978). Vicariant patterns and historical explanation in biogeography. *Systematic Zoology* **27**, 159–88.
ROSS J H (1972). The Flora of Natal. *Memoirs of the Botanical Survey of South Africa* **39**.
ROURKE J P (1980). *The proteas of southern Africa.* Purnell, Cape Town.
SCHALKE H J W G (1973). The Upper Quaternary of the Cape Flats area (Cape Province, South Africa). *Scripta Geologica* **15**, 1–57.
SCHOLTZ A (1985). The palynology of the upper lacustrine sediments of the Arnot Pipe, Banke, Namaqualand. *Annals of the South African Museum* **95**, 1–109.
SCHOLTZ A (1986). Palynological and palaeobotanical studies in the southern Cape. MA thesis, University of Stellenbosch.
SIESSER W G (1980). Late Miocene origin of the Benguela upwelling system off northern Namibia. *Science* **208**, 283–5.
SIESSER W G and DINGLE R V (1981). Tertiary sea level movement around southern Africa. *Journal of Geology* **89**, 83–96.
SORENSON T (1948). A method of establishing groups of equal amplitude in plant sociology based on similarity in species content. *Biologiske Skrifter Kongelige Danske Videnskabernes Selskab* **5**, 1–34.
STEBBINS G L and MAJOR J (1965). Endemism and speciation in the Californian flora. *Ecological Monographs* **35**, 1–35.
STOTT P (1981). *Historical plant geography.* George Allen and Unwin, London.
STOVER L E and PARTRIDGE A D (1973). Tertiary Late Cretaceous spores and pollen from the Gippsland Basin, southeastern Australia. *Proceedings of the Royal Society of Victoria* **85**, 273–86.
STREY R G (1981). Observations on the morphology of the Araliaceae in Southern Africa. *Journal of Dendrology* **1**, 66–83.
STRID A K (1972). Revision of the genus *Adenandra* (Rutaceae). *Opera Botanica* **32**, 1–112.
TAKHTAJAN A (1986). *Floristic regions of the world.* University of California Press, Berkeley.
THIERGART F, FRANZ U and RAUKOPF K (1963). Palynologische Untersuchungen von Tertiarkohlen und einer oberflachen Probe nahe Knysna, Sued-Afrika. *Advancing Frontiers of Plant Science* **4**, 151–78.
VAN WYK B-E (1990). A synopsis of the genus *Lotononis* (Fabaceae — Crotalarieae). *Contributions from the Bolus Herbarium* **13**, 1–292.
WANNTORP H-E, BROOKS D R, NILSSON T, NYLIN S, RONQUIST F, STEARNS S C and WEDELL N (1990). Phylogenetic approaches in ecology. *Oikos* **57**, 119–32.
WEIMARCK H (1933). Die Verbreitung einiger Afrikanisch-montanen Pflanzengruppen, I–II. *Svensk Botanisk Tidskrift* **27**, 400–19.
WEIMARCK H (1934). Monograph of the genus *Cliffortia.* Haka Ohlsson, Lund. 1–229.
WEIMARCK H (1936). Die Verbreitung einiger Afrikanisch-montanen Pflanzengruppen, III–IV. *Svensk Botanisk Tidskrift* **30**, 36–56.
WEIMARCK H (1940). Monograph of the genus *Aristea. Lunds Universitets Arsskrift* N F Avd. 2, Bd 36 nr 1, 1–140.
WEIMARCK H (1941). Phytogeographical groups, centres and intervals within the Cape Flora. *Lunds Universitets Arsskrift* N F Avd. 2, Bd 37, nr 5, 1–143.
WHITE F (1983). *The vegetation of Africa.* Unesco, Switzerland.

6

Plant reproductive ecology

D C le Maitre and J J Midgley

INTRODUCTION

The aim of this chapter is to review explanations for the frequency of some reproductive traits in fynbos plants. We adopt an explicitly comparative and evolutionary approach and focus on traits — e.g. canopy-stored seeds, the abundance of fire-killed (seeding) shrubs — which distinguish fynbos from adjacent vegetation formations in southern Africa and from other mediterranean shrublands. We use a broad definition of reproductive ecology which includes aspects of all stages, from pollination to seedling recruitment, as they are subject to significant selective pressures and are components of the regeneration niche (Grubb 1977). We ask to what extent physical factors — particularly fire and nutrient-poor soils — rather than historical accidents or unique organisms (Westoby 1988) have determined the patterns of reproductive ecology found in fynbos. The first section briefly covers the roles of soil nutrients, climates with winter rainfall and summer drought, and fire as selective pressures on plant reproductive ecology. Next we review patterns in the importance of reproductive traits of fynbos plants and compare them with those in other southern African vegetation formations and other mediterranean shrublands. Finally, we review information on the functioning and correlates of these traits and current hypotheses and explanations for their incidence. In conclusion we suggest questions for research that may lead to satisfactory explanations for the many enigmas in fynbos plant reproductive ecology.

Cape fynbos is generally distinguished from other vegetation formations in southern Africa by its rich flora, many endemic taxa, and the abundance of shrubs and evergreen graminoids (see Cowling and Holmes this volume). Fynbos also has many plant species with reproductive traits — such as seeds which are dispersed by ants — that are rare or absent in other vegetation formations on the subcontinent. Perhaps this is to be expected, since fynbos occurs on nutrient-poor soils and under a climate which ranges from winter to bimodal rainfall, a combination which is not found elsewhere in southern Africa. The theory of convergent evolution predicts that true analogues should be found in regions with similar climates and on similar soils elsewhere in the world. Pioneering studies of this theory, which originated in the northern hemisphere, concentrated on the role of climate. They suggested that the physiological constraints of summer drought select for similar growth forms and morphological traits in widely separated geographic regions, regardless of species ancestry (Mooney and Dunn 1970; Cody and Mooney 1978). Early southern hemisphere studies suggested that the influences of nutrient-poor soils, often parallel to those of climate, can override climatic factors as selective forces (Beadle 1954; Specht 1969). Syntheses of these different viewpoints show that traits such as plant physiognomy and sclerophylly and the spectrum of plant growth forms in different mediterranean shrublands can be explained as a product of ecophysiological adaptations to combinations of moisture and nutrient stress in these shrublands (Kruger et al. 1983; Tenhunen et al. 1987; Stock et al. this volume).

Although the comparative approach has provided satisfactory explanations in the field of plant ecophysiology, it has been less successful in the field of reproductive ecology. This is partly because the field is still in its infancy and partly because reproductive ecology must integrate biotic interactions which are complex and subject to co-evolutionary interactions. In addition, reproductive traits of ecologically and physiognomically convergent species may differ in similar environments. For example, many shrubs which dominate chaparral communities are killed by fire and have fire-cued seed germination and recruitment (Keeley 1986). The analogous matorral shrubs are sprouters and seed germination and seedling recruitment are not dependent on fire (Di Castri 1981; Fuentes et al. 1984, 1986). Some reproductive traits show remarkable convergence. The discovery of the importance of myrmecochory (seed dispersal by ants) in fynbos was the result of a test of convergence theory arising from Berg's (1975) findings in Australia (Slingsby and Bond 1981; Milewski and Bond 1982; Bond and Slingsby 1983). A few studies have included reproductive traits and emphasized the selective importance of physical environments (e.g. Berg 1981; Westoby et al. 1982; Kruger 1983; Beattie 1985; Keeley 1986). Most of these included only a subset of the mediterranean regions and none examined adjacent vegetation formations in southern Africa. In this review we take the first steps towards compiling a synthesis of data from all these areas. While most sprouters also produce seedlings, the terms 'seeders' and 'seeding' are used only for plants which are killed by fire and regenerate exclusively from seeds.

Nutrients, climate, and fire as selective factors

LOW SOIL NUTRIENTS

Low nutrient availability may alter the relative costs of reproductive structures and thus influence the selective advantages of different reproductive traits. Energetically expensive constituents such as lignin, lipid, or resin are inversely correlated with protein content and, therefore, more common on nutrient-poor soils (Bloom et al. 1985). Energetically expensive reproductive structures such as woody fruits, lipid-rich elaiosomes, or pollinator attractants like copious sugary nectar are relatively inexpensive in terms of nutrients. However, the cost of *nutrient* allocation to seeds is relatively high and not recoverable by the parent plant. The benefits of this high maternal investment lie in equipping seeds with the resources required during the crucial regeneration phase (Grubb 1977; Stock et al. 1990). Thus, Kuo et al. (1982) found that species in the Proteaceae allocate as much as 40% of their annual phosphorus uptake to a few high quality seeds. On nutrient-poor soils, increased parental care of seeds will give a high return at a low cost, provided the necessary structures — such as woody cones or elaiosomes — carry a low nutrient cost and provide substantial benefits through reduced predation (Bond and Breytenbach 1985; Campbell 1986). Similarly the maintenance of dormant buds in sprouters requires a supply of carbohydrates but should have a relatively low nutrient cost.

WINTER RAINFALL AND SUMMER DROUGHT

Cape fynbos experiences a variable rainfall regime. In the western Cape more than 60% of the rain occurs in winter and there is a gradual transition eastwards to a bimodal regime with spring and autumn peaks in the south-eastern Cape (Kruger 1979; Deacon et al. this volume). Winter rains are linked to the passage of extensive cold fronts and have a strong orographic component, so they are spatially predictable. This differs markedly from the convectional rains typical of summer rainfall climates where each rainstorm may be independent and has a limited areal coverage (Westoby 1980). Where rainfall patterns are not seasonal, the probability that one rainstorm will be followed by another is low. In contrast, where rainfall is seasonal, good rains after the dry season are likely to be followed by further rainfall. Thus, the onset of the wet season is temporally predictable and this promotes the evolution of appropriate cues for seed germination and growth (Westoby 1980). Except in the north-west, summer rains provide an appreciable percentage of the rainfall and alleviate summer moisture stress, especially in shallow-rooted plants (Miller et al. 1983; Specht and Moll 1983; Stock et al. this volume).

FIRE REGIME

Fynbos is a fire-prone vegetation. Recurrent fires have been a significant selective force in the evolution of plant reproductive ecology because they create open space and increase the availability of resources for recruitment (Gill 1975; Stock and Allsop this volume). Selection for exploitation of this episodic flush of resources may have led to the evolution of several reproductive traits, including fire stimulated seed release, seed germination, and flowering (Gill 1975; Kruger 1983; Zedler and Zammit 1989). Fire regimes in fynbos differ from those in adjacent grassland and savanna where fires occur at intervals of 1–4 years (Figure 6.1) in winter, and in forest where fires are rare (Granger 1984). Fynbos fire regimes resemble most closely those of chaparral, kwongan, maquis, and non-mediterranean heathlands with fires occurring mainly during summer and with fire recurrence intervals from 4–40 years, although intervals in chaparral frequently exceed 50 years (Kruger 1983; Keeley 1986, this volume).

The key components of a fire regime are the frequency, seasonality, and intensity of the fires (Gill 1975). Fire recurrence intervals and seasonality are relatively predictable (Figures 6.1, 6.2) (Van Wilgen 1984; Johnson and Van Wagner 1985) but fire intensity depends primarily on fuel moisture, air temperature, and wind speed (e.g. Cheney 1981; Van Wilgen et al. 1985), all relatively unpredictable factors. Fynbos is most likely to burn in summer, grassy fynbos in spring and autumn, and grassland in winter (Figure 6.2). Lightning is the primary natural source of ignition for fynbos fires and increases in frequency from 0.2 in the west to 3.4 ground strikes per square kilometre per year inland and in the east (Edwards 1984). The effects of fires on plants are also determined by the influence of post-fire climatic conditions (Christensen 1985). Thus, from the plant's perspective the general fire regime is predictable but the particular circumstances associated with each fire are unique (Kruger 1983; Cowling 1987).

PATTERNS

Dominance and species richness of serotinous shrubs

Many fynbos species retain their seeds in persistent fruits and release them only after fire — a trait termed serotiny (Table 6.1; Lamont et al. 1991). Seeds of these species are dispersed by wind after being released (Williams 1972; Bond 1988; Table 6.1). Most of these species are tall fire-killed shrubs of the Proteaceae (Table 6.2) which dominate the overstorey of many fynbos communities in the western and southern Cape. These species are rare in grassy fynbos, where they are confined to habitats which burn less frequently than the surrounding grasslands, and are absent from dune fynbos (Table 6.3; Cowling and Pierce 1988; see also Cowling and Holmes this volume). As rainfall decreases the proportion of serotinous species in the Proteaceae decreases, while that of passively dispersed (telochorous) species increases (Table 6.4).

Serotinous species are absent from the adjacent vegetation formations in southern Africa, even the physiognomically similar renoster shrubland (Table 6.3), except where rainfall is relatively high (Levyns 1929; Boucher and Moll 1981).

Australian kwongan has the greatest number of serotinous species (Tables 6.2, 6.5) where it is found in dominant overstorey shrubs of the Proteaceae and mid-storey Myrtaceae (Milewski and Cowling 1985; Lamont et al. 1991; Table 6.2). The degree of serotiny in three kwongan *Banksia* species increased with decreasing rainfall and vegetation height (Cowling and Lamont 1985). Serotiny is a rare trait in the other mediterranean shrublands where it is confined to gymnosperms (Table 6.2). In California these species are generally associated with nutrient-poor soils (Vogl et al. 1977) but in the Mediterranean Basin the three serotinous *Pinus* species have wide distributions (Table 6.2). In North America serotinous *Pinus* species reach their greatest importance in the evergreen coniferous forests under a cold temperate rather than a mediterranean climate (Lamont et al. 1991).

Abundance and species richness of myrmecochorous plants

Myrmecochorous species are found in a wide range of plant families and growth forms from sedges to woody shrubs and hemiparasitic herbs (Bond and Slingsby 1983; Table 6.1). Unlike the serotinous Proteaceae, myrmecochorous Proteaceae rarely dominate fynbos communities. Few myrmecochorous Proteaceae are found in dry fynbos communities (Table 6.4) and some species are confined to very moist habitats on mountain crests (Rourke 1984). Myrmecochorous mid-stratum and understorey species occur in a wide range of fynbos communities where they can account

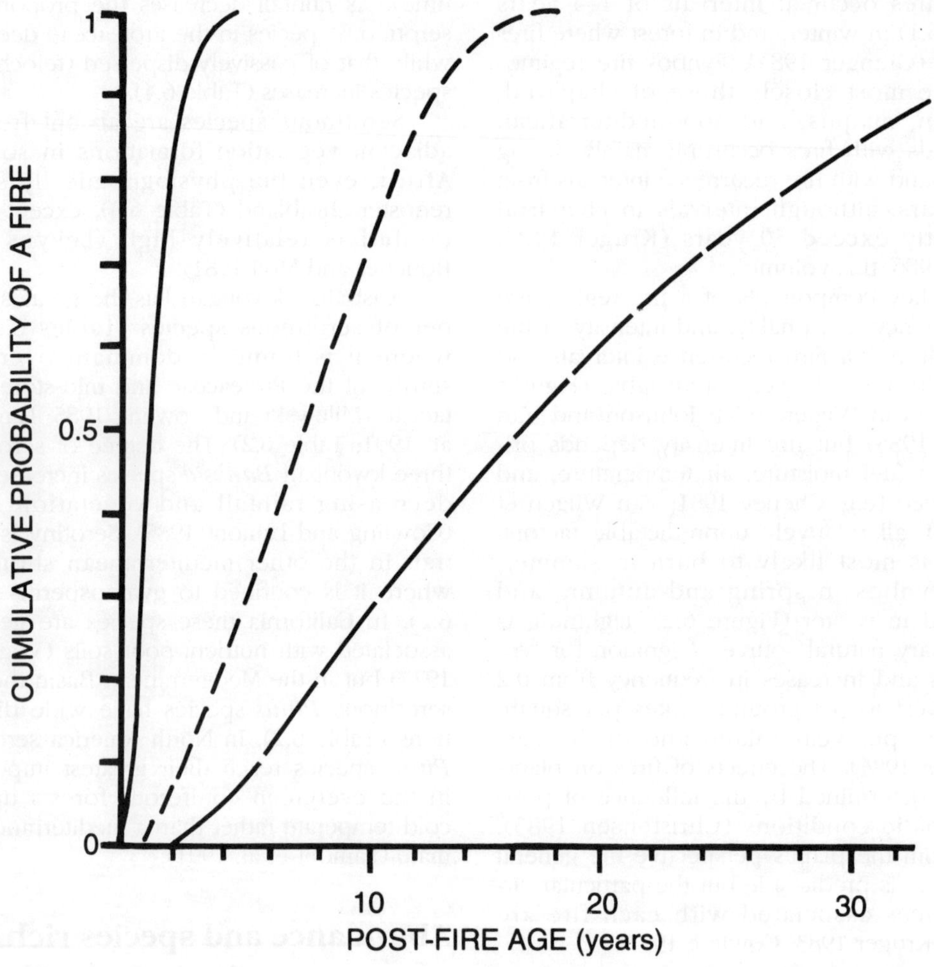

FIGURE 6.1 The generalized cumulative probability of a fire recurring at a point for western Cape mountain fynbos, grassy fynbos, and mesic grassland or savanna. The only specific investigation in southern Africa is that for the Cederberg Mountains in the western Cape by Brown et al. (1991). Recurrence intervals for mesic savanna, grassland, and grassy fynbos were inferred from comments in the literature e.g. Edwards (1984), Cowling (1984).

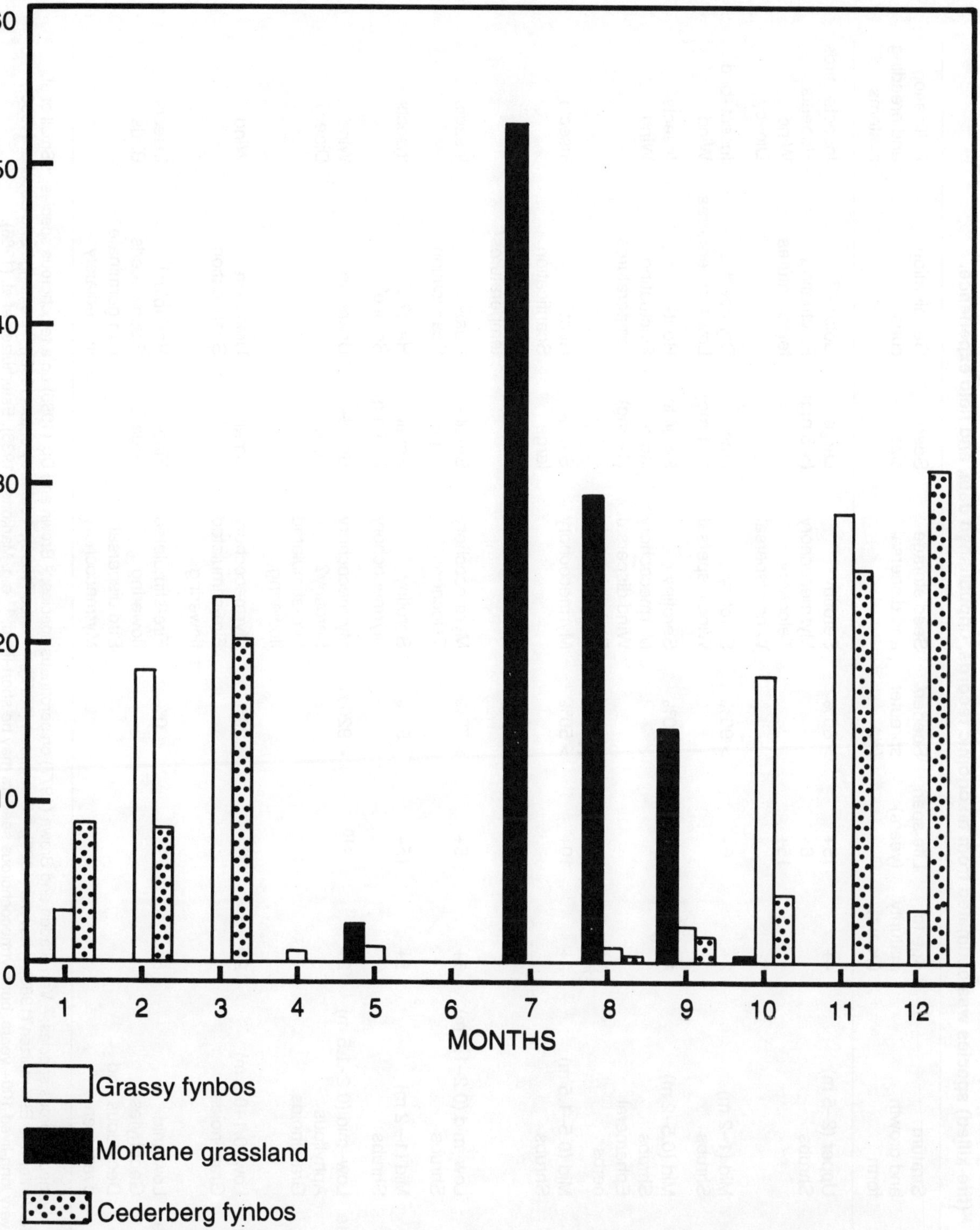

FIGURE 6.2 The distribution of the total area burnt in wildfires ignited by natural causes over the different months of the year (1 = Jan). Western Cape fynbos from data used by Brown et al. (1991), grassy fynbos from unpublished records of the Department of Environment Affairs for catchment areas where the vegetation was mainly grassy fynbos, and montane grassland from data used by Everson et al. (1988).

TABLE 6.1 The growth forms and generalized reproductive traits of typical fynbos families and genera. The percentage of seeders (fire-killed) species was estimated from taxonomic records, unpublished data, and field experience.

Taxon	Stratum and growth form	Age to maturity (years)	Life span (years)	Seeder/ sprouter (%)	Seed storage and dispersal	Seed size	Germination cues	Pollination and breeding systems
Proteaceae 320 spp.	Upper (2–5 m) Shrubs	4+	15+ 6+ 15+	> 80%	Serotiny Myrmecochory Telochory Wind dispersal	Large (> 5 mg)	Cold[1,2,3] Fluctuating temperatures	Insects–birds Rodents Wind Dioecy
Ericaceae 680 spp.	Mid (1–2 m) Shrubs	2+	6+	> 90%	Serotiny Wind dispersal	Fine (< 1 mg)	Dry heat[4] Low temperatures	Insect–bird Wind
Asteraceae 990 spp.	Mid (0.5–2 m) Shrubs Ephemeral herbs	1+	1+	> 50%	Serotiny Myrmecochory Wind dispersal Bird dispersal	Small to large (1+ mg)	Heat[5] Fluctuating temperatures	Insects Wind
Rutaceae 260 spp.	Mid (0.5–1.5 m) Shrubs	2+	10+	> 50%	Myrmecochory	Small–large	Heat[6] Scarification Fluctuating temperatures[7]	Insects
Fabaceae 640 spp.	Low–mid (0.2–1.5 m) Shrubs	1+	5+	> 75%	Myrmecochory Telochory	Small–large	Heat[8] Scarification	Insects
Bruniaceae 75 spp.	Mid (1– 2 m) Shrubs	5+	15+	50%	Serotiny Myrmecochory	Small (< 5 mg)	Heat? Smoke[9]	Insects
Restionaceae 310 spp.	Low–mid (0.2–1.5 m) Aphyllous Graminoids	2+	?[10]	< 920%	Myrmecochory Serotiny? Fire-stimulated flowering	Small–large	Unknown	Wind Dioecy
Cyperaceae 203 spp.	Low (0.1–0.5 m) Graminoids	2+	?	< 20%	Myrmecochory Fire-stimulated flowering	Small	Unknown Scarification	Wind
Orchidaceae Iridaceae Liliaceae Amaryllid.	Low–mid Geophytes Deciduous and evergreen	5+	?	0%	Fire-stimulated flowering Bird dispersal Myrmecochory	Fine–large	Variable[11] Fleshy seeds can germinate immediately	Insects Birds

[1] Brits (1986) myrmecochorous species; [2] Van Staden and Brown (1977) on serotinous species; [3] Brown and Dix (1985) on a telochorous species; [4] Small et al. (1982), Van de Venter and Esterhuizen (1988); [5] Levyns (1935); [6] Blommaert (1972); [7] Pierce (1990); [8] Jefferey et al. (1988); [9] De Lange and Boucher (1990); [10] potentially very long-lived 100+ years, but myrmecochorous seeders may be short-lived; [11] e.g. Markotter (1936), Esterhuizen et al. (1986).

TABLE 6.2 The relative dominance of serotiny in woody plants in the different mediterranean regions of the world. After Lamont et al. (1990) with habitat descriptions, abundance, and additional data from Dallimore and Jackson (1966), Naveh (1975), Vogl et al. (1977), Rundel (1981a), Beard (1984), Bell et al. (1984), and Richardson and Bond (1990).

Region and vegetation	Importance Family (genera/species[1])	Habitat and abundance
Mediterranean shrub and woodland	Cupressaceae (2/2) Pinaceae (1/3)	Primarily on low nutrient or shallow rocky soils. Present but rarely dominant under the current anthropogenic fire regime. Can form a closed forest with sufficient rainfall and less frequent fires, but usually an open forest. *Pinus* species invasive in disturbed areas and old fields.
Californian chaparral	Cupressaceae (1/9)	Primarily on low nutrient, serpentine-derived, and rocky soils, or both. *Pinus radiata* invades disturbed areas in its native habitat.
Chilean matorral	Cupressaceae (1/1)	*Austrocedrus* occurs in temperate forest remnants adjacent to matorral and is more abundant in the south. *Pinus* planted but not invasive.
Cape fynbos	Cupressaceae (1/3) Proteaceae (3/83) Bruniaceae (3/15) Ericaceae (1/1) Asteraceae (2/?)	Two *Widdringtonia* species are confined to restricted habitats, but the third, a lignotuberous sprouting species, has a wide range from the SW Cape to Malawi. Serotinous Proteaceae are widespread and frequently dominant overstorey species over extensive areas on the coastal lowlands and low to middle slopes of the mountains. The Bruniaceae and *Erica sessiliflora* are confined to relatively moist habitats. Serotinous *Pinus* species and serotinous *Hakea* (Proteaceae, Australia) species highly invasive (Richardson et al. this volume).
Australian kwongan and heath	Cupressaceae (2/6) Proteaceae (4/230) Myrtaceae (13/260)	*Callitris* and *Actinostrobus* occur in kwongan but have restricted distributions. The Proteaceae and Myrtaceae are widespread and generally dominant in the overstorey of kwongan and mallee-heath communities. Serotinous *Pinus* species not invasive.

[1] Number of species not necessarily exact but demonstrates relative species richness.

TABLE 6.3 **The percentage of flora/percentage incidence of species with different seed dispersal traits in 0.1 ha plots in different communities (see Cowling and Holmes this volume). Seeds of serotinous species are released after fire and are wind-dispersed; fine seeds like spores are probably wind-dispersed; fleshy fruit = bird-dispersed; small seeds with wings/pappus = wind-dispersed; other = the remainder including Poaceae, details are given in Appendix 1. The data were derived from species lists in Kruger (1979 Appendix 3, incidence = per cent of canopy volume) and the unpublished lists of R M Cowling (incidence = percentage cover). The abbreviations in the sites represent the regions, e.g. SW = south-western Cape.**

Site	Serotinous	Myrmecochorous	Fine seeds (< 1 mg) 'erica'	Fine seeds (< 1 mg) other	Bird-dispersed	Small+ wings, pappus	Other	Number of species
SW ericaceous fynbos	9/10	31/25	11/37	7/.6	2/.4	11/1	29/26	70
SW restioid fynbos	3/.1	26/66	10/2	2/.1	0/0	7/.2	52/32	31
SW ericaceous fynbos	8/1	32/26	10/56	6/.3	2/.2	12/.4	30/16	52
SW proteoid fynbos	8/91	24/2	11/2	6/.2	7/.3	11/.5	33/4	72
SW asteraceous fynbos	.8/7	20/20	9/1	8/1	6/4	20/10	36/57	125
SE proteoid fynbos	10/17	30/36	10/25	7/1	2/.01	20/5	21/16	41
E grassy fynbos	3/15	22/32	2/9	13/3	7/2	20/18	33/21	60
SW proteoid fynbos (lowland)	5/68	31/11	10/6	5/.1	5/.4	10/8	34/6	96
SE dune asteraceous fynbos	0/0	27/27	3/20	5/2	19/4	24/29	22/18	59
C asteraceous fynbos (dry)	0/0	13/3	0/0	19/4	3/.4	26/28	39/65	69
NW asteraceous fynbos (dry)	0/0	19/19	0/0	13/1	2/3	27/3	39/74	52
E renoster shrubland	0/0	17/3	0/0	0/0	13/7	17/40	53/50	87
W renoster shrubland	0/0	7/2	3/.3	4/.5	16/5	21/53	49/39	101
Succulent karroid shrubland	0/0	7/6	4/4	28/28	5/8	28/28	28/26	109
SE dune thicket	0/0	4/9	1/.1	11/3	44/72	12/5	28/11	93
SE succulent thicket	0/0	12/8	1/.01	16/9	30/60	15/9	26/14	98
Afromontane forest	0/0	8/.6	2/.01`	0/0	57/54	19/20	14/25	63
Protea savanna	0/0	8/6	1/.03	5/.2	21/3	23/6	42/85	96

TABLE 6.4 The percentage of Proteaceae species in different seed dispersal and re-generation categories over a moisture gradient based on the mean monthly summer rainfall and evaporation (Midgley 1987). Class 1 = rainfall < 10 mm, evaporation < 300 mm; 2 = rainfall 10–30 mm, evaporation < 200 mm; 3 = rainfall 31–40 mm, evaporation < 200 mm; 4 = rainfall > 30 mm, evaporation < 200 mm. The values are the percentages of the row totals.

Category	Moisture regime class				Number of species
	1	2	3	4	
Serotiny	3.6	28.8	45.0	22.5	111
Myrmecochory	6.5	26.2	50.0	17.3	168
Telochory	26.3	50.0	18.4	5.3	38
Seeder	9.1	29.2	42.8	18.9	264
Sprouter	1.9	34.0	52.8	11.3	53
Seed:sprout	24:1	4.3:1	4:1	8.3:1	

TABLE 6.5 The percentage of species in 0.1 ha plots with different seed bank types in analogous Agulhas Plain fynbos and Barrens (SW Australia) kwongan communities on a range of soil types (R M Cowling unpublished data).

Region	Agulhas Plain fynbos			Barrens kwongan		
Soil type	Calcareous	Laterite	Quartzite	Calcareous	Laterite	Quartzite
Soil-stored – myrmecochorous	13	30	32	42	16	29
Soil-stored – other	40	55	46	33	23	11
Canopy-stored	4	13	20	8	50	47
Neither	43	2	2	17	11	13
Number of species	30	53	41	12	50	45

for a large proportion of the cover (Table 6.3).

Adjacent vegetation formations in southern Africa have few myrmecochorous species (Table 6.3). Members of genera that cross edaphic boundaries, such as *Osteospermum* and *Zygophyllum*, are often myrmecochorous in fynbos but have winged, wind-dispersed fruits in renoster or karroid shrublands (Bond and Slingsby 1983). The relative cover of myrmecochorous species only exceeds 10% in fynbos (Table 6.3) and it is rare in temperate Afromontane forest in contrast to the northern hemisphere temperate forests (Berg 1975; Beattie 1985).

The number and proportion of myrmecochorous species in Australian kwongan are remarkably similar at both the community (Table 6.5) and the regional scale (Table 6.6). Like fynbos, kwongan communities are rarely dominated by myrmecochorous species, and myrmecochory is found in a range of families and life-forms (Berg 1975, 1981). In the other mediterranean regions myrmecochorous species are rare and most species are herbaceous forbs (Table 6.6).

Abundance and species richness of seeding woody plants

The sprouting habit is characteristic of most woody dicotyledons and is found in a range of habitats where it is not associated with recurrent fires (Wells 1969; Gill 1975; Boucher et al. 1990). Fynbos, kwongan, and chaparral are unusual in having communities dominated by seeding woody shrubs and fewer than half the species in the characteristic fynbos families sprout (Table 6.1). Virtually all the seeding *Protea* species are serotinous, as are the Australian *Banksia* species (Table 6.7). Both genera only have canopy-seed storage (serotiny) and short-lived seeds (Lamont et al. 1985), while *Leucadendron* has many non-serotinous species with soil-stored seeds (Williams 1972; Lamont et al. 1985).

There are no fynbos analogues of kwongan, chaparral, maquis, or matorral communities dominated by sprouting, tall, overstorey shrubs (Table 6.8). Most sprouting Proteaceae species in fynbos are found in the intermediate regions of a gradient of increasing summer drought (Table 6.4). There is also a marked gradient in the proportion of sprouting species from west to east in the biome, with eastern grassy fynbos species of *Protea*, *Leucadendron*, and *Leucospermum* (myrmecochorous) species all sprouters (Table 6.7). The same trends are found in the Australian genus *Banksia* (Table 6.7).

The dominant woody species in savanna (Rutherford 1981) and subtropical thicket (Cowling and Pierce 1988) are sprouters. Proteaceae in grassland and savanna all sprout except for *Protea subvestita* which is serotinous and confined to high altitude grassland and Afromontane fynbos communities which burn less frequently (Rourke 1983; Table 6.7).

The only tall, epicormically sprouting shrub in fynbos is *Protea nitida* which can form an open savanna-like community in certain habitats (Kruger 1979). In contrast, epicormic sprouting is found in all the dominant species in riverine (Van der Merwe 1966) and Afromontane forest (Phillips 1931) which are an integral part of fynbos landscapes. Many kwongan *Banksia* species are tall shrubs with epicormic sprouting (Table 6.7). It is also a common trait of species in other mediterranean regions e.g. the evergreen *Quercus* woodlands in California and Europe (Table 6.8). Species with epicormic sprouting are rare in most chaparral communities but are common in riverine forest communities in chaparral landscapes (J E Keeley personal communication 1990).

Rarity of fleshy fruited plants

Fynbos has few bird-dispersed species compared with the adjacent thicket and forest formations (Siegfried 1983; Knight 1988) and these species are not dominant (Table 6.3). Bird dispersal is a key process in savanna and forest communities (Table 6.3; Bews 1917; Phillips 1931; Johnson this volume). More than 80% of the tree species in succulent karroid shrubland, dune thicket, and Afromontane forest are bird-dispersed, as are about 50% of the shrubs and 70% of the vines (compiled from data for Table 6.3). From 3–28% of fynbos shrubs and 26–46% of renoster shrubland shrub species are bird-dispersed. Bird-dispersed forbs were only found in grassy fynbos (25% of forb species), dune asteraceous fynbos (22%), and in dune (17%) and succulent thicket (17%) communities.

TABLE 6.6 The importance of myrmecochory in the different mediterranean regions of the world. Berg (1975) estimated that there were fewer than 300 species of myrmecochorous plants in the world, excluding Australia and the Cape; virtually all of these were temperate forest floor herbs. In Australia and the Cape the species belong to both Gondwanan families e.g. Proteaceae, Restionaceae, and non-Gondwanan families e.g. Asteraceae, Euphorbiaceae, Liliaceae (*sensu lato*), with myrmecochorous species in the northern hemisphere.

Region and vegetation	Importance	Notes
Mediterranean shrub and woodlands	Families ? Species < 50 ?	Data sparse; confined to herbs; 10 species (3.6% of flora) in garrigue (see Berg 1981)
Californian chaparral	Families ? Species < 50 ?	Data sparse; described in the woody shrub *Dendromecon rigida* (Berg 1966; Bullock 1989). Potentially present in *Fremontodendron californicum* and *Montia perfoliata* (J E Keeley personal communication 1990).
Chilean matorral	Families ? Species ?	No data from Chile but some of the genera listed by Berg (1966) present.
Cape fynbos	Families 29 Species *c.* 1 200 (*c.* 14% of flora in Bond and Goldblatt 1984)	Present in several endemic families (Bond and Slingsby 1983); local floras range from 29% on infertile soils to 24% on fertile soils (Milewski and Bond 1982). Present in dicotyledons and monocotyledons, graminoids, geophytes, and woody plants; at least one invasive Australian *Acacia* species is myrmecochorous (Richardson et al. this volume). See also Table 6.3.
Australian kwongan	Families 20 Species *c.* 500 (*c.* 20% of kwongan flora)	Present in some endemic kwongan families (collated from Lamont et al. 1984 and Berg 1975). Berg (1975) estimated 1 500 species for Australia, most of them in heathlands; importance in local floras ranges from 29% on infertile soils to 13% on fertile soils (Milewski and Bond 1982). Present in a similar range of growth forms. See also Table 6.3.

TABLE 6.7 The post-fire recovery modes of the genera *Protea* (Rourke 1982), *Leucadendron* (Williams 1972), *Leucospermum* (southern Africa, Rourke 1972), and *Banksia* (Australia, George 1981). The African genera are grouped according to their occurrence in different vegetation formations. Fynbos lowland species are only those predominantly on the coastal plains in the south-western Cape. Each species was recorded separately for fynbos, grassland, savanna, and alpine heath. The values in parentheses are the percentages of species which are serotinous.

Genus	Biome	Number of species	Post-fire recovery mode (%) Seed	Sprout: Epicormic	Sprout: Rootstock
Protea[1]	Fynbos – upland	57	70.2 (80)	1.8 (0)	29.0 (13)
	– lowland	12	66.7 (100)	8.3 (0)	25.0 (0)
	– grassy	6	33.3 (100)	16.7 (0)	50.0 (100)
	Grassland	8	12.5 (100)	50.0 (0)	37.5 (0)
	Savanna	7	12.5 (100)	87.5 (0)	0.0
	Afro-alpine heath	2	50.0 (100)	0.0	50.0 (0)
Leucadendron[2]	Fynbos – upland	69	89.9 (47)	0.0	10.1 (71)
	– lowland	16	87.5 (71)	0.0	12.5 (0)
	– grassy	2	0.0	0.0	100.0 (100)
	Grassland	2	0.0	0.0	100.0 (100)
Leucospermum[3]	Fynbos – upland	29	86.2	10.3[4]	3.5
	– lowland	15	60.0	20.0	20.0
	– grassy	1	0.0	0.0	100.0
	Grassland	4	0.0	0.0	100.0
Banksia	Heath[5]	32	64.0 (100)	4.0	32.0 (88)
	Open forest	43	46.5 (95)	25.6 (36)[6]	38.1 (100)
	Tall open forest	25	40.0 (90)	36.0 (44)[6]	24.0 (100)
	Mallee	7	28.6 (100)	0.0	71.4 (100)
	Savanna	21	38.1 (100)	14.3 (0)	47.6 (100)

1 No *Protea* or *Banksia* species have persistent seed banks in the soil.
2 Many *Leucadendron* species have persistent seed banks in the soil and six species are myrmecochorous.
3 *Leucospermum* species are all myrmecochorous.
4 These species are not true epicormic sprouters but are able to continue growing from active buds protected by living leaves and bud scales. They are relatively easily killed during fires (Rourke 1972).
5 Vegetation formations after Specht (1981a) supplemented with descriptions in George (1981, 1985); most mallee species occur in communities with a heath understorey.
6 All weakly serotinous species.

TABLE 6.8 The relative dominance of different kinds of sprouting species in the different mediterranean regions of the world.

Region and vegetation	Notes
Mediterranean shrub and woodland	Most of the trees in the mediterranean region (e.g. *Quercus, Betula*) sprout either from the base or from epicormic buds (Trabaud 1981; data from Specht 1988), the notable exceptions being *Pinus* species which are serotinous (see Table 6.5). Lignotuberous woody and herbaceous sprouters are dominant or prominent in maqui and garrigue and are prominent in open shrublands (Naveh 1973; Trabaud 1981).
Californian chaparral	Many *Quercus* species are epicormic sprouters (Hanes 1971; data from Specht 1988). Vigorous sprouting tree and shrub species with seed and sprout recruitment between fires widespread and sometimes dominant, lignotuberous sprouters (e.g. *Adenostoma*) dominant in some communities (Keeley and Keeley 1988). Sprouting herbs rare except shortly after fires (Keeley 1986; Keeley et al. 1986). Epicormic sprouters common in riverine vegetation (J E Keeley personal communication 1990).
Chilean matorral	The dominant woody tree and shrub species have lignotubers and some species also sprout epicormically (Trabaud 1981; Montenegro et al. 1983; data from Specht 1988). Sprouting herbs more prominent than California (Keeley 1986).
Cape fynbos	Only *Protea nitida* is an epicormic sprouter but it does not form closed stands (Kruger 1983). This type of sprouting is found in trees of forest, thicket, and riverine forest (e.g. *Cunonia, Maytenus, Olea, Rapanea*) which can also invade fynbos in certain situations (Manders and Richardson 1991). Lignotuberous woody species common but rarely dominant except in restricted habitats (e.g. *Brunia, Widdringtonia nodiflora*). Herbaceous sprouting perennials common (e.g. Restionaceae) and remain prominent (Kruger 1983).
Australian kwongan and heath	Epicormically sprouting trees can form 'closed' woodland (e.g. *Banksia, Eucalyptus*). Lignotuberous woody sprouters (e.g. *Banksia, Hakea*) are also dominant over extensive areas (Beard 1984; data from Specht 1988). Pate et al. (1984): 13 of 429 (3%) species epicormic and 59.9% lignotuber sprouters. Sprouting herbaceous species common and remain prominent (Kruger 1983).

Species with fleshy fruits are also rare in kwongan communities but common in maquis, chaparral, and matorral communities (Table 6.9). In chaparral, bird-dispersed species (obligate sprouters of Zedler 1981; Keeley 1986) can be important in some communities e.g. *Heteromeles arbutifolia*, and seedling recruitment is linked to long fire-free intervals (Keeley and Keeley 1988). Bird dispersal is a key process in woody shrub recruitment in Chilean matorral (Fuentes et al. 1984, 1986).

Fire stimulated flowering

Fire stimulates flowering in many species of fynbos geophytes, herbaceous perennials — including grasses, sedges, and Restionaceae — and some shrub species (Kruger 1983). Most fynbos communities have many species of geophytes, Cyperaceae, and Restionaceae, and few grass species. Grasses are most abundant on granite and shale-derived soils which have a higher nutrient status (Kruger 1979; Cowling 1983a, b). Geophytes are abundant in asteraceous fynbos in the north-west and uncommon in grassy fynbos (Kruger 1979; Cowling 1983a).

In South Africa, geophytes are also prominent in renoster shrubland and succulent karroid shrubland but are relatively rare in grassland and savanna (Cowling 1983; Gibbs Russel 1987). Fire stimulates the flowering of geophytes in grassland, savanna (Frost 1984), and renoster shrubland (Levyns 1929) but fire is not required for flowering in succulent karoo or dry asteraceous fynbos.

Perennial herbs with fire stimulated flowering are rarely abundant in chaparral (Keeley 1986; Zedler and Zammit 1989), maquis, or matorral and fires may even inhibit the flowering of perennial herbs in maquis and matorral (Trabaud 1981). Kwongan communities have a diverse flora of geophytes but species with bulbs or corms — notably the Iridaceae — are rare compared with fynbos (Pate and Dixon 1982; Milewski 1983). Fynbos also has many hysteranthous geophytes — i.e. flowering at a time when functional leaves are absent — but these species are rare in kwongan (Pate and Dixon 1982). Fynbos also has a richer flora of Cyperaceae (203 spp.) and Restionaceae (310 spp.) than kwongan with 70 and 32 species respectively (Lamont et al. 1984). Species of these families, especially Restionaceae, are also more prominent and widespread in fynbos than in kwongan communities (Levyns 1961; Milewski and Cowling 1985). Fire stimulated flowering of grasses is found in all mediterranean shrublands (Gill and Groves 1981). Grasses are most abundant in maquis, chaparral, and matorral but are relatively rare in kwongan (Trabaud 1981; Kruger 1983; Beard 1984).

Fire stimulated germination

Species with seed germination which is directly or indirectly stimulated by fire are prominent in the flora of chaparral, fynbos, and kwongan but are apparently rare in maquis and matorral (Trabaud 1981; Kruger 1983). Fire annuals and seed-regenerating subshrubs are prominent in chaparral but are generally rare after fires in maquis, although occasionally abundant in frequently burnt areas e.g. Cistaceae (Trabaud 1981; Kruger 1983; Keeley 1986).

Fynbos has a number of short-lived fire ephemerals but few if any annuals with, or without, fire stimulated germination. Annuals are most abundant in dry asteraceous fynbos in the north-western fynbos biome where most species are not fire dependent (Kruger 1979). Annuals are rare in other vegetation formations in South Africa except in the succulent karoo and desert biomes (Gibbs Russel 1987). The abundance of annuals in fynbos increases after fires (Kruger 1979, 1983) but fire stimulated germination of herbaceous species has not been demonstrated experimentally.

Abundance of bird-, mammal-, and wind-pollinated plants

Seventy-five per cent of all the bird-pollinated species in South Africa are found in the Cape flora, where about 430 species (5%) — judging by morphological criteria — are bird-pollinated (Rebelo 1987). A large proportion of the species are in the genus *Erica* (96 spp.) and the Proteaceae (79 spp.) (Rebelo 1987). Bird-pollinated Proteaceae and Ericaceae are most prominent in fynbos communities on the coastal plains, the south-western mountains, and the upper regions of the inland mountains.

Five (3.5%) of 141 Chilean matorral

TABLE 6.9 The prominence of plant species with fleshy fruits in the different mediterranean regions of the world.

Region and vegetation	Notes
Mediterranean shrub and woodland	Fleshy fruits are common in woody plants (e.g. *Arbutus, Juniperus, Olea*) and also in 'semi-woodys' (e.g. *Smilax*); 9% of garrigue species (Willson et al. 1989), 40% of *Quercus ilex* forest species (Braun-Blanquet 1936 from Debussche et al. 1982); Debussche et al. (1982) estimate 110 species in provincial flora; 14% of Israeli shrubland flora (Izhaki and Safriel 1985). Müller (1933 from Willson et al. 1989): 68% of tree species, 89% of vines, 24% of shrub species and 0% of forbs in garrigue.
Californian chaparral	Fleshy fruits are common in chaparral shrubs. Percentage of woody shrub species with < 15 mm wide fleshy fruits was 24% in coastal sage, 32% in chaparral, and 52% in evergreen conifer forest (Hoffman et al. 1989). The 'obligate resprouters' of Zedler (1981), Keeley (1986) are bird-dispersed (e.g. *Heteromeles arbutifolia*) and can be dominant in some chaparral communities (e.g. Keeley and Keeley 1988).
Chilean matorral	Data at the flora level not available. Bird dispersal is common in the dominant shrubs and trees (Fuentes et al. 1984, 1986). Species with < 15 mm wide fleshy fruits comprised 20% of species in coast matorral, 31% in interior matorral, and 65% in evergreen broad-leaved forest (Hoffman et al. 1989).
Cape fynbos	Few species are bird-dispersed (Cowling and Holmes this volume) but bird dispersal is common in thicket, forest, and savanna formations (Table 6.4). The aerial parasites *Cassytha* and *Viscum* are bird-dispersed. Somewhat fleshy fruits are found in *Anomalanthus* (Ericaceae), Grubbiaceae, *Cliffortia baccans*, *Colpoon* (Bean 1990) but the dispersers are unknown. Bird dispersal is a key process in invasions by exotic Australian *Acacia* species (Richardson et al. this volume).
Australian kwongan and heath	Few species are bird-dispersed: 12% in eastern heathlands (Clifford and Drake 1979), Victorian heaths 2–5% (from Willson et al. 1989) and 6.5% of 429 kwongan species (Pate et al. 1984). One *Cassytha* species (1%) versus 13 species (32%) on infertile and fertile soils, respectively, in Western Australian Barrens kwongan (Milewski and Bond 1982). No trees at this site were bird-dispersed (Milewski and Cowling 1985). Bird dispersal is important in the invasion of the Cape species *Chrysanthemoides monilifera* (Weiss and Milton 1984).

species and 19 (10%) of 193 chaparral species are bird-pollinated (Moldenke 1977). Fifteen per cent of the temperate flora of western Australia is bird-pollinated compared with 7.4% of its tropical flora and 5.6% of its desert flora (Keighery 1982). However, none of the dominant woody shrubs of chaparral and matorral are bird-pollinated, unlike the Proteaceae, Ericaceae, and Myrtaceae which dominate fynbos or kwongan communities (Clifford and Drake 1979; Collins and Rebelo 1987). In South Africa bird-pollinated species also occur in karroid shrubland, renoster shrubland, thicket, and forest (Rebelo 1987) but these species are rarely abundant.

Small mammals (four rodent species and one insectivore, *Elephantulus*) act as pollen vectors for some 36 fynbos shrub species, mainly from the genus *Protea* (Wiens et al. 1983; Rebelo and Breytenbach 1987; Johnson this volume; Table 6.1). A convergent syndrome is found in Australian kwongan and heathland where small marsupials act as pollinators (Wiens et al. 1983; Collins and Rebelo 1987; Rebelo and Breytenbach 1987; Johnson this volume). Rodent pollination is apparently absent elsewhere in southern Africa and chaparral and matorral (Rebelo and Breytenbach 1987).

About 12% of the Cape flora is wind-pollinated and the species range from tall *Leucadendron* shrubs to the aphyllous reeds of the Restionaceae and the graminoid Cyperaceae (Table 6.1) (Koutnik 1987). This is slightly lower than the percentages in Chile and about equal with California (excluding conifers) (Koutnik 1987). No Australian heathland Proteaceae are wind-pollinated (Collins and Rebelo 1987) and kwongan has few Restionaceae (32 spp.) and Cyperaceae (70 spp.) (Lamont et al. 1984) compared with fynbos (Table 6.1).

Breeding systems

Approximately 6.6% of the Cape flora, 9.8% of the Cape Peninsula flora, and 11.7% of the Cape Hangklip flora is dioecious (Steiner 1987, 1988). This is higher than the floras of south Australia (3.9%), kwongan (4.4%), and California (2.8%) (Steiner 1988). More than half the dioecious species in fynbos belong to the single family Restionaceae, and 15% come from two genera, *Leucadendron* and *Aulax* (Proteaceae). Dioecious *Leucadendron* species and Restionaceae are dominant in many fynbos communities (Kruger 1979) but dioecious species are apparently rare in kwongan and in chaparral.

Heterostyly is relatively common in fynbos with 162 species in eight genera (Ornduff 1974; Steiner 1987). Enantiomorphy is a rare form of heteromorphy where plants have flowers with left and right handed forms. It is known from only 10 genera in five families worldwide with four of these genera (three families) being represented in the Cape flora (Ornduff 1974; Dulberger and Ornduff 1980).

SUMMARY OF REPRODUCTIVE DIFFERENCES AND SIMILARITIES

Fynbos and kwongan differ from other mediterranean shrublands — and fynbos from other vegetation formations in southern Africa — in the following ways:

- They have many serotinous and myrmecochorous plant species, and serotinous species are widespread and frequently dominant.
- Myrmecochory is most common in woody rather than herbaceous plants.
- They have few bird-dispersed, woody species and these species are rarely abundant.
- They have woody sprouting species with fire stimulated flowering and these species can be abundant.
- They have a relatively poor flora of fire ephemeral herbaceous species, especially annuals, with fire stimulated germination compared with chaparral.
- They have few Poaceae but a diverse and abundant flora of perennial herbs, particularly geophytes, with fire stimulated flowering.
- They have many small non-flying mammal-, bird- and wind-pollinated species (especially the last in fynbos).

Fynbos differs from kwongan in the following ways:

- Kwongan has more serotinous species versus a greater range of families with serotinous species in fynbos.
- Many kwongan Epacridaceae are fleshy fruited and possibly vertebrate-dispersed (R M Cowling unpublished data), although some species are apparently myrmecochorous in contrast to the closely related Cape Ericaceae-Ericeae with only the genus *Acrostemon* having somewhat fleshy fruit.

• Kwongan communities have more serotinous woody sprouting species.
• Kwongan has few hysteranthous geophytes.
• Only Cape Proteaceae have dioecious and wind-pollinated genera.

The following are unique to fynbos:
• The rarity of sprouting shrubs and trees in the overstorey — notably those capable of sprouting epicormically — and the absence of communities with a closed canopy of these species.
• The prominence of wind pollination and dioecy, largely due to the high species richness and abundance of Restionaceae.
• The number of heterostylous and enantiomorphic species.

EXPLANATIONS AND ENIGMAS

Serotiny and myrmecochory

Serotiny leads to accumulation of a seed bank in the canopy, protects seeds from both predispersal predators and fire, and times seed release (*en masse*) into the post-fire environment (Lamont et al. 1991). Seeds of serotinous species do not require fires, or post-fire environmental conditions for germination and germinate best under cold conditions (< 15°C) (Deall and Brown 1981; Mitchell et al. 1986; Table 6.1). This differs from classical cold stratification where seeds germinate under relatively warm conditions after a period of cold stratification — usually at less than 5°C (Harper 1977). Seeds of telochorous and myrmecochorous species require fires for optimal germination both to scarify the hard seeds and to provide appropriate soil temperature regimes for germination (Brits 1986). Therefore — unlike myrmecochorous and telochorous species — serotinous species do not have seed banks that persist after they die. They must maintain their position in the canopy and be relatively long-lived in order to survive from fire to fire (Kruger 1983). Mortality rates of both *Protea neriifolia* (serotinous) and *Orothamnus zeyheri* (myrmecochorous) increase with increasing age and their lifespans are about 50 and 30 years respectively (Figure 6.3). This fits with observations that serotinous fynbos Proteaceae are generally taller (Table 6.10), more robust, and longer lived than telochorous and myrmecochorous species. These lifespans and mortality patterns differ markedly from those observed in *Ceanothus* and *Banksia* species where mortality rates were constant or decreased with increasing age (Figure 6.3). There was no evidence of density-dependent mortality in the only study of serotinous Cape Proteaceae (Kruger 1987) but it had significant effects on, for example, *Ceanothus megacarpus* in chaparral (Schlesinger et al. 1982) and apparently also on *Banksia ornata* (Specht 1981b) in Australian heath. Relatively short lifespans are also found in myrmecochorous species in chaparral e.g. *Dendromecon rigida* (Bullock 1989) and kwongan e.g. *Grevillea leucopteris* (Lamont 1982). Few, if any, of the short-lived fire ephemeral species are myrmecochorous in fynbos (Bond and Slingsby 1983) but the Gyrostemonaceae in kwongan are (Berg 1975; Pate et al. 1985). The links between these traits are evidence that serotiny and myrmecochory are associated with distinct and alternative life histories which have evolved in response to similar selective factors.

NUTRIENT-POOR SOILS AND THE RELATIVE COSTS OF SEEDS

Woody serotinous structures are constructed from carbohydrate-rich compounds and have a low nutrient cost compared with the costly investment in protein-rich seed embryos (Esler et al. 1989). The returns from this low cost investment are two-fold: seeds are protected from heat during fires (e.g. Beaufait 1960) and seeds are partially protected from pre-dispersal granivores (Scott 1982; Coetzee and Giliomee 1987; Johnson this volume). Lipid-rich elaiosomes are also rich in carbohydrates and low in nutrients compared with embryos (Milewski and Bond 1982; Westoby et al. 1982). The annual seed production of mature shrubs of the Proteaceae seems to be markedly lower than that of analogous chaparral species (Keeley 1986; Table 6.11), except in the case of *Paranomus bracteolaris* (Table 6.1; Le Maitre 1988). Most nutrient allocation studies have been made on species with large seeds. It is not clear whether similar arguments apply to species with small to minute seeds such as kwongan Myrtaceae and fynbos Bruniaceae and *Erica sessiliflora*. As these species seem to produce many seeds, the net cost to the parent might be the same.

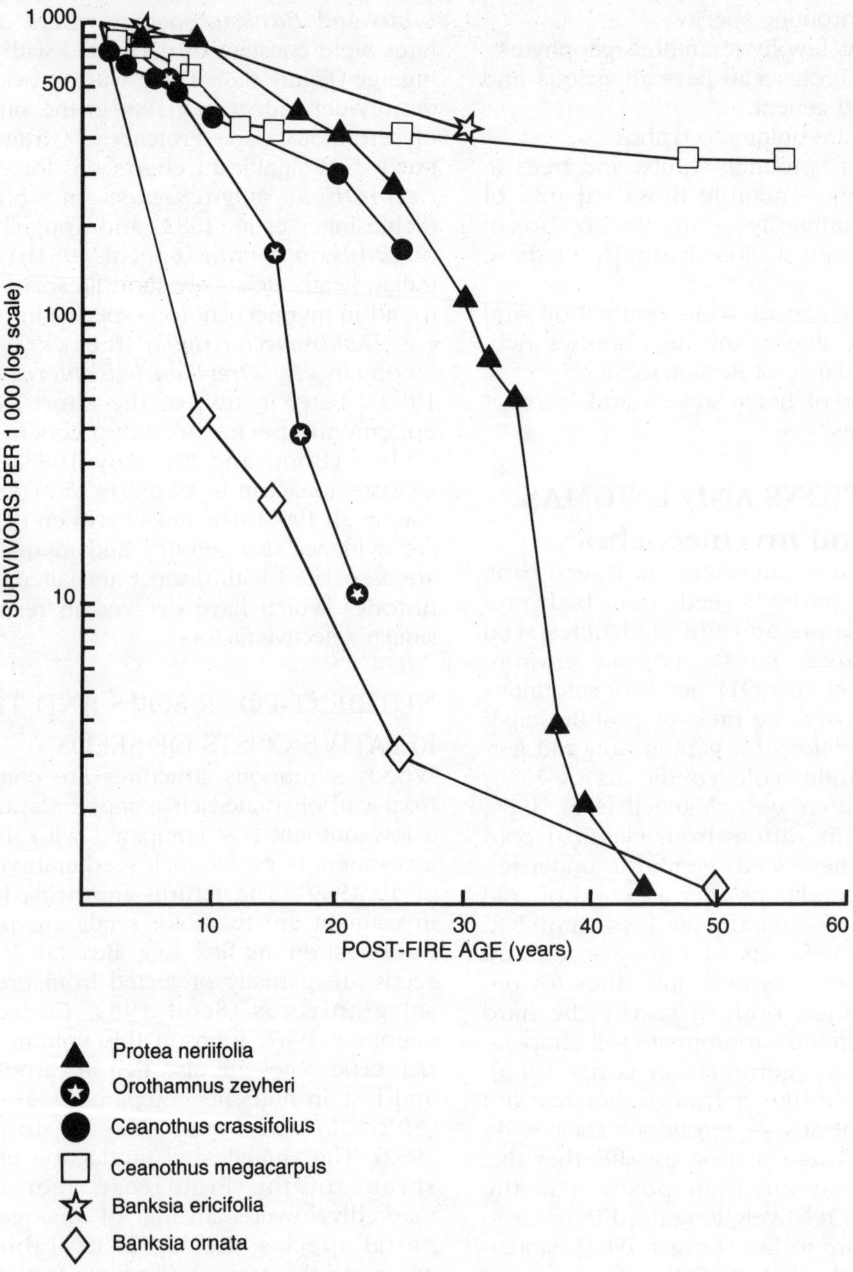

FIGURE 6.3 The survivorship curves for obligate seed-regenerating woody shrub species from the different mediterranean regions of the world. Fynbos: *Orothamnus zeyheri* from unpublished records, Jonkershoek FRC; *Protea neriifolia* from Higgins and Manders (1991) and additional unpublished data, Jonkershoek FRC. Australian heath: *Banksia ornata* from Specht (1981); *B. ericifolia* from Bradstock and O'Connell (1988) and Morris and Myerscough (1988). Californian chaparral: *Ceanothus megacarpus* from Schlesinger et al. (1982) and Montygierd-Loyba and Keeley (1987); *C. crassifolius* from Horton and Kraebel (1955). Note that the survivorship axis uses a log scale.

TABLE 6.10 The percentage of Proteaceae species in different height classes divided according to seed dispersal and regeneration categories (Midgley 1987). The values are the percentages of row totals except for the ratio of seeder to sprouter species.

Category	Height class (m) < 0.5	0.5–1	1.1–2.4	2.5–5	> 5	Number of Species
Serotiny	10.7	16.7	32.1	34.5	6.0	84
Myrmecochory	24.5	36.1	25.2	14.3	0.0	147
Telochory	3.2	12.9	71.0	12.9	0.0	31
Seeder	8.2	30.3	36.1	23.6	1.9	208
Sprouter	53.7	14.8	20.4	9.3	1.8	54
Seed:sprout	0.6:1	7.8:1	6.8:1	9.8:1	4.0:1	

However, the costs to the seedling are high because small seeds can only provide a limited proportion of the nutrients needed for establishment. In fynbos and kwongan most, if not all, small-seeded species have some form of mycorrhizal symbiosis. This could be an alternative method of providing for offspring at a low cost, per seed, to the parent. Small seeds may also escape predation through their small size (Thompson 1987).

SEROTINY: LOW NUTRIENTS AND WIND-DISPERSED SEEDS

Seeds of serotinous species are released into relatively open post-fire environments where long-range dispersal is possible. Wind can tumble *Protea* seeds for 100 metres or more until they are trapped by vegetation remnants or rocky outcrops. Winged and hairy *Leucadendron* seeds are generally dispersed for shorter distances (Bond 1988). It is unlikely that wind dispersal in serotinous Proteaceae evolved as a way of minimizing competition from the parent plant (Janzen 1971) as most shrubs are killed by fires. But wind dispersal could have evolved in response to strong selection to reduce seed predation — as argued above — by escaping from the seed shadow around parent plants. Long-range seed dispersal may also be advantageous for colonizing new sites, or recolonizing sites where local extinction has occurred (Bond 1988; Lamont et al. 1991). Wind dispersal has also been a major factor in invasions by alien *Pinus* and *Hakea* species (Richardson et al. this volume). However, not all serotinous species are wind-dispersed and many non-serotinous species are wind-dispersed (Lamont et al. 1991). As seed wings and hairs which facilitate dispersal also have a low nutrient cost (Wes-toby et al. 1982), wind dispersal, at least in *Protea* species, may simply be a legacy of the evolution of serotiny in lineages with wind-dispersed, non-serotinous ancestors.

MYRMECOCHORY: LOW NUTRIENTS AND DIRECTED DISPERSAL

Myrmecochory should be favoured in nutrient-poor environments because of the advantages of directed dispersal to nutrient-rich ant nests (Beattie 1985). Research in kwongan and fynbos has not supported these predictions because soil nutrient levels in ant nests, or those in which seedlings of myrmecochores are found, are not higher and may even be lower than those in adjacent soils (Rice and Westoby 1986; Bond and Stock 1990; Johnson this volume). Seeds of a myrmecochorous *Leucospermum* were dispersed to open patches by ants because their nests were located there, while wind-dispersed seeds of a serotinous *Protea* tended to cluster around vegetation remnants (Yeaton and Bond 1990). Directed dispersal patterns like these could promote the coexistence of myrmecochorous and serotinous Proteaceae (Yeaton and Bond 1990; see also Cowling and Gxaba 1990).

TABLE 6.11 The viable seed production, seed banks, and seedling recruitment in species from chaparral, kwongan, and fynbos. Data for chaparral: Keeley (1977), Keeley and Zedler (1978); kwongan: Enright and Lamont (1989), Lamont and Barker (1988), Lamont et al. (1990), Cowling et al. (1986), Cowling and Lamont (1987), Bradstock and O'Connel (1988); fynbos: Le Maitre (1988, unpublished data). Data for recruitment are for the first winter after the fire and, where possible, for autumn fires which result in the smallest losses to seed predators (Bond 1984). All the kwongan species and *Protea neriifolia* are serotinous, the remainder have soil-stored seeds.

Region and species (regeneration s = seed v = sprout)	Stand age (yrs)	Stand density (per ha)	Annual seed crop per plant	Seed bank/ plant	Seedlings per seed	Seedlings per parent	First summer mortality (%)
Chaparral							
Ceanothus greggii (s)	90	7 900	3 758	332	0.006	1.82	2–5
C. leucodermis (v)	90	3 540	3 518	236	0.003	0.58	0–6
Arctostaphylos glauca (s)	90	860	3 322	2 174	0.005	9.88	33–44
A. glandulosa (v)	90	1 560	4 678	1 846	0.001	0.96	55–71
Kwongan							
Banksia hookeriana (s)	15	—	—	362	0.14	73.5	68
B. leptophylla (s)	15	—	—	217	0.44	96.0	10
B. attenuata (v)	15	—	—	58	0.25	14.6	60
B. menziesii (v)	15	—	—	5	0.20	1.0	67
B. candolleana (v)	15	—	—	31	0.10	3.2	100
B. burdettii (s)	16	—	240	832	.01–.03	12–26	—
B. attenuata (v)	15	—	11	56	0.21	12	58
B. leptophylla (s)	15	—	147	1 344	0.11	7	95
B. menziesii (v)	15	—	6	2	0.05	1	100
B. prionotes (s)	15	—	33	124	0.18	22	14
B. ericifolia (s)	13	—	—	18	0.23	0.5–9.8	—
Petrophile pulchella (s)	13	—	—	31	0.17	0.6–7.2	—
Fynbos							
Protea neriifolia (s)	27	4 400	9	44	0.27	18.71	11
Leucadendron pubescens[1] (s)	16	1 690	181	272	0.013	3.55	10
Paranomus bracteolaris[1] (s)	16	465	3 086	4 630	0.001	6.60	15

[1] Both species have soil-stored seeds. The seed banks in the soil were estimated as 1.5 times the annual crop as few seeds appear to persist (Le Maitre 1988). There were 5–20 times as many seedlings inside herbivore exclosures in this study (Le Maitre 1988).

CONSTRAINTS ON REPRODUCTION BETWEEN FIRES

In all mediterranean shrublands reproduction of most of the dominant woody shrubs between fires is limited by a variety of factors (Kruger 1983; Keeley 1986). Thus, there is a strong selection for persistent seed banks in order to maximize recruitment in open environments after fires (Givnish 1981; Lamont et al. 1991). Fuel accumulation rates in nutrient-poor environments are slow relative to those in grasslands, resulting in relatively long minimum (4+ years) and modal intervals (8–20+ years) between fires (Gill and Groves 1981; Kruger 1983; Keeley 1986; Figure 6.1). These longer intervals allow slow maturing woody seeding species to accumulate sufficient seeds for population replacement, compared with frequently burnt habitats like grasslands where seeding Proteaceae are generally absent (Table 6.7).

Additional support for the hypothesis that intermediate fire recurrence intervals promote the evolution of seed banks comes from studies of temperate forest *Pinus* species in which the degree of serotiny is polymorphic and genetically determined (Perry and Lotan 1979; Givnish 1981). Non-serotiny is favoured where there is gap by gap replacement and serotiny where stand-replacing fires recur within the lifespan of the species. The same trends are found in Proteaceae where the importance and number of species with serotiny and myrmecochory are highest in fynbos and kwongan and markedly lower or absent in environments with shorter or longer intervals between fires e.g. savanna and forest (Tables 6.4, 6.8). Seedling recruitment of myrmecochorous species is associated with cleanly burnt sites, suggesting that stand-replacing fires could also favour myrmecochory.

WINTER RAINFALL AND SUMMER DROUGHT

In mediterranean climates fires are most likely during summer and autumn (Figure 6.2) and they are, predictably, followed by cool, wet conditions which provide reliable cues for seed germination (Westoby 1980). Seeds of serotinous species germinate readily once released, a trait which could result in recruitment failure if droughts follow good rains (Zammit and Westoby 1987; Le Maitre 1990). As total rainfall decreases, it generally becomes more erratic and unreliable. This increases the risk of mortality during germination and recruitment and may explain why few serotinous Proteaceae occur in arid environments (Table 6.4). Seeds of serotinous species which are released after winter or spring fires may only germinate in the following winter but are then exposed to intense predation (Bond 1984). Serotinous pines dominate in areas which also have a strongly seasonal climate with intermediate fire intervals; fires occur in summer and seed germination cues restrict germination to spring (Lamont et al. 1991).

Soil-stored seeds are hard-coated and relatively safe from predation in the soil (Thompson 1987), so deferred germination is not associated with a high risk of mortality. This reduces the risk of failure because seeds can remain dormant in the soil and only germinate during wet winters. Seeds of serotinous species also germinate on the surface and have to establish a root system. Soil-stored seeds germinate below the surface and roots are established before the cotyledons emerge. These traits favour species with soil-stored seeds in arid environments but these hypotheses cannot explain why there is a greater proportion of telochorous species than myrmecochorous species in arid environments (Table 6.4), as seed of both kinds are stored in the soil.

FIRE INTENSITY

Canopy-stored seeds are generally well protected and appear to experience little mortality during fires compared with soil-stored seeds (Lamont et al. 1991). However, seeds of myrmecochorous species are apparently buried at greater depths than those of telochorous species (Manders 1990a). The only soil-stored seeds which survived an exceptionally intense fire in fynbos were those of myrmecochorous species (Richardson and Van Wilgen 1986). The net recruitment (established seedlings per seed in the seed banks) of chaparral and fynbos shrub species with soil-stored seeds is an order of magnitude lower than the net recruitment of serotinous kwongan and fynbos shrub species (Table

6.11). This suggests that seed losses during recruitment are lower for serotinous species than species with soil-stored seeds. The higher net recruitment rates of serotinous species (Table 6.11) may explain their greater importance and abundance on nutrient-poor soils where seed production is more expensive.

Fires in chaparral are more intense than in fynbos but the fuel is concentrated in the shrub canopy in contrast to the high understorey fuel loadings in fynbos (B W van Wilgen unpublished data). Thus, intense ground fires may select for serotiny with seeds in the canopy, or myrmecochory with active seed burial, rather than telochory in fynbos. The intense canopy fires in chaparral could kill seeds stored in the canopy and would favour species with seeds in the soil.

PREDATOR AVOIDANCE AND SATIATION

Both myrmecochory and serotiny minimize post-dispersal predation (O'Dowd and Hay 1980; Bond 1984; O'Dowd and Gill 1984; Beattie 1985; Bond and Breytenbach 1985; Johnson this volume), although the *modus operandi* differs: serotiny maximizes predator avoidance between — and satiation after — fires, while myrmecochory maximizes predator avoidance between fires. Predation pressure could be a significant factor in nutrient-poor environments because the high nutrient cost of seeds limits seed production (Kuo et al. 1982) and provides strong selection for traits that minimize seed predation. The evidence for the predation hypothesis is compelling but many species of *Leucadendron* are neither serotinous nor myrmecochorous (Tables 6.5, 6.8). Granivores feed intensively on seeds of these species (Le Maitre 1988) and they are preferred to serotinous species in cafetaria trials (Midgley 1989). The predictions of the predation hypothesis are also contradicted by the virtual absence of traits such as myrmecochory which could reduce the high rates of seed predation recorded in chaparral species (Mills and Kummerow 1989). Studies on the costs of seed production and seed predation in ecologically similar species in different environments are needed to resolve these disparities.

Seeders versus sprouters

This section focuses on 'facultative' sprouters, which recruit seedlings primarily after fires (Keeley 1986), and comprise most of the sprouting woody species of fynbos and kwongan. 'Obligate' sprouters, which have a different life history involving traits such as bird dispersal and recruitment between fires (Keeley 1986), are discussed in the section on fleshy fruits. Sprouting is ubiquitous in woody dico-tyledonous plants in all environments, including tropical forest and savanna, and has clearly not evolved only in response to fire (Gill 1975; James 1984; Zedler and Zammit 1989; Boucher et al. 1990). The question of whether sprouting or seeding are derived traits may differ between taxa. Sprouting appears to be a derived trait in *Leucadendron* (Midgley 1987) while epicormic sprouting is primitive, and lignotuber sprouting (in adults) and obligate seeding are derived traits in *Banksia* (George 1981). In chaparral genera the consensus is that obligate seeding is a derived trait which is reflected in a number of features e.g. drought tolerance (Keeley 1986; Davis 1989). Either way, the existence of divergent patterns of traits in sprouters and seeders (Keeley and Zedler 1978; Carpenter and Recher 1979; Parker 1984; Zammit and Westoby 1987) suggests that sprouting and seeding represent different life histories rather than differences involving single traits.

FIRE FREQUENCY AND SEVERITY

The abundance and dominance of seeders is linked to fire frequency. In California, lightning fire frequency decreases from north to south, from high to low elevations, and from inland to the coast. The lowest fire frequencies occur in the coast ranges of southern California which is where the greatest abundance and number of species of seeding shrubs is found (Keeley and Zedler 1978). A similar trend is evident in fynbos where the proportion of seeding Proteaceae species increases with decreasing fire frequency e.g. from grassy to montane fynbos (Table 6.7; Figure 6.1).

Lamont et al. (1985) argued that sprouting Proteaceae are more abundant and diverse in kwongan than in fynbos because kwongan is drier and thus burns more frequently. This contradicts evidence that the cover of sprouting species increases with increasing rainfall (Specht 1981b) and that moisture availability (Moisture Index 6 Specht and Moll 1983) is higher in kwongan than fynbos (Kruger 1983). Seeder dominance in chaparral may be associated with reliable rainfall rather than fire frequency because fire-cued germination results in a high risk of failure during germination and establishment if winter rains are erratic (Minnich and Howard 1984).

COMPETITION BETWEEN ADULT SPROUTERS AND SEEDLINGS

Competition from sprouter adults could exclude seedlings (Keeley 1977; Specht 1981b; Keeley 1986). In Keeley's model the key factor is variable intervals between fires: with frequent fires sprouter adults exclude seedlings; with long intervals between fires sprouters undergo density-dependent mortality and can experience high adult mortality during the intense fires which follow, thus leaving gaps for seedling recruitment. Seeder seedlings occur in greater numbers (Keeley and Zedler 1978), are more drought resistant, and grow faster, and thus outcompete sprouter seedlings in these gaps (Parker 1984). In Specht's model the key factor is rainfall: higher rainfall favours sprouters as inter-fire and post-fire seedling regeneration are excluded by the dense canopy and rapid regeneration of sprouters. Keeley's model assumes that the seed of seeders is always abundant but most seed regenerating species in fynbos are shorter lived than co-occurring sprouting species (Kruger 1983; Figure 6.3). The dominance of sprouters does not increase with increasing rainfall in fynbos (Kruger 1983), contrary to the predictions of Specht's (1981b) model, perhaps because the moisture balance in fynbos does not reach the levels needed for sprouter dominance (Kruger 1983).

COMPETITION BETWEEN WOODY AND HERBACEOUS PLANTS

Competition between seedlings of woody and herbaceous plants limits densities of woody plant seedlings (Kruger 1983; Bond 1987). The dense growth of herbaceous plants after fires in chaparral may suppress seedlings of woody seeders (Bond 1987). Herbaceous fire ephemerals actually facilitated woody seedling survival on warm slopes, possibly by shading them. On cooler slopes competition from herbaceous plants reduced seedling survival but not to the same extent as the facilitation on warm slopes (Bond 1987). Thus, the abundance and distribution of woody seeder species on different slopes in chaparral may be related to interactions with herbaceous plants.

Low soil nutrients may limit the relative growth rate and competitiveness of herbaceous species compared with seedlings of woody species (Bond 1989). This hypothesis predicts that slow growing woody seeders would be more prevalent in fynbos and kwongan and on nutrient-poor soils in chaparral. This prediction is supported in fynbos and kwongan and in chaparral where the dominant seeders on nutrient-poor chaparral soils generally are serotinous conifers (see Vogl et al. 1977) whose seedlings are highly sensitive to competition from herbs (Bond 1989).

RESOURCE ALLOCATION TO BUDS RATHER THAN SEEDS: TRADE-OFFS

Seedling recruitment of sprouters is limited primarily by low fecundity (Midgley 1987). If seeding is to have a selective advantage over sprouting, then seeders must produce, on average, more seedlings than the sum of sprouter adult survival and seedling production (Keeley 1986; Bond 1987; Hilbert 1987). As fire recurrence intervals in chaparral and kwongan are apparently *generally* greater than the time required by seeders to accumulate adequate seed banks for population replacement, seed regeneration should be expected in these communities (Kruger 1983; Keeley 1986; Zedler and Zammit 1989). However, both chaparral and kwongan have extensive communities dominated by sprouting species. This occurs despite the generally lower seed production and/or seedling recruitment and/or higher first summer seedling mortality of sprouting species (Table 6.11).

Lignotuberous sprouters have to devote resources to maintaining a woody lignotuber and their multi-stemmed growth form precludes sustained height growth (Midgley 1987). Therefore, they are frequently overtopped and suppressed by the taller seeders in fynbos (Table 6.10). This is not a problem for epicormic sprouters or in more open vegetation such as kwongan where epicormic sprouters are prominent (Tables 6.7, 6.8). In general epicormic sprouters appear to be relatively fire sensitive and will sprout primarily from the base after intense fires (Gill 1981). Kwongan vegetation is taller but more open than analogous fynbos communities (Milewski 1981, 1983) and thus has lower fuel loads. Epicormic sprouting may, therefore, be more important in kwongan because fire intensities are generally lower.

Fleshy fruited species

All the mediterranean regions have a flora of bird-dispersed, primarily woody, species (Table 6.9) with the following traits: vigorous sprouting, no persistent seed banks, and seedling recruitment between fires (Zedler 1981; Keeley 1986; Manders 1990a). Species with the same suite of traits are common in savanna (Bews 1917), in dune thicket (Cowling and Pierce 1988), and Afromontane forest (Phillips 1931, Van der Merwe 1966) which are intermingled with fynbos in many areas.

SOIL NUTRIENTS

Bird dispersal is thought to be rare in fynbos and kwongan because the low levels of soil nutrients — notably phosphorus and potassium — limit the production of fleshy fruits (Milewski 1982; Milewski and Bond 1982; see also Westoby et al. 1990). Bird-dispersed species are prominent in renoster shrubland and thicket communities on relatively nutrient-rich soils, but are rare in fynbos and important in Afromontane forest (Table 6.3) which both occur on soils derived from nutrient-poor sandstones. Although the concentrations of some nutrients in forest soils may be higher than those in adjacent fynbos (Van Daalen 1984; Manders 1990b; Cowling and Holmes this volume), this may be a consequence of different nutrient cycling patterns in forest environments rather than a prerequisite for the establishment of forest species (Manders 1990b). The successful establishment of an artificial forest of mainly fleshy fruited species on a site in fynbos in the south-western Cape provides additional evidence that soil nutrient levels are not the limiting factor (Knight 1988). Bird-dispersed plants can also establish themselves in a wide variety of fynbos communities e.g. asteraceous fynbos on coastal dunes (Table 6.3; Cowling and Pierce 1988) and fynbos islands (Bond et al. 1988). The predictions of the nutrient hypothesis are not supported by the available evidence.

DIRECTED DISPERSAL

Plants should disperse propagules to escape from parent and parental seed shadow, to colonize new sites (e.g. gaps), and to disperse to favourable microsites or away from intra- or inter-specific competition (Howe and Smallwood 1982). Seedling recruitment occurs during long fire-free intervals (Zedler 1981; Keeley 1986; Knight 1988) and recruitment after fires in fynbos appears to be constrained by lack of suitable microsites as sufficient seeds are being brought in by birds or wind (Manders 1990a, b). The seedlings of obligate seeders must be tolerant of competition, particularly for light, so the primary function of bird dispersal in fynbos apparently is to colonize new and favourable microsites. If bird dispersal is a viable option for forest and thicket species in the same environment, why is it rare in fynbos species? Why do fynbos communities lack bird-dispersed species — like the highly invasive weed *Acacia cyclops* (Glyphis et al. 1981; Richardson et al. this volume) — with persistent seed banks and fire stimulated seed germination?

PHYLOGENY: WHERE ARE THE BIRD-DISPERSED TAXA?

Typical tree and typical fleshy fruited families (e.g. from Roth 1987) — Apocynaceae, Araliaceae, Burseraceae, Euphorbiaceae, Lauraceae, Loranthaceae, Melastomaceae, Moraceae, Myristicaceae Myrtaceae, Palmae, Rubiaceae, Sapindaceae, Sapotaceae, Ulmaceae — are absent from fynbos but many are common in adjacent forest and thicket communities (Dyer 1975). In families where some taxa have fleshy fruits, the subfamilies with fleshy fruits are absent from fynbos (e.g. Ericaceae, Rutaceae, Rosaceae, Proteaceae). Similarly, the western Australian families with fleshy fruits — Myoporaceae, Elaeocarpaceae, Epacridaceae, Casuarinaceae, and Halogaceae (Clifford and Drake 1979) — are not found in fynbos. The (sub)families of bird-dispersed species in mediterranean Spain (Herrera

1984) e.g. Caprifoliaceae, Rosaceae, Ericaceae, Vitaceae, are also absent from fynbos. However, species of these families are present in thicket and forest communities adjacent to fynbos. *Maytenus* (Celastraceae), *Rhus* (Anacardiaceae), and *Diospyros* (Ebenaceae) occur in many fynbos communities but fynbos *Rhus* and *Diospyros* species have dry rather than fleshy fruit. Some typical fynbos genera (e.g. *Passerina*) have both dry and fleshy fruited members. Therefore it is unlikely that phylogeny is a major factor limiting bird dispersal in fynbos.

REPRODUCTIVE DYNAMICS OF BIRD-DISPERSED SPECIES

The reproductive syndromes of obligate sprouters — sprouting, bird dispersal, recruitment between fires (Zedler 1981; Keeley 1986) — are adapted to forest and thicket dynamics and the tree habit. These species are typically gap-phase recruiters. Epicormic sprouting is a means of retaining and reoccupying a site when disturbances, such as wind throw or toppling canopy trees, do not result in death but only in the loss of branches or other damage (Boucher et al. 1990).

The fire-prone nature of fynbos is the key factor limiting extensive invasion by bird-dispersed species. Frequent fires generally kill young trees before they reach a size that enables them to survive fires (Le Maitre 1989) and also reduce the availability of suitably shady micro-climates for seedling establishment and survival of forest species (Manders 1990b). Therefore, while the invasion of forest species is more evident in post-mature fynbos (e.g. Kruger 1983), old dune fynbos (Cowling and Pierce 1988; Pierce 1990), and old chaparral stands (Keeley et al. 1986), seed dispersal and recruitment of forest species are continuing processes even in young fynbos (Manders 1990a). Forest and thicket communities are, consequently, dominated by species which have an alternative life history of persisting by sprouting and recruiting seedlings between fires. The same syndromes are found in matorral species where fires are rare, bird dispersal is important in gap colonization and many shrub species act as nurses for their own and other species seedlings (Fuentes et al. 1984, 1986).

Fires, seed germination, and flowering

FIRES AND SEED GERMINATION

The strong link between seed germination and the occurrence of fires is well known but few studies have examined the mechanisms involved.

Allelopathy Allelopathic substances, primarily in litter, were thought to inhibit germination between fires. Heat from fires denatured these substances and broke seed dormancy (McPherson and Muller 1969), but there is no convincing evidence for allelopathic inhibition in natural environments (Keeley et al. 1985). In pot trials, 6% of *Widdringtonia cedarbergensis* seeds germinated in soil from a *Widdringtonia* stand (with or without litter) compared with 46% on the same soil with the litter burnt off (Manders 1987). However, *Widdringtonia* seedlings occurred in both burnt and unburnt sites in the field. Leachate solutions obtained from foliage inhibited germination of *Agathosma* and *Muraltia* species (Pierce 1990), but no field tests have been conducted.

Heating by fire Seed germination may be stimulated when heat from fires chars or fractures the seed coat. This phenomenon is well known in hard-seeded species (Gill 1975; Keeley 1987; Trabaud and Oustric 1989) including fynbos species in the Fabaceae — *Podalyria* and *Virgilia* (Jefferey et al. 1988) — and telochorous Proteaceae — *Leucadendron tinctum* (Brown and Dix 1985). Dry heat stimulated germination of *Agathosma betulina* and *A. crenulata* (Rutaceae) (Blommaert 1972) but not the six fynbos species studied by Pierce (1990), including two *Agathosma* species. Dry heat and simulated fire effects (release of ammonia and ethylene) stimulated germination of *Erica hebecalyx* (Ericaceae), but not the serotinous *E. sessiliflora* (Van de Venter and Esterhuizen 1988).

Charate Organic substances leached from heated or charred wood (charate) stimulate germination of several chaparral species, especially fire ephemerals, although germination of some of these species was also increased by fluctuating temperatures (Wicklow 1977; Keeley et al. 1985; Keeley and Pizzorno 1986). Germination of these species is not stimulated by wood ash (Wicklow 1977; Keeley et al.

1985). Seeds of the chaparral sub-shrub *Artemisia* germinated readily in light but required charate to stimulate germination in the dark (Keeley 1987). Charate did not significantly increase germination in six fynbos species (Pierce 1990), none of which were true fire ephemerals. No fynbos or kwongan fire ephemerals (see Bell et al. 1984; Pate et al. 1985; Kruger 1987) have yet been studied and species in the same or closely related families — Apiaceae (Hydrophyllaceae), Asteraceae, Crassulaceae, Geraniaceae, Poaceae, Portulaceae, and Scrophulariaceae (Table 6.12) — are the most likely candidates.

Smoke Chemicals in smoke stimulated seed germination of *Adouinia* (Bruniaceae, myrmecochorous) in field experiments and water extracts from smoke gave the highest germination in laboratory experiments (De Lange and Boucher 1990). The active principle in charred wood extracts — probably an oligosaccharin produced by a hemicellulose compound — is effective at low concentrations (Keeley and Pizzorno 1986) and the same, or similar, chemical may be present in smoke particles and smoke extracts.

Indirect interactions Fire-induced alterations in soil microclimate, especially the greater fluctuations in soil temperature, also stimulate seed germination. Examples in fynbos are *Stoebe* (Asteraceae) (Levyns 1935), *Drosera* (Droseraceae) (Ferreira and Small 1974), *Agathosma* (Rutaceae), *Metalasia*, *Felicia* (Asteraceae), *Muraltia* (Polygalaceae) (Pierce 1990), and *Leucospermum* and *Serruria* (Proteaceae) (Brits 1986, 1987) (Table 6.1). Increased oxygen concentrations in soil after fires may also stimulate germination (Brits 1986). Soil storage, or incubation in darkness, followed by exposure to light — especially light with a high red/far-red ratio typical of an open environment — increased germination in European *Calluna* and *Erica* (Pons 1989). Light reduced germination rates or percentages, or both, in most species in a sample of 29 Cape Asteraceae (Schutte and Parkhurst 1961). Cold conditions during the wet season after a fire also provide reliable cues for seed germination (Table 6.1; Keeley et al. 1985; Keeley 1986). Stimulation of germination by cold conditions (< 15°C) is found in fynbos Proteaceae (e.g. Deall and Brown 1981; Mitchell et al. 1986) and in *Watsonia* (Esterhuizen et al. 1986). These requirements are not present in *Banksia* species growing in kwongan or heath where optimal temperatures for germination range between 15°C and 30°C (Zammit and Westoby 1987); 10°C may inhibit germination (Cowling and Lamont 1987).

All these germination cues are linked to fire, either directly through heat or smoke or indirectly through changed environments or charate. They are also not isolated traits (Keeley et al. 1985). Serotinous species and geophytes, which both release seeds into the post-fire environment, have simple germination cues — e.g. a low temperature stimulus — and germinate during the first wet season after a fire (Keeley et al. 1985; Table 6.1). Myrmecochorous and telochorous species — with soil-stored seeds — have more complex germination requirements and will germinate only after fires provide the appropriate cue — direct heating of the seeds and soil temperature fluctuations (Keeley 1987). Germination of some of these species — notably fire ephemerals — requires a stimulus derived from charred wood or smoke and is, therefore, directly dependent on fires (Wicklow 1977; Keeley et al. 1985; De Lange and Boucher 1990).

FIRE STIMULATED FLOWERING

Fire stimulated flowering is common in all fire-prone vegetation types. It has been reported in Iridaceae, Orchidaceae, Liliaceae, Amaryllidaceae, Poaceae, and Restionaceae in fynbos (Kruger 1983; Frost 1984) and kwongan (Bell et al. 1984). Geophytes in fynbos appear to respond primarily to the indirect effects of the changed post-fire environment i.e. changes in temperature or insolation (Stone 1951; Gill 1981; Rundel 1981b). Cues directly related to fire — e.g. chemicals in smoke — have also been proposed as stimuli for flowering in geophytes — e.g. *Watsonia* (Iridaceae) — but the experimental design did not exclude the indirect influence of the altered environment after a fire (Bean 1962). Gill and Ingwersen (1976) showed that fire stimulated flowering in the grass-plant *Xanthorrhea* via the ethylene produced by fire damaged leaf bases, but this mechanism is unlikely in species with subterranean bulbs.

Some kwongan or heathland woody shrub species — e.g. *Telopea* and *Lambertia* (Pro-

TABLE 6.12 The taxonomic distribution of monocarpic and polycarpic fire ephemeral seeders in chaparral, kwongan, and fynbos. The importance of the trait in the family is indicated by upper case = important, lower case = relatively unimportant. Data from Mooney et al. (1977), Keeley et al. (1985), Bell et al. (1984), Pate et al. (1985), Trabaud and Oustric (1989), Musil and De Witt (1990), Van Wilgen and Forsyth (unpublished data). M = monocarpic, P = polycarpic, N = not recorded as having fire ephemerals, ? = family represented but fire responses uncertain or unknown, — = no data.

Family	Region		
	California	Australia	Cape
Aizoaceae	—	m	mp
Apiaceae[1]	Mp	M	p
Asteraceae	MP	M	mP
Boraginaceae	M	?	N
Cistaceae	P	—	—
Crassulaceae	m	—	P
Euphorbiaceae	?	m	m
Fabaceae (Papil)	MP	N	P
Geraniaceae	?	?	p
Goodeniaceae	—	p	—
Gyrostemonaceae	—	P	—
Hydrophyllaceae[1]	Mp	—	?[2]
Liliaceae	N	m	N
Lobeliaceae	?	m	mP
Loganiaceae	?	m	N
Malvaceae	?	p	P
Mesembryanthemaceae	?	—	mp
Onagraceae	M	—	?
Poaceae	?	m	mp
Polemoniaceae	M	—	—
Polygalaceae	—	N	P
Polygonaceae	Mp	N	?
Portulacaceae	M	M	m
Santalaceae[3]	—	N	P
Scrophulariaceae	MP	?	MP
Solanaceae	p	p	p[4]

[1] Apiaceae and Hydrophyllaceae are closely related. *Centella* (common in fynbos, myrmecochorous and not short-lived) was formerly in the Hydrophyllaceae.
[2] Two species of *Codon* occur but are not listed in Bond and Goldblatt (1984).
[3] All Cape species are root hemiparasites of the genus *Thesium*.
[4] Most *Solanum* species are long-lived.

teaceae) (Pyke 1983) and *Angophora hispida* (Myrtaceae) (Auld 1987) — also exhibit fire stimulated flowering but apparently do not have persistent seed banks. Similar responses are found in several woody fynbos sprouters e.g. rhizomatous *Protea* species which do not have pre-fire seed banks, but other species — e.g. *Leucadendron salignum* and *Mimetes cucullatus* — accumulate seed banks between fires.

WHY HAVE SPECIES EVOLVED THESE SPECIALIZED RESPONSES?

Fire stimulated germination is generally seen as an evolutionary response to increased availability of resources (light, space, nutrients) and reduced competition after fires (Rundel 1981b; Kruger 1983). Fire stimulated flowering may also be a response to these same conditions and the increased opportunities for seedling recruitment (Gill 1981; Rundel 1981b). Herbivory and seed predation are also generally lower in the post-fire environment. Mass flowering and seed production of sprouters are analogous to mast seeding, attracting pollinators and saturating seed predators (Salisbury 1942; Waller 1979; Silvertown 1980; Gill 1981). Fire stimulated mass germination may also saturate herbivores and reduce seedling herbivory after fires (Pierce 1990).

Selection will not favour mass seeding if predators can track the increased seed output or if episodic seed production results in a lower net recruitment than consistent seed production (Waller 1979). The changed environment after fires results in declines in the populations of at least the small mammalian granivores (Fox et al. 1985; Midgley and Clayton 1989) and, as the conditions between fires are unfavourable for recruitment, there is no loss of fitness due to episodic rather than regular seed production. Le Maitre (1984) found evidence that mass flowering of *Watsonia* satiated pre- and post-dispersal seed predators and resulted in recruitment of seedlings. In contrast, inter-fire flowering resulted in few seedlings (Kruger and Bigalke 1984; Le Maitre 1984). Bond (1987) studied the post-fire flowering of *Nolina parryi* (Agavaceae) in chaparral and found that fire reduced pre-dispersal predation and increased seedling recruitment. There was no evidence of post-dispersal predator satiation, however, as seed predation was low, but an influx of rodent herbivores resulted in the loss of all the seedlings. All these studies support the hypothesis that seed predation is a significant selective force (Johnson this volume).

Fire annuals

Three explanations have been proposed for the rarity of fire annuals in fynbos and kwongan versus chaparral:

- differences in soil nutrient availability (Kruger 1983; Keeley 1986);
- competition between growth forms for light and space (Keeley 1986); and
- the relative importance of summer rainfall (Milewski 1981).

SOIL NUTRIENTS

Annuals have to accumulate most of the nutrient reserves required for growth and reproduction each generation and only a small fraction of this accumulated capital can be passed on in each seed (Harper 1977). Thus, the low levels of nutrients in fynbos soils, except for a short-lived increase immediately after fires (Stock and Allsopp this volume), limit annuals. Perennial herbs and short-lived fire ephemerals have a competitive advantage because they can assimilate and recycle reserves more efficiently and so achieve a greater reproductive output (Harper 1977). Thus, annuals are important in chaparral but relatively unimportant even after fires in kwongan and fynbos. As efficient nutrient cycling is typical of geophytes (e.g. Pate and Dixon 1982; Hocking 1984), this hypothesis could also explain the abundance of geophytes in fynbos but not their rarity in kwongan (Milewski 1983), where fynbos geophytes have invaded a variety of communities (Pate and Dixon 1982).

COMPETITION BETWEEN GROWTH FORMS

Sprouting herbaceous perennials and subshrubs are rare in chaparral because of the dense canopy of woody plants which can develop on chaparral soils compared with the more open canopy in nutrient-limited fynbos and kwongan (Keeley 1986). Thus competition for light and space are important in chaparral (Keeley 1986), and there is a selective advantage in persisting between fires as seeds rather than dormant or suppressed perennial herbs. The relatively open canopy in fynbos allows perennial herbaceous plants such as Restionaceae to persist between fires. After fires these plants sprout rapidly and occupy space (Kruger 1983) thus limiting the abundance of annuals, especially in the second and subsequent years after fires. This prediction of the hypothesis is supported by the prominence of annuals in arid fynbos which has open patches between the shrub clumps (Kruger 1979), but is contradicted by the fact that kwongan stands with a similar structure have few annuals (see Milewski and Bond 1982; Milewski and Cowling 1985).

RELATIVE IMPORTANCE OF SUMMER RAINFALL

Fynbos, and to a lesser extent kwongan, have sufficient rainfall in summer to alleviate drought stress and to flush out nutrients, enabling shallow rooted, evergreen, herbaceous growth forms such as Restionaceae to persist (Milewski 1983; Stock et al. this volume). This gives herbaceous perennials a selective and competitive advantage over annuals. The hypothesis is supported by the tendency for the cover of tall woody shrubs and herbaceous perennials to decrease and the cover of annuals and deciduous geophytes to increase with increasing summer drought stress (Kruger 1979; Campbell 1985; Cowling and Holmes this volume). Similar gradients exist in chaparral where there is a transition to coastal chaparral (coastal sage) and desert chaparral with increasing summer drought. These drier vegetation types are characterized by clumps of woody species and gaps in which herbaceous species may flourish (Mooney et al. 1977; Keeley and Keeley 1988).

Geophyte life histories

There are no explanations for the diversity of the geophytic flora with bulbs or corms in fynbos compared with analogous kwongan communities. Neither is there any explanation for why most kwongan Liliaceae (*sensu lato*) and Iridaceae are evergreen and grass-like, while fynbos communities are dominated by deciduous, non-graminoid geophytes. The rarity of the hysteranthous habit (flowering and leafing temporally separated) in kwongan communities may be related to the rarity of deciduous geophytes as evergreen geophytes have a synanthous habit (leaves and flowers concurrent). The rarity of geophytes in other mediterranean shrublands appears to be due to competition for light and water from woody plants as they are more common in the drier, more open shrublands. Geophytic plants may be less common in southern African grassland and savanna because of competition from grasses: there are many geophytic species in succulent karroid shrublands which have a sparse grass component. There is evidence that competition from invasive exotic annuals, including grasses, results in a reduction in species richness and density of geophytes in fynbos habitats (Vlok 1988).

The general evolutionary trend among geophytic species appears to be from evergreenness to drought deciduousness and from a synanthous to a hysteranthous habit (Burtt 1970; Dafni et al. 1981a, b). Hysteranthy has advantages over synanthy in certain kinds of environments. Firstly, an ability to flower during the period when growth is not possible frees the plant from the morphological and developmental constraint that leaf growth must precede flower development (Burtt 1970). This may reduce competition for pollinators or dispersers and results in the availability of ripe seed for seedling recruitment at the onset of the wet season (Burtt 1970; Dafni et al. 1981a, b; Snijman 1984; Johnson this volume). Secondly, hysteranthy provides greater flexibility in the allocation of resources (Dafni et al. 1981a, b). Synanthous species must provide for leaf growth and seed production from current reserves and inputs from the current season, as well as retaining sufficient reserves to survive the following drought. Unfavourable conditions can result in flower or seed abortion and the loss of that investment. Hysteranthous species accumulate the capital for both reproduction and survival prior to flowering, so they are less dependent on a favourable growing season than synanthous species. They can therefore invest in flowering with a low risk of premature abortion due to unfavourable conditions. This model predicts that there will be a gradient from synanthy under high (reliable) rainfall to hysteranthy under low (unreliable) rainfall. The Cape genus *Haemanthus* (Amaryllidaceae) shows this trend from evergreen in the eastern Cape (e.g. *H. albiflos*) to deciduous and hysteranthous in the western Cape and Namaqualand.

Hysteranthy and synanthy are also linked to fire responses of geophytes. Fynbos Amaryllidaceae all have fleshy, often 'viviparous' fruit with no dormancy (Markotter 1936; Snijman 1984), except for the genus *Cyrtanthus* which has dry seeds with strong dormancy (Figure 6.4; Olivier and Werner 1980). Fleshy fruited amaryllids have fire stimulated flowering, but this occurs during the normal fire season i.e. late summer and autumn (Snijman 1984). The classic fire-lily response of flowering within days of a fire even in spring or early summer, is found in a few *Cyrtanthus* (Amaryllidaceae) species which are known not to flower between fires (Olivier and Werner 1980; D C le Maitre and P J Brown unpublished data). If fleshy fruited amaryllids were to flower after mid-summer fires, the seeds would die during the summer drought. However, the seeds of

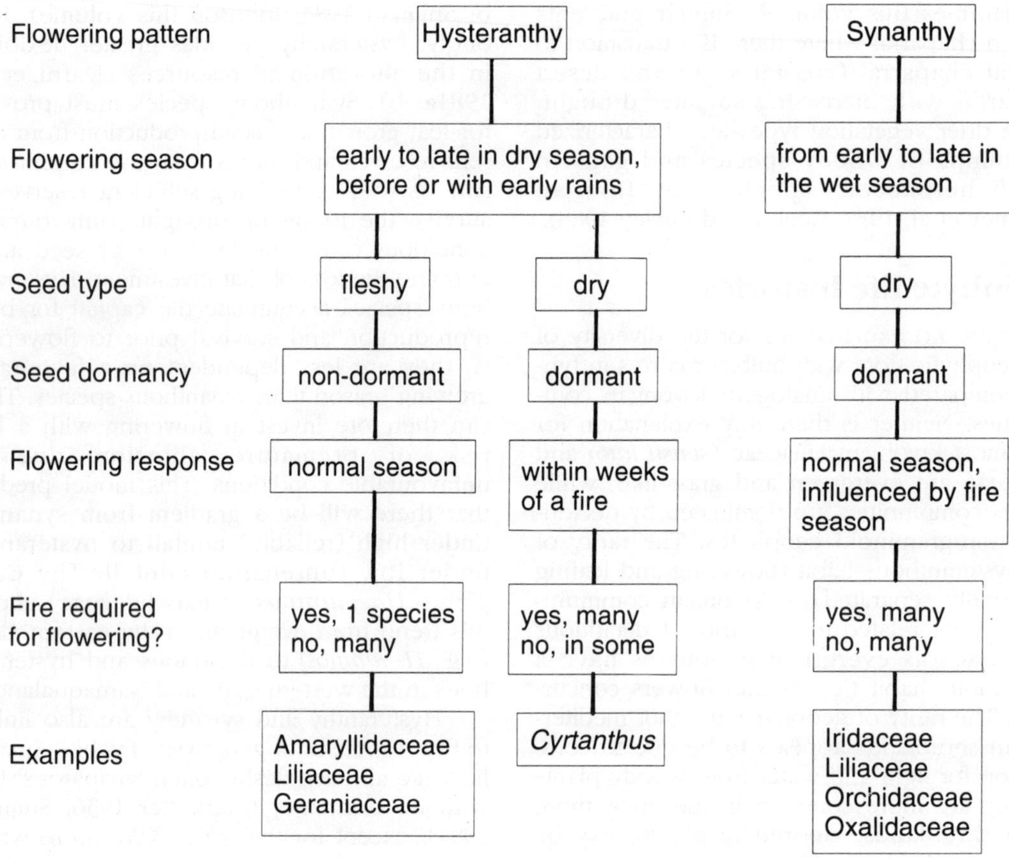

FIGURE 6.4 The relationship between flowering patterns and reproductive traits in typical fynbos geophytes with examples of families or genera with the combination of traits. *Haemanthus canaliculatus* appears to be the only fleshy fruited amaryllid which requires fire to stimulate flowering (Snijman 1984).

Cyrtanthus species are dormant. Thus, hysteranthy and dormant seeds were prerequisites for the evolution of the unique fire response of *Cyrtanthus*.

Pollination modes

BIRD POLLINATION

Birds require large quantities of sugar-rich nectar to meet their energy needs (Rebelo 1987). Plants can derive considerable benefits from the potential for long-range dispersal of pollen by birds compared with the short ranges typical of entomophily (Rebelo 1987). These needs can be met at a low cost to the plant because carbohydrate production is relatively cheap on nutrient-poor soils (Milewski 1983; Bloom et al. 1985; Rebelo 1987) and pollen is also relatively inexpensive. Thus, low soil nutrients favour bird pollination (Rebelo 1987). However, Afromontane forest on the same soils has few bird-pollinated species, with only *Halleria* and *Burchellia* being truly bird-pollinated.

Cool wet conditions are generally unfavourable for insect pollinators as many species only become active at temperatures over 20°C. Cool wet conditions in temperate Australia may, therefore, have favoured bird pollination over insect pollination (Keighery 1982). However, the southern Cape forest climates are also cool and summer temperatures are cool at high altitudes in fynbos where insect-pollinated species are common. Another hypothesis suggests that birds will tend to move more often between plants than between inflorescences on plants than insects, thus enhancing outcrossing, but there is no definite evidence for this in the Proteaceae (Collins and Rebelo 1987). At this stage there are no convincing hypotheses which can explain why bird pollination has evolved in insect-pollinated lineages. Long-range pollen movement has often been suggested but remains putative.

MAMMAL POLLINATION

Mammal pollination in Proteaceae is associated with a number of morphological traits e.g. sugar-rich nectar (Wiens et al. 1983) or inflorescences hidden in the canopy to minimize nectar thieving (Rebelo and Breytenbach 1987). Many species also have alternative pollinators (Rebelo and Breytenbach 1987; Johnson this volume). Mammal-pollinated species flower mainly in spring. This may be an alternative to competing for bird or insect pollination when many other species are in flower and also coincides with the high energy requirements of the mammals during their breeding season (Wiens et al. 1983; Rebelo and Breytenbach 1987). Wiens et al. (1983) noted that mammal-pollinated Proteaceae occurred in small fragmented populations and restricted habitats and suggested that these factors favoured the evolution of mammal pollination because of the more reliable services of the ubiquitous, unspecialized mammals. Rebelo and Breytenbach (1987) reject this explanation because the same fragmented distributions and small population sizes are found in some insect- and bird-pollinated Proteaceae. Thus, there are still no satisfactory explanations for the evolution of rodent pollination from bird/insect pollination in fynbos. Turner (1982) suggests that mammal and lorikeet pollination preceded the advent of honeyeaters in Australia. Bird pollination syndromes of, for example, Australian *Banksia*, have evolved from mammal/lorikeet pollination in contrast to the evolution of mammal from bird/insect pollination in the Cape (Wiens and Rourke 1977). Thus, the strongly convergent mammal pollination syndromes of Cape and Australian Proteaceae may have different evolutionary histories.

WIND POLLINATION

Wind pollination is generally seen simply as an alternative to zoophily in species-poor habitats (where pollen panmixia is not a problem) or cool habitats (where insect pollination is seasonally restricted or not possible) (e.g. Whitehead 1983). Although fynbos is species-rich and structurally dense, wind pollination has evolved in many typically zoophilous families e.g. several *Leucadendron* (Proteaceae) species, several species and genera in the Ericaceae, *Empleurum* (Rutaceae), *Cliffortia* (Rosaceae), and *Tarchonanthus* (Asteraceae). The predictions of this hypothesis are also contradicted by the evolution of wind pollination from insect pollination — e.g. *Leucadendron* (Williams 1972; Midgley 1987) — in species-rich fynbos.

Breeding systems

Dioecy is commonly associated with woodiness, fleshy fruits, and small insect-pollinated flowers (Bawa 1980; Givnish 1982; Fox 1985; Muenchow 1987). These correlations are weak or absent in fynbos but the analysis was heavily biased by the large number of non-woody, dry seeded, wind-pollinated Restionaceae (Steiner 1988; Table 6.1). When the Restionaceae were excluded, dioecy in fynbos was correlated with fleshy fruits and woodiness (Steiner 1988). Steiner suggests that the problem of dioecy in fynbos is one of explaining the species richness and dominance of Restionaceae which may have little to do with the effects of fire or nutrients on reproductive biology. Alternatively, the key might be the constraints that limited nutrients place on seed production. Limitations on seed production would favour outbreeding with strong maternal selection to ensure that the genetic quality of the offspring is maximized (Ayre and Whelan 1989). Thus outbreeding systems such as dioecy and heterostyly would be favoured on nutrient-poor soils. Steiner's (1987) list of species shows that dioecy, woodiness, and fleshy fruits are only found in tree and shrub species which are rare in fynbos but common in the adjacent forest and thicket communities. Dioecy and woodiness in fynbos are found primarily in the genera *Leucadendron*, *Aulax*, and *Anthospermum* which have dry seeds or fruit. None of the current hypotheses can explain why dioecy and fleshy fruit are common in forest, and dioecy and dry fruit in fynbos communities.

SUMMARY OF EXPLANATION AND ENIGMAS

Serotiny and myrmecochory The strongest hypothesis is that both traits have evolved in response to selection to reduce losses of seeds which are expensive to produce on nutrient-poor soils. But its predictions are contradicted by the abundance of telochorous (passively dispersed) species which lose many seeds to seed predators. Both traits attain maximum importance in environments with intermediate fire frequencies and these fire frequencies are linked to fuel accumulation patterns on nutrient-poor soils. Intermediate fire frequencies also select for seed banks in environments in which recruitment between fires is limited by seed and seedling predation and competition. Serotiny is also coupled to a predictable time span between fires and favourable conditions for germination and reliable rainfall. Coexistence of myrmecochorous and serotinous species may be maintained by different seed dispersal patterns but this does not explain the evolution of these traits in separate phylogenetic lineages. Fire, nutrients, predictable rainfall, and seed/seedling predation are all factors but a satisfactory explanation remains elusive.

Seeders versus sprouters If fire frequencies are appropriate for seeders, the most compelling hypotheses are those involving competitive interactions, either between sprouter adults and seedlings of any kind, or between seeder and sprouter seedlings, or between seedlings of woody and herbaceous plants. There are too few studies to identify the strongest contender and the hypotheses not mutually exclusive.

Fleshy fruit The evidence does not support the hypothesis that fleshy fruited species are rare in fynbos and kwongan because of the costs of fruit production on nutrient-poor soils. It does support the hypothesis that fleshy fruited species are rare in fynbos and kwongan because the species' life histories are geared to reproduction in a closed, shady environment. Fire-free intervals are rarely long enough to provide appropriate conditions for seedling establishment over extensive areas. There is no explanation for the relative abundance of bird-dispersed species in other mediterranean shrublands.

Fires, seed germination, and flowering The evidence supports the hypothesis that both traits have evolved in response to the increased availability of resources and relatively low seed predation and herbivory levels after fires, compared with the limited opportunities for reproduction between fires.

Fire annuals The strongest hypothesis is that fire annuals are sparse in fynbos and kwongan because the low nutrient soils favour perennial herbs — including geophytes — and short-lived (3–10 years) fire ephemerals which are able to recycle nutrients more efficiently. A complementary explanation is that summer rains in fynbos enable shallow-rooted perennials to survive the summer drought and increase their abundance compared with similar kwongan communities.

Geophyte life histories There are no hypotheses explaining why fynbos communities have more geophytes with corms and bulbs and a hysteranthous habit than analogous kwongan communities. The rarity of geophytes in other mediterranean shrublands and southern African grassland and savanna appears to be due to competition for light from woody plants or grasses as they are more common in drier areas which have more open vegetation.

Pollination syndromes There is little support for the hypothesis that bird pollination is prominent in fynbos and kwongan because of the nutrient-poor soils or because of the cool climates. There are also no convincing explanations for the evolution of small mammal pollination from bird/insect-pollinated ancestors, or wind pollination in insect-pollinated genera in fynbos.

Breeding systems The hypothesis that dioecy, woodiness, and fleshy fruit are correlated is not supported by data from fynbos. Species with this syndrome of traits are rare in fynbos but are common in adjacent forest and thicket communities, while fynbos species have dry fruits. The most compelling hypothesis is that dioecy and heterostyly are prominent because of the genetic benefits of outcrossing.

CONCLUSIONS

Reproductive traits in fynbos are more similar to those in other mediterranean shrublands than to those in communities on relatively nutrient-rich soils in southern Africa. The marked similarities between mediterranean shrublands are linked to mediterranean climates and intermediate intervals between fires. Differences in soil nutrient status underlie many of the differences between fynbos and kwongan and chaparral, maquis, and matorral, but the linkage is complex and often indirect. Fires, or, more correctly, particular fire regimes have been a significant driving force in the evolution of many reproductive traits. Fire is not simply a first order environmental factor as the fire regime is determined by climate, ignition sources, fuel dynamics, and community structure (Gill 1975) — all factors with significant stochastic and biotic components. Another significant driving force has been biotic interactions, particularly the effects of competition between plants and the influence of granivores and herbivores. There are, however, few really satisfying explanations which can account for patterns such as the abundance of myrmecochory and the rarity of fleshy fruit on nutrient-poor soils. This review has also emphasized the importance of viewing reproductive traits in the context of the regeneration niche (*sensu* Grubb 1977). Seeds of serotinous species and geophytes have similar, non fire dependent cues for germination, a similarity which makes sense because seeds of both kinds are released to germinate in the favourable post-fire environment.

Reproductive traits do not appear to be strongly linked to vegetative traits. For example, many serotinous and myrmecochorous shrubs are indistinguishable in classifications based on physiognomy and ecophysiology but their life histories differ fundamentally. This may be due to the different ways in which abiotic and biotic factors influence the evolution of traits. To quote Harper (1977):

> *It is in the nature of adaptation to the physical environment that a degree of ultimate adaptation is possible, provided the environment does not change . . . In contrast, in an environment dominated by the biotic forces of competition, parasitism and predation, the evolutionary game is potentially unending . . . evolution is an existential game in which success is measured by continuing to play the game rather than by winning.*

Are reproductive traits completely independent of vegetative traits? Flower or seed size may be constrained by leaf size or stem thickness because they are allometrically related (Bond and Midgley 1988). *Protea aristata* has smaller leaves and larger inflorescences than predicted by the relationship between leaf and inflorescence size in the genus *Protea*, evidence of a response to selective pressure to escape biomechanical constraints (Le Maitre and Midgley 1991). The evolution of physiognomic characteristics such as evergreenness and woodiness may, therefore, be balanced with selection for reproductive traits such as large flowers and vice versa.

We have focused largely on the question of why single traits have evolved but this can lead to difficulties in explaining alternative traits. For example, how do telochorous species coexist with serotinous and myrmecochorous ones when they apparently have no traits which could minimize seed predation — apparently a major selective pressure for the evolution of serotiny and myrmecochory? There are actually four separate but connected questions:

• Why has a trait evolved from another?
• Why are there so many/few species with a trait (see also Cowling et al. this volume)?
• Why is a trait important/rare?
• How do species with different traits coexist (Bond et al. this volume)?

WHAT CAN BE DONE?

• These analyses have been hindered by a lack of quantitative information on environmental factors and the incidence of reproductive traits. Studies at the community level and of the reproductive ecology of divergent species pairs — congeneric sprouters and seeders — and enigmatic species such as the few fleshy fruited species in fynbos — e.g. *Anomalanthus* and *Grubbia* — are required.
• More experimental studies are needed to expand our understanding of the reproductive ecology of fynbos plants. Why has bird pollination evolved from insect pollination? Fynbos, with its abundance of closely related species with divergent pollination syndromes e.g. *Protea* and *Erica* species, is an ideal testing ground for these ideas. Intra- and interspecific competition have been invoked in several explanations but the experiments of Bond (1987) are the only ones we are aware of.
• Little is known about the real costs of reproduction. How does nutrient scarcity influence the relative costs of nutrient acquisition and allocation? How expensive are carbohydrate-rich structures compared with seeds? What are the relative costs of maternity and paternity in monoecy and dioecy? How do these vary with soil nutrient status? What are the advantages and disadvantages of the minute seeds of Ericaceae and Orchidaceae compared with larger seeds in seedling competition? The Ericaceae in particular seem to be confined to communities on nutrient-poor soils (Table 6.3): is this due to requirements relating to their regeneration niche?
• As always more ingenuity and creative speculation are required and we hope that this review will stimulate research and lead to a deeper understanding of plant reproductive ecology.

This synthesis has shown that biotic interactions are significant factors. To use Hutchinson's theatrical analogy: while climate, nutrients, and fires have provided the theatre, set, and stage lights, we will only really begin to understand why things are the way they are when we observe the interactions between the living actors. And we have to watch closely enough to catch both the spoken words and the unspoken nuances that mark a great play, and great acting.

ACKNOWLEDGEMENTS

We thank R M Cowling and C Hilton-Taylor for allowing us to use unpublished data for our analyses, and W Bond for the invaluable effort and time he put into this chapter. J Keeley and T O'Connor made useful comments on an earlier draft. Both of us were supported in our research by the Conservation Research Programme of the Forestry Branch of the Department of Environment Affairs and by the Division of Forest Science and Technology of the CSIR.

REFERENCES

AULD T D (1986). Population dynamics of the shrub *Acacia suaveolens* (Sm.) Willd.: fire and the transition to seedlings. *Australian Journal of Ecology* **12**, 373–85.

AULD T D (1987). Post-fire demography in the resprouting shrub *Angophora hispida* (Sm.) Blaxell: flowering, seed production, dispersal, seedling establishment and survival. *Proceedings of the Linnean Society of New South Wales* **109**, 259–69.

AYRE D J and WHELAN R J (1989). Factors controlling fruit set in hermaphroditic plants: studies with the Australian Proteaceae. *Trends in Ecology and Evolution* **4**, 267–72.

BEADLE N C W (1954). Soil phosphate and the delimitation of plant communities in eastern Australia. *Ecology* **35**, 370–5.

BEAN P A (1962). An enquiry into the effect of veld fires on certain geophytes. MSc Thesis, University of Cape Town.

BEAN P A (1990). The identity of *Osyris abyssinica* Hochst. var speciosa A. W. Hill. (Santalaceae). *South African Journal of Botany* **56**, 665–9.

BEARD J M (1984). Biogeography of the kwongan. In *Kwongan. Plant life of the sandplain.* (eds Pate J S and Beard J S) University of Western Australia Press, Nedlands WA. 1–26.

BEATTIE A J (1985). *The evolutionary ecology of ant-plant mutualisms.* Cambridge University Press, Cambridge.

BEAUFAIT W R (1960). Some effects of high temperatures on the cones and seeds of jack pine. *Forest Science* **6**, 194–9.

BELL D T, HOPKINS A J M and PATE J S (1984). Fire in the kwongan. In *Kwongan. Plant life of the sand plain.* (eds Pate J S and Beard J S) University of Western Australia Press, Ned-

lands, WA. 178–204.
BELL D T, VLAHOS S and WATSON L E (1987). Stimulation of seed germination of understorey species of the northern jarrah forest of Western Australia. *Australian Journal of Botany* **35**, 593–9.
BERG R (1966). Seed dispersal of *Dendromecon rigida*: its ecologic, evolutionary, and taxonomic significance. *American Journal of Botany* **53**, 61–73.
BERG R (1975). Myrmecochorous plants in Australia and their dispersal by ants. *Australian Journal of Botany* **23**, 475–508.
BERG R (1981). The role of ants in seed dispersal in Australian lowland heath. In *Heathlands and related shrublands of the world: analytical studies.* (ed Specht R L) Elsevier, Amsterdam. 41–50.
BEWS J W (1917). The plant succession in the thornveld. *South African Journal of Science* **14**, 153–72.
BLOMMAERT K L J (1972). Buchu seed germination. *Journal of South African Botany* **38**, 237–9.
BLOOM A J, CHAPIN F S III and MOONEY H A (1985). Resource limitation in plants — an economic analogy. *Annual Review of Ecology and Systematics* **16**, 363–92.
BOND P and GOLDBLATT P (1984). Plants of the Cape Flora : A descriptive catalogue. *Journal of South African Botany* Suppl. Vol. **13**, 1–455.
BOND W J (1984). Fire survival of Cape Proteaceae — influence of fire season and seed predators. *Vegetatio* **56**, 65–74.
BOND W J (1985). Canopy-stored seed reserves (serotiny) in Cape Proteaceae. *South African Journal of Botany* **51**, 181–6.
BOND W J (1987). Regeneration and its importance in the distribution of woody plants. PhD Thesis, University of California, Los Angeles.
BOND W J (1988). Proteas as 'tumbleseeds': wind dispersal through the air and over the soil. *South African Journal of Botany* **54**, 455–60.
BOND W J (1989). The tortoise and the hare: the ecology of angiosperm dominance and gymnosperm persistence. *Biological Journal of the Linnean Society* **36**, 227–49.
BOND W J and BREYTENBACH G J (1985). Ants, rodents and seed predation in Proteaceae. *South African Journal of Zoology* **20**, 150–4.
BOND W J and MIDGLEY J J (1988). Allometry and sexual differences in leaf size. *American Naturalist* **131**, 901–10.
BOND W J, MIDGLEY J AND VLOK J (1988). When is an island not an island? Insular effects and their causes in fynbos shrublands. *Oecologia* **77**, 515–21.
BOND W J and SLINGSBY P (1983). Seed dispersal by ants in shrublands of the Cape Province and its evolutionary implications. *South African Journal of Science* **79**, 213–33.
BOND W J and STOCK W D (1990). The costs of leaving home: ants disperse seeds to low nutrient sites. *Oecologia* **81**, 412–17.
BOUCHER C and MOLL E J (1981). South African mediterranean shrublands. In *Mediterranean-type shrublands of the World.* (eds Di Castri F, Goodall D W and Specht R L) Elsevier Scientific Publishing Company, Amsterdam. 233–48.
BOUCHER D H, VANDERMEER J H, YIH K AND ZAMORA N (1990). Contrasting hurricane damage in tropical rain forest and pine forest. *Ecology* **71**, 2 022–4.
BRADSTOCK R A and O'CONNELL (1988). Demography of woody plants in relation to fire. *Banksia ericifolia* L.f. and *Petrophile pulchella* (Schrad.) R.Br. *Australian Journal of Ecology* **13**, 505–18.
BRITS G J (1986). Influence of fluctuating temperatures and H2O2 treatment on the germination of *Leucospermum cordifolium* and *Serruria florida* (Proteaceae) seeds. *South African Journal of Botany* **52**, 286–90.
BRITS G J (1987). Germination depth vs temperature requirements in naturally dispersed seeds of *Leucospermum cordifolium* and *L. cuneiforme* (Proteaceae). *South African Journal of Botany* **53**, 119–24.
BROWN N A C and DIX L (1985). Germination of the fruits of *Leucadendron tinctum. South African Journal of Botany* **51**, 448–52.
BROWN P J, MANDERS P T, BANDS D P, KRUGER F J and ANDRAG R H (1991). Prescribed burning as a conservation management practice; a case history from the Cederberg, South Africa. *Biological Conservation* (in press).
BULLOCK S H (1989). Life-history and seed dispersal of the short-lived chaparral shrub *Dendromecon rigida* (Papaveraceae). *American Journal of Botany* **76**, 1 506–17.
BURTT B L (1970). The evolution and taxonomic significance of a subterranean ovary in certain monocotyledons. *Israeli Journal of Botany* **19**, 77–90.
CAMPBELL B M (1985). A classification of the mountain vegetation of the fynbos biome. *Memoirs of the Botanical Survey of South Africa* **50**, 1–115.
CAMPBELL B M (1986). Plant spinescence and herbivory in a nutrient-poor ecosystem. *Oikos* **47**, 168—72.
CARPENTER F L and RECHER H F (1979). Pollination, reproduction and fire. *American Naturalist* **113**, 871–9.
CHENEY N P (1981). Fire behaviour. In *Fire and the Australian biota.* (eds Gill A M, Groves R H and Noble I R) Australian Academy of Science, Canberra.
CHRISTENSEN N L (1985). Shrubland fire regimes and their evolutionary consequences. In *The ecology of natural disturbance and patch dynamics.* (eds Pickett S T A and White P S) Academic Press, New York. 86–100.
CLIFFORD H T and DRAKE W E (1979). Pollination and dispersal in eastern Australian heathlands. In *Heathlands and related shrublands: analytical studies.* (ed Specht R L) Elsevier Scientific Publishers, Amsterdam, 39–49.
COATES PALGRAVE K (1977). Trees of southern Africa. C. Struik Publishers, Cape Town.
CODY M L and MOONEY H A (1978). Convergence versus non-convergence in mediterranean-climate ecosystems. *Annual Review of Ecology and Systematics* **9**, 265–321.
COETZEE J H and GILIOMEE J H (1987). Seed predation and survival in the infructescences of *Protea repens* (Proteaceae). *South African Journal of Botany* **53**, 61–4.
COLLINS B G and REBELO A G (1987). Pollination biology of the Proteaceae in Australia and southern Africa. *Australian Journal of Ecology* **12**, 387–421.
COWLING R M (1983a). Diversity relations in Cape shrublands and other vegetation in the south-eastern Cape, South Africa. *Vegetatio* **54**, 103–27.
COWLING R M (1983b). The occurrence of C3 and C4 grasses in fynbos and allied shrublands in the south-eastern Cape, South Africa. *Oecologia* **58**, 121–7.
COWLING R M (1984). A syntaxonomic and synecological study in the Humansdorp region of the fynbos biome. *Bothalia* **15**, 175–227.
COWLING R M (1987). Fire and its role in coexistence and speciation in Gondwana shrublands. *South African Journal of Science* **83**, 106–11.
COWLING R M and GXABA T (1990). Variation in post-fire recruitment of a fynbos overstorey shrub: effects on understorey community structure and implications for community-wide species richness. *South African Journal of Ecology* **1**, 1–7.
COWLING R M and LAMONT B B (1985). Variation in serotiny of three *Banksia* species along a climatic gradient. *Australian Journal of Ecology* **10**, 345–50.
COWLING R M and LAMONT B B (1987). Post-fire recruitment of four co-occurring *Banksia* species. *Journal of Applied Ecology* **24**, 645–58.
COWLING R M, LAMONT B B and PIERCE S M (1986). Seed bank dynamics of four co-occurring *Banksia* species. *Journal*

of Ecology **75**, 289–302.
COWLING R M and PIERCE S M (1988). Secondary succession in coastal dune fynbos: variation due to site and disturbance. *Vegetatio* **76**, 131–9.
DAFNI A, COHEN D and NOY-MEIR I (1981a). Life cycle variation in geophytes. *Annals of the Missouri Botanical Garden* **68**, 653–60.
DAFNI A, SCHMIDA A and AVISHAI M (1981b). Leafless autumnal-flowering geophytes in the Mediterranean region — phytogeographical, ecological and evolutionary aspects. *Plant Systematics and Evolution* **137**, 118–93.
DALLIMORE W and JACKSON A B (1966). A handbook of Coniferae and Ginkgoaceae. 4th Edition (revised by Harrison S G), Edward Arnold Publishers Ltd., London.
DAVIS S D (1989). Patterns in mixed chaparral stands. Differential water status and seedling survival during summer drought. In *The California chaparral. Paradigms reexamined.* (ed Keeley S C) Science Series **34**, Natural History Museum of Los Angeles County, Los Angeles, California. 97–105.
DEALL G B and BROWN N A C (1981). Seed germination in *Protea magnifica* Link. *South African Journal of Science* **77**, 175–6.
DE BUSSCHE M, ESCARRE J and LEPART J (1982). Ornithochory and plant succession in mediterranean abandoned orchards. *Vegetatio* **48**, 255–66.
DE LANGE J H and BOUCHER C (1990). Autecological studies on *Audouinia capitata* (Bruniaceae). I. Plant-derived smoke as a germination cue. *South African Journal of Botany* **56**, 700–3.
DI CASTRI F (1981). Mediterranean-type shrublands of the World. In *Mediterranean-type shrublands of the World.* (eds Di Castri F, Goodall D W and Specht R L) Elsevier Scientific Publishing Company, Amsterdam. 1–52.
DULBERGER R and ORNDUFF R (1980). Floral morphology and the reproductive biology of four species of *Cyanella* (Tecophileaceae). *New Phytologist* **86**, 45–54.
DYER R A (1975). *The genera of southern African flowering plants. Volume 1. Dicotyledons.* Department of Agricultural Services, Pretoria.
EDWARDS D (1984). Fire regimes in the biomes of South Africa. In *Ecological effects of fire in South African ecosystems.* (eds Booysen P de V and Tainton N M) Springer-Verlag Berlin. 19–37.
ENRIGHT N J and LAMONT B B (1989). Seed banks, fire season, safe sites and seedling recruitment in five co-occurring *Banksia* species. *Journal of Ecology* **77**, 111–22.
ESLER K J, COWLING R M, WITKOWSKI E T F and MUSTART P J (1990). Reproductive traits and accumulation of nitrogen and phosphorus during the development of fruits of *Protea compacta* (calcifuge) and *P. obtusifolia* Buek ex Meisn (calcicole). *New Phytologist* **112**, 109–15.
ESTERHUIZEN A D, VAN DE VENTER H A and ROBBERTSE P J (1986). A preliminary study of seed germination of *Watsonia fourcadei. South African Journal of Botany* **52**, 221–5.
EVERSON T M, VAN WILGEN B W and EVERSON C S (1988). Adaptation of a model for rating fire danger in the Natal Drakensberg. *South African Journal of Science* **84**, 44–9.
FERRIERA D P and SMALL J G C (1974). Preliminary studies on seed germination of *Drosera aliciae* Hamet. *Journal of South African Botany* **40**, 65–73.
FOX B J, QUINN R D and BREYTENBACH G J (1985). A comparison of small mammal succession following fire in Australia, California and South Africa. *Proceedings of the Ecological Society of Australia* **14**, 179–97.
FOX J F (1985). Incidence of dioecy in relation to growth form, pollination and dispersal. *Oecologia* **67**, 244–9.
FROST P G H (1984). The responses and survival of organisms in fire-prone environments. In *Ecological effects of fire in South African ecosystems.* (eds Booysen P de V and Tainton N M) Springer-Verlag Berlin. 273–309.
FUENTES E R, OTAIZA R D, ALLIENDE M C, HOFFMAN A and POIANI A (1984). Shrub clumps of the Chilean matorral vegetation: structure and possible maintenance mechanisms. *Oecologia* **62**, 405–11.
FUENTES E R, HOFFMAN A J, POIANI A and ALBIERDE M C (1986). Vegetation change in large clearings: patterns in the Chilean matorral. *Oecologia* **68**, 358–66.
GEORGE A S (1981). The genus *Banksia. Nuytsia* **3**, 239–473.
GIBBS RUSSEL G E (1987). Preliminary floristic analysis of the major biomes in southern Africa. *Bothalia* **17**, 213–27.
GILL A M (1975). Fire and the Australia flora: a review. *Australian Forestry* **38**, 4–25.
GILL A M (1981). Adaptive responses of vascular plant species to fires. In *Fire and the Australian biota.* (eds Gill A M, Groves R H and Noble I R) Australian Academy of Science, Canberra, Australia. 243–71.
GILL A M and GROVES R H (1981). Fire regimes in heathlands and their plant ecological effects. In *Heathlands and related shrublands of the world: analytical studies.* (ed Specht R L) Elsevier, Amsterdam. 61–84.
GILL A M and INGWERSEN F (1976). Growth of *Xanthorrhea australis* R.Br. in relation to fire. *Journal of Applied Ecology* **13**, 195–203.
GIVNISH T J (1981). Serotiny, geography and fire in the pine barrens of New Jersey. *Evolution* **35**, 101–23.
GIVNISH T J (1982). Outcrossing versus ecological constraints in the evolution of dioecy. *American Naturalist* **119**, 849–65.
GLYPHIS J L, MILTON S J and SIEGFRIED W R (1981). Dispersal of *Acacia cyclops* by birds. *Oecologia* **48**, 138–41.
GOLDBLATT P (1989). The genus *Watsonia.* A systematic monograph. *Annals of Kirstenbosch Gardens* **19**.
GRANGER E (1984). Fire in forest. In *Ecological effects of fire in South African Ecosystems.* (eds Booysen P de V and Tainton N M) Springer-Verlag Berlin. 177–97.
GRUBB P J (1977). The maintenance of species richness in plant communities: the importance of the regeneration niche. *Biological Review* **52**, 107–45.
HARPER J L (1977). *The population biology of plants.* Academic Press, New York.
HERRERA C M (1984). A study of avian frugivores, bird-dispersed plants, and their interaction in Mediterranean scrublands. *Ecological Monographs* **54**, 1–23.
HIGGINS K B and MANDERS P T (submitted). Population dynamics of two fynbos shrubs (*Protea neriifolia* and *Protea nitida* Proteaceae) on permanent plots in the Jonkershoek Valley, Stellenbosch, Cape Province. *South African Journal of Ecology.*
HILBERT D W (1987). A model of life-history strategies of chaparral shrubs in relation to fire frequency. In *Plant response to stress — functional analysis in mediterranean ecosystems.* (eds Tenhunen J D, Catarino F M, Lange O L and Oechel W C) Springer-Verlag, Berlin.
HOCKING P J (1984). Accumulation, partitioning and redistribution of dry matter and mineral nutrients in *Ixia flexuosa* L., with special reference to its cormaceous habit. *Annals of Botany* **53**, 489–501.
HOFFMANN A J, TEILLER S and FUENTES E R (1989). Fruit and seed-characteristics of woody species in mediterranean-type regions of Chile and California. *Revista Chilena de Historia Natural* **62**, 43–60.
HORTON J S and KRAEBEL C J (1955). Development of vegetation after fire in the chamise chaparral of southern California. *Ecology* **36**, 244–62.
HOWE H F and SMALLWOOD J (1982). Ecology of seed dispersal. *Annual Review of Ecology and Systematics* **13**, 201–28.
IMS R A (1989). The ecology and evolution of reproductive synchrony. *Trends in Ecology and Evolution* **5**, 135–40.
IZHAKI I and SAFRIEL U N (1985). Why do fleshy-fruit plants

of the Mediterranean scrub intercept fall- but not spring-passage of seed-dispersing migratory birds? *Oecologia* **67**, 40–3.
JAMES S (1984). Lignotubers and burls — their structure, function and ecological significance in mediterranean ecosystems. *Botanical Review* **50**, 225–66.
JANZEN D H (1971). Seed predation by animals. *Annual Review of Ecology and Systematics* **2**, 465–92.
JEFFEREY D J, HOLMES P M and REBELO A G (1988). Effects of dry heat on seed germination in selected indigenous and alien legume species in South Africa. *South African Journal of Botany* **54**, 28–34.
JOHNSON E A and VAN WAGNER C E (1985). The theory and use of two fire history models. *Canadian Journal of Forest Research* **15**, 214–20.
KEELEY J E (1977). Seed production, seed populations in soil, and seedling production after fire for two congeneric pairs of sprouting and seeding shrubs. *Ecology* **58**, 820–9.
KEELEY J E (1986). Resilience of mediterranean shrub communities to fires. In *Resilience in mediterranean-type ecosystems.* (eds Dell B, Hopkins A J M and Lamont B B) Dr W Junk Publishers, Dordrecht. 95–112.
KEELEY J E (1987). Role of fire in seed germination of woody taxa in California chaparral. *Ecology* **68**, 434–43.
KEELEY J E, BROOKS A, BIRD T, CORY S, PARKER H and USINGER E (1986). Demographic structure of chaparral under extended fire-free conditions. In *Proceedings of the chaparral ecosystems research conference.* (ed De Vries J J) Santa Barbara, California. *California Water Resources Centre Report* **62**, University of California, Davis. 133–7.
KEELEY J E and KEELEY S C (1988). Chaparral. In *North American terrestrial vegetation.* (eds Barbour M G and Billings W D) Cambridge University Press, Cambridge. 165–207.
KEELEY J E, MORTON B A, PEDROSA A and TROTTER P (1985). Role of allelopathy, heat and charred wood in the germination of the chaparral herbs and suffrutescents. *Journal of Ecology* **73**, 445–8.
KEELEY J E and ZEDLER P H (1978). Reproduction of chaparral shrubs after fire: a comparison of sprouting and seeding strategies. *American Midland Naturalist* **99**, 142–61.
KEELEY S C and PIZZORNO M (1986). Charred wood stimulated germination of two fire-following herbs of the California chaparral and the role of hemicellulose. *American Journal of Botany* **73**, 1 289–97.
KEELEY S C, KEELEY J E, HUTCHINSON S M and JOHNSON A W (1981). Postfire succession of the herbaceous flora in southern California chaparral. *Ecology* **62**, 1 608–21.
KEIGHERY G J (1982). Bird-pollinated plants in Western Australia. In *Bird pollinated plants in Western Australia and their breeding systems.* (eds Armstrong J A, Powell J M and Richards A J) Royal Botanic Gardens, Sydney. 77–89.
KNIGHT R S (1988). Aspects of plant dispersal in the south-western Cape with particular reference to the role of birds as dispersal agents. PhD Thesis, University of Cape Town.
KOUTNIK D (1987). Wind pollination in the Cape Flora. In *A preliminary synthesis of pollination biology in the Cape Flora.* (ed Rebelo A G) *South African National Scientific Programmes Report* **141**, CSIR, Pretoria. 126–33.
KRUGER F J (1979). South African heathlands. In *Heathlands and related shrublands-descriptive studies.* (ed Specht R L) Elsevier Scientific Publishing Company, Amsterdam. 19–80.
KRUGER F J (1983). Plant community diversity and dynamics in relation to fire. In *Mediterranean-type ecosystems: the role of nutrients.* (eds Kruger F J, Mitchell D T and Jarvis J U M) Springer-Verlag, Berlin. 466–72.
KRUGER F J (1987). Succession after fire in selected fynbos communities of the south-western Cape. PhD thesis, University of the Witwatersrand.
KRUGER F J and BIGALKE R C (1984). Fire in fynbos. In *Ecological effects of fire in South African ecosystems.* (eds Booysen P de V and Tainton N M) Springer-Verlag, Berlin. 67–114.
KRUGER F J, MITCHELL D T and JARVIS J U M (eds) (1983). *Mediterranean-type ecosystems: the role of nutrients.* Springer-Verlag, Berlin.
KUO J, HOCKING P J AND PATE J S (1982). Nutrient reserves in seeds of selected Proteaceous species from south-western Australia. *Australian Journal of Botany* **30**, 231–49.
LAMONT B B (1982). The reproductive biology of *Grevillea leucopteris* (Proteaceae), including reference to its glandular hairs and colonising potential. *Flora* **172**, 1–20.
LAMONT B B and BARKER M J (1988). Seed bank dynamics of a serotinous, fire-sensitive *Banksia* species. *Australian Journal of Botany* **36**, 192–203.
LAMONT B B, COLLINS B G and COWLING R M (1985). Reproductive biology of the Proteaceae in Australia and South Africa. *Proceedings of the Ecological Society of Australia* **14**, 213–24.
LAMONT B B, CONNEL S and BERGL S (1990). Seed bank and population dynamics of *Banksia cuneata. Botanical Gazette* **50** (in press).
LAMONT B B, HOPKINS A J M and HNATIUK R J (1984). The flora — composition, diversity and origins. In *Kwongan. Plant life of the sandplain.* (eds Pate J S and Beard J S) University of Western Australia Press, Nedlands WA. 27–50.
LAMONT B B, LE MAITRE D C, COWLING R M and ENRIGHT N J (1991). Canopy seed storage in woody plants. *Botanical Review* (in press).
LE MAITRE D C (1984). A short note on seed predation in *Watsonia pyramidata* (Andr.) Stapf in relation to season of burn. *Journal of South African Botany* **50**, 407–15.
LE MAITRE D C (1988). Effects of season of burn on the regeneration of two Proteaceae with soil-stored seed. *South African Journal of Botany* **54**, 575–80.
LE MAITRE D C (1989). Mortality of selected sprouting species in Langrivier. In *The short-term effects of fire in Langrivier, Jonkershoek State Forest, Stellenbosch.* (ed Richardson D M) Jonkershoek Forestry Research Centre Report J 2/89. 51–9.
LE MAITRE D C and MIDGLEY J J (1991). Allometric relationships between leaf and inflorescence mass in the genus *Protea* (Proteaceae): an analysis of the exceptions to the rule. *Functional Ecology* **5**, 476–84.
LEVYNS M R (1929). Veld burning experiments at Ida's Valley, Stellenbosch. *Transactions of the Royal Society of South Africa* **17**, 61–92.
LEVYNS M R (1935). Seed germination in some South African seeds. *Journal of South African Botany* **1**, 161–70.
LEVYNS M R (1961). Some impressions of a South African botanist in temperate Western Australia. *Journal of South African Botany* **27**, 87–97.
LINDER H P (1989). Grasses in the Cape Floristic Region: phytogeographical implications. *South African Journal of Science* **85**, 502–5.
LINDER H P and ELLIS R P (1990). Vegetative morphology and interfire survival strategies in the Cape fynbos grasses. *Bothalia* **20**, 91–103.
MANDERS P T (1987). Is there allelopathic self-inhibition of regeneration within *Widdringtonia cedarbergensis* stands. *South African Journal of Botany* **53**, 408–10.
MANDERS P T (1990a). Soil seed banks and post-fire seed deposition across a forest-fynbos ecotone in the Cape Province. *Journal of Vegetation Science* **1**, 491–8.
MANDERS P T (1990b). Fire and other variables as determinants of forest/fynbos boundaries in the Cape Province. *Journal of Vegetation Science* **1**, 483–90.
MANDERS P T and RICHARDSON D M (in press). Colonisation of Cape fynbos communities by forest components. *For-*

est Ecology and Management (in press).
MARKOTTER E I (1936). Die lewensgeskiedenis van sekere geslagte van die Amaryllidaceae. *Annals of the University of Stellenbosch 14 (Series A(2)).*
McPHERSON J K and MULLER C H (1969). Allelopathic effects of *Adenostoma fasciculatum*, 'chamise' in the California chaparral. *Ecological Monographs* **39**, 177–98.
MIDGLEY J J (1987). Aspects of the evolutionary biology of the Proteaceae, with emphasis on the genus *Leucadendron* and its phylogeny. PhD Thesis, University of Cape Town.
MIDGLEY J J (1989). Season of burn and serotinous Proteaceae: a critical review and further data. *South African Journal of Botany* **55**, 165–70.
MIDGLEY J J and CLAYTON P (1989). Short-term effects of an autumn fire on small mammal populations in southern Cape coastal mountain fynbos. *South African Forestry Journal* **153**, 27–30.
MILEWSKI A V (1981). A comparison of vegetation height in relation to the effectiveness of rainfall in the mediterranean and adjacent arid parts of Australia and South Africa. *Journal of Biogeography* **8**, 107–16.
MILEWSKI A V (1982). The occurrence of seeds and fruits taken by ants versus birds in mediterranean Australia and southern Africa, in relation to the availability of soil potassium. *Journal of Biogeography* **9**, 505–16.
MILEWSKI A V (1983). A comparison of ecosystems in mediterranean Australia and southern Africa: nutrient-poor sites at the barrens and the Caledon coast. *Annual Review of Ecology and Systematics* **14**, 57–76.
MILEWSKI A V and BOND W J (1982). Convergence of myrmecochory in mediterranean Australia and South Africa. In *Ant-plant interactions in Australia.* (ed Buckley R C) Dr.W.Junk, The Hague. 89–98.
MILEWSKI A V and COWLING R M (1985). Anomalies in the plant and animal communities in similar environments at the Barrens, Western Australia, and the Caledon coast, South Africa. *Proceedings of the Ecological Society of Australia* **14**, 199–212.
MILLER P C, MILLER J M and MILLER P M (1983). Seasonal progression of plant water relations in fynbos in the Western Cape Province, South Africa. *Oecologia* **56**, 392–6.
MILLS J N and KUMMEROW J (1989). Herbivores, seed predators, and chaparral succession. In *The California chaparral. Paradigms reexamined.* (ed Keeley S C) Science Series **34**, Natural History Museum of Los Angeles County, Los Angeles, California. 49–55.
MINNICH R and HOWARD L (1984). Biogeography and prehistory of shrublands. In *Shrublands in California: literature review and research needed for management.* (ed De Vries J) Water Resources Centre Contribution **191**, University of California, Davis. 8–24.
MITCHELL J J, VAN STADEN J and BROWN N A C (1986). Germination of *Protea compacta* achenes: the relationship between incubation temperature and endogenous cytokinin levels. *Acta Horticulturae* **185**, 31–7.
MOLDENKE A (1977). Insect-plant relations. In *Chile-California Mediterranean scrub atlas. A comparative analysis.* (eds Thrower N J W and Bradbury D E) Dowden, Hutchinson & Ross, Inc., Pennsylvania. 199–218.
MONTENEGRO G, AVILA G and SCHATTE P (1983). Presence and development of lignotubers in shrubs of the Chilean matorral. *Canadian Journal of Botany* **61**, 1 804–8.
MONTYGIERD-LOYBA T M and KEELEY J E (1987). Demographic structure of Ceanothus megacarpus chaparral in the long absence of fire. *Ecology* **68**, 211–13.
MOONEY H A and DUNN E (1970). Convergent evolution of mediterranean climate sclerophyll shrubs. *Evolution* **24**, 292–303.
MOONEY H A, KUMMEROW J, JOHNSON A W, PARSONS D J, KEELEY S, HOFFMAN A, HAYS R I, GILIBERTO J and CHU C (1977). The producers — their resources and adaptive responses. In *Convergent evolution in Chile and California: mediterranean-type ecosystems.* (ed Mooney H A) Dowden, Hutchinson and Ross, Stroudsberg, Pennsylvania.
MORRIS E C and MYERSCOUGH P J (1988). Survivorship, growth and self-thinning in *Banksia ericifolia. Australian Journal of Ecology* **13**, 181–9.
MUENCHOW G E (1987). Is dioecy associated with fleshy fruit. *American Journal of Botany* **74**, 287–93.
MUSIL C F and DE WITT D M (1990). Post-fire regeneration in a sand plain lowland fynbos community. *South African Journal of Botany* **56**, 167–84.
NAVEH Z (1973). The ecology of fire in Israel. In *Proceedings of the 13th Annual Tall Timbers Fire Ecology Conference.* Tall Timbers Research Station, Tallahasee, Florida. 131–70.
NAVEH Z (1975). The evolutionary significance of fire in the mediterranean region. *Vegetatio* **29**, 199–208.
O'DOWD D J and GILL A M (1984). Predator satiation and site amelioration following fires: mass reproduction of alpine ash (*Eucalyptus delegatensis*) in southeastern Australia. *Ecology* **65**, 1 052–66.
O'DOWD D J and HAY M E (1980). Mutualism between harvester ants and a desert ephemeral: seed escape from rodents. *Ecology* **61**, 531–40.
OLIVIER W and WERNER W (1980). The genus *Cyrtanthus* Ait. *Veld & Flora* **68**, 78–81.
ORNDUFF R (1974). Heterostyly in South African plants: a conspectus. *Journal of South African Botany* **40**, 169–87.
PARKER V T (1984). Correlation of physiological divergence with reproductive mode in chaparral shrubs. *Madrono* **31**, 231–42.
PATE J S and DIXON K S (1982). *Tuberous, cormous and bulbous plants — biology of an adaptive strategy in Western Australia.* University of Western Australia Press, Nedlands WA.
PATE J S, DIXON K W and ORSHAN G (1984). Growth and life form characteristics of kwongan species. In *Kwongan-plant life of the sandplain.* (eds Pate J S and Beard J S) University of Western Australia Press, Perth. 84–100.
PATE J S, CASSON N E, RULLO J and KUO J (1985). Biology of fire ephemerals of the sandplains of the kwongan of southwestern Australia. *Australian Journal of Plant Physiology* **12**, 641–55.
PERRY D A and LOTAN J E (1979). A model of fire selection for serotiny in lodgepole pine. *Evolution* **33**, 958–68.
PHILLIPS J F V (1931). Forest-succession and ecology in the Knysna region. *Memoirs of the Botanical survey of South Africa* **14**.
PIERCE S M (1990). Pattern and process in south coast dune fynbos: population, community and landscape level studies. PhD thesis, University of Cape Town.
PONS T L (1989). Dormancy and germination of *Calluna vulgaris* (L.) Hull and *Erica tetralix* L. seeds. *Acta Oecologia. Oecologia Plantarum* **10**, 35–43.
PYKE G H (1983). Relationship between time since the last fire and flowering in *Telopea speciosissima* and *Lambertia formosa. Australian Journal of Botany* **31**, 293–6.
READ D J (1983). The biology of mycorrhiza in the Ericales. *Canadian Journal of Botany* **61**, 985–1 004.
REBELO A G (1987). Bird pollination in the Cape flora. In *A preliminary synthesis of pollination biology in the Cape flora.* (ed Rebelo A G) *South African National Scientific Programmes Report* **141**, CSIR, Pretoria. 83–108.
REBELO A G and BREYTENBACH G J (1987). Mammal pollination in the Cape flora. In *A preliminary synthesis of pollination biology in the Cape flora.* (ed Rebelo A G) *South African National Scientific Programmes Report* **141**, CSIR, Pretoria. 109–25.
RICHARDSON D M and BOND W J (1991). Determinants of

plant distribution: evidence from pine invasions. *American Naturalist* **136** (in press).
RICHARDSON D M and VAN WILGEN B W (1986). The effects of fire in felled *Hakea sericea* and natural fynbos and implications for weed control in mountain catchments. *South African Forestry Journal* **139**, 4–14.
RICE B and WESTOBY M (1986). Evidence against the hypothesis that ant-dispersed seeds reach nutrient-enriched microsites. *Ecology* **67**, 1 270–4.
ROTH I (1987). *Stratification of a tropical forest as seen in dispersal types.* Tasks for vegetation science **17**. Dr W Junk Publishers, Dordrecht.
ROURKE J P (1972). Taxonomic studies on *Leucospermum* R.Br. *Journal of South African Botany Supplementary Volume* **8**.
ROURKE J P (1982). *The proteas of southern Africa.* 2nd Edition, Centaur Publishers, Johannesburg.
ROURKE J P (1984). A revision of the genus *Mimetes* Salisb. (Proteaceae). *Journal of South African Botany* **50**, 171–236.
ROURKE J P (1987). The inflorescence morphology and systematics of Aulax (Proteaceae). *South African Journal of Botany* **53**, 464–80.
ROURKE J P and WIENS D (1977). Convergent floral evolution in South African and Australian Proteaceae and its possible bearing on pollination by non-flying mammals. *Annals of the Missouri Botanical Garden* **64**, 1–17.
RUNDEL P W (1981a). The matorral zone of central Chile. In *Mediterranean-type shrublands.* (eds Di Castri F, Goodall D W and Specht R L) Elsevier Scientific Publishing Company, Amsterdam. 175–201.
RUNDEL P W (1981b). Fire as an ecological factor. In *Physiological Plant Ecology I: Response to the physical environment. Encyclopaedia of Plant Physiology Vol.* **12A**. (eds Lange O L, Nobel P S, Osmond C B and Ziegler H) Springer-Verlag, Berlin. 501–38.
RUTHERFORD M C (1981). Survival, regeneration and leaf biomass changes in woody plants following spring burns in *Burkea africana — Ochna pulchra* savannah. *Bothalia* **13**, 531–52.
SALISBURY E J (1942). *The reproductive capacity of plants.* Bell, London.
SCHLESINGER W H, GRAY J T, GILL D S and MAHALL B E (1982). *Ceanothus megacarpus* chaparral: a synthesis of ecosystem properties during development and annual growth. *Botanical Review* **48**, 71–117.
SCHUTTE K D and PARKHURST D (1960). The influence of light upon the germination of some Compositae seeds. *Journal of South African Botany* **26**, 149–53.
SCOTT J K (1982). The impact of destructive insects on reproduction in six species of *Banksia* L.f. (Proteaceae). *Australian Journal of Zoology* **30**, 901–21.
SIEGFRIED W R (1983). Trophic structure of some communities of fynbos birds. *Journal of South African Botany* **49**, 1–43.
SILVERTOWN J (1980). The evolutionary ecology of mast seeding in trees. *Biological Journal of the Linnaean Society* **14**, 235–50.
SLINGSBY P and BOND W J (1981). Ants — friends of the fynbos. *Veld & Flora* **67**, 39–45.
SMALL J G C, ROBBERTSE P J, GROBBELAAR N and BADENHORST C M (1982). The effect of time of application and sterilization method of gibberelic acid and temperature on seed germination of *Erica junonia*, an endangered species. *South African Journal of Botany* **1**, 139–41.
SNIJMAN D (1984). A revision of the genus *Haemanthus* (Amaryllidaceae). *Journal of South African Botany* Supplementary Volume **12**.
SPECHT R L (1969). A comparison of the sclerophyllous vegetation characteristic of mediterranean type climates in France, California and southern Australia. I. Structure, morphology and succession. *Australian Journal of Botany* **17**, 277–92.
SPECHT R L (1981a). Major vegetation formations in Australia. In *Ecological biogeography of Australia.* (ed Keast A) Dr W Junk Publishers, The Hague. 163–291.
SPECHT R L (1981b). Responses to fires in heathlands and related shrublands. In *Fire and the Australian Biota.* (eds Gill A M, Groves R H and Noble I R) Australian Academy of Science, Canberra. 395–415.
SPECHT R L (ed) (1988). *Mediterranean-type ecosystems. A data source book.* Kluwer Academic Publishers, Dordrecht.
SPECHT R L and MOLL E J (1983). Mediterranean-type heathlands and sclerophyllous shrublands of the world: an overview. In *Mediterranean-type ecosystems: the role of nutrients.* (eds Kruger F J, Mitchell D T and Jarvis J U M) Springer-Verlag, Berlin. 41–65.
STEINER K E (1987). Breeding systems in the Cape flora. In *A preliminary synthesis of pollination biology in the Cape flora.* (ed Rebelo A G) *South African National Scientific Programmes Report* **141**, CSIR, Pretoria. 22–51.
STEINER K E (1988). Dioecism and its correlates in the Cape Flora of South Africa. *American Journal of Botany* **75**, 1 742–54.
STOCK W D, PATE J S and DELPHS J (1990). Influence of seed size and quality on seedling development under low nutrient conditions in five Australian and South African members of the Proteaceae. *Journal of Ecology* **78**, 1 005–20.
STONE E C (1951). The stimulative effect of fire on the flowering of the golden brodiaea (*Brodiaea ixioides* Wats. var. *lugens* Jeps.). *Ecology* **32**, 534–7.
TENHUNEN J D, CATARINO F M, LANGE O L and OECHEL W C (eds) (1987). *Plant response to stress. Functional analysis in Mediterranean Ecosystems.* Springer-Verlag, Berlin.
THOMPSON K (1987). Seeds and seed banks. *New Phytologist* (Supplement) **106**, 23–34.
TRABAUD L (1981). Man and fire: impacts on mediterranean vegetation. In *Mediterranean-type shrublands of the World.* (eds Di Castri F, Goodall D W and Specht R L) Elsevier Scientific Publishing Company, Amsterdam. 523–37.
TRABAUD L and OUSTRIC J (1989). Heat requirements for seed germination of three Cistus species in the garrigue of southern France. *Flora* **183**, 321–5.
TURNER V (1982). Marsupials as pollinators in Australia. In *Bird pollinated plants in Western Australia and their breeding systems.* (eds Armstrong J A, Powell J M and Richards A J) Royal Botanic Gardens, Sydney. 55–66.
VAN DAALEN J C (1984). Distinguishing features of forest species on nutrient-poor soils in the southern Cape. *Bothalia* **15**, 229–39.
VAN DE VENTER H A and ESTERHUIZEN A D (1988). The effect of factors associated with fire on seed germination of *Erica sessiliflora* and *E. hebecalyx* (Ericaceae). *South African Journal of Botany* **54**, 301–4.
VAN DER MERWE P (1966). Die flora van Swartboskloof en Stellenbosch en die herstel van soorte na 'n brand. *Annals of the University of Stellenbosch* **41** (Series A).
VAN STADEN J and BROWN N A C (1977). Studies on the seed germination of South African Proteaceae — a review. *Seed Science and Technology* **5**, 633–43.
VAN WILGEN B W (1981). Some effects of fire frequency on fynbos plant community composition and structure at Jonkershoek, Stellenbosch. *South African Forestry Journal* **118**, 42–55.
VAN WILGEN B W (1984). Fire climates in the southern and western Cape and their potential use in fire control and management. *South African Journal of Science* **80**, 358–62.
VAN WILGEN B W, LE MAITRE D C and KRUGER F J (1985). Fire behaviour in South African fynbos (macchia) vegetation

and predictions from Rothermel's fire model. *Journal of Applied Ecology* **22**, 955–66.
VAN WYK A E (1989). Floristics of the Natal/Pondoland sandstones. In *Biogeography of the mixed evergreen forests of southern Africa.* (ed Geldenhuys C J) Occasional Report **45**, Terrestrial Ecosystems Section, Ecosystem Programmes, CSIR, Pretoria.
VLOK J H J (1988). Alpha-diversity of lowland fynbos herbs and various levels of infestation by alien annuals. *South African Journal of Botany* **54**, 623–7.
VOGL R J, ARMSTRONG W P, WHITE K L and COLE K L (1977). The closed-cone pines and cypresses. In *Terrestrial vegetation of California.* (eds Barbour M and Major J) Wiley, New York.
WALKER J (1981). Fuel dynamics in Australian vegetation. In *Fire and the Australian biota.* (eds Gill A M, Groves R H and Noble I R) Australian Academy of Science, Canberra. 101–27.
WALLER D M (1979). Models of mast seeding in trees. *Journal of Theoretical Biology* **50**, 223–32.
WEISS P W and MILTON S J (1984). *Chrysanthemoides monilifera* and *Acacia longifolia* in Australia and South Africa. In *Medecos 4 Proceedings* (ed. Bell D), University of Western Australia, Perth. 2.
WELLS P V (1969). The relation between mode of reproduction and extent of speciation in woody genera of the California chaparral. *Evolution* **23**, 264–7.
WESTOBY M (1980). Elements of a theory of vegetation dynamics in arid rangelands. *Israel Journal of Botany* **28**, 169–94.
WESTOBY M (1988). Comparing Australian ecosystems to those elsewhere. *Bio Science* **38**, 549–56.
WESTOBY M, RICE B and HOWELL J (1990). Seed size and plant growth form as factors in dispersal spectra. *Ecology* **71**, 1 307–22.
WESTOBY M, RICE B, SHELLEY J M, HAIG D and KOHEN J L (1982). Plant's use of ants for dispersal at West Head New South Wales. In *Ant-plant interactions in Australia* (ed Buckley R C) Dr W. Junk, The Hague. 5–87.
WHITEHEAD D R (1983). Wind-pollination in the angiosperms: evolutionary and environmental considerations. In *Pollination biology.* (ed Real L) Academic Press, New York. 97–108.
WICKLOW D T (1977). Germination response in *Emmananthe pendulifolia* (Hydrophyllaceae). *Ecology* **58**, 201–5.
WIENS D, ROURKE J P, CASPER B B, RICKARDT E A, LAPINE T R, PETERSON C J and CHANNING A (1983). Nonflying mammal pollination of southern African proteas: a non-coevolved system. *Annals of the Missouri Botanical Garden* **70**, 1–31.
WILLIAMS I J M (1972). A revision of the genus *Leucadendron* (Proteaceae). *Contributions of the Bolus Herbarium* **3**.
WILLSON M M, IRVINE A K and WALSH N G (1989). Vertebrate dispersal syndromes in some Australian and New Zealand plant communities, with geographic comparisons. *Biotropica* **21**, 133–47.
YEATON R I and BOND W J (1991). Competition between two shrub species: directed dispersal by ants promotes coexistence. *American Naturalist* (in press).
ZAMMIT C A and WESTOBY M (1987). Seedling recruitment strategies in obligate seeding and sprouting *Banksia* shrubs. *Ecology* **68**, 1 984–92.
ZAMMIT C A and ZEDLER P H (1988). The influence of dominant shrubs, fire, and time since fire on soil seed banks in mixed chaparral. *Vegetatio* **75**, 175–87.
ZEDLER P H (1981). Vegetation change in chaparral and desert communities in San Diego county, California. In *Forest succession: concepts and applications.* (eds West D C, Shugart H H and Botkin D B) Springer-Verlag, Berlin. 387–402.
ZEDLER P H and ZAMMIT C A (1989). A population-based critique of concepts of change in the chaparral. In *The California chaparral. Paradigms reexamined.* (ed Keeley S C) Science Series **34**, Natural History Museum of Los Angeles County, Los Angeles, California. 73–83.

APPENDIX 6.1 The system used to classify plants for Table 6.3.

This classification was based primarily on information derived from descriptions for the genera in Dyer (1975), supplemented with information on woody vine, shrub, and tree species from Coates-Palgrave (1977). Succulent karoo species were classified with assistance from Craig Hilton-Taylor. Each class represents a distinct seed dispersal and/or seed biology syndrome.

Serotiny Seeds dispersed only after fire; all plant species which belonged to the families listed in Table 6.2 and known from the literature or field observation to be serotinous. The Acanthaceae and Mesembryanthemaceae were not included because seed release may be independent of fire (see below).

Myrmecochory Seed dispersed by ants; all plant species belonging to genera listed by Slingsby and Bond (1983) and known to be myrmecochorous from field observation. This includes genera with myrmecochorous species in fynbos and non-myrmecochorous species in other vegetation formations (Slingsby and Bond 1983).

Fine seeds Species with minute wind- or water-dispersed seeds without wings or a pappus. Ericaceae and Orchidaceae were separated as 'erica' because these plants require mycorrhizae for proper growth and development and are most common in habitats with nutrient-poor soils. The remaining species may also have mycorrhizal symbionts and include Scrophulariaceae, Crassulaceae, and Mesembryanthemaceae.

Bird-dispersed Species with fleshy fruits known to be dispersed by birds including Liliaceae (Asparagaceae), Sapindaceae, Podocarpaceae, Ebenaceae, Celastraceae, and Myrsinaceae (see also Knight 1988).

Small with wings or pappus Species known to be wind-dispersed e.g. small seeds with pappus, winged seeds including Asteraceae, and Asclepediaceae.

Other Species included here either could not be classified — including many species of geophytes, Cucurbitaceae (probably vertebrate-dispersed), and Poaceae (many of which may have epizoochorous dispersal) — or the dispersal agent was unknown.

The classification is not precise but it is adequate for this analysis of the relative species richness and importance of dispersal traits as the general patterns are unlikely to change significantly in a more refined classification.

7

Plant-animal relationships

S D Johnson

INTRODUCTION

In this chapter I review research in fynbos on pollination, seed dispersal, seed predation, and herbivory. My intention is not to provide an exhaustive catalogue, but rather a compilation that gives a broad overview of research on plant-animal interactions in the fynbos.

Successful reproduction in most flowering plants depends on complex interactions with animal pollinators, dispersers, herbivores, and seed predators. It is thought that in some cases plants and animals may share an intimate history of 'coevolution' (Ehrlich and Raven 1964) which involves reciprocal adaptation between interacting species (Thompson 1989). This theory of coevolution has provided plant-animal interaction research with a central framework, although some have been too quick to see coevolution behind every phylogenetic tree (Janzen 1980; Thompson 1982; Feinsinger 1983; Schemske 1983)! In spite of what one might expect from the rich floral diversity of fynbos (Goldblatt 1978; Cowling et al. this volume), there is an apparent lack of close coevolution in plant-animal relationships. These relationships appear to be unilaterally evolved; that is, fynbos plants show numerous adaptations to animals, especially in their reproductive traits, without any obvious reciprocal adaptations by animals.

The basis for all plant-animal interaction research remains sound natural history based on field observations. Rudolf Marloth (1855–1931), an amateur botanist, pioneered this research in the Cape Floristic Region. In the four volumes of his monumental *Flora of South Africa* (1913–1932) Marloth treated plants not as isolated taxonomic entities, but rather as components of a living, interacting environment. The period between Marloth and the late 1970s was uneventful, with most research effort concentrating on describing the immense floristic diversity of the Cape flora. The 1980s saw a renewed appreciation for the role that animals play in fynbos ecology with a publication on pollination biology (Rebelo 1987a), the rediscovery of seed dispersal by ants (Slingsby and Bond 1981), and the realization that seed predation by rodents has a major impact on the population dynamics of many fynbos plants (Bond 1984; Bond and Breytenbach 1985).

POLLINATION

This section covers some of the natural history aspects of pollination — namely 'what pollinates what?' — followed by a discussion of various aspects of pollination ecology, including factors controlling fruit and seed set, mimicry, phenology, and gene flow. The problem of coevolution in pollination relationships is discussed in detail in a general section at the end of the chapter.

Unfortunately, pollination biology has never been an important research theme in fynbos. Early studies consist of the detailed observations of Marloth (1896, 1903, 1908, 1913–1932) and an extensive study by Vogel (1954) which described 'pollination syndromes' of South African plants. Information on insect pollination in South Africa is mostly scattered throughout the entomological literature (Usher 1972; Watmough 1974; Michener 1979; Claassens and Dickson 1980). Only recently have there been a few experimental insect pollination studies (Louw and Nicolson 1983; Coetzee and Giliomee 1985; Hattingh and Giliomee 1989). Bird and mammal pollination have, in contrast, been well researched (Mostert et al. 1980; Collins 1983a, b; Wiens et al. 1983; Rebelo et al. 1984; Collins and Rebelo 1987). Much of the information on pollination in fynbos was summarized in a review volume (Rebelo 1987a).

Insect pollination

> *Local botanists as well as entomologists have repeatedly noticed that often there seems to be an entire absence of insect life, although the fields or the hillsides may be aglow with flowers.*
>
> R Marloth (1908)

Anyone expecting to find fynbos teeming with insect life is bound to be disappointed. Perhaps it is for this reason, or because humans have a greater interest in vertebrates, that bird and mammal pollination have received far more attention than insect pollination in fynbos. Despite the apparent scarcity of pollinating insects, the majority of fynbos plants are insect-pollinated (Whitehead et al. 1987). A rough breakdown of the Cape flora, based on floral characteristics, is as follows: ornithophily (bird pollination) *c.* 4% (Rebelo 1987b), therophily (mammal pollination) < 1% (Rebelo and Breytenbach 1987), and anemophily (wind pollination) *c.* 12% (Koutnick 1987), with the remainder (*c.* 83%) comprising mostly entomophilous (insect-pollinated) plants and a small percentage of autogamous species (Steiner 1987). As in most regions of the world, four insect orders — Coleoptera, Diptera, Hymenoptera, and Lepidoptera — dominate the insect pollinator spectrum (Kevan and Baker 1983).

COLEOPTERA

Beetles are an important and conspicuous component of the insect pollinator fauna in fynbos. A large beetle fauna has been identified in the inflorescences of members of the genus *Protea* (Proteaceae) (Gess 1968; Coetzee and Giliomee 1985). The exclusion of birds, but not insects, from inflorescences of *Protea* species (traditionally considered to be bird-pollinated) has been shown to have no effect on seed set. This indicates that insects (mostly small beetles) may pollinate species such as *Protea repens* even though they receive regular visits by birds (Coetzee and Giliomee 1985). Gess (1968) showed that the attraction of the green *Protea* beetle, *Trichostetha fascicularis* (Cetoniinae), to *Protea* inflorescences was primarily visual, but that the impulse to enter a flower was stimulated by the scent of the flower. This beetle is closely associated with the genus *Protea* and feeds on nectar by means of specialized brush-tipped maxillae (Gess 1968). Small beetles (Nitulidae, Alticidae, Curculionidae) have been shown to carry pollen grains of some species of *Leucadendron* (Proteaceae) (Hattingh and Giliomee 1989). Interestingly, the direction of evolution in *Leucadendron* seems to be towards wind pollination which is a derived trait in the genus (Midgley 1987).

Monkey beetles (Hopliini: Rutelinae), a group endemic to southern Africa, are ubiquitous on numerous fynbos Asteraceae and Bruniaceae (Whitehead et al. 1987) (Plate 7.1a). These beetles are, however, most abundant in karoo areas north of the fynbos region. The curious way in which they burrow into inflorescences with their elongated hind legs protruding above the florets is reminiscent of 'ducks feeding on a pond' (Giliomee 1986). Many monkey beetles are as active and agile as bees, contradicting the oft-held assertion that beetles are clumsy pollinators. Monkey beetles often visit red flowers; this observation, together with the work of Dafni et al. (1990) in Israel, seems to confirm that long wavelength colours are visible to some beetles.

Blister beetles, especially *Mylabris* species (Meloidae), are usually observed eating petals of flowers of the Iridaceae including *Aristea*, *Dietes*, and *Ixia* (Plate 7.1b). These brightly coloured, toxic beetles may, in some cases, prove to be important pollinators with petals as their unique reward! Another group of blister beetles, the Zonitinae, feed on nectar using modified maxillae (De Moor 1985).

DIPTERA

A conspicuous group of floral visitors in fynbos is the long-proboscid flies which include horse flies (Tabanidae), tangle-wing flies (Nemestrinidae), and bee flies (Bombyliidae) (Plate 7.1c, d). The Cape Floristic Region appears to be a centre for speciation in these taxa (Usher 1972; Bowden 1978). Important fynbos genera such as *Pelargonium* (Geraniaceae); *Nivenia*, *Aristea*, *Gladiolus*, *Watsonia* (Iridaceae); *Erica* (Ericaceae); and *Agapanthus* (Alliaceae) contain species which are visited regularly by these flies (Marloth 1913–1932; Bezzi 1924; Vogel 1954; Rebelo et al. 1985; Goldblatt 1989; Goldblatt and Bernhardt 1990; S D Johnson unpublished data). Flowers visited by long-proboscid flies usually have a deep corolla tube or, alternatively, anthers which project a distance from the flower. There is a functional difference in the way that most Tabanidae and Nemestrinidae feed from flowers — most Tabanidae have a proboscis projecting forwards, whereas the proboscis of Nemestrinidae tends to project downwards. Also, in contrast to the Tabanidae and Nemestrinidae which tend to hover while feeding, the Bombyliidae often settle on the corolla of a flower. These morphological and behavioural traits would have a strong selective influence on the morphological characteristics of host flowers. The most extreme proboscis length among the Cape dipterans is found in *Moegistorrhynchus longirostris* (Nemestrinidae) which has a proboscis extending 60–70 mm (Plate 7.1c). According to Vogel (1954) this fly visits *Lapeirousia anceps* (Iridaceae) which has an exceptionally long corolla tube.

Other less specialized dipterans, e.g. Syrphidae and Muscidae, do not seem to play as prominent a role as the long-proboscid flies. Fly-pollinated plants with flowers which resemble and smell like decaying meat, e.g. *Orbea* (Asclepiadaceae) and *Ferraria crispa* (Iridaceae), occur in fynbos, but these plants are more common in drier karroid shrublands.

HYMENOPTERA

Despite the floral diversity of fynbos, the region does not appear to have a particularly rich bee fauna (Michener 1979). The Cape honey-bee, *Apis mellifera capensis* (Apidae), is indigenous to fynbos regions and probably plays an important pollinating role. A direct comparison of foraging efficiency of the two subspecies in a fynbos region indicated that *A. m. capensis* foragers were more successful than those of *A. m. adansonii* (W-Worswick 1988). The long, cold, wet winters of the fynbos region are clearly unsuitable for insect activity, yet *A. m. capensis* copes with this period by forming compact clusters, abandoning brood rearing, storing sufficient honey, and having a conservative pattern of thermoregulation (W-Worswick 1987; Hepburn and Jacot-Guillarmod 1991).

Since fynbos proper contains practically no trees, it has been argued that lack of tree nesting sites may limit the population densities of bees (Whitehead et al. 1987). However, many streams in fynbos are wooded and kloofs containing Afromontane forest often project fingerlike into fynbos; bee nests are usually common in these forests.

Carpenter bees, *Xylocopa* species (Anthophoridae), are quite common in fynbos. In some respects the 13 species of *Xylocopa* in fynbos (Watmough 1974) perform the ecological role of bumblebees (*Bombus*) which are absent from southern Africa. Carpenter bees have both the strength and 'intelligence' to manipulate the large flowers of many legumes. They are not restricted to legumes, however, and visit a wide range of flower types including the bell-like flowers of *Amaryllis belladonna* (Amaryllidaceae) and the brushlike flowers of *Haemanthus coccineus* (Amaryllidaceae) (Plate 7.1e). *Orphium frutescens* (Gentianaceae), a species with poricidal anther dehiscence, is buzz-pollinated by this bee. *Xylocopa* is quick to recolonize burnt areas where it utilizes burnt shrubs as nesting sites and also pollinates many petaloid monocotyledons that flower shortly after fire.

Solitary oil-collecting bees of the genus *Rediviva* (Melittidae) have been found to visit many fynbos orchids of the section Coryciinae e.g. *Pterygodium*, *Disperis*, and *Ceretandra* (Steiner 1989; K Steiner personal communication). Bees of the same genus also collect oil from some *Diascia* species (Scrophulariaceae), although mainly in non-fynbos regions (Whitehead et al. 1984; Steiner and Whitehead 1988).

Although wasps are occasionally seen visiting flowers of fynbos plants, their role in pollination is unknown. Wasps have been found with pollinia of the fynbos orchids *Disa bivalvata* and *Disa atricapilla* (S D Johnson personal observation; K Steiner personal communication). Pollination by ants has not been recorded in the fynbos region.

LEPIDOPTERA

Although of similar morphology, moths and butterflies have different flower preferences and will be considered separately.

Butterflies are not common in fynbos, probably because the vegetation is sclerophyllous and its low nitrogen content is unsuitable for phytophagous larvae (Cottrell 1985). Browns (Satyridae) and blues (Lycaenidae) are the best represented groups, although neither group is renowned for its flower-visiting habits. An exception among the Satyridae is the large *Meneris (Aeropetes) tulbaghia* which is an important pollinator of high altitude geophytes which flower in autumn. *M. tulbaghia* appears to visit red flowers exclusively — when offered a choice of artificial flowers of different colours, this butterfly chose red in more than 90% of cases (S D Johnson unpublished data).

Other fynbos butterflies do not appear to be as specialized as *M. tulbaghia* and show few floral preferences. Many of these butterflies, including the prominent *Papilio demodocus* (Papilionidae), *Catopsilio florella* (Pieridae), and *Acrea horta* (Acreidae), are opportunistic, often visiting flowers which are clearly adapted to other vectors. According to Wiklund et al. (1979) this behavioural pattern reflects the parasitic nature of many butterflies, their long pliable probosces enabling them to rob nectar from flowers adapted for other pollinators. Fynbos plants which are pollinated by butterflies either have narrow corolla tubes with inserted stamens which deposit pollen on to the proboscis, e.g. *Crassula coccinea* (Crassulaceae), or have exserted stamens which place pollen on to the body and wings, e.g. *Nerine sarniensis*, *Cyrtanthus guthrei*, and *Brunsvigia marginata* (Amaryllidaceae) (S D Johnson unpublished data). The latter pollination mechanism is probably more efficient (Cruden and Herman-Parker 1979). The proboscis and legs of butterflies are ideal for attachment of orchid pollinia because they lack scales (Plate 7.1f).

Moth pollination is totally unstudied in fynbos, although nocturnal fragrances in many species with pale coloured flowers, including *Gnidia* and *Struthiola* (Thymelaeaceae) and *Amaryllis belladonna* (Amaryllidaceae), as well as nocturnal anthesis in others e.g. *Zaluzianskya* (Scrophulariaceae) suggest moth pollination. Like butterflies, moths are probably limited by a scarcity of suitable larval host plants in fynbos vegetation. Hawkmoths (Sphingidae), an important component of the pollinator spectrum elsewhere in Africa (Skaife 1979), are not often seen in fynbos. High wind velocities which are frequent in the fynbos region are known to be unfavourable for sphingids (Eisikowitch and Galil 1971).

OTHER INSECTS

Earwigs (Dermaptera), together with a small beetle, *Pria cinerascens* (Nitulidae), seem to play a role in the pollination of *Witsenia maura* (Iridaceae), a curious plant that appears otherwise adapted to bird pollination (Wright et al. 1989). Marloth (1903) observed small hemipterans (*Pameridae roridulae*) being showered in pollen from the anthers of another unusual endemic plant, *Roridula gorgonias* (Roridulaceae). These hemipterans, which live on the phloem sap of *Roridula*, do not appear to be impeded by the sticky balsam-like substance covering the leaves (Marloth 1903).

Bird pollination

Ornithophilous plant species comprise about 4% of the Cape flora (Rebelo et al. 1984). This is a higher percentage than for the South African flora as a whole (2.5%), but is much lower than in regions of the tropics and Australia (Keighery 1982; Rebelo 1987b and references therein). The flower-bird fauna in fynbos consists of a single species of *Promerops* (sugar bird), about seven *Nectarinia* species, and one *Anthreptes* species (both sunbird genera).

Birds have several features which make them effective pollinators. They are large, highly mobile, and have feathers which are ideal for transporting pollen. They are active in cold weather and, unlike some insects, can learn to associate long wavelength colours such as red with a reward. In addition to the classical bird-red, a wide range of flower colours occurs in fynbos ornithophilous species. In the genus *Erica* (Ericaceae), ornithophilous species show a higher incidence of colour polymorphism than insect-visited species (Rebelo and Siegfried 1985).

The disadvantages of attracting birds as pollinators include the expense to plants of providing large amounts of nectar. Since fynbos flower birds perch while feeding, another cost might also be the provision of a rigid perch. For instance, bird-pollinated *Erica* species in fynbos have thicker stems than insect-pollinated species (Siegfried et al. 1985). Most Proteaceae have very robust inflorescences with woody bracts that easily support the weight of flower birds. *Babiana (Antholyza) ringens* (Iridaceae) is a curious plant with a sterile spike which is used by birds as a perch while feeding. Flowers visited by fynbos flower birds are usually arranged in brushlike inflorescences, e.g. *Leucospermum*, *Protea* (Proteaceae), or are tubular, e.g. *Erica* (Eri-

7.1a

7.1b

7.1c

PLATE 7.1

PLATE 7.1a Monkey beetles (Hopliini: Rutelinae) foraging on *Berzelia intermedia* (Bruniaceae) (Photo: A V Hall).
PLATE 7.1b Blister beetles (Meloidae) consuming petals of *Dietes iridoides* (Iridaceae) (Photo: S Johnson).
PLATE 7.1c Long-proboscid flies occurring in the fynbos region. Clockwise from top left: *Philoliche rostrata* (Tabanidae), *Philoliche gulosa* (Tabanidae), *Prosoeca nitidula* (Nemestrinidae), and *Moegistorrhynchus longirostris* (Nemestrinidae) (South African Museum collection) (Photo: S Johnson).

7.1d

7.1e

7.1f

PLATE 7.1d Long-proboscid fly (Bombyliidae) taking nectar from *Selago serrata* (Selaginaceae) (Photo: S Johnson).
PLATE 7.1e Carpenter bee (*Xylocopa* sp.) on *Haemanthus coccineus* (Amaryllidaceae) (Photo: S Johnson).
PLATE 7.1f *Meneris Aeropetes tulbaghia* (Satyridae) carrying pollinia of *Disa uniflora* (Orchidaceae) on its legs (Photo: S Johnson).

caceae); *Tritoniopsis* section Anapalina, *Watsonia* (both Iridaceae), with some intermediates, e.g. *Mimetes* (Proteaceae).

The Cape sugarbird, *Promerops cafer*, has a close association with ornithophilous members of the Proteaceae. It is the largest fynbos flower bird (*c.* 36 g, Maclean 1985) and has a distribution corresponding almost exactly with the boundaries of the Cape Floristic Region (Maclean 1985). Insects found in the bird's gut, as well as the low amino acid content of nectar found in one of its host plants (*Protea repens*), led Mostert et al. (1980) to conclude that the bird supplements its nectar diet with insects. Males of *P. cafer* are strongly territorial — the size of their territory increasing as the availability of nectar declines (Collins 1983b).

Sunbirds are much smaller than the Cape sugarbird, ranging in mass from *c.* 7–15 g (Maclean 1985). The orange-breasted sunbird, *Nectarinia famosa* (Plate 7.2a), is the most common sunbird in fynbos at higher altitudes and is usually observed visiting those members of the genus *Erica* which possess long tubular corollas (Rebelo 1983; Rebelo et al. 1984). The lesser-double collared sunbird, *Nectarinia chalybea*, is more commonly found at lower altitudes, particularly in dune thicket and fynbos. It is an important visitor, not only to *Erica*, but also to various geophyte species including *Haemanthus coccineus*, *Brunsvigia orientalis* (both Amaryllidaceae), and *Babiana (Antholyza) ringens* (Iridaceae). Rebelo (1987c) recorded this species visiting and removing pollinia from *Satyrium odorum* (Orchidaceae), which is the only record of a bird visiting an orchid species in the Cape Floristic Region. The malachite sunbird *Nectarinia amethystina* occurs in low numbers in fynbos. This large sunbird has no particular affinities with any plant group except perhaps *Aloe* species (Asphodelaceae). A few other sunbirds — *Nectarinia fusca*, *N. veroxi*, *N. afra*, and *Anthreptes collaris* — occur in a small part of the Cape Floristic Region, but are also widespread in other regions (Maclean 1985).

Flower birds require a year-round nectar supply and it is essential that there is a suite of plants which meet these requirements. The mediterranean environment of the south-western Cape does not, however, favour flowering during much of the summer. Some high altitude sites do receive summer precipitation in the form of mist. *Erica* species characteristic of high altitude ericaceous fynbos (Cowling and Holmes this volume) are often the only plants in flower in mid-summer (Rebelo et al. 1984). In contrast, lower lying areas are often dominated by large stands of Proteaceae which exhibit a winter flowering peak (Pierce 1984). It has been suggested that sunbirds migrate during mid-summer to high lying regions where *Erica* may be in flower (Siegfried 1983; Rebelo et al. 1984; Rebelo 1987b). A group of fynbos geophytes flower in a leafless state in late summer/autumn, using resources stored in bulbs or corms (Le Maitre and Midgley this volume). Many of these geophytes, e.g. *Cyrtanthus*, *Haemanthus*, and *Brunsvigia* (all Amaryllidaceae), are important food sources for birds at a time when few other plants are in flower.

A feature of bird pollination in both Australia and South Africa is that it appears to be most strongly associated with plant communities on nutrient-poor soils e.g. fynbos and kwongan (Keighery 1982; Rebelo 1987b). It has been argued (Milewski 1983; Rebelo 1987b) that bird pollination may be more economically viable in nitrogen-poor environments as 'excess' carbon unavailable for growth can be used to produce copious amounts of nutritionally inexpensive nectar. Another possible explanation is that conditions in kwongan and fynbos are not as favourable for insects as adjacent regions, resulting in plants shifting to bird pollination. Environmental factors unfavourable to insects in fynbos could include high wind velocities (Deacon et al. this volume), frequent mists and rain in mountainous regions, and the unpalatability of sclerophyllous foliage. Clearly neither of these explanations (excess carbon or insect shortage) is fully satisfactory, and more data and insight are needed on the relative merits of birds versus other pollinators in different ecosystems.

Mammal pollination

Bat pollination is unknown in fynbos, but non-flying mammals have been shown to pollinate some ground-flowering members of the Proteaceae (Rourke and Wiens 1977; Wiens et al. 1983) (Plate 7.2b). Rodents are common in fynbos, where they are responsible for extensive predation of seeds and seedlings (see below), and it is perhaps not surprising that certain plants utilize them as pollinators.

Protea species which are visited by small mammals usually have the following characteristics: bowl-shaped inflorescences of the brush type situated near the ground and hidden beneath foliage, a yeasty odour, and copious production of sucrose-rich nectar (Cowling and Mitchell 1981) with a high total carbohydrate content (> 30%) and few amino acids (Wiens et al. 1983). When small mammals were given a choice of typically mammal- versus bird-pollinated *Protea* inflorescences, they showed a significant preference for the former (Wiens et al. 1983). The primary attractant seems to be the yeasty odour.

It appears that small mammal pollination has evolved from bird or insect pollination on separate occasions in five lineages of *Protea* (Wiens et al. 1983) and possibly one lineage of *Leucospermum* (Rebelo and Breytenbach 1987). This begs the question — what stimulated the shift to mammal pollination? Most of the mammal-pollinated Proteaceae occur in small isolated populations and consequently may have been neglected by birds in favour of species which form large monospecific stands (Wiens et al. 1983). In contrast to birds, which migrate to superior food sources, relatively homebound rodents may have offered a more dependable service. Another explanation is that cryptic inflorescences beneath foliage might be less visible to harmful insects than the prominent, brightly coloured bird-visited inflorescences and would thus suffer less insect damage to flowers (Rebelo and Breytenbach 1987). It should be noted, however, that many of the beetles previously considered harmful to *Protea* inflorescences have been shown to play a role in pollination (Coetzee and Giliomee 1985).

Ecological questions

Having surveyed the spectrum of pollinating animals that occur in fynbos I now ask, more precisely, how they affect the reproductive success of their host plants. The description of pollination categories (cf. Vogel 1954; Faegri and Van der Pijl 1979), while still important, has now largely been replaced by the rapidly growing field of experimental pollination ecology. Unfortunately, fynbos pollination biologists are still largely at the stage of identifying pollinators and hence some of the theory which has recently been developed in the northern hemisphere (Lovett Doust and Lovett Doust 1988; Barret and Eckert 1990) remains to be tested locally. In this section limits to seed production, phenology, floral mimicry, and gene flow are discussed.

DO POLLINATORS DETERMINE SEED OUTPUT?

Low fruit and seed set is often observed in natural plant populations and it is generally thought that the limitation of either pollinators or nutrients is partly responsible for this phenomenon (Bierzychudeck 1981). Hand pollination or addition of nutrients to plants may boost seed set, suggesting either pollinator or nutrient limitation respectively. Even so, it is very rare that full reproductive potential is reached in outcrossing plants.

The question of why plants seem to produce an excess of flowers and ovules is complex. Flowers which do not develop fruits may still contribute to the fitness of the plant by donating pollen (male fitness); extra flowers may also increase the attractiveness of the floral display. Having a large number of flowers may enable a plant to benefit from years in which nutrients or pollinators are available in abundance (bet-hedging). Finally, it may be selectively advantageous for nutrient-limited plants to have more fertilized ovules than can be translated into seeds, as selective abortion may allow the plant to ‘choose’ the highest quality seed (Ayre and Whelan 1989; Stock et al. 1989).

The pollinator/nutrient limitation problem can be illustrated using the Proteaceae, a conspicuous fynbos family, as an example. Fynbos biologists have generally assumed that seed production in hermaphrodite members of the Proteaceae is limited by nutrients and not pollinators (cf. Collins and Rebelo 1987). There have been five major reasons behind this assumption:

- Fynbos generally occurs on low nutrient soils.
- The large seeds of many Proteaceae contain high levels of nutrients, especially nitrogen and phosphorus (Kuo et al. 1982; Esler et al. 1990).

- Observed levels of fruit set in the Proteaceae are very low (Horn 1962; Rebelo and Rourke 1986; Collins and Rebelo 1987; Stock et al. 1989). Collins and Rebelo (1987) presented data which show that mean fruit set in a number of *Leucospermum* and *Protea* species may be as low as 5.6% and 9.2% respectively. A similar situation is found in the Australian Proteaceae — mean seed set for a group of *Banksia* species was 2.3% (Collins and Rebelo 1987). Interesting exceptions are the dioecious fynbos Proteaceae, *Leucadendron* and *Aulax*, which have relatively high levels of fruit set (Rebelo and Rourke 1986; Collins and Rebelo 1987; Hattingh and Giliomee 1989). Low levels of seed set in hermaphrodite Proteaceae may, however, equally reflect pollinator limitation and cannot be used as evidence for nutrient limitation *per se*.
- Studies of Australian Proteaceae in similar environments to fynbos have not, until recently, found an increase in fruit set with additional hand pollination, suggesting that pollinators were not limiting fruit set (e.g. Pyke 1982). However, more recent Australian studies have found that addition of pollen may increase fruit set of Proteaceae, even in species previously considered nutrient-limited (e.g. Whelan and Goldingay 1989).
- Proteaceae which are fertilized with nutrient solutions may in some cases show higher fruit set. For example, follicle production of the Australian *Banksia laricina* doubled from 7.1 per plant to 14.3 per plant in plants given additional nitrogen and phosphorus (Stock et al. 1989). Based on fertilization experiments, Witkowski (1990) inferred nutrient limitation of seed production of the fynbos shrub *Leucospermum parile* (Proteaceae), but the results of this study are equivocal as most seed was released prior to harvest of the inflorescences.

It is clear that no general conclusions about the role of pollinators and nutrients in determining fruit set of fynbos Proteaceae can yet be reached. Experiments which simultaneously test a number of hypotheses about limits to fruit set will have to be performed. Other alternatives should also be explored. Coetzee and Giliomee (1987), for instance, suggested that low seed set in *Protea repens* (Proteaceae) may function as a predator avoidance mechanism, as a seed predator would be forced to eat through large numbers of nutritionally poor, infertile seeds. Distinction should also be made between the number of inflorescences setting fruit and the number of fruits per inflorescence — these two parameters may be under different control in the Proteaceae (Ayre and Whelan 1989). Experiments should ideally be performed over at least two years as evidence from some plant groups (e.g. tropical orchids) indicates that high fruit set in one year may deplete resources to the extent that growth and flowering are depressed in the following year (Zimmerman and Mitchell Aide 1989).

In studies with fynbos petaloid monocots, the addition of pollen seems mostly to increase fruit and seed set (Table 7.1), suggesting that availability of pollinators should have a strong effect on female fitness of these plants. Pollinator limitation is probably particularly acute for geophytes which flower in the immediate post-fire period when most pollinator populations are depleted. For instance, hand pollination resulted in an 800% increase in seed production of *Cyrtanthus ventricosus* (Amaryllidaceae), a fire lily which flowers about two weeks following fire (Table 7.1).

Pollinator limitation may be especially severe in small nature reserves. Plant populations in these 'islands' of natural vegetation may be too small to support viable populations of pollinating animals. Hand pollination experiments could be carried out in these small reserves to ascertain whether isolated plants are more pollinator-limited than plants in large undisturbed regions. This method might allow extinction to be predicted before it occurs. Specialist species which rely on a single pollinator would be most prone to pollination-related extinction. Bond et al. (1988) found no significant evidence that plants with specialized flowers (e.g. those with deep corolla tubes) were less common in isolated patches of fynbos. They did notice, however, that orchids, which are thought to have specialized pollination mechanisms, were scarce in isolated fynbos fragments.

FLORAL MIMICRY

Floral mimicry is fairly common in the floras of the mediterranean-type climate regions of Europe and Australia (Dafni 1984; Dafni and Bernhardt 1990), but relatively scarce in fynbos. Bolus (1888) noted a striking similarity in

TABLE 7.1 **A comparison of fruit set (%) and seed set ($\bar{x} \pm$ SD) in hand-pollinated and untreated (naturally pollinated) flowers of six fynbos geophytes. The fruit set values (raw data prior to transformation into percentages) were analysed using the chi-square test of independence with Yates' correction for continuity. The seed set data (when obtainable) were analysed using the Mann-Whitney *U* test. Unpublished data of S D Johnson.**

Species	Locality (year)	Percentage fruit set		χ^2	*P*	Seed set/capsule		*Z*	*P*
		Untreated	Hand-pollinated			Untreated	Hand-pollinated		
Amaryllis belladonna (Amaryllidaceae)	Millers Point (1989)	66 (51)[1]	82 (51)[1]	2.5	0.11	7.1 ± 10.5	18.8 ± 14.1	4.7	< 0.001
Cyrtanthus ventricosus (Amaryllidaceae)	Constantia Nek (1990)	11 (112)	41 (296)	32.3	< 0.001	1 ± 4	8.3 ± 12.2	5.9	< 0.001
Cyrtanthus guthrei (Amaryllidaceae)	Bredasdorp (1991)	6 (17)	44 (18)	4.9	< 0.05	–	–	–	–
Haemanthus rotundifolius (Amaryllidaceae)	Cape Point (1991)	3 (4 844)	19 (1 275)	302.7	< 0.001	–	–	–	–
Disa fasciata (Orchidaceae)	Constantiaberg (1990)	46 (26)	84 (25)	6.4	< 0.05	–	–	–	–
Disa uniflora (Orchidaceae)	Table Mountain (1990)	44 (18)	71 (41)	2.7	0.1	–	–	–	–
	Table Mountain (1991)	39 (138)	84 (25)	15.4	< 0.001	–	–	–	–

[1] Number of observations.

morphology between the rare orchid *Disa fasciata* and the more abundant species *Adenandra villosa* (Rutaceae) (Plate 7.2c), and suggested they may share a common pollinator. Another orchid, *Disa ferruginea* which is nectarless, closely resembles a sympatric nectar-producing species *Tritoniopsis (Anapalina) triticea* (Iridaceae). Both species flower together in mature fynbos and are pollinated by the butterfly *Meneris tulbaghia* (Satyridae) which does not appear capable of discriminating between the species in the field (S D Johnson unpublished data). There are no confirmed examples of sexual mimicry in fynbos orchids.

PHENOLOGY

Research on the phenology of flowering in fynbos in relation to pollination has been neglected. Biologists elsewhere have vigorously debated whether flowering patterns can be explained in terms of competition for pollinators, seasonal pollinator abundance, or other factors (Waser 1979; Rathcke and Lacey 1985; Zimmerman et al. 1989). Because seed ecology has been an important fynbos research theme, flowering times have sometimes been interpreted in terms of optimal timing for seed release, but rarely in terms of pollination. For instance, spring flowering in ant-dispersed Proteaceae has been related to subsequent seed release in summer when ants are most active (Pierce 1984).

Most of the important insect pollinators, including Diptera, Coleoptera, Lepidoptera, and some Hymenoptera, have a seasonal adult winged stage. The seasonal availability of insect pollinators should correspond with the flowering times of nectar plants (Waser 1979). For instance, an analysis of flowering patterns of fynbos petaloid monocots shows an autumn peak for red flowered species. This corresponds to the seasonal availability of the butterfly *Meneris tulbaghia* which is on the wing from December through to April/May (Figure 7.1). This butterfly visits red flowers exclusively (S D Johnson unpublished data). It is not always obvious whether the flowering times of plants have evolved in response to pollinators or whether the seasonal emergence of insect pollinators has responded to flowering times. The suite of red flowered geophytes in fynbos probably flower in autumn in response to *M. tulbaghia*, as the butterfly has a similar phenology in non-fynbos regions.

The phenology of the Cape honey-bee seems to be synchronized closely with the different patterns of flowering found in the winter rainfall (western) and all-year rainfall (eastern) areas of the Cape Floristic Region. Hepburn and Jacot-Guillarmod (1991) collated evidence which indicates that reproductive swarming takes place around the time of peak flowering in fynbos — September in the western Cape and November in the southern Cape. Likewise, bees tend to abandon colonies (abscond) during the low flowering period which again differs by about two months between the western and southern Cape (Hepburn and Jacot-Guillarmod 1991).

GENE FLOW

The large number of species which are packed into the small area of fynbos suggests a history of rampant speciation (see Cowling et al.; Linder et al. this volume). A combination of environmental heterogeneity and limited gene flow could have facilitated this process (speciation in fynbos plants is discussed in the final section of this chapter).

Gene flow occurs through pollen and seed dispersal; their relative importance in determining population genetic structure will differ from species to species. Seed dispersal potentially contributes more to the genetic structure of populations as seeds have twice the allelic content of pollen. Gene flow is often thought to be restricted in fynbos (Linder 1985; Cowling 1987). This is a reasonable assumption for seed dispersal since fynbos is characterized by an absence of taxa with seeds dispersed long distances by birds or mammals (see below). However, there is no reason at this stage to assume that pollen dispersal distances of fynbos plants are any different to other regions.

SEED DISPERSAL

If the latter (plants) did not possess the means of dissemination, and their offspring were consequently to germinate in close proximity to their parents, it would lead to deadly competition between the individuals most similar to each other, it would favour close fertilization, and weaken the reproductive power of the species and it would

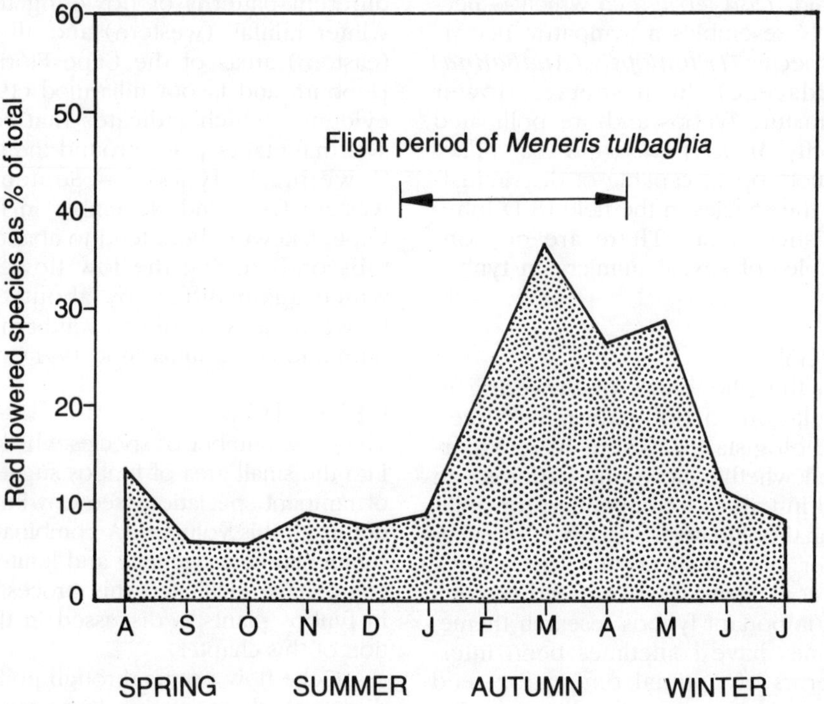

FIGURE 7.1 The distribution of the flowering times of red flowered petaloid monocots expressed as a percentage of the total number of petaloid monocots in flower in a particular month. The horizontal line depicts the flight period of the butterfly *Meneris (Aeropetes) tulbaghia* (Satyridae) which visits red flowers exclusively. The flowering times were calculated from the ranges given in Bond and Goldblatt (1984).

> *further danger its existence by restricting it to a limited area, which local dangers may render unfit for it, while many other suitable localities would remain unoccupied. Effective arrangements for securing the spreading of its seeds are consequently one of the most necessary outfits of a plant in the general struggle for existence.*
>
> R Marloth (1894)

Marloth correctly realized that there are many advantages of long-distance seed dispersal, including avoidance of competition and inbreeding as well as allowing colonization of new localities. Marloth must have been surprised, therefore, when some years later, he found that many fynbos plants have seeds dispersed short distances by ants (Marloth 1913). The dispersal of seeds by ants (myrmecochory) appears to be the only biotic seed dispersal mechanism of any importance in fynbos; plants with dispersal of seeds by birds and by attachment to animals (Van der Pijl 1982) are not found in significant numbers (Le Maitre and Midgley this volume).

Seed dispersal by birds

Species with fleshy fruits attractive to birds are rare in fynbos, but are relatively common in adjacent renoster shrubland, coastal thicket, and Afromontane forest (Knight 1988). This is reflected in the scarcity in fynbos regions of frugivorous birds such as bulbuls (Pycnonotidae), mousebirds (Coliidae), doves (Columbidae), and white-eyes (Zosteropidae) (Siegfried 1983). Possible reasons for the absence of bird-dispersed plants in fynbos, discussed by Le Maitre and Midgley (this volume), include:

- the high nutrient costs of producing fruits;
- long age to maturity of bird-dispersed plants;
- non-importance of microsites for recruitment; and
- phylogenetic legacy.

The nutrient argument (Milewski 1983) is based on the observation that vegetation types on nutrient-rich soils adjacent to fynbos have many bird-dispersed species (Le Maitre and Midgley this volume). The age to maturity argument effectively explains why bird-dispersed Afromontane trees do not invade fynbos: these trees may be killed by fire before they reach the age at which they produce fruits. Nonetheless, bird-dispersed species with a short age to maturity, e.g. *Chironia baccifera* (Gentianaceae), do occur in fynbos, implying that age to maturity is probably not the ultimate (evolutionary) explanation for the scarcity of bird-dispersed species in fynbos. The microsite argument is that since fynbos recruitment occurs shortly after fire, microsites are not important. Bird dispersal is most effective in vegetation not prone to fires, where directed dispersal to microsites below bird perches is important. Finally, coastal thicket and Afromontane forest have tropical affinities and the high proportion of bird-dispersed species in those vegetation types may simply reflect a phylogenetic legacy of fleshy fruits.

Seed dispersal by ants

Although Marloth had described myrmecochory in fynbos taxa as early as 1913, it was only recently that the importance of this dispersal mode became apparent (Slingsby and Bond 1981). There are now thought to be about 1 300 myrmecochorous taxa in fynbos (20% of the total of *c.* 6 500 strictly fynbos spp.). Only Australia has a comparable number (1 500) of myrmecochorous taxa (Berg 1975), but Australia also has a larger flora (*c.* 18 000 spp.). Ecological factors common to fynbos and environmentally analogous regions in Australia may explain the convergent evolution for this dispersal mode in both regions (Milewski and Bond 1982; Milewski 1982, 1983; Bond and Slingsby 1983; Westoby et al. 1990). Myrmecochory appears to have evolved independently in at least 20 families in the Cape flora (Bond and Slingsby 1983), suggesting powerful selective pressures for this mode of dispersal (Plate 7.2d).

Ant dispersal and the burial of seeds appear to be essential to the survival of the myrmecochorous taxa which have been studied. Studies with *Leucospermum conocarpodendron* (Proteaceae) have shown that, although removal of elaiosomes (food reward for ants) does not affect germination, seeds with elaiosomes removed are quickly eaten by rodents and consequently have little chance of success in the field (Slingsby and Bond 1985). The most important ants involved in dispersing Proteaceae seeds appear to be *Anaplolepis steingroeveri* and, to a lesser extent, *Pheidole capensis* (Slingsby and Bond 1985; Bond et al. 1991) (Plate 7.2e). Areas of fynbos close to human habitation have been invaded by the Argentine ant, *Iridomyrmex humilis*, which poses a threat to myrmecochorous species as it consumes elaiosomes without burying seed (Bond and Slingsby 1984a, b). Bond and Slingsby (1984b) found that the myrmecochorous species *Mimetes cucullatus* (Proteaceae) seldom occurred in areas infested by Argentine ants and that only 0.7% of seeds emerged as seedlings in these areas compared with 35% emergence in uninfested areas. Certain local ant species in the genus *Crematogaster* also remove elaiosomes without burying seed. These ants are indigenous and are not a threat to the flora, although they may preclude colonization by myrmecochorous taxa in areas where they are common.

While alien ants pose a threat to indigenous fynbos plants, indigenous ants appear to be assisting alien plants to invade natural tracts of fynbos. Holmes (1990) showed that in low density stands of *Acacia cyclops* and *A. saligna*, seed removal by ants may exceed the rate of predation by rodents. In this way, seeds achieve protection from rodent predation until they germinate following fire. As *Acacia* stands thicken, seed production is probably sufficient to saturate seed predators (Holmes 1990).

Seed release by myrmecochorous plants is thought to correspond with periods of maximum ant activity. It appears as if seeds of myrmecochorous Proteaceae are released in spring and summer when high temperatures are favourable for ant activity (Pierce 1984; Lamont et al. 1985; Breytenbach et al. 1986).

The germination of fynbos myrmecochorous seeds seems to be controlled largely by temperature fluctuations (Brits 1987; Pierce 1990). Bond et al. (1990) found that the regeneration of myrmecochorous taxa was greatest following intense, hot fires. This may reflect either that vegetation cover is removed more completely by hot fires resulting in greater soil temperature fluctuation, or that a greater heat pulse, essential for germination of some species, penetrates the soil. Effective germination would depend largely on a combination of depth of seed burial and the depth to which cues are effective. Galleries of *Anaplolepis custodiens* are usually found 4–7 cm below ground (Bond and Slingsby 1983) and the hypocotyl length of Proteaceae seedlings (a good measure of burial depth) are usually 2–5 cm (Brits 1987; Bond and Stock 1989).

Although the importance of ants to the success of myrmecochorous taxa cannot be doubted, it has proved frustratingly difficult to elucidate exactly why this means of dispersal has evolved in preference to others (Bond et al. 1991). Current explanations for the adaptiveness of myrmecochory in fynbos vegetation centre around seed escape from fire, escape from seed predators, directed dispersal to nutrient-enriched sites, and directed dispersal to escape competition. It is difficult to separate ecological and evolutionary hypotheses in the examples which follow. In some cases it is possible that the present-day ecological benefits of myrmecochory were not the reasons for the evolution of the dispersal mode.

DIRECTED DISPERSAL TO NUTRIENT-ENRICHED SITES

In some northern hemisphere studies ants have been shown to disperse seeds to nests which are nutrient-enriched relative to surrounding soil (Beattie and Culver 1983). The idea that directed dispersal to nutrient-enriched sites has been an important factor in the evolution of myrmecochory in nutrient-poor areas of South Africa and Australia is an attractive one, but lacks experimental support. In Australia, Rice and Westoby (1986) found that soil around seedlings from ant-dispersed seeds was no different in nitrogen and phosphorous from soil around other seedlings. Bond and Stock (1989) showed that ants dispersed seeds of the fynbos shrub *Leucospermum conocarpodendron* (Proteaceae) to sites which were even lower in nutrients than sites where passively dispersed seed had germinated. This was because soil nutrients were inversely correlated with both depth and distance from parents (Bond and Stock 1990).

DIRECTED DISPERSAL AVOIDS COMPETITION

Directed dispersal by ants may help myrmecochorous fynbos shrubs to avoid competition with sympatric *Protea* species. Yeaton and Bond (1990) found that wind-dispersed seeds of the highly competitive serotinous shrub *Protea lepidocarpodendron* accumulate against burnt skeletons of *L. conocarpodendron* following their release after fire. Seedlings of *L. conocarpodendron* would thus be rapidly outperformed if they were not dispersed to open sites by ants. Coexistence of these two species is discussed by Bond et al. (this volume). Directed dispersal to open sites may also help seedlings to avoid competition with parents in species which produce dense sprouts following fire e.g. *Mimetes cucullatus* (Proteaceae).

AVOIDANCE OF PREDATION

A number of studies show that myrmecochorous seeds which are not rapidly buried by ants are quickly eaten by granivorous rodents (Bond and Slingsby 1984b; Bond and Breytenbach 1985; cf. O'Dowd and Hay 1980). Bond and Breytenbach (1985) found that up to 100% of seeds of *Mimetes pauciflorus* and *Leucospermum glabrum* (Proteaceae) in ant-proof exclosures were removed by rodents, although when ants had access to seeds, they removed and buried them within minutes. Interestingly, seeds with intact elaiosomes were discovered more readily by both ants and rodents than seeds with elaiosomes removed, suggesting that the mutualist and the predator may utilize a common cue (Bond and Breytenbach 1985).

Predation pressure by rodents would provide a powerful selective force for the evolution of myrmecochory in fynbos plants. This hypothesis is less persuasive, however, in the light of evidence which suggests that post-dispersal vertebrate predation does not seem to be as important in south-west Australian kwongan (Cowling and Lamont 1987; Andersen 1988; Hughes and Westoby 1990) where myrmecochory is at least as common as in fynbos (Milewski and Bond 1982). Many myrmecochorous fynbos shrubs, e.g. *Muraltia* (Polygalaceae) and *Agathosma* (Rutaceae), have tiny seeds, possibly as an allometric consequence of having small leaves (Pierce 1990). It appears that these seeds are eaten not so much by rodents as by ants (Pierce and Cowling 1991). Myrmecochory in these species may have evolved as a mechanism to avoid predation by granivorous ant species which do not bury seeds.

AVOIDANCE OF FIRE DAMAGE

Seed burial by ants may have evolved as a means of avoiding heat damage to seeds. This may explain why myrmecochory is prevalent in fynbos as well as in Australian vegetation which is also fire-prone. The hypothesis is not easy to test, however — merely showing that myrmecochorous seeds do not survive fire on the soil surface is inappropriate as the ability to survive high temperatures may have been lost through evolutionary time.

MYRMECOCHORY VERSUS OTHER DISPERSAL MODES

Wind and ballistic dispersal of seed appear to be the only successful alternatives to myrmecochory in fynbos; bird dispersal of seeds is found in only a handful of taxa.

The most striking difference between wind and ant dispersal is the distance over which seeds are dispersed. Whereas ants seldom disperse seeds more than five metres (Slingsby and Bond 1985), wind-dispersed seeds may travel hundreds of metres. Bond (1988) found that *Protea* seeds may travel 30 metres through the air and then far greater distances over soil until impeded by an obstruction (e.g. burnt plant skeletons). Wind-dispersed seeds are typical of serotinous species which release seed after fire, as well as the numerous geophytes and herbs which flower in the immediate post-fire period. By contrast ant dispersal takes place mostly in mature unburnt vegetation.

It is interesting that many families, e.g. Proteaceae, Asteraceae, and Restionaceae, have both wind- and ant-dispersed representatives in fynbos. These families would be ideal subjects for studying the ecological implications of dispersal mode. Particularly valuable would be comparisons between closely related species which share a common pollination mode but which differ in means of dispersal. Expectations would be that ant-dispersed species would have smaller breeding neighbourhoods and reduced colonization ability. Thus, both speciation and extinction rates should be higher in myrmecochorous taxa (Cowling et al. this volume). Differences in dispersal distance probably explain why many myrmecochorous Proteaceae are confined to small populations, whereas wind-dispersed *Protea* species are often widely dispersed and cover-abundant.

In conclusion, the evolution of myrmecochory must be understood in terms of the alternative options. Myrmecochory in fynbos probably evolved under a combination of selective pressures of which seed predation was the strongest. Serotiny in the Proteaceae, which is associated with wind-dispersed seed, seems to be an alternative strategy to myrmecochory and probably involves a greater risk of seed predation (see below), but also has advantages in terms of increased gene flow and colonizing ability. Myrmecochory has evolved independently in numerous families and in some cases the reasons may have been quite different; for instance, burial by ants of seeds of the root parasite *Myrstapetalon* (Balanophoraceae) may well help in the establishment of a parasitic relationship with its proteoid hosts (Marloth 1913).

SEED PREDATION

Research on fynbos plants, particularly Proteaceae, has shown that pre- and post-dispersal seed predation has a major impact on reproductive success (Myburgh et al. 1974; Bond et al. 1984; Bond and Breytenbach 1985; Coetzee and Giliomee 1987; Pierce 1990; Van Hensbergen et al. 1991).

Pre-dispersal predation by insects

Fynbos species with canopy-stored seed (serotiny) appear to be particularly vulnerable to pre-dispersal insect predation. The Proteaceae have been well studied in this regard and a large number of insect seed predators (mostly larvae of Lepidoptera and Coleoptera) have been identified (Gess 1968; Myburgh and Rust 1975). These insects can have a devastating impact on the seed bank: for instance, 84% of *Protea repens* (Proteaceae) seeds were damaged by insects within 102 weeks after flowering (Coetzee and Giliomee 1987) (Figure 7.2). Slightly lower levels of predation were reported for *Protea magnifica (barbigera)* by Myburgh et al. (1974) and Wright and Giliomee (1991). However, levels of seed predation for *Protea laurifolia* in the same communities as the predation-susceptible *P. magnifica* were much lower (*c.* 10%) (Wright and Giliomee 1991).

Seed stored in the canopy is vulnerable as it can be readily located by insects. As a first mechanical line of defence, serotinous Proteaceae have thick, hard, woody cones (Stock et al. 1990). Furthermore, it has been suggested that low seed set in Proteaceae may be an adaptive strategy to frustrate seed predators (Coetzee and Giliomee 1987; Wright and Giliomee submitted; P J Mustart, M Wright and R M Cowling unpublished data): most insect larvae appear to bore through the infructescence in a random fashion and would have less chance of encountering the few fertile seeds. Mustart et al. (unpublished data) have found that the few fertile seeds in the infructescences of *Protea susannae* and *P. obtusifolia* are clustered and they suggest that this may further reduce the probability of an insect encountering fertile seeds. This hypothesis (low seed set in the Proteaceae is a mechanism to avoid seed predation) must remain tentative until alternative hypotheses such as pollinator limitation have been tested. Seed clustering itself may possibly result when pollen is deposited on a number of adjacent florets during a single probe by a bird searching for nectar or insects.

Predation by small mammals

Fynbos appears to have denser populations of small mammals than are found in similar shrublands in Australia and California (Fox et al. 1985). Predation of seeds by small mammals has an enormous impact in fynbos. To illustrate, attempts to re-establish the Clanwilliam cedar *Widdringtonia cedarbergensis* (Cupressaceae) have been frustrated largely due to seed predation by granivorous rodents (Botha 1988).

PREDATION AND SINGLE-AGEDNESS OF FYNBOS

Although fire reduces rodent populations considerably, some rodents inevitably survive by sheltering in streamside vegetation or among rocky outcrops (Willan and Bigalke 1982). The post-fire environment lacks cover and is especially unsuitable for recolonization by rodents such as the otherwise ubiquitous striped field mouse, *Rhabdomys pumilio*, which avoids open areas and is normally associated with thick grassy vegetation (Bond et al. 1980). Less timid species such as *Aethomys namaquensis* may colonize a burnt area sooner. The general trend is that small mammal populations increase as vegetation matures (Willan and Bigalke 1982; Fox et al. 1985; Breytenbach 1987; Midgley and Clayton 1990; Van Hensbergen et al. 1991). The consequence for plants may be that the immediate post-fire environment is a relatively predator-free 'window' for recruitment by seed.

Both Bond (1984) and Breytenbach (1984) have argued that single-agedness of many fynbos plants results from lack of inter-fire recruitment, owing to seed and seedling predation by rodents. This appears to be true for weakly serotinous species such as *Protea repens* (Proteaceae) which may release seed prematurely into mature vegetation (Bond 1984). It should be borne in mind, however, that reproductive traits of most fynbos plants show adaptation to recruitment in the immediate post-fire period (Le Maitre and Midgley this volume). Post-dispersal predation in mature fynbos is of such high levels (Bond 1984; Pierce and Cowling 1991) that it would have provided a powerful selective force for the evolution of traits such as serotiny and myrmecochory, as well as post-fire flowering and seed release by geophytes. Seed predation is not the only factor which may prevent recruitment in mature fynbos — competition for water, shading of seedlings, and depletion of nutrients should also be considered and weighed up against seed predation when trying to explain post-fire recruitment strategies.

PREDATION AND SEASON OF BURN

Poor recruitment of serotinous Proteaceae is often observed following spring burns (Bond et al. 1984; Van Wilgen and Viviers 1985; Midgley 1989). Bond (1984) suggested that the

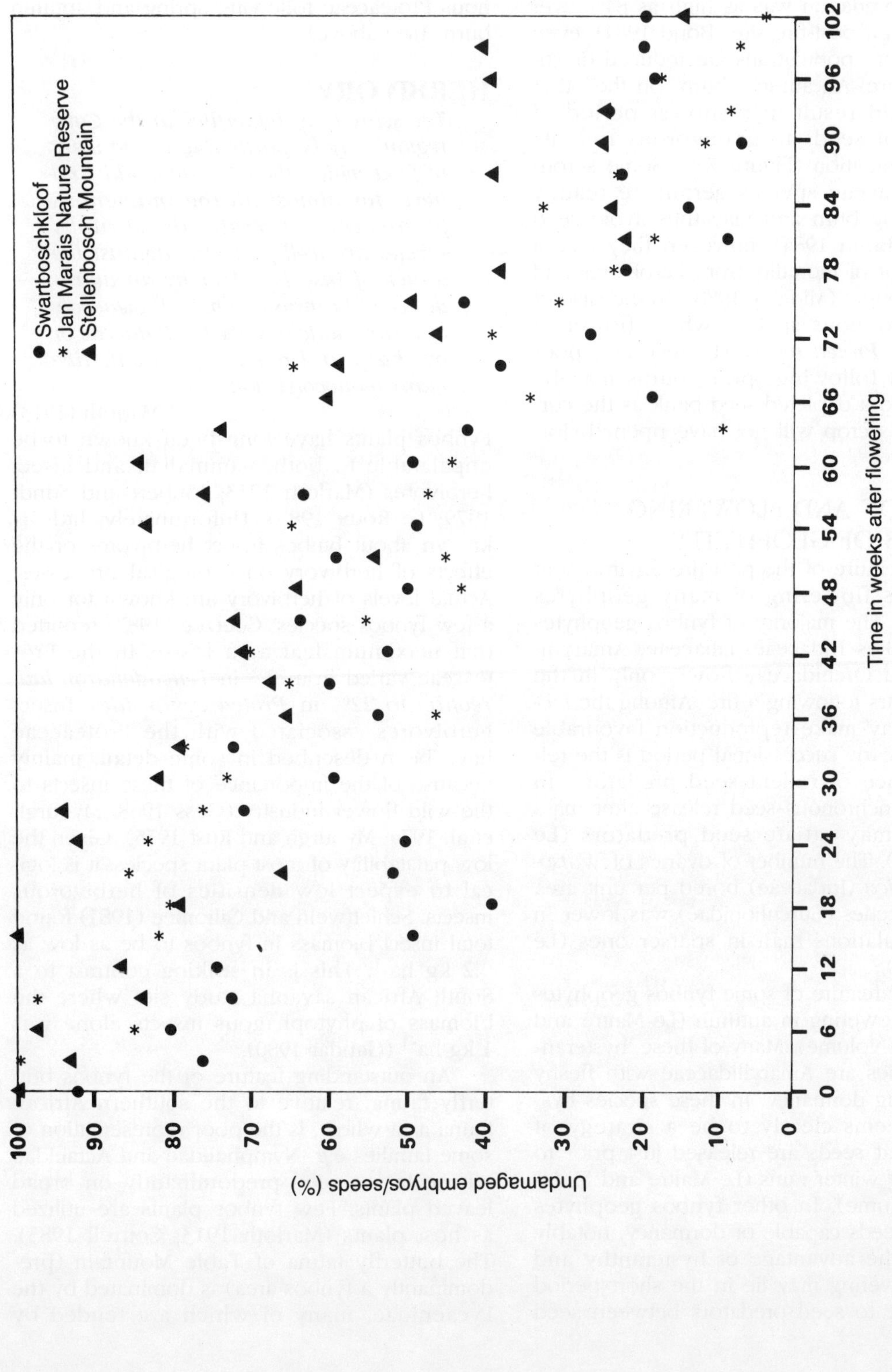

FIGURE 7.2 The percentage of undamaged seeds of *Protea repens* related to the time after flowering for three areas. The fitted regression line is $y = 89.935 - 0.725x$, $R^2 = 0.663$ for the three areas. The peak of the flowering season, mid-May, was calculated as zero weeks after flowering. Reprinted from Coetzee and Giliomee (1987) with the permission of the National Botanical Institute.

long period between seed release in spring and germination at the onset of winter results in heavy seed predation by rodents (Figure 7.3). Seed predation was as high as 84% over 15 weeks in a post-fire site (Bond 1984), even though rodent populations are reduced drastically after fire. An autumn burn, on the other hand, would result in a shorter period of exposure of seeds to granivorous rodents before germination (Figure 7.3). Some serotinous Proteaceae species germinate readily after a spring burn and may thus avoid seed predation (Bond 1984); however, they face a further threat of mortality from herbivores and summer drought (Midgley 1988). In the case of weakly serotinous species which flower in winter, e.g. *Protea repens* (Proteaceae), poor recruitment following spring burns may be due partly to a depleted seed bank as the current season's crop will not have ripened (Jordaan 1949).

PREDATION AND FLOWERING PATTERNS OF GEOPHYTES

A striking feature of the post-fire environment is the mass flowering of many geophytes (Plate 7.2f). The majority of fynbos geophytes in the families Iridaceae, Liliaceae, Amaryllidaceae, and Orchidaceae flower only in the first few years following a fire. Among the factors that may make reproduction favourable during this early successional period is the relative absence of rodent seed predators. In addition, synchronous seed release after mass flowering may satiate seed predators (Le Maitre 1984). The number of ovaries of *Watsonia borbonica* (Iridaceae) bored per unit area by snout beetles (Curculionidae) was lower in dense populations than in sparser ones (Le Maitre 1984).

Another feature of some fynbos geophytes is leafless flowering in autumn (Le Maitre and Midgley this volume). Many of these 'hysteranthous' species are Amaryllidaceae with fleshy seeds lacking dormancy. In these species hysteranthy seems clearly to be a strategy of ensuring that seeds are released just prior to the onset of winter rains (Le Maitre and Midgley this volume). In other fynbos geophytes with hard seeds capable of dormancy, notably Iridaceae, the advantage of hysteranthy and autumn flowering may lie in the short period of exposure to seed predators between seed release and germination. This is a similar argument to the one used by Bond (1984) to explain differences in recruitment of serotinous Proteaceae following spring and autumn burns (see above).

HERBIVORY

> *The scarcity of butterflies in the Cape region may be partly due to the want of food which their larvae could find here, for almost all the indigenous plants and especially the shrubby species are well protected against the attack of insects, either by an abundance of tannins like most Proteaceae, or by aromatic oils like the Rutaceae or by sharp and acrid ingredients like many monocotyledons.*
>
> R Marloth (1913)

Fynbos plants have long been known to be unpalatable to both mammalian and insect herbivores (Marloth 1913; Joubert and Stindt 1979; Le Roux 1988). Unfortunately, little is known about fynbos insect herbivores or the effects of herbivory on ecological processes. Actual levels of herbivory are known for only a few fynbos species. Coetzee (1987) reported that maximum leaf area losses in the Proteaceae varied from 2% in *Leucadendron laureolum* to 22% in *Protea cynaroides.* Insect herbivores associated with the Proteaceae have been described in some detail, mainly because of the importance of these insects to the wild flower industry (Gess 1968; Myburgh et al. 1974; Myburgh and Rust 1975). Given the low palatability of most plant species, it is logical to expect low densities of herbivorous insects. Schlettwein and Giliomee (1987) found total insect biomass in fynbos to be as low as 0.2 kg ha^{-1}. This is in striking contrast to a South African savanna study site where the biomass of phytophagous insects alone was 1 kg ha^{-1} (Gandar 1980).

An outstanding feature of the fynbos butterfly fauna, relative to the southern African fauna as a whole, is the poor representation of some families e.g. Nymphalidae and Acraeidae whose larvae feed predominantly on broad leaved plants. Few fynbos plants are utilized as host plants (Marloth 1913; Cottrell 1985). The butterfly fauna of Table Mountain (predominantly a fynbos area) is dominated by the Lycaenidae, many of which are tended by

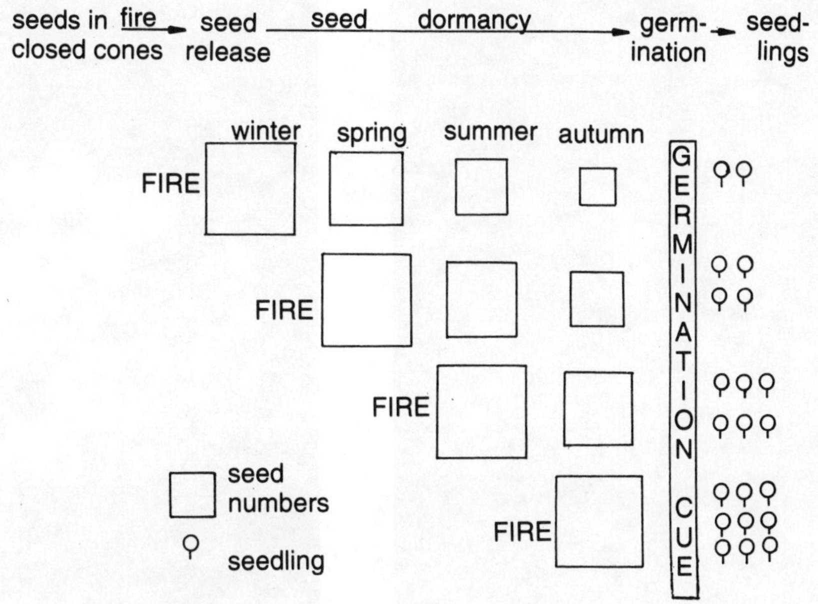

FIGURE 7.3 A diagrammatic representation of the effects of fire season, seed dormancy, and predation on the recruitment of serotinous Cape Proteaceae. Squares represent seed reserves, large after fires have stimulated seed release from cones and becoming progressively smaller from predation before they germinate in autumn. Reprinted from Bond (1984), with the permission of Kluwer Academic Publishers.

ants, and the Satyridae which are mainly grass feeders (Table 7.2). Relative to southern Africa as a whole, Table Mountain has a distinct deficit of Acraeidae, Nymphalidae, and Papilionidae, the larvae of which usually feed on broad leaved plants. Little is known about herbivory by moth larvae in fynbos, except for those of economical relevance to the export of wild flowers. Because of the apparent unpalatability of leaves of fynbos plants, Cottrell (1985) predicted that fynbos might contain a relatively high proportion of phloem sap-suckers, root feeders, and pollen and seed feeders among the insect fauna.

Theories of anti-herbivore defence

There are now, broadly speaking, two theories of anti-herbivore defence (Edwards 1989), both of which would predict high levels of carbon-based defence in fynbos. Apparency theory (Feeny 1975; Rhoades and Cates 1976) distinguishes between plants (or tissues) which are 'apparent' to herbivores, such as long-lived evergreen perennials, and those which are 'unapparent' such as early successional species or deciduous species. It is argued that apparent plants are inevitably discovered by specialist herbivores and that toxins would eventually be overcome through a biochemical coevolutionary 'arms race' between plant and herbivore (Ehrlich and Raven 1964). 'Apparent' plants should, therefore, employ 'quantitative digestibility reducing substances', e.g. tannins, which cannot be detoxified. 'Unapparent' plants, on the other hand, escape specialist herbivores and hence employ 'qualitative toxins' such as alkaloids against generalist herbivores. Two major problems with this theory are firstly the difficulty in defining 'apparency' and secondly the growing realization that some tannins may be detoxified by insects (Bernays 1981; Zucker 1983).

7.2a

7.2c

7.2b

PLATE 7.2

PLATE 7.2a Female orangebreasted sunbird (*Nectarinia famosa*) taking nectar from *Erica* sp. (Photo courtesy of Fitz-Patrick Institute).

PLATE 7.2b *Aethomys namaquensis* foraging on *Protea humiflora* (Proteaceae) (Photo courtesy of J Rourke).

PLATE 7.2c Mimicry or coincidence? *Disa fasciata* (Orchidaceae) (left) and *Adenandra villosa* (Rutaceae) (right) (Photo: S Johnson).

7.2d

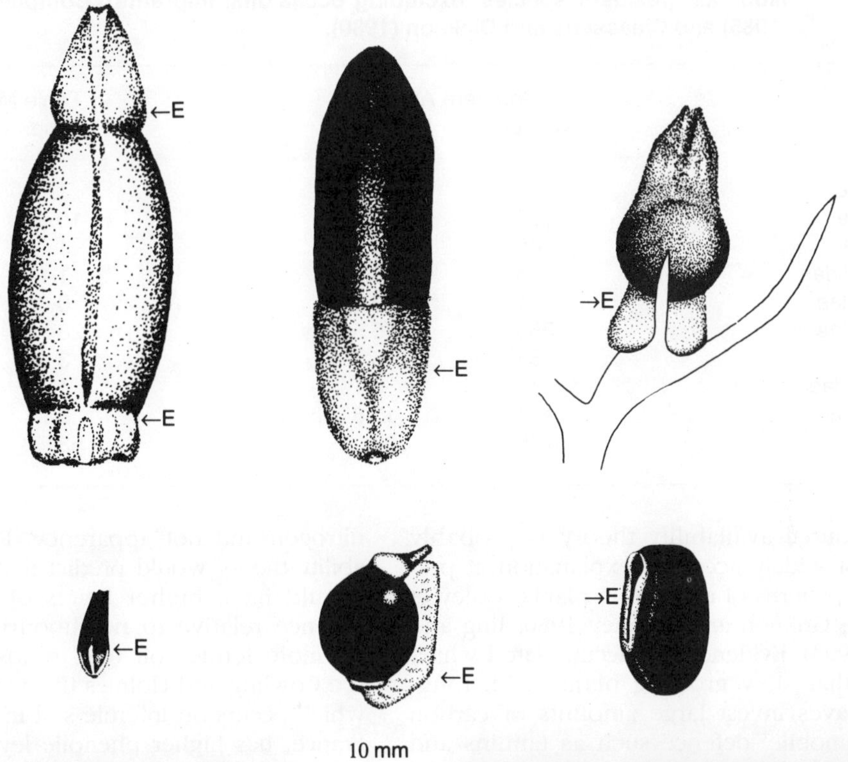

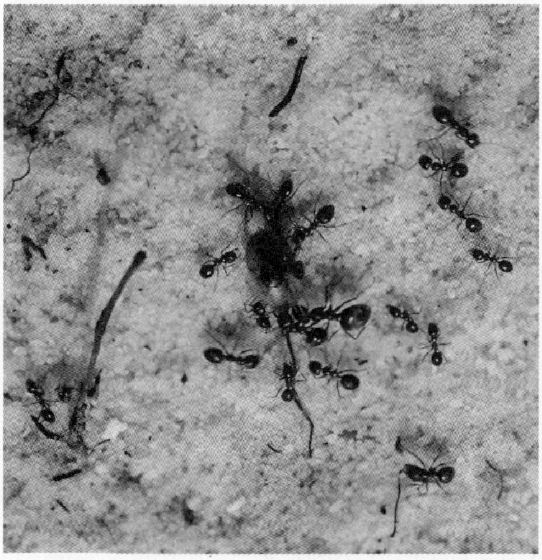

7.2e

7.2f

PLATE 7.2d Typical fynbos seeds bearing an elaiosome (arrowed). Clockwise from top left: *Mimetes cucullatus* (Proteaceae), *Hypodiscus aristatus* (Restionaceae), *Thesium strictum* (Santalaceae) — still attached to branchlet, *Podalyria glauca* (Fabaceae), *Zygophyllum fulcrum* (Zygophyllaceae), and *Stilbe* sp. (Stilbaceae) (Drawings by P Slingsby).

PLATE 7.2e Ants *anoplolepis steingroeveri* manoeuvring elaiosome-bearing seeds of *mimetes cucullatus* (Photo: J C Paterson-Jones).

PLATE 7.2f *Bulbinella nutans* (Asphodelaceae) flowering *en masse* following a fire (Photo: S Johnson).

TABLE 7.2 The number of species per family for the butterfly fauna of southern Africa and Table Mountain (resident species, excluding occasional migrants). Compiled from Cottrell (1985) and Claassens and Dickson (1980).

	Southern Africa		Table Mountain	
	No. spp.	% of total	No. spp.	% of total
Danaidae	7	0.8	1	1.8
Satyridae	77	9.7	10	18.8
Acreidae	46	5.8	1	1.8
Nymphalidae	103	13.1	1	1.8
Libytheidae	1	0.1	0	0
Lycaenidae	358	45.5	28	52.8
Pieridae	53	6.7	3	5.7
Papilionidae	17	2.1	1	1.8
Hesperidae	124	15.8	8	15
Total	786	100	53	100

Resource availability theory is probably the most widely accepted explanation at present for patterns of defence in plants (Coley et al. 1985; Gulmon and Mooney 1986; Jing and Coley 1990). Evidence has accumulated which shows that slow growing plants with long-lived leaves invest large amounts of carbon into 'immobile' defence such as tannins and lignins. Plants with fast growth rates and short-lived leaves, on the other hand, generally invest relatively small amounts of nitrogen into 'mobile' defence compounds such as alkaloids and cardiac glycosides. Since these 'mobile' compounds have a rapid turnover, their production cost accumulates over the leaf's lifetime. The compounds are affordable, however, as the leaves in which they are found are typically short-lived (Gulmon and Mooney 1986). For nutrient-limited fynbos plants, the difficulty in obtaining nitrogen to replace a leaf lost to a herbivore should result in selection for high levels of defence (Janzen 1974).

Patterns of anti-herbivore defence in fynbos

The mosaic of shrublands in the Cape Floristic Region, which occurs on soils of differing fertility, provides an ideal testing ground for the theories outlined above. These shrublands consist mainly of evergreen plants of similar apparency, hence any differences in defence would have to be attributed to availability of nitrogen and not apparency. Resource availability theory would predict that fynbos plants should have higher levels of carbon-based defence relative to neighbouring shrublands on more fertile soil (e.g. renoster shrubland; see Cowling and Holmes this volume). Maquis, which occurs on infertile soil in mediterranean France, has higher phenolic levels, lower leaf nitrogen levels, and less grazing damage than adjacent garrigue on more fertile, calcareous soil (Glyphis and Puttick 1989).

Campbell (1986) tested the prediction that levels of spinescence should be higher in fynbos relative to adjacent vegetation on more fertile soils, since fynbos leaves contain valuable long-term stores of nitrogen which are costly to replace (Stock and Allsopp this volume). Contrary to prediction, Campbell (1986) found that levels of spinescence were in fact lower in fynbos and this was attributed to exceptionally low leaf nitrogen levels which are often below the critical point for supporting herbivores (Mattson 1980). Fynbos provides very poor grazing for both domestic livestock (Le Roux 1988) and indigenous ungulates (Norton 1980; Rebelo this volume). Fynbos plants have therefore never been under selective pressure to develop spinescence. It is interesting that one of the few fynbos taxa to possess spines, *Aspalathus* (Fabaceae), also has nitrogen-fixing symbionts and may therefore have higher leaf nitrogen levels and palatability. The small leaf sizes typical of fynbos plants would mean that mammalian herbivores would be forced to chew a mouthful of leaves and twigs of little sustenance (Campbell 1986; Norton 1980). Thus the only mammalian herbivores that occur in fynbos, albeit in low numbers, are small specialist herbivores such as the klipspringer antelope *Oreotragus oreatragus* (Norton 1980) and the rock hyrax *Procavia capensis* (Morrow et al. 1983). By contrast, adjacent renoster shrubland on more nutrient-rich soils previously supported a large spectrum of mammalian herbivores, including elephant *Loxodonta africana* and black rhinoceros *Diceros bicornis*, prior to intense exploitation by humans (Hendey 1987; Rebelo et al. this volume). The palatable grass component of renoster shrubland was probably maintained at higher levels than at present by regular burning by pastoralists (Novellie 1987).

Mole rats and geophytes

More than 16% of species in the Cape flora are geophytes, mostly petaloid monocots in the families Iridaceae, Liliaceae, and Orchidaceae (Goldblatt 1978). Nutrients are stored in bulbs and corms, hence it is not surprising that many Cape geophytes e.g. *Boophane*, *Amaryllis* (Amaryllidaceae), and *Morea* (Iridaceae) are defended by potent toxins. Nevertheless, mole rats (Bathyergidae) are important consumers of corms and bulbs of geophytes in the Cape Floristic Region. Mole rats even consume geophytes, such as *Ornithogalum thyrsoides* (Hyacinthaceae), which are known to contain toxins capable of causing death to humans and livestock (Lovegrove and Jarvis 1986). Some highly palatable geophytes e.g. *Micranthus junceus* (Iridaceae) have segmented corms much like garlic which are broken apart by mole rats; uneaten segments may be strewn in burrows, thereby functioning as dispersal units (Lovegrove and Jarvis 1986). In addition, many geophytes have clusters of cormlets which remain behind if the parent corm is removed by mole rats. Renoster shrubland is particularly rich in geophytes (Le Maitre and Midgley this volume) as well as mole rats, suggesting a possible mutualistic interaction. Raitt (1986), on the other hand, found a negative relationship between the presence of mole rats and orchids at a site on the Cape Flats.

DISCUSSION: HOW MUCH COEVOLUTION IN FYNBOS?

In previous sections I covered some of the natural history and ecological aspects of fynbos plant-animal relationships. To complete the story, I consider the following two questions. Have these relationships coevolved? Can the floral richness of fynbos be linked to interactions with animals, though not necessarily in the coevolutionary sense?

Coevolution, or 'reciprocal change in interacting species' (Thompson 1989), has become the central concept in the study of plant-animal relationships. Yet there is still much confusion as to what constitutes coevolution (Janzen 1980). Generally speaking 'pair-wise coevolution' refers to reciprocal change between pairs of interacting species, whereas 'diffuse coevolution' describes reciprocal change between a number of species. If evolutionary change has occurred on one side of a relationship only, the situation is one of 'unilateral evolution' which is not coevolution at all.

Pollination

There are no clear cases, as yet, of pair-wise coevolved plant-pollinator relationships in fynbos. There are in fact few cases of pair-wise coevolved pollination relationships anywhere; those that have been documented in *Ficus* and possibly *Yucca* are unusual in that the pollinating wasps are also seed parasites, making the association more intimate than usual pollination relationships (Schemske 1983; Kiester et al. 1984). Perhaps the best candidates for diffusely coevolved relationships in fynbos are those between long proboscid flies and plants with long-tubed flowers (Plate 7.1c). The evolution of very deep corolla tubes and long insect probosces can be explained by runaway coevolution: longer probosces enable the insect to reach more nectar, but in response, selection occurs for a deeper corolla tube which forces the insect to press against the reproductive organs of a flower (thereby increasing pollination efficiency) which in turn results in selection for longer probosces (Nilsson 1988).

There are no butterfly groups that are closely associated with fynbos (Cottrell 1985), hence coevolution between butterflies and flowers is unlikely. While many fynbos plants have evolved large red flowers which are attractive to the butterfly *Meneris tulbaghia*, this species occurs throughout southern Africa and cannot be considered to have coevolved with fynbos *per se.* The Hopliini, a group of specialized flower-visiting beetles, are common in fynbos but appear to have radiated most extensively in the succulent karroid shrublands north of the fynbos (Scholtz and Holm 1985). Further work may reveal coevolution in the thousands of undescribed fynbos insect-flower relationships, but at present it seems that few specialized flower-visiting insects are associated with the region.

A few hundred ornithophilous fynbos plant species are pollinated by a handful of sunbirds and one species of sugarbird, implying little specialization on the part of birds (Rebelo et al. 1985). There are two broad bird-flower relationships — between the Cape sugarbird, *Promerops cafer*, and the Proteaceae, and between the orange-breasted sunbird, *Nectarinia famosa*, and long-tubed *Erica* species (Siegfried 1983; Rebelo 1983; Rebelo et al. 1984). These relationships may be diffusely coevolved (Rebelo et al. 1984). Both bird species are restricted to the Cape Floristic Region (Maclean 1985).

Many *Protea* species show clear-cut adaptations to pollination by mammals (see earlier section). However, the rodents themselves show no obvious adaptations to *Protea* — indeed, *Protea* nectar appears to be only a minor part of their diet. The *Protea*-mammal system is thus not coevolved and is instead a good example of unilateral evolution (Wiens et al. 1983). By contrast, the Australian honey possum, *Tarsipes rostratus*, has several adaptations for nectar feeding and pollen digestion and appears to have coevolved with species of *Banksia* and *Dryandra.*

Certain aspects of fynbos ecology would militate against close coevolution:

• Fynbos is subject to recurrent fire. Each fire is an idiosyncratic event varying in intensity and season, following which local extinctions may occur (Cowling 1987). Species which recruit from seed are exterminated from an area if fire occurs before they reach reproductive age. Extinctions may also occur should the interval between fires be too long (Bond et al. 1988). A fire-prone system is thus unsuitable for specialists, particularly those locked into obligate mutualisms, since the extinction of one species could mean the extinction of the other.

• There is a serial successional pattern of flowering which proceeds for several years after fire. It stands to reason that a pollinator is unlikely to specialize on a species such as *Cyrtanthus ventricosus* (Amaryllidaceae) or scores of other plants which only flower in the first year or two following fire.

• Fynbos, like other temperate regions, has a seasonal climate which does not allow year-round flowering. Long-lived insects and birds which rely on nectar for survival cannot specialize on a single species and are forced to forage on a succession of species throughout the year, thus limiting the potential for close coevolution.

• The apparently low insect biomass in the fynbos region (Marloth 1908; Schlettwein and Giliomee 1987) may mean that plants are consistently pollinator-limited, especially after fire (Table 7.1). This would result in intense selection for floral modifications which increase the attractiveness of flowers to insects and which increase the efficiency of pollen transfer. Insects, on the other hand, may be limited more by the availability of larval food than nectar, resulting in few modifications for traits involved with pollination. If the fitness consequences of a mutualism are asymmetrical, we expect little coevolution (Schemske 1983; Kiester et al. 1984).

• Brushlike inflorescences are usually associated with pollination by perching birds in fynbos and in Australia. These inflorescences, unlike tubular flowers, do not allow the plant to discriminate between floral visitors (Stiles 1978). There is, therefore, little opportunity for specialization between perching birds and fynbos plants, with the exception of species with tubular flowers. This might explain why bird pollination systems in fynbos are less closely coevolved than tropical systems involving hummingbird pollination of predominantly tubular flowers (Stiles 1978).

Seed dispersal

There may be good reasons for a myrmecochorous species to evolve a relationship with a single ant species. Germination may be a function of burial depth (Brits 1987; Bond et al. 1990) and, assuming that different ants have different gallery depths, germination success may be related to which ants are involved in dispersal. Myrmecochory does not, however, appear to be a coevolved system in fynbos. There are clear adaptations to ant dispersal on the part of the plants with little evidence that ants have evolved any particular morphological or behavioural traits in response. Pre-viously it was thought that ants with a wide distribution in southern Africa, such as *Anaplolepis custodiens* (Prins 1978), buried seeds with elaiosomes only within the fynbos region, suggesting coevolved behavioural traits in the ants (see Bond and Slingsby 1984b). However, it seems that *Anaplolepis* responds to elaiosome-bearing seed regardless of whether the ants occur in the fynbos region or in the karoo, a region with few or no myrmecochorous plants (Breytenbach 1988). It appears that the elaiosomes of fynbos myrmecochorous seeds are attractive to most granivorous ants, implying a general chemical cue. The main reason for the lack of specialization on the part of ants is probably that elaiosomes are not a major feature of their generalist diet, as evident from the wide distributions of some of the ant species involved in myrmecochory. In addition, seeds have few physical features which allow specialization, a prerequisite for close coevolution. It is hard to imagine a seed feature that could allow only one ant species access to an elaiosome in the same way that a tubular corolla restricts floral visitors (see also Wheelwright and Orians 1982). One possibility is that seed size itself may act as a filtering mechanism. Little is known about the ants involved in dispersing the very small elaiosome-bearing seeds of numerous fynbos species (but see Pierce 1990).

Herbivory

The relationship between *Heliconius* butterflies and their passerine host plants was the inspiration for Ehrlich and Raven's seminal (1964) paper on coevolution. This 'arms race' concept of biochemical coevolution was later built into the 'apparency theory' of defence (Feeny 1975; Rhoades and Cates 1975). Little is known about insect herbivory in fynbos, but it appears to be very limited (Picker 1987; Coetzee 1987). This may be due not so much to the presence of tannins, since Australian *Eucalyptus* species with high tannin levels are heavily browsed (Fox and Macauley 1977), but rather to very low levels of leaf nitrogen (Campbell 1986). Cottrell (1985) analysed the Cape butterflies on a generic level and found that only three genera could be considered to have evolved in the fynbos region and that two of these genera, *Oxychaeta* (Lycaenidae) and *Thestor* (Lycaenidae), probably have aphytophagous, ant-fed larvae. The third genus, *Melampias* (Satyridae), consists of only one species. The genus *Capys* (Lycaenidae) is closely associated with the Proteaceae, a typical fynbos family, but only one of the four species (*C. alpheus*) occurs in the fynbos region. It appears, then, that the unpalatable nature of fynbos has precluded coevolution with butterflies (Cottrell 1985).

Lovegrove and Jarvis (1986) have argued that the cormlets, palatability, and corm segments of *Micranthus junceus* suggest coevolution between this geophyte and mole rats. This suggestion was, however, premature since it was not shown that mole rats had evolved in response to *M. junceus* (B Lovegrove personal communication). The relationship between *M. junceus* and mole rats probably typifies unilateral evolution and not coevolution, although mole rats may have coevolved with geophytes as a whole in a broad (diffuse) sense.

In summary, fynbos has surprisingly few known coevolved plant-animal relationships. Yet it should be emphasized that this does not mean that animals have not influenced the evolution of fynbos plants. In fact, an outstanding characteristic of fynbos plants are the many reproductive features, e.g. diversity of floral morphologies, serotiny, and myrmecochory, that can be attributed to selective pressures from animals. The next question then is: to what extent have animals, particularly pollinators, influenced speciation in the region?

Speciation

The pollination component of any ecosystem is in a sense a major evolutionary steering wheel.

C H Stirton (1981)

Traditional explanations for the high plant diversity of fynbos centre around environmental heterogeneity (e.g. soil diversity, dissected landscape, and moisture gradients) and the reproductive isolation of populations through limited gene flow and transient, fire-created niches (Linder 1985; Cowling 1987; Cowling et al. this volume). The role of pollinators and seed dispersers in the speciation of fynbos plants has not been considered apart from their role in gene flow.

Many fynbos genera, e.g. *Gladiolus*, *Lapeirousia*, *Tritoniopsis* (Iridaceae), *Disa* (Orchidaceae), *Erica* (Ericaceae), and *Pelargonium* (Geraniaceae), exhibit great floral diversity, indicating numerous pollinator shifts which may have promoted speciation in these groups. The extensive adaptive radiation of Cape genera into different pollination syndromes was first described in detail by Vogel (1954). Speciation through pollinator shifts requires strong selection on floral characteristics as well as gradients of pollinator availability. Strong selection on floral characteristics occurs when pollinators limit reproductive success and consequently intra- and inter-specific competition for pollinators is severe. At present there is some limited evidence which supports the idea that pollinators limit seed production in geophytes, particularly after fire (Table 7.1).

Strong pollinator gradients would be expected in the mountainous Cape Floristic Region (cf. Cruden 1972). Such gradients may be on a geographical or a habitat scale. Robertson and Wyatt (1990), for instance, attributed morphological variation in the North American orchid *Platanthera ciliaris* to different butterflies which pollinate the orchid in coastal and mountain sites. Pollinator gradients may also occur on a smaller scale; sub-populations of the fynbos orchid *Disa uniflora* growing in an open area were seldom pollinated (< 10% of stigmas with pollen), while plants growing a few hundred metres away among rocks were visited far more frequently (> 70% of stigmas with pollen) (S D Johnson unpublished data). This sharp discrepancy in reproductive success between sub-populations of the orchid is simply due to the preference of its pollinator, the butterfly *Meneris tulbaghia*, for rocky habitats.

A plant which invades a new area may be pollinated effectively by a pollinator which is different from that for which the plant is adapted; selection will subsequently alter floral features to suit the new pollinator (Stebbins 1970). Such modified floral features may be used by taxonomists to delimit species. For instance, epidermal sculpturing in the Papilionidae appears to function as a foothold for insects and is also a major taxonomic marker (Stirton 1981). Should selection by pollinators alter floral features, other plant parts may also become altered through allometric scaling; larger flowers need to be supported by larger stems and leaves (Midgley and Bond 1989). Thus a pollinator shift may in fact alter the entire plant. Furthermore, should isolated populations undergo a pollinator shift, they would become reproductively isolated and free to explore new habitats. This is a different concept of speciation from the traditional one which holds that isolating mechanisms, e.g. different pollination mechanisms, evolve *subsequent* to habitat specialization in order to prevent hybrid unfitness (Mayr 1963).

The extent of pollinator-induced adaptive radiation in fynbos taxa will become clearer as systematists attempt to create monophyletic genera. Previously, taxonomy based on phenetics tended to result in genera consisting of species which were similar due to convergence for a single pollination syndrome. For instance, many former genera in the Cape, such as *Anapalina*, *Anomalesia*, *Homoglossum*, and *Antholyza* (all Iridaceae), were constructed using characters associated with bird pollination. It has now been realized that such genera were often artificial polyphyletic constructs (Goldblatt and De Vos 1989; Goldblatt 1990; Goldblatt 1991).

At present the ideas on pollination and speciation included in this section are fairly speculative and require testing. I have included them to stimulate consideration of factors, other than the physical landscape alone, which may be responsible for the species richness of fynbos.

CONCLUSION

Biologists visiting the fynbos region are often struck by the apparent lifelessness of the landscape. No large mammals are seen roaming about, few bird calls can be heard, and one does not need insect-proof netting to sleep at night. They may conclude that this corner of Africa is a botanical wonderland and a zoological desert. Yet, as we have seen, many ecological processes in fynbos are sustained by interactions between plants and small, often inconspicuous animals. Rodents, beetles, flies with absurdly long probosces, and ants have all played their part in shaping a unique flora.

ACKNOWLEDGEMENTS

I wish to thank R M Cowling, R Whelan, J Majer, K Johnson, and W Bond for their valuable suggestions on how to improve the manuscript. Financial support included the Foundation for Research Development Special Programme for Evolutionary Biology and the Charles Dixon Award.

REFERENCES

ANDERSEN A N (1988). Dispersal distance as a benefit of myrmecochory. *Oecologia* **75**, 507–11.

AYRE D J and WHELAN R J (1989). Factors controlling fruit set in hermaphroditic plants: studies with the Australian Proteaceae. *Trends in Ecology and Evolution* **4**, 267–72.

BARRET C H and ECKERT C G (1990). Current issues in plant reproductive ecology. *Israel Journal of Botany* **39**, 5–12.

BEATTIE A J and CULVER D C (1983). The nest chemistry of two seed-dispersing ant species. *Oecologia* **56**, 99–103.

BERG R Y (1975). Myrmecochorous plants in Australia and their dispersal by ants. *Australian Journal of Botany* **23**, 475–508.

BERNAYS E A (1981). Plant tannins and insect herbivores: an appraisal. *Ecological Entomology* **6**, 353–60.

BEZZI M (1924). The South African Nemestrinidae (Diptera) as represented in the South African Museum. *Annals of the South African Museum* **19**, 164–90.

BIERZYCHUDEK P (1981). Pollinator limitation of plant reproductive effort. *American Naturalist* **117**, 838–40.

BOLUS H (1888). The Orchids of the Cape Peninsula. *Transactions of the South African Philosophical Society* **5**, 1–201.

BOND P and GOLDBLATT P (1984). Plants of the Cape Flora. *Journal of South African Botany, Supplementary Volume* **13**, 1–455.

BOND W J (1984). Fire survival of Cape Proteaceae — influence of fire season and seed predators. *Vegetatio* **56**, 65–74.

BOND W J (1988). Proteas as tumbleseeds: wind dispersal through the air and over soil. *South African Journal of Botany* **54**, 455–60.

BOND W J and BREYTENBACH G J (1985). Ants, rodents and seed predation in Proteaceae. *South African Journal of Zoology* **20**, 150–4.

BOND W, FERGUSON M and FORSYTH G (1980). Small mammals and habitat structure along altitudinal gradients in the southern Cape mountains. *South African Journal of Zoology* **15**, 34–43.

BOND W J, LE ROUX D and ERNTZEN R (1990). Fire intensity and regeneration of myrmecochorous Proteaceae. *South African Journal of Botany* **56**, 326–30.

BOND W J, MIDGLEY J and VLOK J (1988). When is an island not an island? Insular effects and their causes in fynbos islands. *Oecologia* **77**, 515–21.

BOND W J and SLINGSBY P (1983). Seed dispersal by ants in shrublands of the Cape Province and its evolutionary implications. *South African Journal of Science* **79**, 231–3.

BOND W and SLINGSBY P (1984a). Proteas, ants and invaders: disruption of a delicate dependence. *South African Journal of Science* **80**, 201.

BOND W and SLINGSBY P (1984b). Collapse of an ant-plant mutualism: the Argentine ant (*Iridomyrmex humilis*) and myrmecochorous Proteaceae. *Ecology* **65**, 1 031–7.

BOND W J, VLOK J and VIVIERS M (1984). Variation in seedling recruitment of Cape Proteaceae after fire. *Journal of Ecology* **72**, 209–21.

BOND W J and STOCK W D (1989). The costs of leaving home: ants disperse seeds to low nutrient sites. *Oecologia* **81**, 412–17.

BOND W J, YEATON R and STOCK W D (1991). Myrmecochory in Cape fynbos. In *Ant-plant interactions.* (eds Huxley C and Cutler D) Oxford University Press, Oxford. (in press).

BOTHA S A (1988). Die invloed van diere op die populasiebiologie van die Clanwilliam seder (*Widdringtonia cederbergensis* Marsh) met spesiale verwysing na die rol van klein soogdiere. MSc thesis, University of Stellenbosch.

BOWDEN R N (1978). Diptera. In *Biogeography and ecology of southern Africa.* (ed Werger M J A) W Junk, The Hague. 774–96.

BREYTENBACH G J (1984). Single agedness in fynbos: a predation hypothesis. In *Medecos IV Proceedings of the 4th International Conference on Mediterranean Ecosystems.* (ed Dell B) Nedlands WA Botany Department, University of Western Australia. 14–15.

BREYTENBACH G J (1987). Small mammal dynamics in relation to fire. In *Disturbance and the dynamics of fynbos biome communities.* (eds Cowling R M, Le Maitre D C, McKenzie B, Prys-Jones R P and Van Wilgen B W) *South African National Scientific Programmes Report* **135**, CSIR, Pretoria. 56–67.

BREYTENBACH G J (1988). Why are myrmecochorous plants limited to fynbos (Macchia) vegetation types? *South African Forestry Journal* **144**, 3–5.

BREYTENBACH G J, CUNCLIFFE R N and COWLING R M (1986). Community structure and species interactions in the fynbos biome. *South African National Scientific Programmes Occasional Report* **12**, CSIR, Pretoria.

BRITS G J (1987). Germination depth vs. temperature requirements in naturally dispersed seeds of *Leucospermum cordifolium* and *L. cuneiforme* (Proteaceae). *South African Journal of Botany* **53**, 119–24.

CAMPBELL B M (1986). Plant spinescence and herbivory in a nutrient poor ecosystem. *Oikos* **47**, 168–72.

CLAASSENS A J M and DICKSON C G C (1980). *The butterflies of the Table Mountain range.* C Struik Publishers, Cape Town.

COETZEE J H (1987). Herbivory and defence mechanisms of five Proteaceae species in the fynbos biome. In *Proceedings of the Sixth Entomological Congress,* Entomological Society of southern Africa, Pretoria. 14–15.

COETZEE J H and GILIOMEE J H (1985). Insects in association with the inflorescence of *Protea repens* (L.) (Proteaceae) and their role in pollination. *Journal of the Entomological Society of Southern Africa* **48**, 303–14.

COETZEE J H and GILIOMEE J H (1987). Seed predation and survival in the infructescences of *Protea repens* (Proteaceae). *South African Journal of Botany* **53**, 61–4.

COLEY P D, BRYANT J P and CHAPIN F S (1985). Resource availability and plant antiherbivore defense. *Science* **230**, 895–9.

COLLINS B G (1983a). Seasonal variations in the energetics of territorial Cape sugarbirds. *Ostrich* **54**, 121–5.

COLLINS B G (1983b). Pollination of *Mimetes hirtus* (Proteaceae) by Cape sugarbirds and orange-breasted sunbirds. *Journal of South African Botany* **49**, 125–42.

COLLINS B G and REBELO T (1987). Pollination biology of the Proteaceae in Australia and southern Africa. *Australian Journal of Ecology* **12**, 387–421.

COTTRELL C B (1985). The absence of coevolutionary associations with Capensis floral element plants in the larval/plant relationships of southwestern Cape butterflies. In *Species and speciation.* (ed Vrba E S) Transvaal Museum Monograph No. 4, Transvaal Museum, Pretoria. 115–24.

COWLING R M (1987). Fire and its role in coexistence and speciation in Gondwanan shrublands. *South African Journal of Science* **83**, 106–11.

COWLING R M and MITCHELL D T (1981). Sugar composition, total nitrogen and accumulation of C_{14} assimilates in floral nectaries of Protea species. *Journal of South African Botany* **47**, 743–50.

COWLING R M and LAMONT B B (1987). Post-fire recruitment

of four co-occurring *Banksia* species. *Journal of Applied Ecology* **24**, 645–58.

CRUDEN R W (1972). Pollinators in high altitude ecosystems: Relative effectiveness of birds and bees. *Science* **176**, 1 439–40.

CRUDEN R W and HERMAN-PARKER S M (1979). Butterfly pollination of *Caesalpinia pulcherrima*, with observations on a psychophilous syndrome. *Journal of Ecology* **67**, 155–68.

DAFNI A (1984). Mimicry and deception in pollination. *Annual Review of Ecology and Systematics* **15**, 259–78.

DAFNI A and BERNHARDT P (1990). Pollination of terrestrial orchids of southern Australia and the Mediterranean region. *Evolutionary Biology* **24**, 193–252.

DAFNI A, BERNHARDT P, SHMIDA A, IVRI Y, GREENBAUM S, O'TOOLE C H and LOSITO L (1990). Red bowl-shaped flowers: convergence for beetle pollination in the Mediterranean region. *Israel Journal of Botany* **39**, 81–92.

DE MOOR F C (1985). Meloidae. In *Insects of southern Africa.* (eds Scholtz C H and Holm E) Butterworths, Durban. 260–3.

EDWARDS P J (1989). Insect herbivory and plant defence theory. In *Toward a more exact ecology.* (eds Grubb P J and Whittaker J B). Blackwell Scientific Publications, London. 275–98.

EHRLICH P R and RAVEN P H (1964). Butterflies and plants: a study in coevolution. *Evolution* **18**, 586–608.

EISIKOWITCH D and GALIL J (1971). Effect of wind on the pollination of *Pancratium maritimum* L. (Amaryllidaceae) by hawkmoths (Lepidoptera: Sphingidae). *Journal of Animal Ecology* **40**, 673–8.

ESLER K J, COWLING R M, WITKOWSKI E T F and MUSTART P J (1990). Reproductive traits and accumulation of nitrogen and phosphorus during the development of fruits of *Protea compacta* (calcifuge) and *P. obtusifolia* Buek ex Meisn. (calcicole). *New Phytologist* **112**, 109–15.

FAEGRI K and VAN DER PIJL L (1979). *The principles of pollination ecology.* Pergamon Press, Oxford.

FEENY P (1975). Biochemical coevolution between plants and their insect herbivores. In *Coevolution of animals and plants.* (eds Gilbert L E and Raven P H) University of Texas Press, Austin. 3–20.

FEINSINGER P (1983). Coevolution and pollination. In *Coevolution.* (eds Futuyma P and Slatkin M) Sinauer, Sunderland. 282–310.

FOX B J, QUINN R Q and BREYTENBACH G J (1985). A comparison of small-mammal succession following fire in shrublands of Australia, California and South Africa. *Proceedings of the Ecological Society of Australia* **14**, 179–97.

FOX L R and MACAULEY B J (1977). Insect grazing on *Eucalyptus* in response to variation in leaf tannins and nitrogen. *Oecologia* **29**, 145–62.

GANDAR M V (1980). Short-term effects of the exclusion of large-mammals and insects in broad leaf savanna. *South African Journal of Science* **76**, 29–31.

GESS F W (1968). Insects found on Proteas. *The Journal of the Botanical Society of South Africa* **54**, 29–33.

GILIOMEE J H (1986). Hunting for insect pollinators on fynbos flowers. *Veld & Flora* **72**, 6–7.

GLYPHIS J P and PUTTICK G M (1989). Phenolics, nutrition and insect herbivory in some garrigue and maquis plant species. *Oecologia* **78**, 259–63.

GOLDBLATT P (1978). An analysis of the flora of southern Africa: its characteristics, relationships, and origins. *Annals of the Missouri Botanical Garden* **65**, 369–436.

GOLDBLATT P (1989). *The genus Watsonia.* National Botanic Garden, Cape Town.

GOLDBLATT P (1990). Status of the southern African *Anapalina* and *Antholyza* (Iridaceae) genera, based solely on characters for bird pollination, and a new species of *Tritoniopsis. South African Journal of Botany* **56**, 577–82.

GOLDBLATT P (1991). An overview of the systematics, phylogeny and biology of the African Iridaceae. In *Systematics, biology and evolution of some South African taxa.* (eds Linder H P and Hall A V) *Contributions from the Bolus Herbarium* **13**, 1–74.

GOLDBLATT P and DE VOS M P (1989). The reduction of *Oenostachys*, *Homoglossum* and *Anomalesia*, putative sunbird pollinated genera, in *Gladiolus* L. (Iridaceae-Ixioideae). *Bulletin du Museum National d'Histoire Naturalle: ser. 4, Adansonia* **11**, 417–28.

GOLDBLATT P and BERNHARDT P (1990). Pollination biology of *Nivenia* (Iridaceae) and the presence of heterostylous self-incompatability. *Israel Journal of Botany* **39**, 93–111.

GULMON S L and MOONEY H A (1986). Costs of defense and their effects on plant productivity. In *On the economy of plant form and function.* (ed Givnish T J) Cambridge University Press, Cambridge. 681–98.

HATTINGH V and GILIOMEE J H (1989). Pollination of certain *Leucadendron* species (Proteaceae). *South African Journal of Botany* **55**, 387–93.

HENDEY Q B (1983). Palaeontology and palaeoecology of the fynbos region: an introduction. In *Fynbos palaeoecology: a preliminary synthesis.* (eds Deacon H J, Hendey Q B and Lambrechts J J N) *South African National Scientific Programmes Report* **75**, CSIR, Pretoria. 87–99.

HEPBURN H R and JACOT-GUILLARMOD A (1991). The Cape honeybee and the fynbos biome. *South African Journal of Science* **87**, 70–3.

HOLMES P M (1990). Dispersal and predation in alien *Acacia. Oecologia* **83**, 288–90.

HORN W (1962). Breeding research on South African plants: Fertility of Proteaceae. *The Journal of South African Botany* **28**, 259–68.

HUGHES L and WESTOBY M (1990). Removal rates of seeds adapted to dispersal by ants. *Ecology* **71**, 138–48.

JANZEN D H (1974). Tropical black-water rivers, animals and mast fruiting by the Dipterocarpaceae. *Biotropica* **6**, 69–103.

JANZEN D H (1980). When it is coevolution? *Evolution* **34**, 611–12.

JING S W and COLEY P D (1990). Dioecy and herbivory: the effect of growth rate on plant defense in *Acer negundo. Oikos* **58**, 369–77.

JORDAAN P G (1949). Aantekeninge oor die voorplanting en brandeperiodes van *Protea mellifera* Thunb. *Journal of South African Botany* **15**, 121–5.

JOUBERT J G V and STINDT H W (1979). The nutritive value of natural pastures in the district of Swellendam in the winter rainfall area of the Republic of South Africa. *Technical Communication, Department of Agriculture, Technical Services, Republic of South Africa* **156**, 1–10.

KEIGHERY G J (1982). Bird pollinated plants in Western Australia and their breeding systems. In *Pollination and evolution.* (eds Armstrong J A, Powell J M and Richards A J) Royal Botanic Gardens, Sydney. 77–89.

KEVAN P G and BAKER H G (1983). Insects as flower visitors and pollinators. *Annal Review of Entomology* **28**, 407–53.

KIESTER A R, LANDE R and SCHEMSKE D W (1984). Models of coevolution and speciation in plants and their pollinators. *American Naturalist* **124**, 220–43.

KNIGHT R S (1988). Aspects of plant dispersal in the southwestern Cape with particular reference to the roles of birds as dispersal agents. PhD thesis, University of Cape Town.

KOUTNIK D (1987). Wind pollination in the Cape flora. In *A preliminary synthesis of pollination biology in the Cape flora.* (ed Rebelo A G) *South African National Scientific Programmes Report* **141**, CSIR, Pretoria. 126–33.

KUO J, HOCKING P J and PATE J S (1982). Nutrient reserves

in seeds of selected proteaceous species from south-western Australia. *Australian Journal of Botany* **30**, 231–49.

LAMONT B B, COLLINS B G and COWLING R M (1985). Reproductive biology of the Proteaceae in Australia and South Africa. *Proceedings of the Ecological Society of Australia* **14**, 213–24.

LE ROUX G H (1988). Die gebruik van die Kaapse fynbos vir weiding en die probleme wat dit skep. *Suid-Afrikaanse Bosboutydskrif* **146**, 51–4.

LE MAITRE D C (1984). A short note on seed predation in *Watsonia pyramidata* (Andr.) Stapf. in relation to season of burn. *Journal of South African Botany* **50**, 407–45.

LINDER H P (1985). Gene flow, speciation and species diversity patterns in a species-rich area: the Cape flora. In *Species and speciation.* (ed Vrba E S) Transvaal Museum Monograph No. 4, Transvaal Museum, Pretoria. 53–7.

LOUW G N and NICOLSON S W (1983). Thermal, energetic and nutritional considerations in the foraging and reproduction of the carpenter bee *Xylocopa capitata. Journal of the Entomological Society of Southern Africa* **46**, 227–40.

LOVEGROVE B G and JARVIS J U M (1986). Coevolution between mole-rats (Bathyergidae) and a geophyte, *Micranthus* (Iridaceae). *Cimbebasia* **8**, 79–85.

LOVETT DOUST J and LOVETT DOUST L (1988). *Plant Reproductive Ecology: Patterns and Strategies.* Oxford University Press, London.

MACLEAN G L (1985). *Robert's Birds of Southern Africa.* The Trustees of the John Voelcker Bird Book Fund, Cape Town.

MARLOTH R (1894). On the means of the distribution of seeds in the South African flora. *The Transactions of the South African Philosophical Society* **7**, 74–88.

MARLOTH R (1896). The fertilization of Disa uniflora Berg, by insects. *The Transactions of the South African Philosophical Society* **8**, 93–5.

MARLOTH R (1903). Some recent observations on the biology of *Roridula. Annals of Botany* **17**, 151–7.

MARLOTH R (1908). Some observations on entomophilous flowers. *South African Journal of Science* **4**, 110–13.

MARLOTH R (1913–1932). *The Flora of South Africa,* **1–4**. Darter Bros & Sons, London.

MATTSON W J (1980). Herbivory in relation to plant nitrogen content. *Annual Review of Ecology and Systematics* **11**, 119–61.

MAYR E (1963). *Animal species and evolution.* Harvard University Press, Cambridge.

MICHENER C D (1979). Biogeography of the bees. *Annals of the Missouri Botanical Garden* **66**, 277–347.

MIDGLEY J J (1987). Aspects of the evolutionary biology of the Proteaceae, with emphasis on the genus *Leucadendron* and its phylogeny. PhD thesis, University of Cape Town.

MIDGLEY J J (1988). Mortality of Cape Proteaceae seedlings during their first summer. *South African Forestry Journal* **145**, 9–12.

MIDGLEY J J (1989). Season of burn of serotinous Proteaceae: a critical review and further data. *South African Journal of Botany* **55**, 165–70.

MIDGLEY J and BOND W (1989). Leaf size and inflorescence size may be allometrically related traits. *Oecologia* **78**, 427–9.

MIDGLEY J J and CLAYTON P (1990). Short-term effects of an autumn fire on small mammal populations in southern Cape coastal mountain fynbos. *South African Forestry Journal* **153**, 27–30.

MILEWSKI A V (1982). The occurrence of seeds and fruits taken by ants versus birds in mediterranean Australia and southern Africa, in relation to the availability of soil potassium. *Journal of Biogeography* **9**, 505–16.

MILEWSKI A V (1983). A comparison of ecosystems in mediterranean Australia and southern Africa: nutrient poor sites at the Barrens and the Caledon coast. *Annual Review of Ecology and Systematics* **14**, 57–76.

MILEWSKI A V and BOND W J (1982). Convergence of myrmecochory in mediterranean Australia and South Africa. In *Ant-plant interactions in Australia.* (ed Buckley R C) W Junk, The Hague. 89–98.

MORROW P A, DAY J A, FOX M D, FROST P G H, JARVIS J U M, MILEWSKI A V and NORTON P M (1983). Interactions between plants and animals. In *Mineral nutrition in Mediterranean ecosystems.* (ed Day J A) *South African National Scientific Programmes Report* **71**, CSIR, Pretoria. 111–23.

MOSTERT D P, SIEGFRIED W R and LOUW G N (1980). Protea nectar and satellite fauna in relation to the food requirements and pollinating role of the Cape Sugarbird. *South African Journal of Science* **76**, 409–12.

MYBURGH A C, STARKE L C and RUST D J (1974). Destructive insects in the seed heads of *Protea barbigera* Meisner (Proteaceae). *Journal of the Entomological Society of Southern Africa* **37**, 23–9.

MYBURGH A C and RUST D J (1975). A survey of pests of the Proteaceae in the western and southern Cape Province. *Journal of the Entomological Society of Southern Africa* **38**, 55–60.

NILSSON L A (1988). The evolution of flowers with deep corolla tubes. *Nature* **334**, 147–9.

NORTON P M (1980). The habitat and feeding ecology of the Klipspringer *Oreotragus oreotragus* (Zimmerman, 1783), in two areas of the Cape Province. MSc thesis, University of Pretoria.

NOVELLIE P (1987). Interrelationships between fire, grazing and grass cover at the Bontebok National Park. *Koedoe* **30**, 1–17.

O'DOWD D J and HAY M E (1980). Mutualism between harvester ants and a desert ephemeral: seed escape from rodents. *Ecology* **61**, 531–40.

PICKER M (1987). Determinants of herbivory on *Cliffortia.* (Abstract). In *Proceedings of the Sixth Entomological Congress,* Entomological Society of southern Africa, Pretoria. 62–3.

PIERCE S M (1984). A synthesis of plant phenology in the fynbos biome. *South African National Scientific Programmes Report* **88**, CSIR, Pretoria.

PIERCE S M (1990). Pattern and process in south coast dune fynbos: population, community and landscape level studies. PhD thesis, University of Cape Town.

PIERCE S M and COWLING R M (1991) Dynamics of soil-stored seedbank of six shrubs in fire-prone dune fynbos *Journal of Ecology* (in press).

PRINS A J (1978). Hymenoptera. In *Biogeography and ecology of southern Africa.* (ed Werger M J A) W Junk, The Hague. 823–76.

PYKE G H (1982). Evolution of inflorescence size and height in Waratahs (*Telopea speciosissima*): the difficulties of interpreting correlations between plant traits and fruit set. In *Pollination and Evolution.* (eds Armstrong J A, Powell J M and Richards A J) Royal Botanic Gardens, Sydney. 77–89.

RAITT L M (1986). High orchid densities and a hybrid *Satyrium* at Blue Downs near Blackheath, South Africa. *South African Journal of Botany* **52**, 189–91.

RATHCKE B and LACEY E P (1985). Phenological patterns of terrestrial plants. *Annual Review of Ecology and Systematics* **16**, 179–214.

REBELO A G (1983). Birds, blossoms and beauty. *Veld & Flora* **69**, 24–5.

REBELO A G (ed) (1987a). A preliminary synthesis of pollination biology in the Cape flora. *South African National Scientific Programmes Report* **141**, CSIR, Pretoria.

REBELO A G (1987b). Bird pollination in the Cape flora. In *A preliminary synthesis of pollination biology in the Cape flora.* (ed Rebelo A G) *South African National Scientific Programmes Report* **141**, CSIR, Pretoria. 83–108.

REBELO A G (1987c). Sunbird feeding at *Satyrium odorum* Sond. flowers. *Ostrich* **58**, 185–6.

REBELO A G and BREYTENBACH G J (1987). Mammal pollination in the Cape flora. In *A preliminary synthesis of pollination biology in the Cape flora.* (ed Rebelo A G) *South African National Scientific Programmes Report* **141**, CSIR, Pretoria. 109–25.

REBELO A G and ROURKE J P (1986). Seed germination and seed set in southern African Proteaceae: ecological determinants and horticultural problems. *Acta Horticulturae* **185**, 75–88.

REBELO A G and SIEGFRIED W R (1985). Colour and size of flowers in relation to pollination of *Erica* species. *Oecologia* **65**, 584–90.

REBELO A G, SIEGFRIED W R and CROWE A A (1984). Avian pollinators and the pollinating syndromes of selected Mountain Fynbos plants. *South African Journal of Botany* **3**, 285–96.

REBELO A G, SIEGFRIED W R and OLIVER E G H (1985). Pollination syndromes of *Erica* species in the south-western Cape. *South African Journal of Botany* **51**, 270–80.

RHOADES D F and CATES R G (1976). Toward a general theory of plant anti-herbivore chemistry. *Recent Advances in Phytochemistry* **10**, 168–213.

RICE B and WESTOBY M (1986). Evidence against the hypothesis that ant-dispersed seeds reach nutrient-enriched microsites. *Ecology* **67**, 1 270–4.

ROBERTSON J L and WYATT R (1990). Evidence for pollination ecotypes in the yellow-fringed orchid, *Platanthera ciliaris. Evolution* **44**, 121–33.

ROURKE J and WIENS D (1977). Convergent floral evolution in South African and Australian Proteaceae and its possible bearing on pollination by nonflying mammals. *Annals of the Missouri Botanical Garden* **64**, 1–17.

SCHEMSKE D W (1983). Limits to specialisation and coevolution in plant-animal mutualisms. In *Proceedings of the Fifth Annual Spring Systematics Symposium: Coevolution.* (ed Nitecki M H) The University of Chicago Press, Chicago. 67–110.

SCHLETTWEIN C H G and GILIOMEE J H (1987). Comparison of insect biomass and community structure between fynbos sites of different ages after fire with particular reference to ants, leafhoppers and grasshoppers. *Annale van die Universiteit van Stellenbosch* Serie A 3 (Landbouwetenskappe) **2**, nr 2.

SCHOLTZ C H and HOLM E (1985). Scarabaeoidea. In *Insects of southern Africa.* (eds Scholtz C H and Holm E) Butterworths, Durban. 214–23.

SIEGFRIED W R (1983). Trophic structure of some communities of fynbos birds. *Journal of South African Botany* **49**, 1–43.

SIEGFRIED W R, REBELO A G and PRYS-JONES R P (1985). Stem thickness of *Erica* plants in relation to avian pollination. *Oikos* **45**, 153–5.

SKAIFE S H (1979). *African Insect Life.* C Struik, Cape Town.

SLINGSBY P and BOND W (1981). Ants — friends of the fynbos. *Veld & Flora* **67**, 39–45.

SLINGSBY P and BOND W J (1985). The influence of ants on the dispersal distance and seedling recruitment of *Leucospermum conocarpodendron* (L.) Buek (Proteaceae). *South African Journal of Botany* **51**, 30–4.

STEBBINS G L (1970). Adaptive radiation of reproductive characteristics in Angiosperms, 1: Pollination mechanisms. *Annual Review of Ecology and Systematics* **1**, 307–26.

STEINER K E (1987). Breeding systems in the Cape flora. In *A preliminary synthesis of pollination biology in the Cape flora.* (ed Rebelo A G) *South African National Scientific Programmes Report* **141**, CSIR, Pretoria. 22–51.

STEINER K E (1989). The pollination of *Disperis* (Orchidaceae) by oil collecting bees in southern Africa. *Lindleyana* **4**, 164–83.

STEINER K E and WHITEHEAD V B (1988). The association between oil-producing flowers and oil-collecting bees in the Drakensberg of southern Africa. *Monographs of Systematic Botany of the Missouri Botanical Garden* **25**, 259–77.

STILES F G (1978). Ecological and evolutionary implications of bird pollination. *American Zoologist* **18**, 715–27.

STIRTON C H (1981). Petal sculpturing in Papilionoid legumes. In *Advances in legume systematics.* (eds Polhill R M and Raven P H) Royal Botanic Gardens, Kew. 771–8.

STOCK W D, PATE J S, KUO J and HANSEN A P (1989). Resource control of seed set in *Banksia laricina* C. Gardner (Proteaceae). *Functional Ecology 3*, 453–60.

STOCK W D, PATE J S and RASINS E (1991). Seed developmental patterns in *Banksia attenuata* R. Br. and *B. laricina* C. Gardner in relation to mechanical defence costs. *New Phytologist* (in press).

THOMPSON J N (1982). *Interaction and coevolution.* John Wiley & Sons, New York.

THOMPSON J N (1989). Concepts of coevolution. *Trends in Ecology and Evolution* **4**, 179–83.

USHER P J (1972). A review of the South African horsefly fauna (Diptera: Tabanidae). *Annals of the Natal Museum* **21**, 459–507.

VAN DER PIJL L (1982). *Principles of dispersal in higher plants.* Springer-Verlag, Berlin.

VAN HENSBERGEN B, BOTHA S A and FORSYTH G (1991). Do small mammals govern vegetation recovery in fynbos. In *Fire in South African mountain fynbos: ecosystem, community and species responses at Swartboschkloof* (eds Van Wilgen B J, Richardson D M, Kruger F J and Van Hensbergen B). Springer-Verlag, Berlin (in press).

VAN WILGEN B W and VIVIERS M (1985). The effect of season of fire on serotinous Proteaceae in the western Cape and the implications for fynbos management. *South African Forestry Journal* **133**, 49–53.

VOGEL S (1954). *Blutenbiologische typen als elemente der sippengliederung.* Veb Gustav Fischer Verlag, Jena.

WASER N M (1979). Pollinator availability as a determinant of flowering time in Ocotillo (*Fouquieria splendens*). *Oecologia* **36**, 223–36.

WATMOUGH R (1974). Biology and behaviour of Carpenter bees in southern Africa. *Journal of the Entomological Society of Southern Africa* **37**, 261–81.

WESTOBY M, RICE B and HOWELL J (1990). Seed size and plant growth form as factors in dispersal spectra. *Ecology* **71**, 1 307–15.

WHEELWRIGHT N T and ORIANS G H (1982). Seed dispersal by animals: contrasts with pollen dispersal, problems of terminology, and constraints on coevolution. *American Naturalist* **119**, 402–13.

WHELAN R J and GOLDINGAY R (1989). Factors affecting fruit-set in *Telopea speciosissima* (Proteaceae): the importance of pollen limitation. *Journal of Ecology* **77**, 1 123–34.

WHITEHEAD V B, SCHELPE E A C L E and ANTHONY N C (1984). The bee *Rediviva longimanus* Michener (Apoidea: Melittidue) collecting pollen and oil from *Diascia longicornus* (Thunb.) Pruce (Scrophulariaceae). *South African Journal of Science* **80**, 286.

WHITEHEAD V B, GILIOMEE J H and REBELO A G (1987). Insect pollination in the Cape flora. In *A preliminary synthesis of pollination biology in the Cape flora.* (ed Rebelo A G) *South African National Scientific Programmes Report* **141**, CSIR, Pretoria. 52–82.

WIENS D, ROURKE J P, CASPER B B, RICKART E A, LAPINE T R, PETERSON C J and CHANNING A (1983). Nonflying mammal pollination of southern African Proteas: a non-coevolved system. *Annals of the Missouri Botanical Garden* **70**, 1–31.

WIKLUND C, ERIKSSON T and LUNDBERG H (1979). The wood white butterfly *Leptidea sinapsis* and its nectar plants: a case of mutualism or parasitism? *Oikos* **33**, 358–62.

WILLAN K and BIGALKE R C (1982). The effects of fire regime on small mammals in s.w. Cape montane fynbos

(Cape Macchia). In *Proceedings of the Symposium on Dynamics and Management of Mediterranean-Type Ecosystems.* (eds Conrad C E and Oechel WC) General Technical Report PSW-58 United States Department of Agriculture, Pacific Southwest Forest and Range Experiment Station, Berkeley. 207–12.

WITKOWSKI E T F (1990). Nutrient limitation of inflorescence and seed production in *Leucospermum parile* (Proteaceae) in the Cape fynbos. *Journal of Applied Ecology* **27**, 148–58.

WRIGHT M G, WRIGHT G E P and SMITH P (1989). Entomophily and seed predation of *Witsenia maura* (Iridaceae), a rare fynbos species. *South African Journal of Botany* **55**, 273–7.

WRIGHT M G and GILIOMEE G H (submitted). Pre-dispersal seed predation in *Protea magnifica* and *Protea laurifolia* (Proteaceae).

W-WORSWICK P V (1987). Comparative study of colony thermoregulation in the African honeybee, *Apis mellifera adansonii* Latreille and the Cape honeybee, *Apis mellifera capensis* Escholtz. *Comparative Biochemistry and Physiology* A **86**, 95–102.

W-WORSWICK P V (1988). Comparison of nectar foraging efficiency in the Cape honeybee, *Apis mellifera capensis* Escholtz, and the African honeybee, *Apis mellifera adansonii* Latreille, in the western Cape Province. *South African Journal of Zoology* **23**, 124–7.

YEATON R I and BOND W J (1991). Competition between two shrub species: directed dispersal by ants promotes coexistence. *American Naturalist.* (in press).

ZIMMERMAN J K and MITCHELL AIDE T (1989). Patterns of fruit production in a neotropical orchid: pollinator vs resource limitation. *American Journal of Botany* **76**, 67–73.

ZIMMERMAN J K, ROUBIK D W and ACKERMAN J D (1989). Asynchronous phenologies of a neotropical orchid and its euglossine bee pollinator. *Ecology* **70**, 1 192–5.

ZUCKER W V (1983). Tannins: does structure determine function? An ecological perspective. *American Naturalist* **121**, 335–65.

8

Competition and coexistence

W J Bond, R M Cowling and M B Richards

It is generally accepted that in order to coexist more than transiently (plant) species must differ . . . if species are too similar all but one will be eliminated in competition.

(Newman 1982)

I do not question the existence of the differences among plant species . . . The question is what these differences have to do with coexistence. I do not know the answer to this question and I do not believe that anybody else does either.

(Yodzis 1986)

INTRODUCTION

Cape fynbos is distinguished by its enormous diversity of plant species concentrated in a few large genera packed into a small geographic region. How do these many closely related species coexist? There is a wealth of new theory in plant ecology which addresses this problem (Silvertown and Laws 1987). Older notions inherited from zoologists that predict niche differentiation in coexisting species (see Schoener 1989 for a comprehensive discussion) have been challenged by new theories that predict niche convergence (Yodzis 1986; Pacala 1988). Field biologists have been more concerned with the description of plant associations and their correlation with abiotic variables than with mechanisms determining composition of communities. Thus there is little empirical work to test the new theoretical speculations on what limits membership of plant communities. Here we report recent studies of mechanisms governing coexistence in fynbos, and some first attempts to test these opposing theories of plant coexistence.

There has been a long tradition of vegetation description in South Africa in which the distribution of plant associations is generally related to abiotic growing conditions (Adamson 1927; Acocks 1953). Past disturbance, especially by fire and grazing, has also been considered important in determining plant species distribution. This traditional view — that the location of species can be explained in terms of the match of abiotic site conditions and species physiologies or past disturbance history — was challenged by Cody (1986). He argued that the local distribution of fynbos Proteaceae is determined more by the abundance of other species than by local soil or climate. According to Cody, out of a large regional pool of Proteaceae species only a small subset of species with complementary niches can coexist. In principle, one could develop a set of 'assembly rules' (Diamond 1975) from which vegetation patterns could be predicted if the pertinent niche dimensions were known. In contradiction, Cowling (1987) has dismissed habitat or resource differentiation as a factor in species coexistence in fynbos. Citing a number of recent theoretical papers (Grubb 1977; Chesson and Warner 1981; Warner and Chesson 1985), he argued that spatial and temporal variation in fire regimes is inconsistent with equilibrial explanations of coexistence, and that variation itself is the key to explaining coexistence.

Our central interest in this chapter is whether competition, alone or in combination with other processes, structures species membership in fynbos communities. We emphasize competition since classical theory, based on the competitive exclusion principle, provides a set of 'rules' on how communities can be assembled. We first discuss predictions of classical theory and contrast these with more recent work, much of it designed specifically for sessile organisms. We then compare theoretical predictions on community structure with observed distribution patterns of ecologically and taxonomically similar species in fynbos. Finally we review the few studies which look directly at the processes which underlie species distribution patterns, including alternatives to competition. We conclude by considering the implications of competition theory in veld management.

THEORY

Contemporary theory has split into two sets of contradictory predictions for coexistence in communities. Classical theory predicts that the most similar species cannot coexist, while a variety of more recent models predict the reverse — the most similar species are most likely to coexist.

Classical theory

Classical competition theory, based on the competitive exclusion principle, leads to several predictions for species distributions in nature:

- coexisting potential competitors should exhibit niche differentiation (Newman 1982);
- niche differentiation will often be apparent as morphological differentiation (Cody 1986);
- potential competitors with little or no niche differentiation should not coexist and should, instead, show habitat differentiation (each species occurs where the other is absent) (Cody 1978; Werner 1979).

Another possibility is that apparent coexistence is merely temporary co-occurrence, and the poorer competitor will be displaced given more time for competition to act. This would likely be more common in communities where disturbances, such as fire, occur with some regularity (Huston 1979; Denslow 1985).

The classical theory, based on the Lotka-Volterra equations, has been modified to deal with various departures from basic assumptions (Chesson and Case 1986; Schoener 1989). Plant ecologists have long been sceptical of the applicability of niche theory to plants (for an early critique see Whittaker 1965). Much recent work has attempted to extend competition theory to sessile organisms. It is argued that plants compete for too few and too similar resources for niche theory to explain local diversity (e.g. Silvertown and Laws 1987). Tilman (1982, 1986) ingeniously extended classical niche theory to plants and argued that many species can coexist on a few resources if there is spatial variation in resource supply. He explicitly predicted high diversity in fynbos-type vegetation on nutrient-poor soils (Tilman 1986), but the theory has yet to be tested empirically.

The regeneration niche

Grubb (1977) made an important extension to classical niche theory for plants. He recognized four niche axes as critical for segregation of plant species — growth form, habitat differences, phenological differences, and regeneration. Plants with similar growth forms, shared habitats, and growing seasons ('trophically equivalent' in the jargon of Shmida and Ellner 1984) often occur in the same community. Grubb argued that coexistence can be explained in such species by differences in their regeneration requirements including dispersal, germination, and seedling establishment.

Lottery models

Until recently, competition models largely ignored age structure and regeneration events. Grubb's observations (and those of Sale (1977) for coral reef fishes) led to a spate of new mathematical models testing the plausibility of coexistence of trophically equivalent sessile species which differ only in reproductive traits (Fagerstrom and Agren 1979; Chesson and Warner 1981; Shmida and Ellner 1984; Chesson 1986; Comins and Noble 1985; Fagerstrom 1988). Shmida and Ellner (1984), for example, have shown that an inferior competitor can coexist with a superior one if it is better at colonizing unoccupied sites.

Lottery models are a promising new class of models which encapsulate many recognizable features of fynbos plant life histories. Lottery models assume that competition among recruits is for space, and that space is allocated randomly (as a 'lottery') to propagules of the different species (Fagerstrom and Agren 1979; Chesson and Warner 1981; Shmida and Ellner 1984; Chesson 1986; Fagerstrom 1988). Thus the outcome of competitive interactions will be in proportion to the reproductive success of the competing species. The sensitivity of juvenile stages to environmental fluctuations adds a potential stochastic element to species interactions and lottery models incorporate this stochasticity.

Two basic criteria must be met for long-term coexistence in lottery models (Fagerstrom 1988). First, environmental variation is necessary which leads to occasional periods of strong recruitment in each species. Coexistence is not possible if one species *consistently* has stronger recruitment (or higher survival rates). Secondly, the effects of a good recruitment period must be 'stored' over lean years. For example, a tree species may persist in a forest even though good recruitment is rare and episodic as long as adult trees survive across several episodes and have repeated reproductive opportunities. Warner and Chesson (1985) particularly emphasize this 'storage effect' as crucial for coexistence in lottery models. A species can persist despite low average recruitment if a population can 'store' the effects of good recruitment years across the lean periods. Long-term coexistence is possible when each species has positive average population growth. Warner and Chesson (1985) strongly emphasized the need for long lifespans and overlapping generations for the 'storage effect' to explain coexistence. However, recent work by Laurie (1989) shows that lottery models can readily be applied to the non-overlapping life histories characteristic of many fynbos plants (see below).

Lottery models do not explain away the problem of the coexistence of species utilizing similar resources. On the contrary, they provide a new set of testable predictions on community assembly 'rules' (Warner and Chesson 1985; Fagerstrom 1988). In complete contradiction to classical niche theory, lottery models predict that the most similar species should coexist. The corollary is that dissimilar species should not (e.g. a species with consistently high seed production should exclude a competitively equivalent species with consistently lower production). Convergence, rather than differentiation, is the predicted evolutionary outcome of competitive interaction (Fagerstrom 1988). A second important prediction is that environmental uncertainty promotes coexistence by promoting, for example, species-specific variation in reproductive output (Warner and Chesson 1985; Fagerstrom 1988). At least one result of classical niche theory predicts the reverse with fewer species able to coexist in variable environments (May and MacArthur 1972).

FYNBOS STUDIES

Most fynbos studies have focused on proteoid (Proteaceae) shrubs, particularly the genera *Protea*, *Leucadendron*, and *Leucospermum*. Species in these genera often form the overstorey of fynbos communities (Cowling and Holmes this volume). Many proteoid shrubs are killed by fire and, in the first year after fire, recruit from a seed bank in the soil or in the canopy (serotinous cones). Models based on discrete generations are therefore appropriate for describing population growth. Other proteoids sprout after fire but seedling regeneration is also largely confined to the post-fire period.

Similar fire-dependent life histories are shared by shrubs and trees in many areas of the world including Australian kwongan and Californian chaparral (Gill 1975; Keeley 1977).

Alpha diversity — the richness of species in small areas — is the appropriate scale for the most recent discussion of coexistence in plants (Auerbach and Shmida 1987). Up to nine proteoid shrubs of similar general stature and seed biology may coexist in small areas (50–100 m^2) in parts of the fynbos biome. Other fynbos shrubs are similarly diverse with the Ericaceae producing perhaps the richest mix of species in small sites (Cowling 1987). We concentrate on the alpha diversity of proteoid shrubs in this chapter.

Assembly rules for proteoid communities

Here we examine distributional data for Proteaceae to test the conflicting predictions of lottery and niche models of plant community structure.

COEXISTENCE AND NICHE DIFFERENTIATION

Cody (1986) was the first to search for niche differentiation among coexisting species in the fynbos. He studied patterns of leaf morphology in the dominant tall shrubby proteoids (*Protea*, *Leucadendron*, *Leucospermum*) across four regions. Cody recorded leaf length (size) and leaf width/length (shape) of all species found in five hectare study sites in each region. Figure 8.1 shows that each species (or sex in the case of *Leucadendron*) has a different leaf morphology from the next with little or no overlap. Cody explained a single exception, where two species did show substantial overlap, by habitat segregation within his sample site. Data from a fifth site, Elandskloof, with variable within-site topography, also showed substantial morphological overlap which Cody interpreted similarly as small-scale habitat segregation. Cody also studied the relationship between *Leucadendron* leaf morphology and species distribution and concluded that morphologically similar species 'tend either to be allopatric or to occupy different habitats (such as different slopes, aspects or elevations)'.

Cody explained these segregation patterns both in terms of the sorting of *species* with appropriate non-overlapping leaf forms (other species with the same leaf form would be excluded), and selective changes of leaf morphology within a species so that it can 'fit' different species assemblages. The latter represents the phenomenon of character displacements.

Cody's approach is rare in the plant literature. It implies that proteoid assemblages are a result of interspecific interactions rather than individual species responses to, for example, soil differences or past disturbance regimes. Cody's study leads to a number of predictions and is based on a number of untested assumptions. First, his is an essentially static or equilibrial argument. There is little indication of the processes which produce the pattern. Population densities of serotinous proteoid are known to fluctuate widely from one fire to the next, especially after different fire seasons (Bond et al. 1984; Van Wilgen and Viviers 1985; Midgley 1989). However, niche differentiation assumes that proteoid species are both competing and that the environment is saturated. If densities should drop, then presumably species with overlapping leaf morphologies should coexist more easily. This prediction has yet to be tested. Secondly, although Cody discusses the physiological implications of leaf shape and size, the link between leaf morphology and space pre-emption is not obvious. Canopies of individual proteoid shrubs rarely overlap. If one plant belonging to one species has successfully outcompeted others for space, then why should it not occupy all available spaces? Recent studies show that leaf morphology is allometrically linked to reproductive traits (Bond and Midgley 1988; Midgley and Bond 1989) which greatly expands the range of explanations for why leaf morphology should be associated with coexistence. For example, leaf size is correlated with branch density and this, in turn, with number and size of inflorescences. Leaf size may thus be a surrogate for various reproductive traits which may be more important for coexistence.

Cody's study lacks a formal neutral model for the observed distributions, although the chances of getting contiguous and non-overlapping leaf morphologies would seem to be vanishingly small. Cody argues that, given a regional pool of species, only those species with complementary leaf morphologies will coexist in any one site *or* selection will have altered leaves to better 'fit' the local assemblage. However, to support his case, it is necessary to show that pairs of species with overlapping morphologies occur together less often than species with different leaf morphs from the same regional species pool. This is not a trivial problem (Harvey et al. 1983).

INTRASPECIFIC VARIATION IN LEAF MORPHOLOGY

Protea repens is a very common and widespread serotinous species in fynbos. It has a highly variable leaf shape and size (Rourke 1982). To test for leaf morphology as a niche dimension, we asked whether leaves of this species vary in response to the competitive environment. Character displacement has been widely discussed in the zoological literature (Brown and Wilson 1956; Fenchel 1975) but we know of no botanical examples. According to the niche differentiation argument (Cody 1986), one would predict similar leaf morphologies in monospecific stands but divergent morphologies when *P. repens* co-occurs with other proteoid shrubs. The direction in which leaf morphology shifts should presumably depend on the precise mix of congeners.

We sampled five leaves randomly from each of 10 shrubs of *P. repens* in monospecific and mixed stands. Leaves were sampled at sites

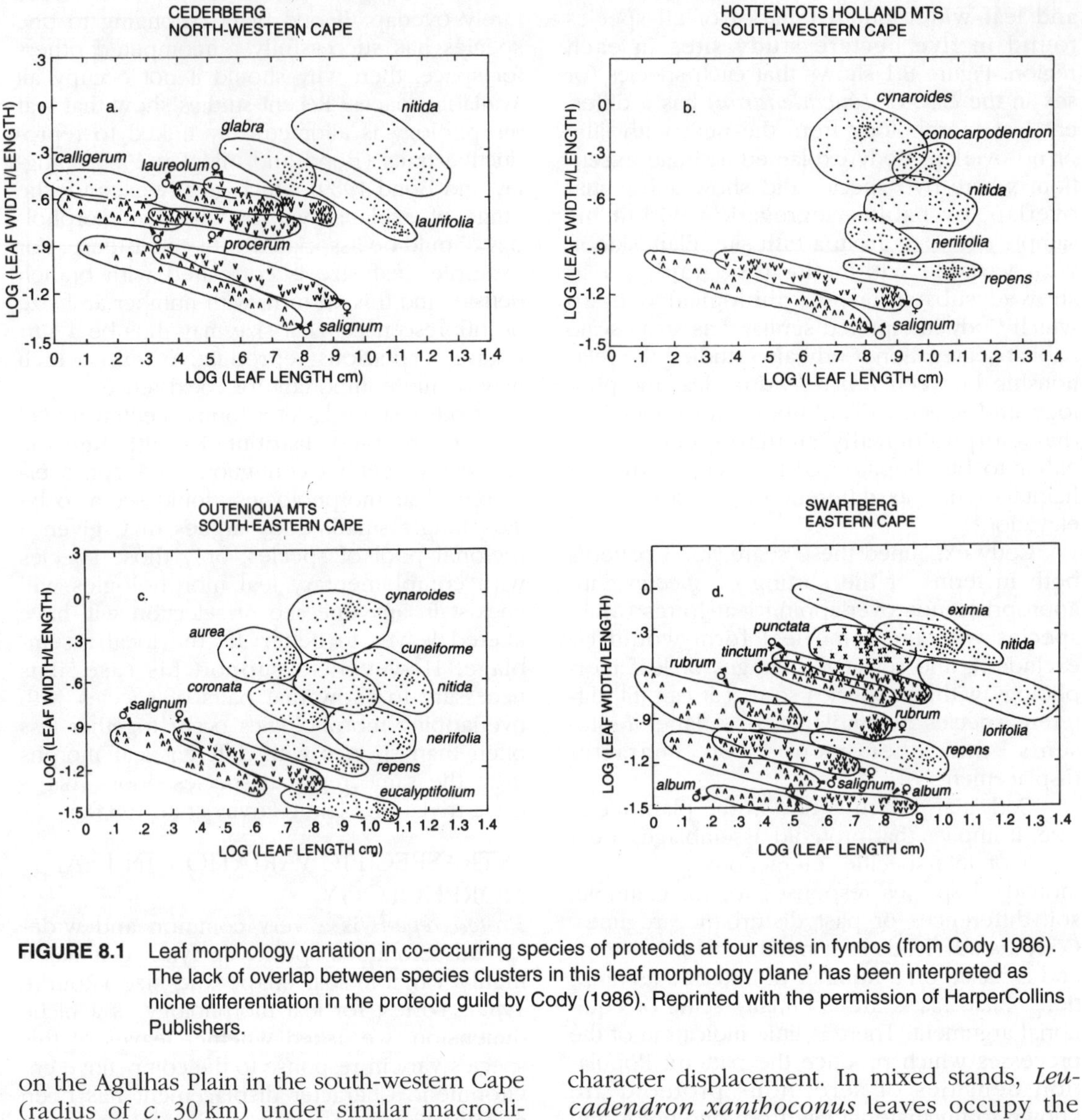

FIGURE 8.1 Leaf morphology variation in co-occurring species of proteoids at four sites in fynbos (from Cody 1986). The lack of overlap between species clusters in this 'leaf morphology plane' has been interpreted as niche differentiation in the proteoid guild by Cody (1986). Reprinted with the permission of HarperCollins Publishers.

on the Agulhas Plain in the south-western Cape (radius of *c.* 30 km) under similar macroclimatic conditions but across very variable soils. Two sample areas had matched sites of mono and multispecies stands on very similar soil in close geographic proximity (< 5 km apart).

The results of discriminant analyses using log leaf length and log (leaf length/leaf width) (Cody 1986) on these populations are shown in Figure 8.2. Monospecific and mixed stands of *P. repens* were significantly different on the basis of these two leaf variables (Wilks Lambda = 0.466, Chi Square = 77.1, $P < 0.001$). The results seem to support the prediction of character displacement. In mixed stands, *Leucadendron xanthoconus* leaves occupy the same position as solitary populations of *P. repens* which now appear to be displaced to a lower, non-overlapping space on the leaf morphology plane (Figure 8.3). At Heuningrug, however, there was no significant discrimination between leaves of mono and mixed populations (Figure 8.2). There was also substantial overlap with *L. coniferum* at this site (although not with *P. susannae*) contrary to the niche differentiation argument (Figure 8.3). Abiotic effects on leaf morphology may further complicate patterns. There was significant dis-

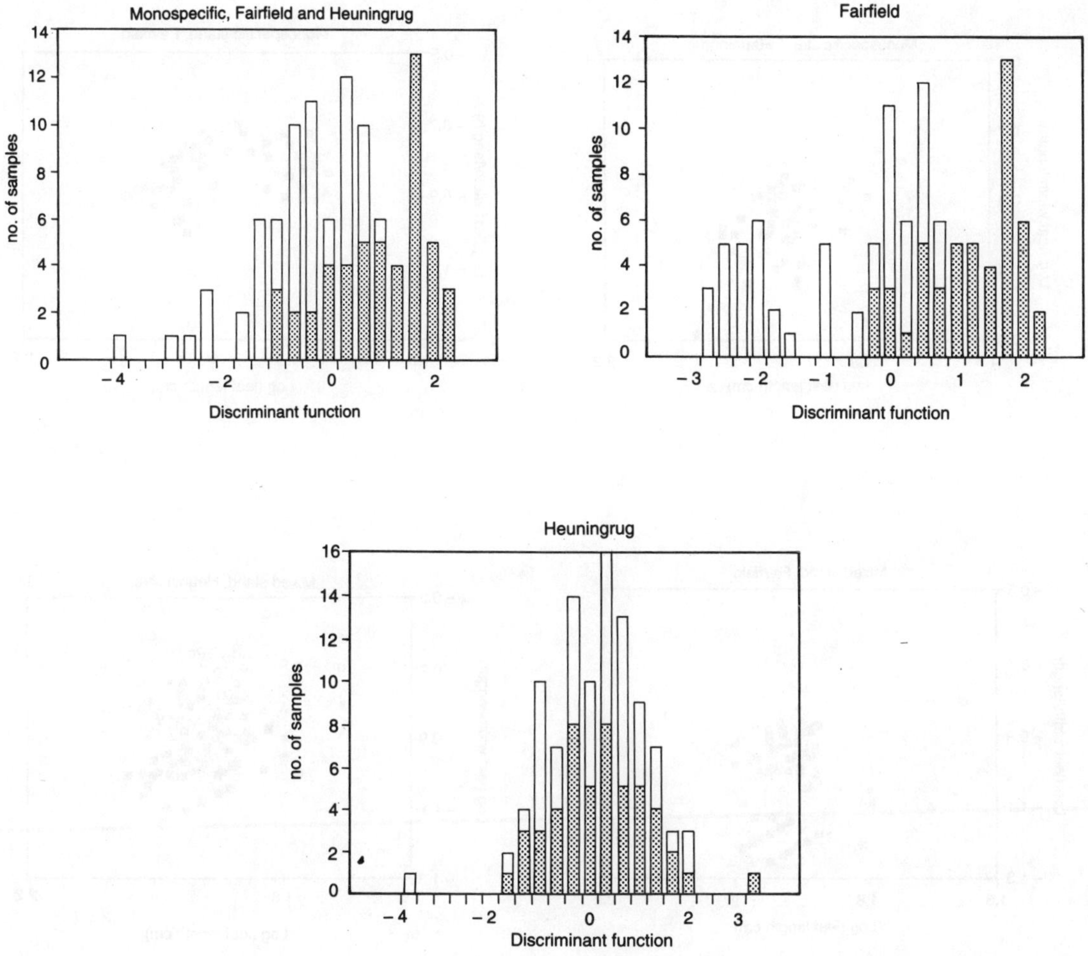

FIGURE 8.2 The results of discriminant analyses on log (leaf length) and log (leaf length/leaf width) of *Protea repens* at Fairfield and Heuningrug on the Agulhas Plain. In each study area, sites were matched for soils (deep sands at Heuningrug, shallow sand over sandstone at Fairfield) but differed in the number of proteoids. Shading represents single or mixed stands for Fairfield and Heuningrug histograms, and the respective study sites for the monospecific comparison.

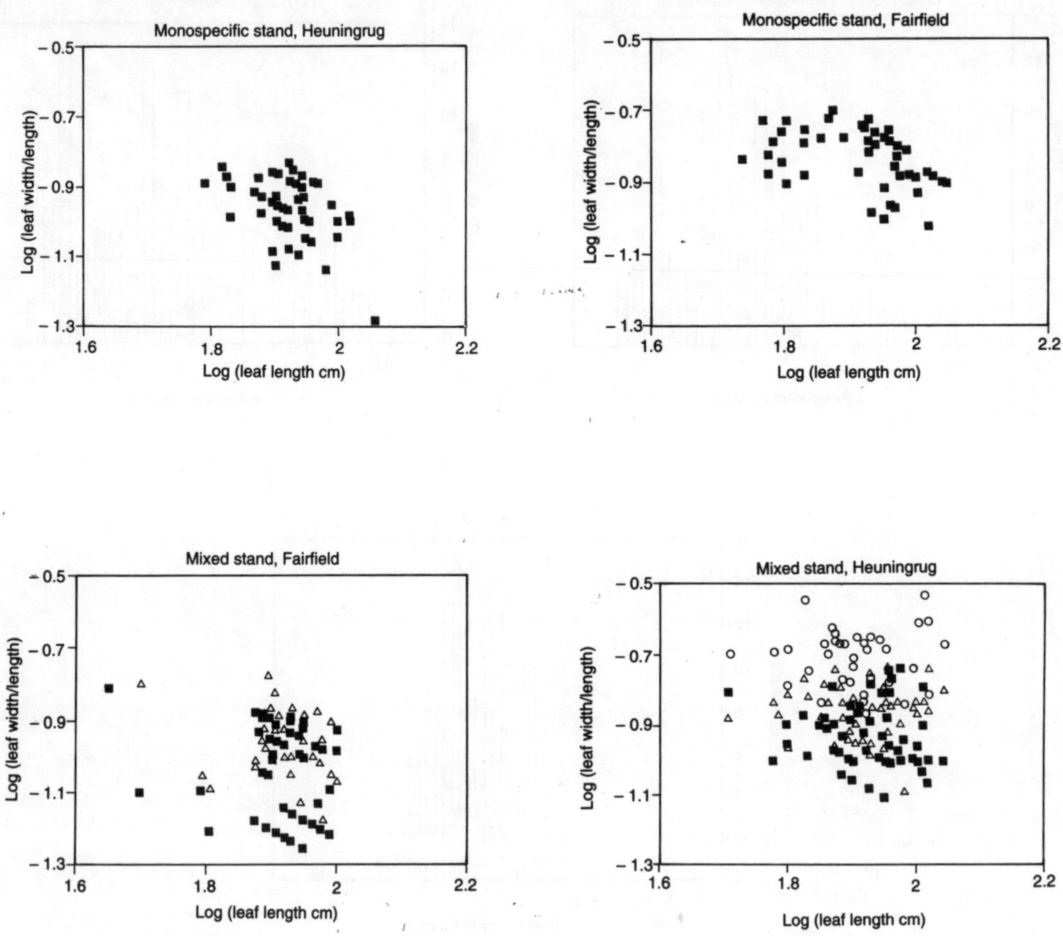

FIGURE 8.3 Leaf morphology variation in *Protea repens* (closed squares) in relation to co-occurring proteoids at matched study sites on the Agulhas Plain. Both Fairfield populations were on shallow sandy soils. The Heuningrug populations were on deep sands. Co-occurring species in the mixed stands were: Fairfield — *L. xanthoconus*; Heuningrug — *P. susannae* (open squares), *L. coniferum* (cross). *Leucadendron* samples include both males and females.

crimination between monospecific stands at Heuningrug and Fairfield (Wilks Lambda = 0.592, Chi Square = 50.9, $P < 0.001$) although neither had proteoid competitors. However, soils were deep at Heuningrug and shallow over sandstone at Fairfield.

The ideal situation of replicated populations in monospecific and multispecific stands matched for soil and climate has still to be found but the preliminary results suggest the pursuit is worthwhile. Besides the need for more replication, these results further emphasize the problem of explaining how and when in the life of a shrub leaf morphology could possibly influence coexistence.

LOTTERY MODELS: CONVERGENCE OR DIVERGENCE?

Warner and Chesson (1985) have warned that 'the storage idea is overly seductive. It is liable to be used either as a last resort explanation after attempts to demonstrate other hypotheses have failed or as an unsupported and convenient explanation without benefit of critical testing'. Here we attempt to test the lottery model by its wider implications for community structure. We are painfully aware that implications of the theory for field situations are still elusive while new refinements (Chesson and Huntly 1989; Chesson 1991) weaken its generality and therefore the value of our tests.

A simple test of lottery model predictions analogous to Cody's (1986) study is to ask whether the most similar species occur together or apart. Here similarity refers to both competitive equivalence and similar mean reproduction (Fagerstrom 1988). Similarity should be at least partly reflected by taxonomy with sister species sharing the most traits. Do closely related species share the same habitats more often than distantly related species? Clearly allopatric speciation may bias the analysis if species occur across geographic barriers. However, in the species-rich Cape flora, many closely related species occur on the same mountain slope or within easy reach of propagule dispersal (Cowling et al. this volume). Several closely related species with overlapping distribution ranges seem *not* to coexist in the same stands e.g. *Leucadendron spissifolium* and *L. salignum*; *L. rubrum* and *L. album*; *L. meridianum* and *L. coniferum*; *Protea neriifolia* and *P. coronata*; *P. repens* and *P. lanceolata*; *P. aurea* and *P. mundii*; *P. susannae* and *P. obtusifolia* (W J Bond and R M Cowling personal observation; J Midgley personal communication). However, we do not have adequate data for formally comparing frequency of co-occurrence of closely related and distantly related species. Also, the appropriate variable for predicting coexistence is reproductive similarity rather than the surrogate of taxonomic relatedness.

COEXISTENCE AND ENVIRONMENTAL VARIABILITY

In a completely uniform environment, seed and seedling production would become more predictably deterministic; there would be no storage effect; and coexistence via lottery would fall away. We tested the implications of environmental variation for species coexistence by comparing species numbers in vegetation samples from phytosociological surveys from the eastern to western extremes of the fynbos biome.

Variation in winter rainfall, especially periodic drought, can influence seed production and seedling survival (e.g. Keeley and Keeley 1988 for *Arctostaphylos* in chaparral). The reliability of winter rainfall in the fynbos biome diminishes from west to east (Table 8.1). According to lottery theory, we would predict higher average alpha species richness in areas with higher rainfall variability since this should increase variance in seedling production.

A comparison of mean species numbers of proteoids in 10 x 10 metre plots in vegetation surveys from eastern and western areas appears to support theoretical predictions since eastern areas have significantly higher mean species richness. This is particularly striking since regional richness of Proteaceae declines markedly from west to east across the fynbos biome in opposition to these diversity trends (Oliver et al. 1983). These results contradict Rickler's (1987) suggestion that alpha diversity is coupled to regional species richness, implying instead local control on richness (see also Cowling et al. this volume). However, the results are not predicted exclusively by lottery theory since niche arguments could also be developed (e.g. greater phenological differentiation is possible in the nonseasonal rainfall areas of the east).

TABLE 8.1 The species richness of the overstorey Proteaceae in phytosociological plots from vegetation surveys in western and eastern fynbos. The coefficient of variation of winter rainfall (June–August for 30 years) for stations near the study area has been calculated from Zucchini et al.'s (1984) rainfall simulation model. Only non-sprouting species are included.

Western	Rainfall cv (%)	Mean no. spp.	SE	No. plots	Total spp.
Cape of Good Hope[1]	21	1.47	0.097	55	7
Swartboschkloof[2]	30	1.29	0.050	85	3
Cape Hangklip[3]	29	1.46	0.054	129	7
Eastern					
Langeberg[4]	38	2.15	0.165	53	7
Moordkuil's[5]	43	2.61	0.272	23	8
Elandsberg[6]	42	2.08	0.288	12	6

[1] R M Cowling (unpublished data)
[2] McDonald (1988)
[3] Boucher (1978)
[4] D J McDonald (unpublished data)
[5] J Midgley and J Vlok (unpublished data)
[6] Cowling (1983)

Species interactions and coexistence

Although statistical studies of species distributions give insights into community structure, they are essentially descriptive. What processes underlie pattern and what is the evidence for the role of competition?

COEXISTENCE AND THE REGENERATION NICHE

The dominant proteoid shrubs in many fynbos stands share similar growth forms, habitat requirements, and growing seasons. However, there are often interspecific differences in seed mass, dispersal properties, seed number, seed bank size, etc. Yeaton and Bond (1991) studied a pair of co-occurring 'trophically equivalent' species differing primarily in seed biology. *Protea lepidocarpodendron* is a serotinous, wind-dispersed species, and *Leucospermum conocarpodendron*, a shrub with soil stored, ant-dispersed seed. By comparing *Leucospermum* performance in the open (i.e. without neighbouring proteoids) with performance near *Protea* shrubs, Yeaton and Bond (1991) showed strongly asymmetric competition with the *Protea* completely dominating the *Leucospermum* (Table 8.2) at all stages of its life history. *Protea* can even influence the likelihood of *Leucospermum* surviving a fire since individuals near other proteoids (canopies < .25 m apart) are more likely to be killed by fire (Table 8.2).

Because of dispersal differences, seedlings of the two species occupy different, and complementary, microsites after fire (Table 8.3). To test the long-term implications of dispersal differences for coexistence, Yeaton and Bond (1991) used a Markov model with transition probabilities based on seedling frequencies in the different microsites (Table 8.4). The Markov model predicts that dispersal differences are not sufficient, in this case, to ensure coexistence. Unless fires reduce *Protea* densities and create open space for colonization by *Leucospermum* every third or fourth fire cycle, *Leucospermum* would be eliminated from the community. Dispersal by ants does, however, reduce the rate of exclusion (Table 8.4). This study, therefore, predicts that coexistence of these species is temporary and that the two species should occupy different habitats — *Leucospermum* should persist where *Protea* does not. Distribution patterns of the two species support this prediction since *Leu-*

Table 8.2 The competitive effects of *Protea lepidocarpodendron* on *Leucospermum conocarpodendron* (from Yeaton and Bond 1991).

1) Seedling growth

The height of neighbouring plants (within 50 cm) three years after a fire at Cape Point. ($P < 0.001$)

	n	Height (cm)
Protea	50	29.5 ± 1.0
Leucospermum	50	15.6 ± 0.6

2) Shrub size at maturity

The effect of *Protea* canopies on stem diameter of *Leucospermum* at Steenberg Plateau. Vegetation was 15–16 years old. ($P < 0.001$)

	n	Stem diameter (cm)
In the open	25	6.9 ± 0.2
Under *Protea*	25	3.3 ± 0.2

3) Reproductive performance

The effect of *Protea* canopies on the flowering of *Leucospermum* at Steenberg Plateau during spring 1988. ($P < 0.001$)

	n	Mean no. of inflorescences
In the open	25	8.9 ± 1.7
Under *Protea*	25	0.2 ± 0.2

4) Fire survival

The neighbour effects on the survival of *Leucospermum* after a light burn at Rooihoogte, Cape Point. *n* is the number of plants censused. ($\chi^2$10.6, $P < 0.001$)

	n	% survival
In the open	62	56.5
Next to *Protea*	8	21.1

Table 8.3 The mean density (± SE) of seedlings m^{-2} of *Protea lepidocarpodendron* and *Leucospermum conocarpodendron* in three subhabitats in fynbos at Miller's Point, Cape Peninsula. *n* = 50 for each condition (from Yeaton and Bond 1991).

Subhabitats	*Protea*	*Leucospermum*
Open	0.3 (± 0.09)	1.7 (± 0.37)
Under *Leucospermum*	3.5 (± 0.69)	2.6 (± 0.35)
Under *Protea*	12.8 (± 7.07)	0.6 (± 0.14)

Table 8.4 The transition matrices derived from the seedling distributions of myrmecochorous *Leucospermum conocarpodendron* and serotinous *Protea lepidocarpodendron*. The values of the matrices are derived under the assumption that if both *Protea* and *Leucospermum* occur in the same patch, *Protea* will dominate by the next fire. The values in the lower matrix assume, in addition, no dispersal of *Leucospermum* seeds beyond its canopy radius. The predicted stationary states are the proportions of the two species (and open space) in the community after many fire cycles with the same transition characteristics (from Yeaton and Bond 1991).

	Probability of subhabitat becoming			
Subhabitat	Open	*Protea*	*Leucospermum*	Stationary state
With ants				
Open	0.30	0.16	0.54	0.08
Under *Protea*	0.06	0.92	0.02	0.84
Under *Leucosp.*	0.08	0.66	0.26	0.08
No ants				
Open	0.84	0.16	0.00	0.33
Under *Protea*	0.08	0.92	0.00	0.67
Under *Leucosp.*	0.08	0.66	0.26	0.00

cospermum is largely confined to areas without *Protea* (Taylor 1984; W J Bond personal observation).

This study cannot readily be extrapolated to other species pairs since *Leucospermum conocarpodendron* is exceptional in its genus, being neither fast growing nor short-lived and therefore more likely to be outcompeted by *Protea* in the face of recurrent fire. However, the implications of serotiny and myrmecochory for seedling distribution may be quite general according to a recent study by J Midgley and G von Maltitz (in preparation). They showed that seedlings of myrmecochorous and serotinous co-occurring *Protea*, *Leucadendron*, and *Leucospermum* species are distributed differently. Serotinous seeds are highly aggregated, irrespective of species, whereas seedlings of myrmecochorous species are more isolated. Dispersal by ants apparently minimizes interspecific competition and this may promote coexistence with wind-dispersed serotinous species. Interestingly, Midgley and Von Maltitz found little evidence for seedling distribution differences among serotinous species, despite differences in seed morphology. This implies a lottery to reach an establishment site and the potential for intense interactive competition among seedlings which reach the same site.

COEXISTENCE AND LOTTERY MODELS

A central assumption of lottery models is that generations overlap. This is necessary to ensure the 'storage effect' of a period of good recruitment persisting through the lean years (Warner and Chesson 1985). Laurie (1989) first applied lottery models in fynbos. He argued that, even where each generation is killed by fire, the lottery assumption is met if the number of successful recruits following fire is proportional to pre-burn seed reserves. Laurie (personal communication 1990) is currently developing simulation models of proteoid seed bank dynamics based on empirical information on inflorescence production and seed mortality from two study sites. Evidence from several mediterranean regions indicates substantial year to year variance in seed production with peaks and troughs sometimes falling at different periods for different species (see e.g. Cowling and Lamont 1985; Copland and Whelan 1989 for Australian examples, Keeley and Keeley 1988 for Californian). Comparable data are needed for fynbos proteoids.

To illustrate the use of lottery models, a simple example for Proteaceae seedlings is given in Table 8.5. Here adults are killed by fire, produce seedlings only after fire, and acquire space in proportion to the number of seedlings as a fraction of all proteoid seedlings. Seedling production was generated from a negative exponential distribution (with mean proportional to variance) which approximates field census data (Bond 1989). The process was simulated for 20 fires or until one species dropped to < 0.001 of the population when it was considered locally extinct. This model assumes the habitat is always saturated.

The results of a series of simulations are shown in Figure 8.4. These clearly show the qualitative results of lottery models — the most similar species coexist the longest and the least similar the shortest. However, the simulation also shows that even identical species will not coexist indefinitely. Table 8.4 indicates how this can happen due to a random series of successive poor recruitment episodes. To persist indefinitely, populations would have to be replenished from elsewhere (i.e. a spatial 'storage effect'), or have a 'rare species advantage' allowing them to bounce back from near extinction. Recently Chesson (1991) has noted that species obeying the lottery model are most likely to coexist if they share similar *average* birth and death rates and similar *average* variances over all environmental conditions. However, long-term coexistence is not possible unless each species is favoured by some specific conditions under which it performs better than its competitor. In this sense, lottery models still require 'niches' for long-term coexistence (Chesson 1991).

Lottery models have the attraction of realistically including variation in reproduction and variation in the timing of regeneration. Fynbos proteoids offer excellent opportunities for testing implications of lottery models for species coexistence. It should also be relatively easy to test the validity of the biological assumptions and the implications of relaxing the assumptions (e.g. Yeaton and Bond 1991; Midgley and Von Maltitz in preparation). The most difficult problem seems to be how to phrase lottery model predictions in testable ways.

Table 8.5 A lottery model simulating the population growth of two proteoid competitors (see text). In this example, the frequency distribution of λ is assumed to fit a negative exponential to simulate field data for Proteaceae (see Bond 1989).

Variables

Population growth rate = λ_{ij} where λ = seedlings/parents,
i = species A or B, j = fire number

Proportion of shrubs = p_{ij}

Calculations

1) Estimate λ_{ij} for fire j, given the mean and variance of λ_i

λ_{ij} = -mean λ_i x Ln (RND) where RND is a random number and 0 < = RND < = 1

2) Estimate p_{ij} after fire j

e.g. $$p_{A1} = \frac{\lambda_{A1} \times p_{A1}}{\lambda_{A1} \times p_{A1} + \lambda_{B1} \times p_{B1}}$$

Example

Mean Population growth rates $\lambda_{A1} = 2$
$\lambda_{B1} = 3$

Fire no.	λ_A	λ_B	p_A	p_B
0	2	3	0.5	0.5
1	0.86	6.79	0.11	0.89
2	0.85	7.98	0.01	0.99
3	0.12	1.30	0.001	0.999
4	1.31	0.31	0.005	0.995
5	0.38	6.64	< 0.001 = EXCLUDED	

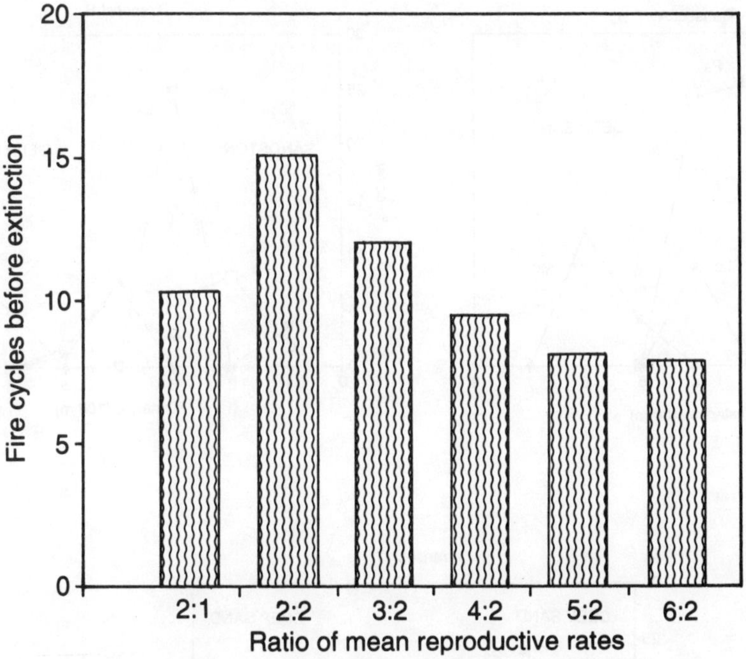

FIGURE 8.4 The outcome of repeated simulations of a lottery model for two competing proteoid species. Each bar represents 50 simulations of a run of 20 fires. Note that species with identical reproductive rates coexist the longest but exclusion still occurs. See text and Table 8.5 for details.

Alternatives to competition

Although competition has underlain theory predicting community structure, it need not be the only process responsible for apparent niche differentiation or habitat segregation. The classical alternative is that plant distribution is limited by the match between the physiological capabilities of a species and the local abiotic environment.

ARE DISTRIBUTIONS LIMITED BY ABIOTIC FACTORS?

Recent studies have attempted to separate the effects of interspecific competition and abiotic factors in determining distribution of species in salt marshes (Snow and Vince 1984; Bertness and Ellison 1987), semi-arid grasslands (Gurevitch 1986), and Australian woodlands (Lamont et al. 1989). M B Richards and R M Cowling (in preparation) have studied edaphic versus competitive controls on seedling survival in order to determine factors responsible for distribution patterns of two species pairs of *Leucadendron* and one species pair of *Protea*. Species in each pair replace each other across transects through the communities (Figure 8.5). Two of the pairs are differentiated by soil type:

- A calcicole/calcifuge comparison with *L. meridianum* on limestone and *L. coniferum* on neutral to acid sand downslope.
- *L. xanthoconus* on shallow rocky acid sandstone soil with *L. laureolum* on deeper acid sands downslope.
- The third pair has no obvious soil or topographic differences but *P. compacta* changes abruptly to *P. susannae* on deep neutral to slightly acid sands (Figure 8.6).

Reciprocal plantings of seed in monoculture and two-species mixtures were made to test for the effects of competition versus edaphic factors on seedling growth and survival (see Figure 8.6 for experimental layout). Along each transect, seeds of the respective species pairs were planted in cleared experimental plots at three sites. Each plot was divided into 15 sub-plots, 10 of which contained monocultures (five replicates per species), and five of which contained 1:1 mixtures. Treatments were randomly allocated to the sub-plots.

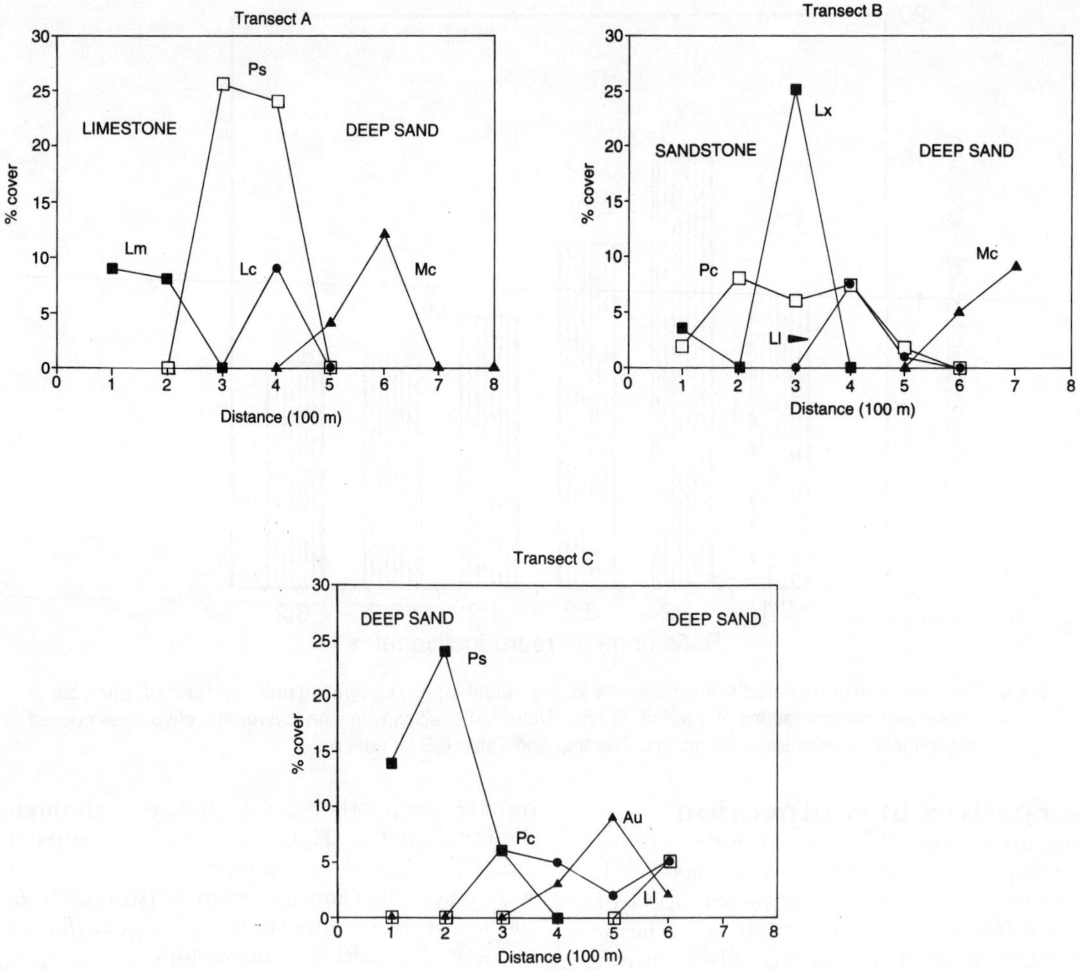

FIGURE 8.5 The distribution of proteoid shrubs across edaphic gradients in lowland fynbos, Soetanysberg. The distribution of the transects in the landscape is indicated in Figure 8.6. Au = *Aulax umbellata*, Lc = *Leucadendron coniferum*, Ll = *L. laureolum*, Lm = *L. meridianum*, Lx = *L. xanthoconus*, Ms = *Mimetes saxatilis*, Mc = *M. cucullatus*, Ps = *Protea susannae*, Pc = *P. compacta*.

After one year, seedling survival exceeded 90% at all the sites in all but one species (Table 8.6). This is perhaps surprising given the apparent edaphic restriction of some of the species pairs. There was little indication of high seedling mortality in seeds germinating in the 'wrong' site. Four of the six species did not show significantly greater mortality in sites where adults had been absent before clearing. *L. meridianum* was one exception in having 28% lower survival in one of the two sites where adults had been absent ($P > 0.001$, Table 8.6). In *P. compacta*, seedling mortality patterns corresponded with adult distribution with 8.5% lower survival at the single site where *P. compacta* adults did not occur ($P < 0.005$). Two species showed small but statistically signi-ficant reductions in survival in monoculture relative to mixtures (*P. compacta* 4.1% lower, $P < 0.05$, *L. xanthoconus* 1.6% lower, $P < 0.05$) possibly indicating intraspecific competition.

With such high overall survival and low (if statistically significant) mortality due to either site or competitive effects, these results suggest that early seedling establishment is not the critical life history stage controlling species distribution. Mustart and Cowling (submitted),

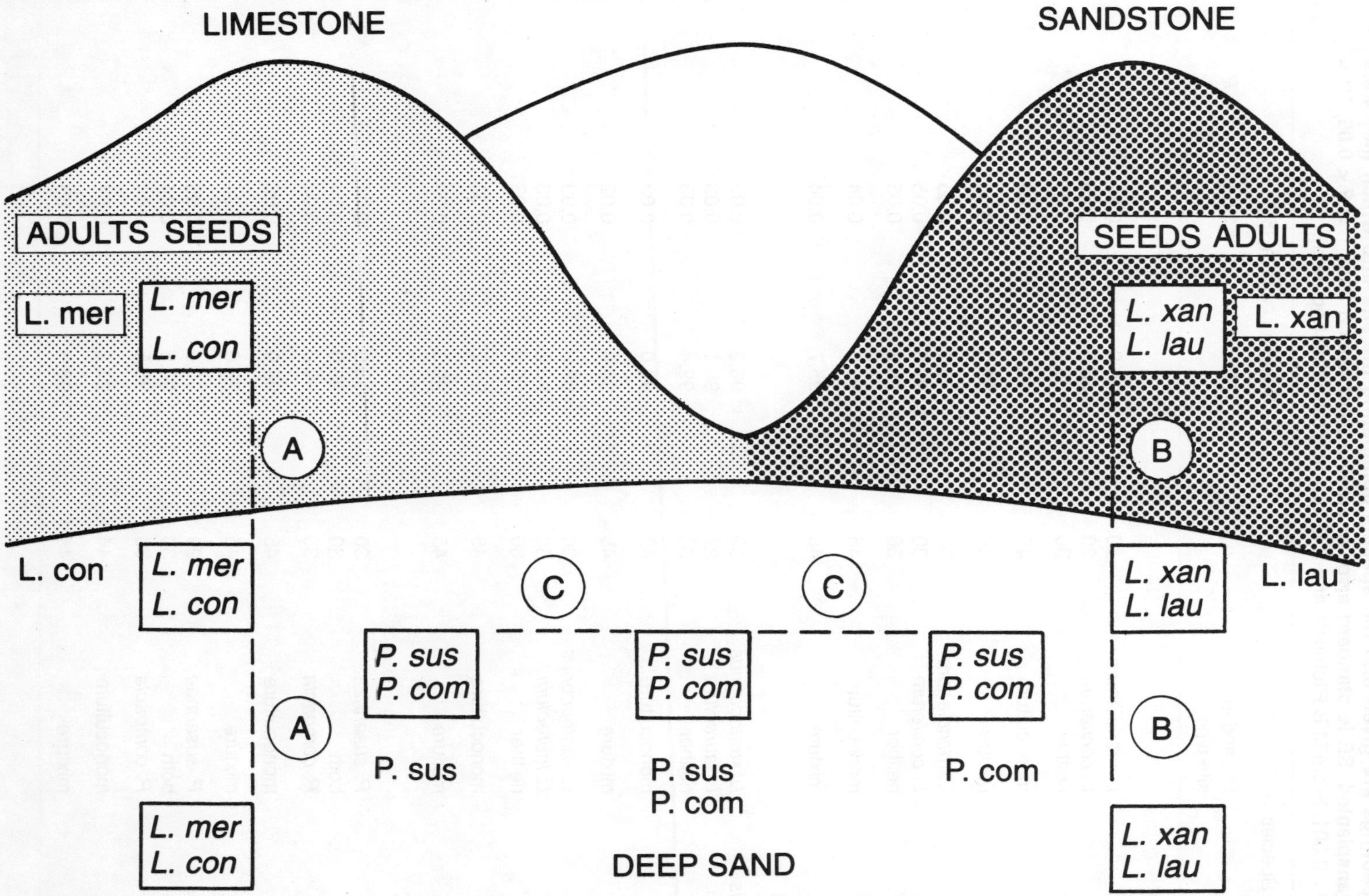

FIGURE 8.6 An experimental design for reciprocal seed plantings in lowland fynbos, Soetanysberg. Seeds were planted into cleared areas of mature fynbos along three transects indicated as A, B, and C. The species used as seed sources at each experimental site are shown in italics, and the dominant proteoid shrubs there in standard text. Limestone derived soils are shallow and calcareous, sandstone soils are rocky and acid, and the sands are deep (> 1.2 m) and neutral. P. com = *Protea compacta*, P. sus = *P. susannae*, L. con = *L. coniferum*, L. mer = *L. meridianum*, L. xan = *L. xanthoconus*, L. lau = *L. laureolum.*

Table 8.6 Seedling survival after transplants to new locations. The species names of seedlings are given and the species dominating the site into which they were transplanted. SE is standard error of arcsine transformed data. * = $P < 0.05$, *** = $P < 0.001$. From M B Richards and R M Cowling (unpublished data).

Species			Seedling survival (%)		
Seed type	Planted in site type	*n*	Mean	SE	*P*
TRANSECT A					
L. meridianum	*L. meridianum*	30	96.9	0.04	
	L. coniferum	30	68.5	0.09	
	neither	30	96.2	0.06	***
	monoculture	45	85.9	0.06	
	mixture	45	92.3	0.06	
L. coniferum	*L. meridianum*	30	95.1	0.05	
	L. coniferum	30	97.6	0.05	
	neither	30	95.2	0.05	
	monoculture	45	96.5	0.04	
	mixture	45	95.7	0.04	
TRANSECT B					
L. xanthoconus	*L. xanthoconus*	30	98.2	0.03	
	L. laureolum	27	99.1	0.03	
	neither	30	99.4	0.03	
	monoculture	42	98.0	0.03	
	mixture	45	99.6	0.02	*
L. laureolum	*L. xanthoconus*	30	98.9	0.03	
	L. laureolum	30	98.2	0.03	
	neither	30	100.0	0.03	
	monoculture	45	99.2	0.03	
	mixture	45	99.3	0.02	
TRANSECT C					
P. susannae	*P. susannae*	30	98.5	0.04	
	both	30	98.0	0.04	
	P. compacta	30	99.6	0.04	
	monoculture	45	98.9	0.03	
	mixture	45	98.8	0.04	
P. compacta	*P. susannae*	29	90.0	0.05	
	both	30	98.5	0.04	
	P. compacta	30	98.5	0.04	*
	monoculture	44	94.2	0.04	
	mixture	45	98.3	0.03	*

however, found significant differences in survival and growth rates in the second year after the establishment of reciprocally transplanted calcicole and calcifuge *Protea* and *Leucadendron* species pairs. Richards and Cowling's study is still in progress and the importance of abiotic factors versus competition may yet emerge in differences in mortality or growth. It is worth noting that soil factors will not change (much) as the seedlings grow but competition obviously will.

EFFECTS OF DISTURBANCE ON SPECIES DISTRIBUTION

Pierce (1990) has argued that life history differences explain apparent habitat differentiation in a suite of small-leaved shrubs in south-eastern Cape dune fynbos. The six species studied have different longevities. Species with short lifespans are excluded from thicket vegetation that seldom burns. At the other extreme, only species with rapid maturation and short lifespans can tolerate the frequent fire regimes occurring in grasslands. A similar explanation seems likely for the restriction of some slow growing Proteaceae to rocky, relatively fire-free high montane habitats (e.g. *Leucadendron dregei*, *Protea venusta*, *P. montana*, and *P. cryophila*). Certain low, spreading, reseeding Proteaceae, such as *Leucospermum secundifolium* and *Leucadendron glaberrimum* subspecies *glaberrimum*, survive fire by being burnt only at the edges of the shrub and appear to be restricted to areas with light fuel loads (W J Bond personal observation). *Leucadendron muirii*, with a similar growth form, survives on bare rocky limestone outcrops, whereas other limestone endemics, such as *L. meridianum*, have an erect growth habit and are burnt regularly (W J Bond and R M Cowling personal observation).

Of course past selection to avoid competition may have led to these life history differences (and to physiological site preferences) but hypotheses invoking the 'ghost of competition past' (Connell 1980) are notoriously difficult to test.

CONCLUSION

Competition studies have barely begun in the fynbos. The discrete generations initiated by fire lend themselves to simple modelling, persuasive field experiments of interactions in the regeneration phase, and unequivocal studies of the effects of neighbours in older phases. Fynbos also has the advantage of large genera such as *Leucadendron*, *Protea*, and *Erica* with many species in a landscape. This facilitates the use of statistical tests in descriptive studies of community structure. All these advantages suggest that substantial insights may be gained from competition studies in fynbos which can balance the largely indirect studies of the other major centres of diversity — the tropical rain forests (e.g. Hubbell and Foster 1986; Gentry 1988).

These studies are also not without an applied interest. Species are currently managed and flowers harvested on the assumption that utilization of one species will have little effect on others. If competition studies indicate that the distribution and abundance of either desirable or undesirable species are a function of competitors, then management prescriptions may be affected. Already in the case of mixed communities of *Leucospermum conocarpodendron* and *Protea lepidocarpodendron*, it is clear that *Protea* densities would have to be reduced either by the use of unseasonal fire or harvesting in order to maintain *Leucospermum* in the community (Yeaton and Bond 1991).

As regards the debate of niche versus non-equilibrial stochastic models, it is clearly too early to judge. Cody's leaf 'niche' does not fit current theories on niche differentiation in sessile organisms (e.g. Yodzis 1986; Pacala 1988). His study is a major challenge to theory that denies the importance of niche differentiation. The mechanism linking leaf size and shape to coexistence is not clear, and both this and community pattern in leaf 'niches' demand further attention. The lottery models risk explaining everything and therefore nothing — as recognized by Warner and Chesson (1985). We have attempted to test some implications of these models, as we see them, for community structure. These are interesting in directly contradicting most traditional theory developed for consumptive competition. Predictions of the theory are supported as regards coexistence in variable environments but not coexistence of closely related species pairs.

The models themselves are poor reflections of reality. Lotka-Volterra models of consumptive competition are inappropriate for plants and deterministic models of competition for space generate opposing predictions (Yodzis 1986; Pacala 1988). The lottery models are closer descriptions of fynbos life histories, although Chesson's models explicitly exclude discrete population growth and Laurie's (1989) modification lacks any explicit reference to dispersal, growth, or competitive differences (but see Shmida and Ellner 1984). There is clearly much scope for modelling the life histories of fynbos plants to explore implications for coexistence. Likewise, too little is known of key stages in plant life history and how species interactions might affect them. Nevertheless, theoretical models have stimulated a series of interesting questions and empirical studies which in turn are beginning to provoke a reassessment of theory. The future in this area looks to be an interesting one.

ACKNOWLEDGEMENTS

We thank D McDonald and J Midgley for providing phytosociological data and S Pierce for assistance in transcribing it. We are very grateful for comments and criticism from M Cody, P Grubb, J Keeley, H Laurie, J Midgley, and R I Yeaton. W J Bond would particularly like to thank H Laurie for helpful discussions on lottery models. We are grateful for financial support from the Foundation for Research Development, the South African Nature Foundation, and the Research Committee of the University of Cape Town.

REFERENCES

ACOCKS J P H (1953). Veld types of South Africa. *Memoirs of the Botanical Survey of South Africa* **28**.

ADAMSON R S (1927). The plant communities of Table Mountain. I. Preliminary Account. *Journal of Ecology* **15**, 278–309.

AUERBACH M and A SHMIDA (1987). Spatial scale and the determinants of plant species richness. *Trends in Ecology and Evolution* **2**, 238–42.

BERTNESS M D and ELLISON A M (1987). Determinants of pattern in a New England salt marsh plant community. *Ecological Monographs* **57**, 129–47.

BOND W J (1989). Describing and conserving biotic diversity. In *Biotic diversity in Southern Africa: concepts and conservation.* (ed Huntley B J) Oxford University Press, Cape Town. 2–18.

BOND W J, VLOK J and VIVIERS M (1984). Variation in seedling recruitment of Cape Proteaceae after fire. *Journal of Ecology* **72**, 209–21.

BOND W J and MIDGLEY J (1988). Allometry and sexual differences in leaf size. *American Naturalist* **131**, 901–10.

BOUCHER C (1978). Cape Hangklip Area 2. The vegetation. *Bothalia* **12**, 455–97.

BROWN W L and WILSON E O (1956). Character displacement. *Systematic Zoology* **7**, 49–64.

CHESSON P L (1986). Environmental variation and the coexistence of species. In *Community ecology.* (eds Diamond J and Case T J) Harper and Row, Cambridge. 240–56.

CHESSON P L (1991). A need for niches? *Trends in Ecology and Evolution* **6**, 26–8.

CHESSON P L and WARNER R R (1981). Environmental variability promotes coexistence in lottery competitive systems. *American Naturalist* **117**, 923–43.

CHESSON P L and CASE T J (1986). Overview: nonequilibrium community theories: chance, variability, history and coexistence. In *Community ecology.* (eds Diamond J and Case T J) Harper and Row, Cambridge. 229–39.

CHESSON P L and HUNTLY N (1989). Short term instabilities and long term community dynamics. *Trends in Ecology and Evolution* **4**, 293–8.

CODY M L (1978). Distributional ecology of *Haplopappus* and *Chrysothamnus* in the Mojave Desert. 1. Niche position and niche shifts on north facing granite slopes. *American Journal of Botany* **65**, 1 107–16.

CODY M L (1986). Structural niches in plant communities. In *Community ecology.* (eds Diamond J and Case T J) Harper and Row, New York. 381–405.

COMINS H N and I R NOBLE. (1985). Dispersal, variability and transient niches: species coexistence in a uniformly variable environment. *American Naturalist* **126**, 706–23.

CONNELL J H (1980). Diversity and the coevolution of competitors, or the ghost of competition past. *Oikos* **35**, 131–8.

COPLAND B J and WHELAN R J (1989). Seasonal variation in flowering intensity and pollination limitation of fruit set in four co-occurring *Banksia* species. *Journal of Ecology* **77**, 509–23.

COWLING R M (1983). Vegetation studies in the Humansdorp region of the fynbos biome. Unpublished PhD thesis, University of Cape Town.

COWLING R M (1987). Fire and its role in coexistence and speciation in Gondwana shrublands. *South African Journal Science* **83**, 106–11.

COWLING R M, LAMONT B (1985). Serotiny in three Western Australian *Banksia* species along a climatic gradient. *Australian Journal of Ecology* **10**, 345–50.

COWLING R M, STRAKER C J, DEIGNAN M T (1990). Does microsymbiont specificity determine plant species turnover and speciation in Gondwana shrublands: a hypothesis. *South African Journal of Science.* (in press).

DENSLOW J S (1985). Disturbance mediated coexistence of species. In *The ecology of natural disturbance and patch dynamics.* (eds Pickett S T A and White T S) Academic Press, New York. 307–24.

DIAMOND J M (1975). Assembly of species communities. In *Ecology and Evolution of Communities.* (eds Cody M L and Diamond J M) Harvard University Press, Cambridge, Mass. 342–444.

FAGERSTROM T and AGREN G I (1979). Theory for coexistence of species differing by regeneration properties. *Oikos* **33**, 1–10.

FAGERSTROM T (1988). Lotteries in communities of sessile organisms. *Trends in Ecology and Evolution* **3**, 303–6.

FENCHEL T (1975). Character displacement and coexistence in mud snails (Hydrobiidae). *Oecologia* **20**, 19–32.

GENTRY A H (1988). Changes in plant community diversity and floristic composition on environmental and geographic gradients. *Annals of Missouri Botanical Garden* **75**, 1–34.

GILL A M (1975). Fire and the Australian flora: a review. *Australian Forestry* **38**, 4–25.

GRUBB P J (1977). The maintenance of species richness in plant communities. The importance of the regeneration niche. *Biological Review* **52**, 107–45.

GUREVITCH J (1986). Competition and the local distribution of the grass *Stipa neomexicana. Ecology* **67**, 46–57.

HARVEY P H, COLWELL R K, SILVERTOWN J V, MAY R M (1983). Null models in ecology. *Annual Review of Ecology and Systematics* **14**, 189–211.

HUBBELL S P and R B FOSTER (1986). Biology, chance, and

history and the structure of tropical rain forest tree communities. In *Community ecology.* (eds Diamond J and Case T J) Harper and Row, Cambridge. 314–29.

HUSTON M (1979). A general hypothesis of species diversity. *American Naturalist* **113**, 81–101.

KEELEY J E (1977). Seed production, seed populations in soil, and seedling production after fire for two congeneric pairs of sprouting and non-sprouting chaparral shrubs. *Ecology* **58**, 820–9.

KEELEY J E and S E KEELEY (1988). Temporal and spatial variation in fruit production by California Chaparral Shrubs. In *Time scales and water stress.* (eds Di Castri F, Floret C, Rambal S, Roy J) Proceedings of the 5th International Conference on Mediterranean Ecosystems, IUBS, Paris. 57–463.

LAMONT B B, ENRIGHT N J, BERGL S M. (1989). Coexistence and competitive exclusion of *Banksia hookeriana* in the presence of congeneric seedlings along a topographic gradient. *Oikos* **56**, 39–42.

LAURIE H (1989). Can lottery models work for populations with discrete generations? Applications in Cape fynbos shrublands. Unpublished manuscript, University of Cape Town.

MAY R M, MACARTHUR R H (1972). Niche overlap as a function of environmental variability. *Proceedings of the National Academy of Sciences* **69**, 1 109–13.

McDONALD D J (1988). A synopsis of the plant communities of Swartboskloof, Jonkershoek, Cape Province. *Bothalia* **18**, 233–60.

MIDGLEY J J (1989). Season of burn of serotinous Proteaceae: a critical review and further data. *South African Journal of Botany* **55**, 165–70.

MIDGLEY J and VON MALTITZ G (1990). Nearest neighbour relations of seedlings of serotinous and myrmecochorous Proteaceae. (in preparation).

MIDGLEY J and BOND W J (1989). Leaf size and inflorescence size may be allometrically related traits. *Oecologia* **78**, 427–9.

MUSTART P J and COWLING R M (submitted). The role of regeneration niche in determining the distribution in fynbos of edaphically restricted *Leucadendron* and *Protea* species. *Journal of Ecology.*

NEWMAN E I (1982). Niche separation and species diversity in terrestrial vegetation. In *The plant community as a working mechanism.* (ed Newman I E) Blackwell Scientific Publications, Oxford. 61–75.

OLIVER E G H, LINDER H P, ROURKE J P (1983). Geographical distribution of present day Cape taxa and their phytogeographical significance. *Bothalia* **14**, 427–40.

PACALA S W (1988). Competitive equivalence: the coevolutionary consequences of sedentary habit. *American Naturalist* **132**, 576–93.

PIERCE S M 1990. Pattern and process in south coast dune fynbos: population, community and landscape level studies. PhD thesis, University of Cape Town.

RICKLEFS R E (1987). Community diversity: relative roles of local and regional processes. *Science* **235**, 167–71.

ROURKE J P (1982). *The Proteas of Southern Africa.* Struik, Cape Town.

SALE P F (1977). Maintenance of high diversity in coral reef fish communities. *American Naturalist* **111**, 337–59.

SCHOENER T W (1989). The ecological niche. In *Ecological concepts, 29th symposium of the British Ecological Society.* (ed Cherret J M) Blackwell, Oxford. 79–113.

SHMIDA A and ELLNER S (1984). Coexistence of plant species with similar niches. *Vegetatio* **58**, 29–55.

SILVERTOWN J and LAW R (1987). Do plants need niches? *Trends in Ecology and Evolution* **2**, 24–6.

SNOW A A and VINCE S W (1984). Plant zonation in an alaskan salt marsh. II. An experimental study of the role of edaphic conditions. *Journal of Ecology* **72**, 669–84.

TAYLOR H C (1984). A vegetation survey of the Cape of Good Hope Nature Reserve. II. Descriptive account. *Bothalia* **15**, 259–91.

TILMAN D (1982). Resource competition and community structure. Princeton University Press, Princeton, New Jersey.

TILMAN D (1986). Evolution and differentiation in terrestrial plant communities: the importance of the soil resource: light gradient. In *Community ecology.* (eds Diamond J and Case T J) Harper and Row, Cambridge. 359–80.

VAN WILGEN B W and VIVIERS M (1985). The effects of season of fire on serotinous Proteaceae and the implications for fynbos management. *South African Forestry Journal* **133**, 49–53.

WARNER R R and P CHESSON (1985). Coexistence mediated by recruitment fluctuations: a field guide to the storage effect. *American Naturalist* **125**, 769–87.

WERNER P A (1979). Competition and coexistence of similar species. In *Topics in plant population biology.* (eds Solbrig O T, Jain S, Johnson G B, Raven P H) Columbia University Press, New York. 287–310.

WHITTAKER R H (1965). Dominance and diversity in land plant communities. *Science* **147**, 250–60.

YEATON R I and BOND W J (1991). Competition between two shrub species: dispersal differences and fire promote coexistence. *American Naturalist.* (in press).

YODZIS P (1986). Competition, mortality, and community structure. In *Community ecology.* (eds Diamond J and Case T J). Harper and Row, Cambridge. 480–91.

ZUCCHINI W, HIEMSTRA L A V, SPARKS R S (1984). Estimating the missing values in rainfall records. Water Research Commission Report No. 91/3/84.

9

Plant structure and function

W D Stock, F van der Heyden and O A M Lewis

INTRODUCTION

Detailed studies of the relationships between vegetation form and patterns of resource use in the fynbos biome are comparatively recent and were begun only in 1979 after the initiation of the Fynbos Biome Project. This programme complemented studies in other mediterranean-type ecosystems which had previously been compared in order to test hypotheses regarding convergent evolution of plant form and function (Di Castri and Mooney 1973; Mooney 1977; Thrower and Bradbury 1977; Cody and Mooney 1978). Miller (1981) followed up these studies by testing whether chaparral and matorral vegetations optimize resource use. Optimality arguments suggest that resources such as water, light, and nutrients are maximally used and this leads to specific plant forms becoming common in the vegetation. There has been considerable theoretical opposition to optimality arguments because of the non-falsifiability of assumptions of adaptation, the untestability of optimality models, and the failure to consider non-selective factors shaping evolution (Lewontin 1977; Gould and Lewontin 1979). Nonetheless, we use the optimality approach in this chapter to develop an understanding of fynbos plant form and function, bearing in mind the limitations and advantages of this approach as argued by Givnish (1986).

The role of nutrients as a selective force promoting convergence has been shown to differ between the relatively nutrient-rich chaparral, matorral, and maquis mediterranean climate ecosystems and the nutrient-poor systems of Australia and South Africa (Kruger et al. 1983). Differences in resource use among species and life forms have not been extensively studied in the fynbos biome, although certain distinct and similar patterns of resource use are evident in both Australia and South Africa (Stock 1988). This chapter synthesizes fynbos research by reviewing available data on resource (water, light, and nutrients) use. The chapter also discusses the functional significance of some distinctive features of fynbos, such as the ubiquitous occurrence of sclerophylly (leathery leaf consistency) and the large numbers of plants producing carbon-rich secondary compounds and mechanical structures.

PLANT WATER RELATIONS

Fynbos plant species have a range of anatomical and morphological features which allow the co-occurrence of a number of growth forms with potentially different water usage patterns (Campbell and Werger 1988). Features include variations with respect to leaf longevity, rooting depth, sclerophylly, and leaf size, all of which had been interpreted initially by ecophysiologists as mechanisms of drought resistance. It is therefore not surprising that the first studies of plant form and function in the fynbos investigated water usage characteristics of various plant growth forms. Initial studies concentrated on seasonal and diurnal progressions of leaf conductances and xylem pressure potential (XPP) in response to climatic gradients (Miller et al. 1983, 1984; Sommerville 1983). Interest was focused upon comparisons of the water relations of fynbos plants with those of other mediterranean-type ecosystems. Shorter periods of water stress were observed during summer months in fynbos, which suggested that water is not a major limiting factor in fynbos, as it is in other mediterranean-type ecosystems. Marked differences were, however, observed between different fynbos growth forms. Shallow-rooted restioids and ericoid shrubs exhibited greater summer declines in XPP values than deeper rooted proteoid shrubs. The latter group experienced high

XPPs which remained relatively constant throughout the year. Subsequent studies by Jeffery et al. (1987), Van der Heyden and Lewis (1989), Davis and Midgley (1990), and Richardson and Kruger (1990) have all found similar patterns of seasonal changes in XPP of these growth forms.

The observed patterns of water usage have been interpreted in terms of various mechanisms of water use optimization in response to summer drought. Proteoid overstorey shrub species are considered to be water spenders as they maintain relatively high and constant XPPs throughout the year because they have access to soil water in deep soil horizons (Van der Heyden and Lewis 1988). A drought tolerant group of plants can be recognized which includes mostly restioids and shallow-rooted understorey species which maintain high stomatal conductances, irrespective of water availability, and consequently experience highly variable seasonal and daily XPPs (Miller 1985). The limited stomatal dis-tribution of the ericoid group suggests that these species use water conservatively (water savers) (Van der Heyden and Lewis 1989) as do succulents. Any analysis of the distribution of water use optimization patterns along aridity gradients is complicated in fynbos regions by associated nutrient gradients. Drought deciduous and succulent species are, however, more prevalent in the westerly arid regions on nutrient-rich soils. Water appears, therefore, to act as a selective pressure in the fynbos region (Miller 1985) only on nutrient-rich soils. On low nutrient soils little evidence of diversity in water use patterns and associated morphological and phenological features is evident across the biome and Campbell and Werger (1988) have suggested that soil nutrient levels are paramount in determining the distribution of plant form features. Thus the generalization that climate (moisture availability) is the major factor determining vegetation patterns does not hold for the fynbos region. Ecological interpretations of plant form and performance have often identified low soil nutrient status as being important (Specht et al. 1983) and throughout this chapter physiological explorations of plant performance attempt to evaluate the relative roles of these selective factors.

CARBON GAIN

Photosynthetic gas exchange studies have been undertaken only relatively recently in the fynbos biome. Results show photosynthetic rates to be generally low, similar to those measured in several mediterranean climate sclerophyllous shrubs (Mooney and Dunn 1970; Odening et al. 1974; Gigon 1979). Photosynthetic rates reported for *Protea* species (Mooney et al. 1983) of 3–14 µmol m^{-2} s^{-1} at 21°C compare well with the photosynthetic rates of mountain fynbos species determined by Von Willert et al. (1989) and Van der Heyden and Lewis (1989). Von Willert et al. (1989) could not separate species nor growth forms on the basis of photosynthetic rates, transpiration rates, or leaf conductances. The geophyte, *Dilatris corymbosa*, showed the highest photosynthetic rates of all 16 fynbos species studied. This geophyte appears to have a carbon gaining strategy comparable to that of a drought-deciduous species as the short duration leaves tended to have high photosynthetic rates allowing them to equal the year-round carbon gain of the evergreens.

Van der Heyden and Lewis (1989) distinguished between two distinct gas exchange guilds in fynbos based on photosynthetic rates. The first guild consisted of the restioid and ericoid species which exhibited relatively low photosynthetic rates and the second guild comprised the proteoid species which experienced comparatively high photosynthetic rates throughout the year. Maximum rates of photosynthesis recorded for proteoid species were double the rates recorded for *Thamnochortus lucens* (restioid) and *Erica plukenetii* (ericoid) (Figure 9.1). In addition to the inherent differences in the photosynthetic carbon gaining capacities of the growth forms, Van der Heyden and Lewis (1989) also identified different degrees of summer decline in photosynthetic rates. For example, midday net photosynthetic rates of *Thamnochortus lucens* (restioid) and *Erica plukenetii* (ericoid) during summer were 3% and 6% respectively of their spring midday net photosynthetic rates. In contrast, the midday net assimilation rates of the proteoid species during summer were 43–95% of their spring rates. The limited adjustment of seasonal photosynthetic rates has also been observed by Richardson and Kruger (1990)

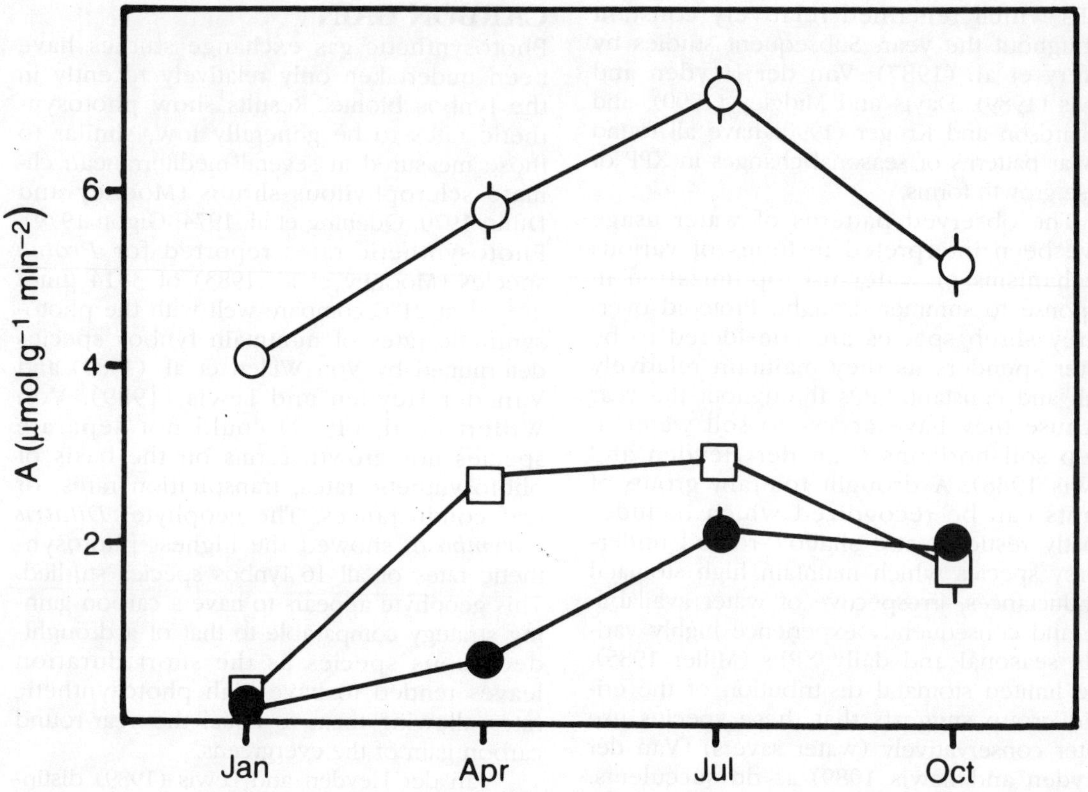

FIGURE 9.1 The midday net photosynthetic rates of *Protea laurifolia* (○——○), *Erica plukenetii* (□——□), and *Thamnochortus lucens* (●——●) over four months in 1986. Each value is the mean of six replicate measurements and the corresponding vertical bar represents the standard error.

and has been attributed to the water gaining ability of deep-rooted proteoids. The inherently higher annual carbon gaining capacity of the proteoids has been interpreted as an adaptive advantage allowing this functional guild to compete favourably during the later successional stages following fire (Van der Heyden and Lewis 1989).

CONTROLS ON PHOTOSYNTHESIS

Diurnal and seasonal patterns of plant water relations should be reflected in photosynthetic carbon gain patterns since both water loss and CO_2 uptake occur through the stomata. However, results obtained for fynbos species do not provide sufficient evidence to confirm that plant water status is the overriding determinant of photosynthetic capacity. Van der Heyden (1988) found no statistically significant correlations between XPPs and net photosynthetic rates. Coefficients of correlation, however, were greater for the shallow-rooted species (r = 0.627–0.744) than for the deep-rooted species (r = 0.355–0.604) which indicated that plant water status is of greater importance in influencing net photosynthetic rates of the shallow-rooted species. Von Willert et al. (1989) also found that midday depressions of photosynthesis, transpiration, and leaf conductance were not accompanied by decreases in XPP for deep-rooted species such as *Rafnia capensis*, *Rhus dissecta*, and *Salvia* species.

Van der Heyden and Lewis (1990) investigated the influence of water availability on photosynthetic responses of proteoid, restioid, and ericoid species during a late summer drought period. Photosynthetic rates of ericoid and proteoid species did not change with the addition of water, which suggested that the rates of these species were not limited by low soil water availability during summer. In contrast, a restioid, *Thamnochortus lucens*, experienced a 20–40% increase in net photosynthetic rates when supplied with water (Figure 9.2). Photosynthetic characteristics of the restioids, exemplified by *T. lucens*, indicate that this growth form is able to occupy a diversity of habitats. The relative inefficiency of its ability to control water loss and the low efficiency of water usage in terms of carbon gain (Van der Heyden and Lewis 1989) are counterbalanced by an ability to take up water as soon as it becomes available, thereby maximizing CO_2 uptake (Van der Heyden and Lewis 1990). An analysis of the vegetation structural characteristics of mountain fynbos (Campbell 1985) showed strong correlations between restioid-dominated fynbos and xeric habitats (areas with shallow and rocky soils) or physiologically water-limited environments (seasonally waterlogged bogs) which represent conditions for which the restioids are well adapted in terms of plant water relations and carbon gaining strategies.

Restionaceae have a number of morphological and anatomical characteristics which allow them to survive periods of low soil water availability. Meney et al. (1990) identified a number of xeromorphic morphological features in *Alexgeorgea* (an Australian member of the Restionaceae) such as heavily lignified culms, non-photosynthetic scale leaves, buried creeping rhizomes, and extensive rhizosheaths. These features also characterize fynbos restioids and are considered as adaptations to avoid desiccation during summer drought. In one species, *Alexgeorgea ganopoda*, an integrated system of aerenchyma was observed which extended through the roots, rhizomes, and the lower regions of the culms. Meney et al. (1990) suggested that the interconnecting aerenchyma in the cortex investments are an adaptation to seasonally flooded habitats. These features also characterize fynbos restioids (Hardcastle and Schtte 1983) which are often associated with waterlogged habitats.

Mooney et al. (1983), Richardson and Kruger (1990), and Van der Heyden and Lewis (1990) reported light saturation for proteoid and restioid species at approximately 1 000 $\mu mol\ m^{-2}\ s^{-1}$ while light saturation of *Erica plukenetii* was recorded at 800 $\mu mol\ m^{-2}\ s^{-1}$ (Van der Heyden and Lewis 1990). The light requirements of mediterranean climate species have been interpreted in terms of their low nitrogen contents which does not allow for high photosynthetic capacities (Mooney et al. 1983). Nitrogen fertilization of a proteoid species, *Protea lepidocarpodendron*, resulted in increased photosynthetic rates (W D Stock and T Manuel unpublished data) which supports the hypothesis that low leaf nitrogen contents probably constrain the photosynthetic capacity of fynbos species.

Temperature optima for photosynthesis of two fynbos species was found to be near 25°C. Van der Heyden and Lewis (1990) found that photosynthetic temperature optima of *Erica plukenetii* and *Protea laurifolia* showed little change during high temperature periods as values remained unchanged from spring to late summer stress periods. Photosynthetic temperature optima of *Thamnochortus lucens* did, however, change from 30°C during November (spring to early summer) to 20°C during February (late summer). The occurrence of a seasonal shift in the photosynthetic temperature optimum in *T. lucens* appears to be related to the absence of physiological or structural means to maintain culm temperatures close to ambient. *Erica* species minimize leaf temperatures by virtue of leaf size and shape. Proteoids, on the other hand, keep leaf temperatures to a minimum by transpirative cooling which is realized by their ability to tap deep water resources unavailable to most species. *T. lucens*, however, has a limited ability to regulate temperatures and therefore requires photosynthetic apparatus which can acclimatize more easily to extremes of temperatures (Van der Heyden and Lewis 1990).

PHOTOSYNTHETIC PATHWAYS

Surveys have been undertaken to determine the geographical distribution of various photosynthetic pathways in southern Africa. Most investigations were aimed at elucidating photosynthetic pathways of specific growth forms or pathways characteristic of specific geographical regions. Fynbos species were only incidentally included in these surveys.

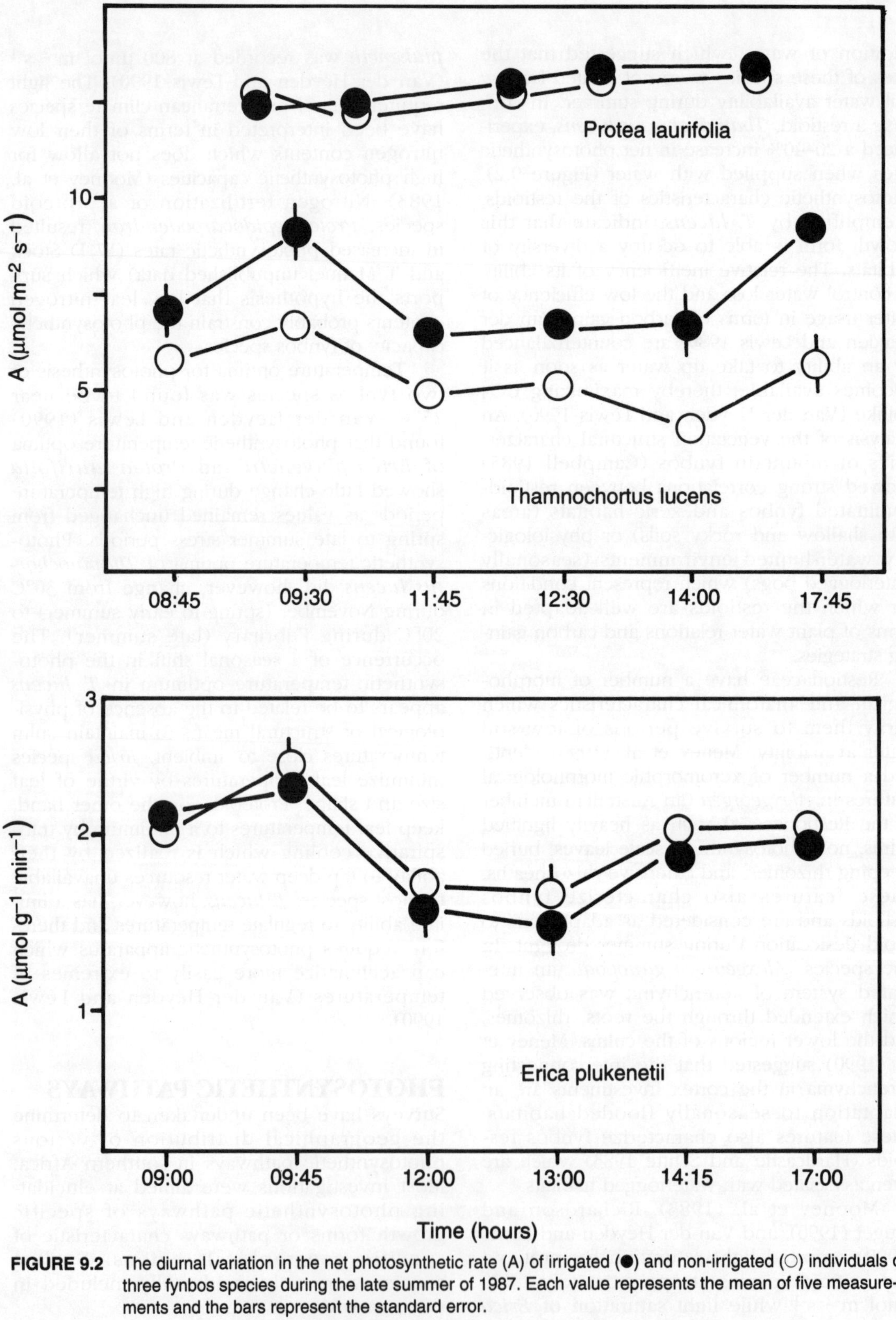

FIGURE 9.2 The diurnal variation in the net photosynthetic rate (A) of irrigated (●) and non-irrigated (○) individuals of three fynbos species during the late summer of 1987. Each value represents the mean of five measurements and the bars represent the standard error.

Schütte et al. (1967) examined the occurrence of the Crassulacean Acid Metabolism (CAM) photosynthetic pathway in 35 South African succulent species. On the basis of the diurnal changes in organic acid concentrations, the authors assigned a degree of CAM activity to each species. The species examined covered a broad spectrum of potentially CAM plants. Fynbos species sampled occurred in the more arid regions and were found to have no or intermediate potential for CAM photosynthesis. This suggested that some succulents in the fynbos have the ability to shift between CAM and C_3 photosynthetic pathways. A succulent (*Ruschia* sp.) from a dry north-western fynbos region showed CAM gas exchange characteristics only towards the end of summer (Miller 1985), indicating that the shift from C_3 to CAM photosynthesis does occur.

Mooney et al. (1977) used carbon isotope ratio technology to determine photosynthetic pathways in succulent plants of southern Africa. Although plant material used in this study was not collected from true mediterranean climate regions, the values obtained for species which occur in the fynbos and the general patterns are comparable with the findings of Schütte et al. (1967) who used titratable acidity measures. Vogel et al. (1978) used both carbon isotope ratios and the presence of Kranz leaf anatomy to distinguish between C_3 and C_4 photosynthetic pathways in approximately 1 000 South African grass species. This survey clearly showed that C_3 grasses predominate in the winter rainfall regions of the western Cape. The authors found that low temperatures prevalent during the growing season, characteristic of fynbos mediterranean climate regions, appear to favour C_3 grasses. A survey of fynbos and allied shrublands in the south-eastern Cape (Cowling 1983) showed that in the non-seasonal rainfall areas of the fynbos biome C_3 and C_4 grasses are equally represented. Cowling (1983) analysed the relationship between the relative cover of C_3 grasses and a range of environmental factors and found that lower temperatures during the growing season correlated significantly with C_3 grass cover.

Photosynthetic pathways of species other than succulents and grasses are typically C_3 and carbon isotope ratios of the ericoid, restioid, and proteoid groups all fall between -25 to -30 per mil (W J Bond unpublished data; D le Roux, W J Bond and W D Stock unpublished data).

CARBON AS AN ABUNDANT RESOURCE

Fynbos shrubs maintain positive carbon balances throughout the year, albeit at relatively low photosynthetic capacities; some other factor appears to limit growth in this ecosystem. Studies of plant response to nutrient additions (Witkowski et al. 1990), litter decomposition processes (Mitchell et al. 1986), and nitrogen allocation patterns (Stock et al. 1987) suggest that nutrient-poor soils impose limitations on plant growth and development. Nutrient-poor systems such as fynbos are often carbon-rich, as plants limited by nutrient availability have surplus carbon which is not allocated to vegetative plant tissue development and maintenance (Baas 1989). A number of fynbos studies support Baas's (1989) carbon/nutrient cycle hypothesis as much of the carbon in the system is utilized in an array of plant secondary compounds and mechanical structures which have profound influences upon abiotic and biotic processes in fynbos ecosystems.

Fire is an example of an abiotic factor which has been shown to play a key role in fynbos nutrient cycling patterns (Stock and Allsopp this volume). Many fynbos adaptations can be interpreted in terms of the carbon/nutrient cycle hypothesis in that carbon produced in excess by a plant can be used for storage or the production of fire resistant structures. Surplus carbon is commonly stored, and resprouting fynbos species allocate considerable quantities of carbon to non-structural carbohydrate reserves in below-ground plant parts. Carbohydrate reserves are utilized for regrowth following fire which enhances the competitive ability of resprouting species. Differences in storage patterns of reseeders and resprouters have been demonstrated in taxa from the mediterranean climate south-western Australia, where resprouters have higher concentrations of starch in their roots than reseeders (Pate et al. 1990). Carbon is also allocated to fire resistant structures such as the cones of many serotinous (canopy-stored seed) Proteaceae (Bond 1985).

Surplus carbon is similarly important in biotic interactions in fynbos environments (see also Johnson this volume) e.g. a large number of taxa in the fynbos produce carbon-rich nectar (Rebelo 1987). The nectar is rich in sucrose, glucose, and fructose (Cowling and Mitchell 1981) and contains low concentrations of nitrogen-based compounds such as proteins and amino acids (Mostert et al. 1980). High sugar content of the nectar attracts nectar-feeding bird and insect pollinators which enhances pollination efficiency (Mostert et al. 1980). Birds also feed on the insect fauna, thereby protecting foliage from nutrient losses through herbivory (Johnson this volume).

Carbon is also used to increase plant defence against herbivory. This is effected by allocating carbon to either secondary metabolites with deterrent or toxic properties or mechanical defence structures. Fynbos plants use both approaches as is evident from the diversity and concentration of polyphenolic compounds present in most taxa (Glyphis and Puttick 1988) and the large number of carbon-rich mechanical structures found protecting seeds not only from fire but also predator damage (e.g. cones of *Leucadendron* spp.). Similar defence structures are found in plants growing in other low nutrient environments of south-western Australia. In Australian Proteaceae, resource allocation to large well defended cones is substantial with defence structure to seed mass ratios in excess of 112:1 (Stock et al. 1991).

Given that fynbos is a carbon-rich ecosystem, why is it that plants which become repositories of substantial quantities of carbon, such as trees, are absent from the system? Moll et al. (1980) hypothesized that a more frequent fire regime in historical times caused the disappearance of trees from fynbos. Other suggestions include a water limitation of forest development. However, investigations of the Afromontane forest-fynbos boundary have shown little support for this hypothesis (Kruger et al. 1988; Manders 1990). We would support the idea that edaphic factors are implicated since patterns of nutrient cycling observed for Afromontane tree species cannot easily be supported by the low nutrient levels found in fynbos soils, especially when compounded by short fire intervals (Cowling and Holmes this volume). Species comparable to *Eucalyptus* (Myrtaceae) in mediterranean south-western Australia, which appear to have a suite of adaptations to cope with low soil nutrients, have not evolved in fynbos. Hence, the absence of trees in fynbos could largely be due to the lack of lineages with characteristics suited for survival in the nutrient-poor and fire-prone fynbos environment (Richardson et al. this volume).

FUNCTIONAL SIGNIFICANCE OF SCLEROPHYLLY IN FYNBOS

A striking feature of fynbos is the dominance of the sclerophyllous leaf (Stock and Allsopp this volume). This carbon-rich leaf is of particular interest and we will discuss it in some detail as this leaf form influences many biotic and abiotic processes governing fynbos dynamics.

Evergreen shrubs possessing sclerophyllous leaves occur in many ecosystems from the tropics to the poles, but are especially dominant in the five mediterranean-type ecosystems of the world. To date the biochemical basis and functional significance of this leaf type is uncertain. Functional interpretations of sclerophylly have emphasized, firstly, its suitability as an adaptation to drought (Schimper 1903; Poole and Miller 1975); secondly, that it is an adaptation improving nutrient use efficiency in low nutrient environments (Beadle 1966; Specht and Rundel 1990); and thirdly, that this modification is a highly successful means of reducing herbivory (Chabot and Hicks 1982).

Considerable differences in interpretation of the phenomenon of sclerophylly have arisen because of the confusion and interchangeability in the use of the term sclerophylly with others, such as xerophytism and xeromorphism. These latter terms are used to describe leaf and plant morphology and physiological performance in terms of water usage (Seddon 1974) and may include characteristics such as a reduced surface area, the presence of leaf hairs, protected stomata, and thick cuticles. Sclerophylly is also an attribute common among xerophytes but the possession of sclerophyllous leaves is not sufficient in itself to predict the water relations of a species.

Kummerow (1973) considered sclerophylls to be a physiologically homogeneous group based primarily on similarities in anatomical structure. Subsequent investigations of the physiological performance of sclerophylls have, however, shown them to be a diverse group (Lo Gullo and Salleo 1988; Rhizopoulou and Mitrakos 1990; Salleo and Lo Gullo 1990). The selective forces responsible for the evolution of sclerophylly and some of its correlates (e.g. leaf longevity and evergreenness) are poorly understood. Below we review recent ecophysiological studies of sclerophylls from various mediterranean-type ecosystems in order to develop our understanding of the functional significance of this leaf type in fynbos.

Water relations of evergreen sclerophylls

Palaeoecological discussions concerning the origins of mediterranean-type ecosystems have traditionally invoked the incidence of summer drought as the primary selective force shaping the characteristics of these floras (Axelrod 1973; Beard 1977; Axelrod and Raven 1978). Deacon (1983) noted that this view has not been critically tested in fynbos mainly because of a paucity of fossils from present-day arid areas adjacent to fynbos. Evidence that sclerophylly arose during the Cretaceous in tropical taxa (Specht 1963) suggests that water scarcity probably has had little influence in the selection for sclerophylly *per se*. This agrees with present distribution of fynbos sclerophylls which show little change in abundance along rainfall gradients (Campbell and Werger 1988; Cowling and Holmes this volume). Sclerophyll abundance does, however, vary along soil fertility and seasonality of rainfall gradients with mesophytic grasses replacing sclerophyllous restioids in the more fertile, non-seasonal rainfall zone in the eastern fynbos biome (Campbell 1983, 1985). Taylor (1980) also observed that sclerophylly was unrelated to climate but strongly associated with nutrient-poor soils derived from the Table Mountain Group Sandstones (Deacon et al. this volume). Oertli et al. (1990) have suggested that sclerophylly is an adaptation to drought only in regions where small but regular inputs of water occur throughout the year so as to overcome extreme water deficits. Areas with prolonged water shortage are dominated by mesophytes with phenological mechanisms to avoid drought (e.g. deciduousness).

No ecophysiological studies have tested the significance of present-day ecological patterns of sclerophyll abundance in the fynbos region. Therefore, we can only infer physiological performance from studies of sclerophylls in other summer arid environments. Present evidence from the Mediterranean Basin (Lo Gullo et al. 1986; Lo Gullo and Salleo 1988; Salleo and Lo Gullo 1990; Rhizopoulou and Mitrakos 1990) shows that contrary to earlier assumptions (Kummerow 1973), sclerophylls are not a physiologically homogeneous group. Lo Gullo and Salleo (1988) and Salleo and Lo Gullo (1990) have shown that the strategies adopted to overcome dry summer periods in phylogenetically unrelated sclerophyll species were substantially different. Some species tolerated substantial water loss (*Olea oleaster*, drought tolerant species), some maintained high water potentials by extracting sufficient water from deeper layers of the soil profile (*Ceratonia siliqua*, drought avoiding water spender species), while *Laurus nobilis* showed an ability to restrict transpirational water loss (water saving species). In addition, comparisons of the water use strategies of phylogenetically related Mediterranean Basin sclerophyll species (*Quercus ilex* and *Q. suber*) showed that the index of sclerophylly (SI = ratio of crude fibre weight/protein weight x 100 (Loveless 1962); cf. sclerophylly often measured by leaf specific weight, LSW in g m^{-2}) gave no indication of a plant's ability to withstand drought stress (Salleo and Lo Gullo 1990). Under conditions of water scarcity *Quercus* species had similar minimum leaf relative water contents (RWC) despite large differences in sclerophylly (130% differences in LSW between *Q. suber* and *Q. ilex*) and duration of peak leaf conductance to water vapour values (*Q. ilex* values twice that of other species) (Salleo and Lo Gullo 1990). The large array of LSW values of species from fynbos and adjacent karoo regions (Table 9.1) does not, therefore, allow the prediction of patterns of water usage by any of the species listed. Rather, other anatomical and physiological properties need to be studied before any relationships between strategies for resisting drought and sclerophylly can be elucidated.

TABLE 9.1 The leaf specific weights (gm^{-2}) of a range of sclerophyllous karoo and fynbos species.

Species	Leaf size	Leaf specific weight (gm^{-2})
Karoo		
Galenia africana[1]	leptophyll	127.2
Osteospermum sinuatum[1]	nanophyll	94.7
Pentzia incana[1]	leptophyll	85.8
Pteronia empetrifolia[1]	leptophyll	272.5
Pteronia pallens[1]	leptophyll	195.5
Rhus undulata[1]	nanophyll	185.9
Rhus lancea[1]	nanophyll	151.3
Rhus longispina[1]	nanophyll	174.5
Fynbos		
Carissa bispinosa[3]	microphyll	84.0
Cassine peragua[3]	microphyll	72.0
Erica phylicaefolia[3]	leptophyll	33.0
Maytenus oleoides[2]	microphyll	181.8
Metalasia muricata[3]	leptophyll	35.0
Olea europeae[3]	microphyll/nanophyll	75.0
Protea repens[2]	microphyll	232.1
P. neriifolia[2]	microphyll	223.4
P. nitida[2]	microphyll	252.6
P. acaulos[2]	microphyll	174.9
P. compacta[3]	microphyll	150.0
Rhus tomentosa[2]	nanophyll	188.9

[1] W D Stock (unpublished data)
[2] Mooney et al. (1983)
[3] Cowling and Campbell (1983)

Sclerophylly and nutrient use efficiency

A nutritional interpretation for the occurrence of sclerophylly has been put forward by Loveless (1961) in which he suggested that sclerophyll leaf metabolism is geared to low nutrient availability (in particular phosphorus). He felt that sclerophylly is the expression of a plant's nutrient conservation strategy and that it produced an answer to the paradox of why this leaf type occurs independently of site water availability (Loveless 1962; Beadle 1966). Sclerophylly, however, is frequently correlated with other leaf characteristics (such as leaf longevity and evergreenness) which have also been interpreted in terms of nutrient conservation strategies (Monk 1966, 1987; Orians and Solbrig 1977; Shaver 1981; Chabot and Hicks 1982; Jonasson 1989; Aerts 1990).

It has been suggested that the suite of interrelated leaf characteristics (longevity, evergreenness, and sclerophylly) is favoured in low nutrient environments because it entails an increased carbon return per unit of nutrient invested (Orians and Solbrig 1977). Leaves are able to photosynthesize for longer periods which allows them a higher NUE than deciduous species (Small 1972). This does, however, have a negative influence upon competitiveness in high nutrient environments (Grime 1977; Chapin 1980) in which faster growing species with higher nutrient requirements would dominate. In discussing the nutrient interpretation of sclerophylly we must be careful not to confuse these interrelated characteristics and draw false inferences and conclusions about their functional significance.

Rundel (1988) and Specht and Rundel (1990) have recently tested hypotheses concerning the relationship between leaf structure and nutrition in sclerophylls from all five mediterranean-type ecosystems. These studies show that very close relationships between leaf N and P, leaf N and Ca, and between index of sclerophylly (SI, see above) and leaf P exist for species from all mediterranean regions. More important, it was found that the fynbos sclerophyll species separated out according to growth form; proteoid and restioid taxa separated from ericoids on the basis of high cellulose and lignin (components of SI) and low ether extractable compounds in the foliage. Proteoids differed from restioids on the basis of high divalent cation content (Mg and Ca) while restioids had higher levels of monovalent potassium ions and lignin and cellulose (Rundel 1988). The nutritional basis underlying sclerophylly thus varies between guilds; as emphasized earlier, sclerophylls cannot be considered as a physiologically homogeneous group.

Studies on the genetic basis of sclerophylly have attempted to establish whether this characteristic is genotypically stable or phenotypically variable. Beadle (1968) applied nutrients to some Australian sclerophylls and found in *Hakea teretifolia* that although leaf length increased some 300%, many leaf characteristics remained the same. For example, leaf diameters, number of vascular bundles, number of fibre groups, sizes of palisade layer cells, cuticle thickness, and wall thickness of fibres and sclereids did not change with nutrient addition (Beadle 1968). The numbers of sclereids in the palisade did, however, decrease (Beadle 1968). This also occurs in fertilized *Hakea suaveolens* (Heide-Jorgensen 1990).

Fynbos sclerophylls show limited plasticity in leaf morphology and physiological performance in response to increased nutrient availability (Witkowski et al. 1990, 1991; W D Stock and T I Abraham unpublished data). Albeit many of the measured responses were significant (W D Stock and T I Abraham unpublished data) as can be seen from field-grown two-year-old *Leucospermum parile* seedlings receiving high, intermediate, and low levels of nutrients (Table 9.2). Plants receiving high levels of nutrients had significantly wider leaves, greater leaf glaucousness, larger numbers of mesophyll cells, and lower leaf specific weights. Similar trends have been observed by F. van der Heyden (unpublished data) who showed larger leaf areas, lower LSW, and higher relative water content in *Leucadendron xanthoconus* seedlings grown at high nutrient levels. Such morphological changes are limited but altered physiological performance is also evident (W D Stock and T Manuel unpublished data). With fertilization, *P. lepidocarpodendron* seedlings showed enhanced stomatal conductances and photosynthetic rates (Figure 9.3), provided water was plentiful. Nutrients, therefore, have a significant effect upon sclerophyll leaf development and characteristics, although the magnitude of induced changes is small.

The sclerophyll leaf appears as a fixed characteristic with little phenotypic variability along gradients of nutrient availability. The evolutionary and ecological evidence accounting for present-day sclerophyll characteristics and distributions appears to support the nutritional interpretation, although further research is necessary to clarify the relationship. In particular, the link between availability of individual elements and sclerophylly needs further attention so that a more precise understanding and possible physiological mechanisms underlying sclerophylly may be developed.

Sclerophylly and herbivory

Plants typical of resource-limited habitats (shaded, arid, or low nutrient environments) are generally unable to acquire sufficient resources to support rapid growth and typically have low maximum potential growth rates (Chapin 1980). When herbivory occurs, low maximum potential growth rates, which are selectively advantageous in resource-limited environments, become a liability because the resource capital and net productivity of slow growing species are more seriously affected at any given rate of herbivory than is the case for faster growing species. Consequently, the intensity of selection for anti-herbivore defences increases as resource availability in the environment declines (Coley et al. 1985) and we would

TABLE 9.2 **The effect of nutrient addition on leaf characteristics of the proteoid sclerophyll, *Leucospermum parile*. Data are means (SE). One-way ANOVAs were used to test for significant differences between treatments; NS = not significant (W D Stock and T I Abraham unpublished data).**

	Treatment						
Characteristic	High nutrient		Low nutrient		No nutrient added		Significance
Leaf length (mm)	36.55	(0.57)	33.71	(1.1)	35.64	(0.92)	NS
Leaf width (mm)	3.4	(0.04)	2.9	(0.04)	2.8	(0.07)	$p < 0.01$
Leaf thickness (mm)	0.58	(0.04)	0.58	(0.02)	0.53	(0.02)	NS
Leaf glaucousness (scale 1–5)	4.3	(0.26)	4	(0.2)	4.9	(0.13)	$p < 0.05$
Leaf angle (^{o})	41.2	(2.16)	42.8	(2.89)	36.5	(2.99)	NS
Diameter of palisade cells (mm)	0.053	(0.015)	0.049	(0.011)	0.053	(0.015)	NS
Diameter of mesophyll cells (mm)	0.046	(0.018)	0.042	(0.15)	0.044	(0.02)	NS
No. of layers of palisade cells	4	(0)	4	(0)	4	(0)	NS
No. of layers of mesophyll cells	3	(0)	2.8	(0.18)	2.5	(0.17)	$p < 0.01$
Leaf specific weight (g m^{-2})	161	(6.9)	178	(12.5)	214	(9.7)	$p < 0.01$
Dry mass of shoot (g)	1.04	(0.18)	0.87	(0.14)	0.91	(0.18)	NS
Dry mass of root (g)	0.31	(0.54)	0.25	(0.04)	0.29	(0.05)	NS
Shoot/root ratio	3.3	(0.28)	3.6	(0.27)	3.5	(0.28)	NS

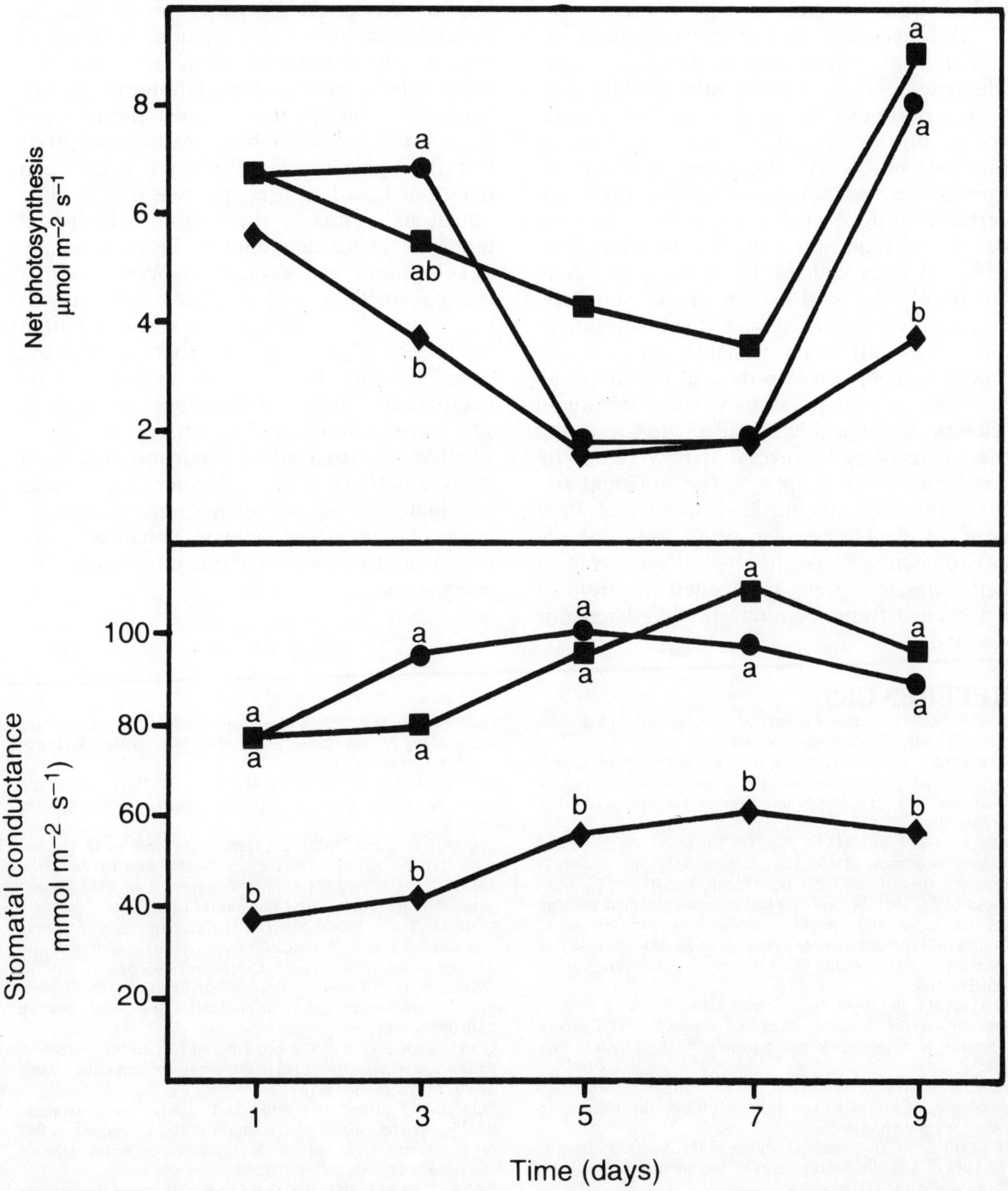

FIGURE 9.3 The net photosynthetic rates and stomatal conductances of *P. lepidocarpodendron* seedlings grown under standard conditions (light intensity 400 µmol m^{-2} s^{-1}, 25°C) at three nutrient levels (■——■ High, ●——● Intermediate, ◆——◆ Low). Significance ($p < 0.05$) was analysed by one-way ANOVAS and multiple range test (SNK) on each sample day.

therefore expect fynbos plants to exhibit a range of modifications reducing herbivory (Johnson this volume).

Defences against herbivory include the allocation of carbon to secondary metabolites (discussed earlier) and structural modifications (e.g. woodiness, seeds protected in woody cones, and sclerophylly) which reduce or preclude herbivory. The sclerophyllous leaf form is one such structural modification which certainly reduces palatability to herbivores because of high fibre content, low water content, and thick cuticles. Fibre, wax, and cutin are poorly digested by mammals especially, and high concentrations of these substances dilute the nutritionally valuable nutrients and energy contained within the leaf (Bryant et al. in press). As yet no studies have interpreted sclerophylly as a primary adaptation reducing the intensity of herbivory. It may rather be considered as an exaptation (*sensu* Gould and Vrba 1982) enhancing leaf survival of slow growing plants characteristic of low resource environments. Thus, the evolutionary role of herbivores in shaping the limited spectrum of fynbos leaf forms currently observed remains uncertain.

CONCLUSIONS

We have reviewed the current ecophysiological understanding of the functioning of fynbos plants and discussed some of the more unusual features of fynbos plant forms. In particular the concept that summer drought has been a major selective force in shaping fynbos form and physiological functioning has been questioned. Rather, it is apparent that the low nutritional status of the soils has been the dominant factor determining the evolutionary development and ecological occurrence of many features of fynbos plants. Low growth rates, low photosynthetic rates, the common occurrence of evergreen sclerophylls, and the varied patterns of carbon allocation can all be interpreted in terms of nutritional constraints. The carbon/nutrient hypothesis has wide applicability in fynbos environments with many abiotic (e.g. slow decomposition rates and high flammability of the vegetation) and biotic (e.g. increased defence, enhanced pollination) implications for the ecology of fynbos ecosystems.

REFERENCES

AERTS R (1990). Nutrient efficiency in evergreen and deciduous species from heathlands. *Oecologia* **84**, 391–7.

AXELROD D I (1973). History of the mediterranean ecosystem in California. In *Mediterranean type ecosystems: origin and structure.* (eds Di Castri F and Mooney H) Springer-Verlag, Berlin, 225–77.

AXELROD D I and RAVEN P H (1978). Late Cretaceous and Tertiary vegetation of Africa. In *Biogeography and ecology of southern Africa.* (ed Werger M J A) Junk, The Hague, 77–130.

BAAS W J (1989). Secondary plant compounds, their ecological significance and consequences for the carbon budget. In *Causes and consequences of variation in growth rate and productivity of higher plants.* (ed Lambers H) Academic Publishing, The Hague.

BEADLE N C W (1966). Soil phosphate and its role in molding segments of the Australian flora and vegetation with special reference to xeromorphy and sclerophylly. *Ecology* **47**, 992–1 007.

BEADLE N C W (1968). Some aspects of the ecology and physiology of Australian xeromorphic plants. *Australian Journal of Science* **30**, 348–55.

BEARD J S (1977). Tertiary evolution of the Australian flora in the light of latitudinal movements of the continent. *Journal of Biogeography* **4**, 111–18.

BRYANT J P, KUROPAT P J, REICHARDT P B and CLAUSEN T P (in press). Controls of the allocation of resources by woody plants to chemical antiherbivore defense. In *Chemical Defense of Plants Against Mammals.* (eds Robbins C and Palo T) CRC Press.

CAMPBELL B M (1983). Montane plant environments in the fynbos biome. *Bothalia* **14**, 283–98.

CAMPBELL B M (1985). A classification of the mountain vegetation of the fynbos biome. *Memoirs of the Botanical Survey of South Africa* **50**, 1–119.

CAMPBELL B M and WERGER M J A (1988). Plant form in the mountains of the Cape, South Africa. *Journal of Ecology* **76**, 637–53.

CHABOT B G and HICKS D J (1982). The ecology of leaf life spans. *Annual Review of Ecology and Systematics* **13**, 229–59.

CHAPIN F S (1980). The mineral nutrition of wild plants. *Annual Review Ecology and Systematics* **11**, 233–60.

CODY M L and MOONEY H A (1978). Convergence versus non-convergence in mediterranean-climate ecosystems. *Annual Review Ecology and Systematics* **9**, 265–321.

COLEY P D, BRYANT J P and CHAPPIN III, FS (1985). Resource availability and plant antiherbivore defense. *Science* **230**, 895–9.

COWLING R M (1983). The occurrence of C_3 and C_4 grasses in fynbos and allied shrublands in the South Eastern Cape, South Africa. *Oecologia* **58**, 121–7.

COWLING R M and MITCHELL D T (1981). Sugar composition, total nitrogen and accumulation of C_{14} assimilates in floral nectaries of *Protea* species. *Journal of South African Botany* **47**, 743–50.

DAVIS G W and MIDGLEY G F (1990). Effects of disturbance by fire and tillage on the water relations of selected mountain fynbos species. *South African Journal of Botany* **56**, 199–205.

DEACON H J (1983). The comparative evolution of mediterranean-type ecosystems: a southern perspective. In *Mediterranean-type ecosystems: the role of nutrients.* (eds Kruger F J, Mitchell D T and Jarvis J U M) Springer-Verlag, Berlin, 3–40.

DI CASTRI F and MOONEY H A (eds) (1973). *Mediterranean*

type ecosystems: origin and structure. Springer-Verlag, New York.

GIGON A (1979). CO^2-gas exchange, water relations and convergence of mediterranean shrub-types from California and Chile. *Oecologia Plantarum* **14**, 129–50.

GOULD S J and LEWONTIN R C (1979). The spandrels of San Marco and the Panglossian paradigm: a critique of the adaptionist programme. *Proceedings of the Royal Society of London, Series B* **205**, 581–98.

GOULD S J and VRBA ES (1982). Exaptation — a missing term in the science of form. *Paleobiology* **8**, 4–15.

GIVNISH T J (ed) (1986). *On the economy of plant form and function.* Cambridge University Press, Cambridge.

GRIME J P (1977). Evidence for the existence of three primary strategies in plants and its relevance to ecological and evolutionary theory. *American Naturalist* **111**, 1 169–94.

GLYPHIS J P and PUTTICK G M (1988). Phenolics in some Southern African Mediterranean shrubland plants. *Phytochemistry* **27**, 743–51.

HARDCASTLE J and SCHUTTE K H (1983). Aspects of an experimental study on root aerenchyma development and the ecological implications thereof. *Bothalia* **14**, 791–4.

HEIDE-JORGENSEN H S (1990). Xeromorphic leaves of *Hakea suaveolens* R. Br. IV. Ontogeny, structure and function of the sclereids. *Australian Journal of Botany* **38**, 25–43.

JEFFERY D, MOLL E J and VAN DER HEYDEN F (1987). Comparative water potentials of 4-month-old and 40-year-old Cape fynbos plants. *South African Journal of Botany* **53**, 32–4.

JONASSON S (1989). Implications of leaf longevity, leaf nutrient re-absorption and translocation for the resource economy of five evergreen species. *Oikos* **56**, 121–31.

KRUGER F J, MITCHELL D T and JARVIS J U M (eds) (1983). *Mediterranean-type ecosystems: the role of nutrients.* Springer-Verlag, Berlin.

KRUGER F J, RICHARDSON D M and SMITH R E (1988). Plant water relations in sclerophyllous trees and shrubs of riparian and hillslope habitats at Swartboskloof south-western Cape Province, South Africa. In *Time scales and water stress.* (eds Di Castri F, Floret C, Rambal S and Roy J) International Union of Biological Sciences, Paris, 575–81.

KUMMEROW J (1973). Tertiary evolution of the Australian flora in the light of latitudinal movements of the continent. *Journal of Biogeography* **4**, 111–18.

LEWONTIN R C (1977). Adaptation. *Scientific American* **239**, 156–69.

LO GULLO M A, SALLEO S and ROSSO R (1986). Drought avoidance strategy in *Ceratonia siliqua* L, a mesomorphic-leaved tree in the xeric Mediterranean area. *Annals of Botany* **58**, 745–56.

LO GULLO M A and SALLEO S (1988). Different strategies of drought resistance in three Mediterranean sclerophyllous trees growing in the same environmental conditions. *New Phytologist* **108**, 267–76.

LOVELESS A R (1961). A nutritional interpretation of sclerophylly based on differences in the chemical composition of sclerophyllous and mesophytic leaves. *Annals of Botany* **25**, 168–84.

LOVELESS A R (1962). Further evidence to support a nutritional interpretation of sclerophylly. *Annals of Botany* **26**, 551–61.

MANDERS P T (1990). Fire and other variables as determinants of forest/fynbos boundaries in the Cape Province. *Journal of Vegetation Science* **1**, 483–90.

MENEY K A, PATE J S and DIXON K W (1990). Comparative morphology, anatomy, phenology and reproductive biology of *Alexgeorgea* spp. (Restionaceae) from south-western Western Australia. *Australian Journal of Botany* **38**, 523–41.

MILLER J M (1985). Plant water relations along a rainfall gradient, between the succulent karoo and mesic mountain fynbos, in the Cederberg Mountains near Clanwilliam, South Africa. MSc thesis, University of Cape Town.

MILLER J M, MILLER P C and MILLER P M (1984). Leaf conductances and xylem pressure potentials in fynbos plant species. *South African Journal of Science* **80**, 381–5.

MILLER P C (ed) (1981). *Resource use by chaparral and matorral.* Springer-Verlag, New York.

MILLER P C, MILLER J M and MILLER P M (1983). Seasonal progression of plant water relations in fynbos in western Cape Province, South Africa. *Oecologia* **56**, 392–6.

MITCHELL D T, COLEY P G F, WEBB S and ALLSOPP N (1986). Litterfall and decomposition processes in the coastal fynbos vegetation, south-western Cape, South Africa. *Journal of Ecology* **74**, 977–93.

MOLL E J, McKENZIE B and McLACHLAN D (1980). A possible explanation for the lack of trees in the fynbos, Cape Province, South Africa. *Biological Conservation* **17**, 221–8.

MONK C D (1966). An ecological significance of evergreenness. *Ecology* **47**, 504–5.

MONK C D (1987). Sclerophylly in *Quercus virginiana* Mill. *Castanea* **52**, 256–61.

MOONEY H A (eds) (1977). *Convergent evolution in Chile and California mediterranean climate ecosystems.* Dowden, Hutchison and Ross, Stroudsburg.

MOONEY H A and DUNN E L (1970). Photosynthetic systems of mediterranean-climate shrubs and trees of California and Chile. *American Naturalist* **104**, 447–53.

MOONEY H A, FIELD C, GULMON S L, RUNDEL P and KRUGER F J (1983). Photosynthetic characteristics of South African sclerophylls. *Oecologia* **58**, 398–401.

MOONEY H A, TROUGHTON J H and BERRY J A (1977). Carbon isotope ratio measurements of succulent plants in Southern Africa. *Oecologia* **30**, 295–305.

MOSTERT D P, SIEGFRIED W R and LOUW G N (1980). *Protea* nectar and satellite fauna in relation to the food requirements and pollinating role of the Cape sugarbird. *South African Journal of Science*, **76**, 409–12.

ODENING W R, STRAIN B R and OECHEL W C (1974). The effect of decreasing water potential on net CO_2 exchange of intact desert shrubs. *Ecology* **55**, 1 086–95.

OERTLI J S, LIPS S H and AGAMI M (1990). The strength of sclerophyllous cells to resist collapse due to negative turgor pressure. *Acta Oecologia* **11**, 281–9.

ORIANS G H and SOLBRIG O T (1977). A cost-income model of leaves and roots with special reference to arid and semi-arid areas. *American Naturalist* **111**, 677–90.

PATE J S, FROEND R H, BOWEN B J, HANSEN A and KUO J (1990). Seedling growth and storage characteristics of seeder and resprouter species of mediterranean-type ecosystems of S.W. Australia. *Annals of Botany* **65**, 585–601.

POOLE D K and MILLER P C (1975). Water relations of selected species of Chaparral and Coastal Sage communities. *Ecology* **56**, 1 118–28.

REBELO A G (ed) (1987). A preliminary synthesis of pollination biology in the Cape flora. *South African National Scientific Programmes Report* **141**, CSIR, Pretoria.

RHIZOPOULOU S and MITRAKOS K (1990). Water relations of evergreen sclerophylls. Seasonal changes in the water relations of eleven species from the same environment. *Annals of Botany* **65**, 171–8.

RICHARDSON D M and KRUGER F J (1990). Water relations and photosynthetic characteristics of selected trees and shrubs of riparian and hillslope habitats in the south western Cape Province, South Africa. *South African Journal of Botany* **56**, 214–25.

RUNDEL P W (1988). Leaf structure and nutrition in mediterranean-climate sclerophylls. In *Mediterranean-type ecosystems.*

A data source book. (ed Specht R L) Kluwer, Dordrecht, 157–67.
SALLEO S and LO GULLO M A (1990). Sclerophylly and plant water relations in three mediterranean *Quercus* species. *Annals of Botany* **65**, 259–70.
SCHIMPER A F W (1903). *Plant-geography upon a physiological basis.* Clarenden Press, Oxford.
SCHUTTE K H, STEYN R and VAN DER WESTHUIZEN M (1967). Crassulacean acid metabolism in South African succulents: a preliminary investigation into its occurrence in various families. *Journal of South African Botany* **33**, 107–10.
SEDDON G (1974). Xerophytes, xeromorphs and sclerophylls: the history of some concepts in ecology. *Biological Journal of the Linnean Society* **6**, 65–87.
SHAVER G R (1981). Mineral nutrition and leaf longevity in an evergreen shrub, *Ledum palustre decumbens. Oecologia* **49**, 362–5.
SMALL E (1972). Photosynthetic rates in relation to nitrogen recycling as an adaptation to nutrient deficiency in peat bog plants. *Canadian Journal of Botany* **50**, 2 227–33.
SOMMERVILLE J E M (1983). Aspects of coastal fynbos phenology. MSc thesis, University of Cape Town.
SPECHT R L (1963). Dark Island heath (Ninety-Mile Plain, South Australia). VII. The effect of fertilizers on composition and growth, 1950–1960. *Australian Journal of Botany* **11**, 67–94.
SPECHT R L, MOLL E J, PRESSINGER F and SOMMERVILLE J E M (1983). Moisture regime and nutrient control of seasonal growth in mediterranean ecosystems. In *Mediterranean-type ecosystems: the role of nutrients.* (eds Kruger F J, Mitchell D T and Jarvis J U M) Springer-Verlag, Berlin, 120–32.
SPECHT R L and RUNDEL P W (1990). Sclerophylly and foliar nutrient status of mediterranean-climate plant communities in southern Australia. *Australian Journal of Botany* **38**, 459–74.
STOCK W D (1988). Nutrients and water stress as selective forces in the evolution of the fynbos and kwongan floras. In *Time Scales and Water Stress.* (eds Di Castri F, Floret C, Rambal S and Roy J). International Union of Biological Sciences, Paris, 623–9.
STOCK W D, SOMMERVILLE J E M and LEWIS O A M (1987). Seasonal allocation of dry mass and nitrogen in a fynbos endemic Restionaceae species *Thamnochortus punctatus* Pil. *Oecologia* **72**, 315–20.
STOCK W D, PATE J S and RASINS E (1991). Seed development patterns in *Banksia attenuata* and *Banksia laricina* in relation to mechanical defense costs. *New Phytologist* **117**, 109–14.
TAYLOR H C (1980). Phytogeography of fynbos. *Bothalia* **13**, 231–5.
THROWER N J W and BRADBURY D E (eds) (1977). *Chile-California mediterranean scrub atlas: a comparative analysis.* Dowden, Hutchison and Ross, Stroudsburg.
VAN DER HEYDEN F (1988). An investigation of photosynthetic carbon fixation in fynbos growth forms and its variation with season and environmental conditions. MSc thesis, University of Cape Town.
VAN DER HEYDEN F and LEWIS O A M (1988). Photosynthetic carbon-fixation in selected fynbos growth forms and its variation with season and environmental conditions. In *Times scales and water stress.* (eds Di Castri F, Floret C, Rambal S and Roy J). The International Union of Biological Sciences, Paris, 589–93.
VAN DER HEYDEN F and LEWIS O A M (1989). Seasonal variation in photosynthetic capacity with respect to plant water status of five species of the mediterranean climate region of South Africa. *South African Journal of Botany* **55**, 509–15.
VAN DER HEYDEN F and LEWIS O A M (1990). Environmental control of photosynthetic gas exchange characteristics of fynbos species representing three growth forms. *South African Journal of Botany* **56**, 654–8.
VOGEL J C, FULS A and ELLIS R P (1978). The geographical distribution of kranz grasses in South Africa. *South African Journal of Science* **74**, 209–15.
VON WILLERT D J, HERPPICH M and MILLER J M (1989). Photosynthetic characteristics and leaf water relations of mountain fynbos vegetation in the Cedarberg area (South Africa). *South African Journal of Botany* **55**, 288–98.
WITKOWSKI E T F (1991). Growth and competition between seedlings of *Protea repens,* (L.) L. and the alien invasive, *Acacia saligna* (Labill.). Wendl. in relation to nutrient availability. *Functional Ecology* **5**, 101–10.
WITKOWSKI E T F, MITCHELL D T and STOCK W D (1990). Response of a Cape fynbos ecosystem to nutrient additons: shoot growth and nutrient contents of a proteoid *Leucospermum parile* and an ericoid (*Phylica cephalanthe*) evergreen shrub. *Acta Oecologia* **11**, 311–26.

10

Functional perspective of ecosystems

W D Stock and N Allsopp

INTRODUCTION

The objective of this chapter is to review results of studies which have used a process-functional approach to study fynbos ecosystems as opposed to the population-community approach adopted in other chapters of this book. A review of process-functional studies is difficult as they operate at a number of spatio-temporal scales, ranging from abiotic input/output nutrient budget studies of total landscape units to localized process studies where abiotic and biotic components are intimately linked, such as in decomposition (O'Neill et al. 1986). However, because the process-functional approach is such a powerful tool for enhancing the understanding of complex ecosystems, where it is often virtually impossible to isolate the functions of individual species, a range of studies will be reviewed which will attempt to cover the hierarchy of functional levels found in fynbos ecosystems. It should be remembered that the functioning ecosystem is the result of the integration and coevolution of biotic communities and abiotic environments and not merely the sum of the properties of individual components.

Odum (1986) likened ecosystems to organisms when he described them as being 'open, far-from-equilibrium thermodynamic systems with input and output environments: but they differ from organisms in the way in which they develop and are controlled'. This concept of ecosystems as quasi-organisms is problematic, however, as it raises the greatest limitation of the process-functional approach, namely that fluxes of nutrients, water, or energy are treated as if they exist independently of the species involved in the system. The holistic viewpoint adopted in process-functional studies is often excessively abstract for many biologists and the conceptualization of ecosystems can often differ according to the nature of the study. In particular, cycles of elements with sedimentary cycles (e.g. P, Ca, K) and those with gaseous cycles (e.g. N, S) have many dissimilar processes. Thus, ecosystem studies tracing these elements will have distinct structural differences. Despite problems with the process-functional approach, the population-community viewpoint also has a number of distinct disadvantages. These are especially associated with studies of below ground processes where it is difficult to isolate and identify individual species and the role they play in ecosystem dynamics. Population-community studies also invariably ignore microbial components of ecosystems because of their complexity. In almost all studies reported in the literature, the population community approach has only considered ecosystem dynamics at the plant-animal community level. Process-functional studies also allow the development of a predictive component which is often a major shortcoming of population-community studies.

The purpose of this chapter is to synthesize the range of process-functional studies undertaken in fynbos over the past decade, highlighting how the results have led to a better understanding of fynbos function (see also earlier review of Groves 1983). These studies have also complemented the range of population-community studies reported in other chapters of this book. The use of these multi-disciplinary and multi-approach studies has led to a substantial increase in the understanding of fynbos ecology which in the long-term should enhance the effective management and conservation of this small but species-rich biome.

NET PRIMARY PRODUCTION AND STANDING PHYTOMASS

Net primary production

The flow of energy and nutrient resources through ecosystems is closely linked to the flow of water through the soil-plant-atmosphere continuum (SPAC). Thus, it is only when soil moisture is adequate that transpirational water-flux from the soil to the atmosphere can occur through open stomata. This in turn allows for the simultaneous diffusion of CO_2 required for the energy trapping process of photosynthesis. With summer drought being a regular occurrence in mediterranean regions, shortage of water in the root zone often occurs and this may induce stomatal closure and concomitant reductions in CO_2 uptake which in turn could act to limit the primary productivity.

Net primary production has been studied in most mediterranean ecosystems. Specht (1969) reported growth rates in Australia and California to range from 64 to 420 g m^{-2} y^{-1}. In fynbos Kruger (1977) found rates of the same order of magnitude with variations from 100 to 400 g m^{-2} y^{-1}. In fynbos the highest rates were recorded in the first two or three post-fire years, after which they fell rapidly as the community aged (Kruger 1977).

Production rates in mediterranean-type ecosystems appear to be only marginally greater than those of adjacent desert scrub biomes (10–250 g m^{-2} y^{-1}) and well below those of nearby savannas and forests (savanna 200–12 000, temperate forest 600–3 000 g m^{-2} y^{-1}, Whittaker 1970). Factors accounting for the low, yet relatively consistent, annual productivity in mediterranean-type ecosystems have been suggested to include low solar energy input (visible radiation required for photosynthesis and radiation energy required to drive water through SPAC), the low volume and seasonality of the water regime, and the low availability of soil nutrients, particularly in the Australian and South African regions (Specht 1969).

Solar radiation in the south-western Cape appears to be lower than in the other regions of South Africa. Monthly mean daily incoming radiation values are highly variable between seasons with the summer period having values of 28–9 MJ m^{-2} d^{-1} and with winter values dropping as low as 9–10 MJ m^{-2} d^{-1} (Reid and De Jager 1988). In comparison the savanna area of the Transvaal highveld has a relatively constant radiation input with values showing small fluctuations from 25 (summer) to 22 MJ m^{-2} d^{-1} (winter). Spring and autumn radiation values are very similar across the whole of South Africa. Thus, although winter radiation inputs are lower than in any other region, annual inputs are high which suggests that radiation *per se* probably does not significantly limit production in the fynbos region. Specht's (1969) studies in France, California, and Australia would also support this contention as he noted that potential plant production is never realized under the prevailing radiation regimes in any of the mediterranean-type ecosystems, probably because of limitations of other resources, such as water and nutrients.

The combination of water shortages during periods of high radiation and plentiful water during low radiation winter periods could possibly account for the low productivity of mediterranean-type ecosystems. However, studies of the water gathering strategies and a consideration of the water relations of species representative of various fynbos guilds reveal that most plants in the system do not suffer from water stress in summer (see Stock et al. this volume). Studies have shown that certain fynbos guilds, such as the proteoid shrub group, are efficient in obtaining water by having a deep tap root system for water acquisition with laterals concentrated in the more nutrient-rich surface layers of the soil (Higgins et al. 1987; Jongens-Roberts and Mitchell 1986). Plants of this guild seldom show great stress and plant water potentials are generally high, never falling below -2.5 MPa (Miller et al. 1983; Van der Heyden and Lewis 1989; Von Willert et al. 1989; Richardson and Kruger 1990). Other plant functional guilds, such as the restioid group, do, however, show summer stress and their productivity might be restricted by water scarcity. In species representative of this group, water potentials as low as -7 MPa have been recorded (Miller et al. 1983; Van der Heyden and Lewis 1989) and in communities dominated by this guild, water scarcity might have a marked influence on productivity. The pronounced flush of spring to summer above ground growth and productivity in many of the different functional guilds (Sommerville 1983; Pierce 1984) of the fynbos regions does suggest that summer water scarcity only has a limited influence on productivity throughout most of the region.

The effect of nutrient availability on ecosystem productivity could be profound. However, to date no studies on net productivity of nutrient-rich ecosystems of the fynbos biome such as thicket, renoster shrubland, and karroid shrubland have been undertaken. From circumstantial evidence it would appear that sclerophylly in fynbos is a functional response to low nutrient environments (Loveless 1962; Small 1972) and that this leaf type places constraints on the overall production. A consequence of the major physiognomic groups of fynbos possessing evergreen, long-lived sclerophyllous leaves is the low structural diversity and resulting low functional diversity of such ecosystems. The need for efficient nutrient acquisition and high nutrient use efficiencies in fynbos appears to have resulted in a reduced number of above ground functional options being exercised in fynbos species despite the high species richness (Cowling et al. this volume).

Below ground functional diversity may, however, be high in fynbos. A mixture of shallow rooted (e.g. Ericaceae and Restionaceae) and moderate to deep rooted (e.g. Proteaceae, Rhamnaceae, and Rutaceae) species results in resource division. In addition, specialized uptake mechanisms such as cluster roots, which increase the surface area for absorption, are prevalent in the Proteaceae, Restionaceae, and Fabaceae (e.g. *Aspalathus*) (Lamont 1981; N. Allsopp unpublished data). Nutrient acquisition is also enhanced by symbiotic associations such as those found in the Ericaceae which are the obligate hosts of ericoid mycorrhizas. Vesicular-arbuscular mycorrhizas (VAM) occur in many common fynbos taxa (e.g. Fabaceae, Asteraceae, Rutaceae, Rhamnaceae, and Bruniaceae) (N Allsopp unpublished data) while nitrogen-fixing rhizobia and bradyrhizobia are found in the Fabaceae (Grobbelaar et al. 1983; Staphorst and Strydom 1975).

These nutrient-acquiring specializations often impose a severe nutrient and carbon drain on the plant. For instance, the ephemeral cluster (proteoid) roots can at any one time make up to 65% of root biomass of Proteaceae (Lamont 1981) while nutritional demands of mycorrhizas (Buwalda and Goh 1982; Stribley and Read 1974), rhizobia, or bradyrhizobia often delay seedling development particularly during the infection process (Atkins et al. 1989).

Fire combined with low nutrient soils may also constrain above ground structural diversity. Because nutrients are scarce, energy may be invested not only in organs enhancing nutrient uptake but considerable biomass may be diverted to below ground tissue so as to enhance regeneration ability after fires. Plants adopting this strategy will have an above ground competitive advantage and will also be able to establish nutrient uptake organs rapidly so as to exploit the post-fire nutrient flush. High investments in nutrient-acquiring structures and in fire survival organs possibly limit allocation to above ground growth and may also restrict the diversity of structural and functional options that could be exercised in the aerial environment of fynbos ecosystems.

Standing phytomass

The phytomass (biomass + necromass) of an ecosystem is similar to primary production in that it is sensitive to moisture and temperature gradients (Whittaker 1970). Under any given climatic conditions, however, these two parameters may be influenced by soil fertility gradients. In studies undertaken in the south-western Cape, the total phytomass and biomass component of fynbos communities ranges extensively and the values are confounded by primarily post-fire stand age (Table 10.1). This makes conclusions concerning the influences of abiotic environmental factors rather difficult. From the results shown in Table 10.1 it is evident that rainfall and elevation (below 1 500 m), with their associated temperature changes, have little effect upon the phytomass of fynbos communities. The relative contribution of the various phytomass components does, however, show interesting trends, with the graminoids becoming an increasingly important percentage of the biomass at higher altitudes (Rutherford 1978) while the reverse was true for shrubs.

TABLE 10.1 The above-ground phytomass of plant communities in the south-western fynbos biome. Rainfall, elevation, and post-fire stand age are given wherever possible as these are often important factors governing above-ground phytomass production and accumulation.

	Rainfall (mm)	Elevation (m)	Stand age (years)	Phytomass (kg m^{-2})
Fynbos Kraaifontein[2]	553[1]	122[1]	11	1.458
Fynbos Pella[3,7]	577	160	20	3.009
			4	0.914
			9	1.468
Fynbos Zachariashoek[4]	1 018	750	12	0.558
Fynbos Jonkershoek[4]	1 700	425	21	3.542
Fynbos Kogelberg[4]	1 150	110	20–2	1.105
Fynbos Jakkalsrivier[5]	655	740	16	1.453
Fynbos Riviersonderend[6]	1 200	1 520	10–14	1.451
Fynbos Riviersonderend[6]	800	910	10	1.142
Renoster shrubland Riviersonderend[6]	500	560	?	1.402
Karroid shrubland Riviersonderend[6]	300	390	?	1.162

[1] Data estimated from Fuggle (1981)
[2] Low (1983)
[3] O'Callaghan (1981)
[4] Van Wilgen and Le Maitre (1981)
[5] Kruger (1977)
[6] Rutherford (1978)
[7] Mitchell et al. (1986)

Phytomass per unit area within the fynbos region of 0.5–4 kg m^{-2} is at the lower range of values reported for shrubland and woodland ecosystems (2–20 kg m^{-2}, Whittaker 1970) and is of the same order of magnitude as desert scrub (0.1–4 kg m^{-2}) and the lower range of savanna ecosystems (0.2–15 kg m^{-2}). The major difference between fynbos and adjacent ecosystems is that the phytomass persists and accumulates each year in the inter-fire period. This is in contrast to both savanna and desert scrub systems where the biomass does not persist for more than one or two years as it is burnt by regular fires (e.g. grass in savannas burned every 1–2 years), consumed by herbivores (e.g. termites in desert communities), or broken down by thermal degradation (e.g. arid and semi-arid regions) (Steinberger et al. 1990).

The major store of carbon compounds in fynbos is the soil which may contain over 88% of organic matter of an 11-year-old post-fire coastal fynbos community (above ground phytomass: 1.73 kg m^{-2}; below ground phytomass > 3 mm: 1.39 kg m^{-2}; soil organic matter in 1.8 m profile: 11.08 kg m^{-2}) (Low 1983). Although values of annual productivity and above ground phytomass in fynbos communities are low, a significant quantity of energy is stored in soils in the form of organic matter which could be used to drive soil processes such as decomposition and mineralization of nutrient elements.

NUTRIENT CYCLING

Achievements and challenges in fynbos nutrient cycling

The range of nutrient cycling patterns in the fynbos biome is expected to be wide because of the diversity of soil types present in the biome, each of which carries a characteristic vegetation type (Deacon et al. this volume). Soils range from relatively high nutrient and well developed clay soils supporting renoster shrubland to extremely nutrient-poor shallow sands on quartzite which support fynbos. Furthermore, these soils occur under a wide range of climatic conditions. This expected high range of nutrient cycling patterns has not been explored as most studies have concentrated on patterns in a coastal lowland fynbos site at Pella, north of Cape Town. The elements N and P were extensively studied in this ecosystem as they have intrinsically different nutrient cycling patterns: N is a globally cycled nutrient as opposed to the local sedimentary cycling patterns typical of P and many cations. N and P are also the two elements suggested to be most likely to limit primary production and other ecosystem processes.

The results of N and P cycling studies in fynbos at Pella (Figure 10.1) show a range of interesting patterns and give some indication of the mechanisms which regulate these patterns. Although limited to a single site, these studies allow for the links between soil fertility and processes important to maintenance of fynbos structure and function to be established. As with carbon, the largest reserves of N and P in the ecosystem are located in the soil (Low 1983; Stock 1985). This is in contrast to other low nutrient ecosystems such as certain tropical rain forests which Whittaker (1970) described as systems with a nutrient-rich economy perched on nutrient-poor substrates. Tropical rain forests on infertile sites do, however, display a range of characteristics in common with fynbos ecosystems. These include a high incidence of sclerophyllous leaves, low tissue nutrient concentrations, and high root/shoot ratios (Vitousek and Sanford 1986). The incidence of sclerophylly and low tissue nutrient levels (Table 10.2) are inextricably linked as a sclerophyllous leaf is rich in structural carbon compounds (Mitchell et al. 1986) and phenolic compounds (Glyphis and Puttick 1988) (presumably to deter herbivores) which lead to a dilution of the nutrients contained per unit weight of leaf. The low foliar nutrient levels and high fibre content may also contribute directly to reducing herbivory which is of a very limited scale in fynbos (Joubert and Stindt 1979; Glyphis 1985; Johnson this volume). Thicket and renoster shrubland ecosystems on the more nutrient-rich soils show higher levels of herbivory which suggests that nutrient availability has a pronounced effect on structure and function at all levels within the ecosystem.

The transfer of nutrients from the standing phytomass to the soil is complex and depends upon the processes of litterfall and decomposition. Litterfall studies in fynbos are scarce and data exist for two lowland and three mountain sites only. Generally litter mass at such sites increases with increasing standing phytomass and age of the community: N and P pool sizes vary accordingly (Table 10.3). The actual rates of litter production are low. Mitchell et al. (1986) observed that an eight-year-old lowland fynbos community produces some 78 g m^{-2} per year and that it would take between 16–19 years for 95% of steady-state to be reached in the litter layer. Steady-state occurs when the composition and average biomass of the community become constant (Olson 1963).

Litter production in fynbos is low when compared to other mediterranean-type ecosystems where values of 409, 355, and 194 g m^{-2} have been recorded for Australian kwongan (Maggs and Pearson 1977), Californian chaparral (Mooney et al. 1977), and Californian coastal sage scrub (Gray and Schlesinger 1981) respectively. Little is known of the rates of nutrient return via litterfall in fynbos. How-

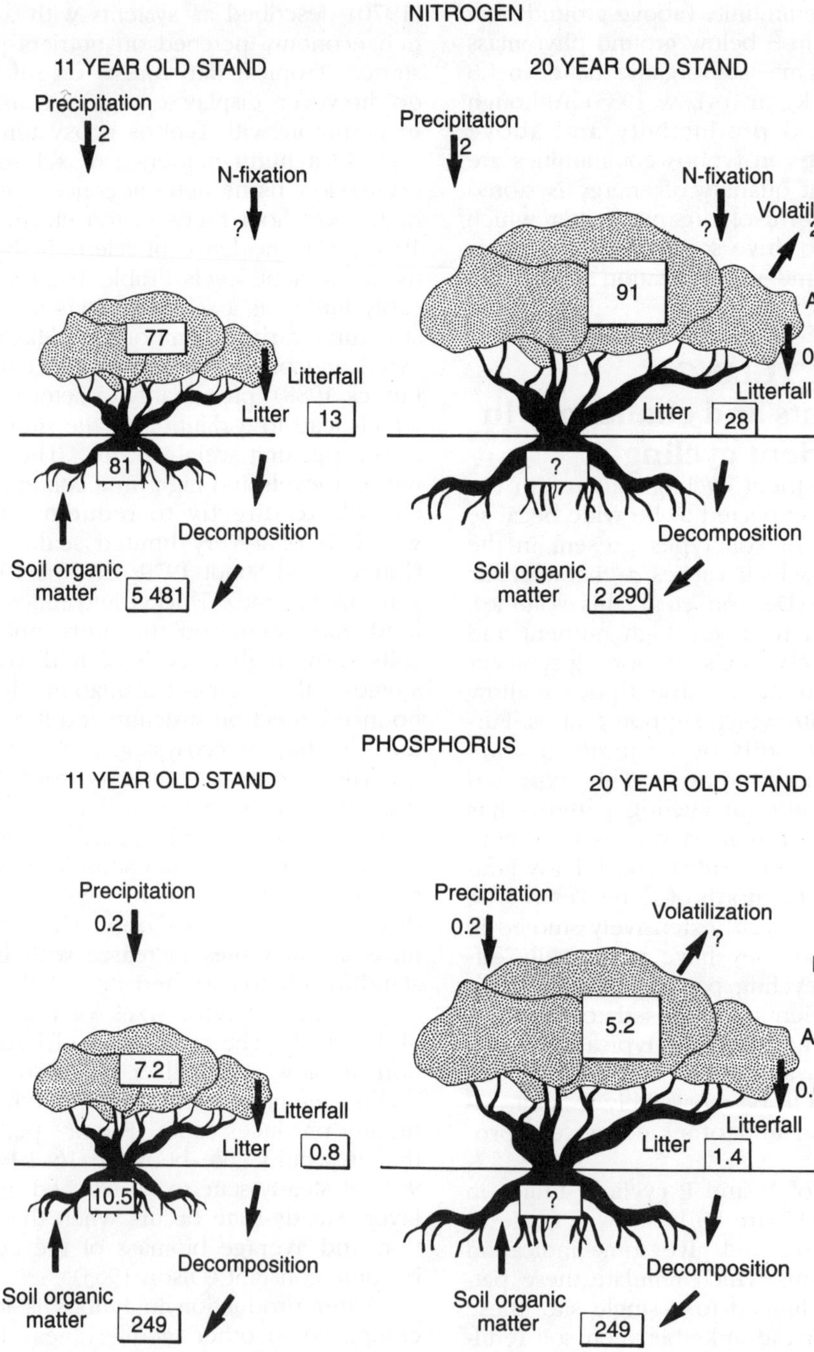

FIGURE 10.1 The major compartments (kg ha^{-1}) and pathways of nitrogen and phosphorus cycling (kg ha^{-1} y^{-1}) in 11 and 20 year old stands of fynbos on deep sands on the western Cape coastal foreland (data from Low 1983; Stock 1985; Mitchell et al. 1986). The volatilization and ashing effects of fire are given in kg ha^{-1} per fire.

TABLE 10.2 The organic and inorganic constituents of proteoid and ericoid shrub leaf litter and restioid culm litter of a fynbos community at Pella, north of Cape Town. Data from Mitchell et al. (1986).

	Physiognomic group		
	Proteoid ($mg\ g^{-1}$)	Ericoid ($mg\ g^{-1}$)	Restioid ($mg\ g^{-1}$)
Carbon	508.3	530.1	494.0
Fats and waxes	8.8	7.6	5.5
Soluble carbohydrates	13.8	18.8	18.7
Soluble phenolic compounds	22.0	22.2	20.8
Holocellulose	119.8	71.3	319.5
Lignin	359.5	348.4	173.3
Calcium	11.0	6.8	0.8
Magnesium	1.9	1.5	0.5
Potassium	0.5	0.7	1.4
Iron	0.9	0.7	0.2
Phosphorus	0.3	0.3	0.1
Nitrogen	5.7	5.9	4.3

TABLE 10.3 The dry mass, nitrogen, and phosphorus pools in fynbos litter at lowland and mountain sites.

	Stand age	Litter mass $g\ m^{-2}$	Total N $g\ m^{-2}$	Total P $g\ m^{-2}$
Lowland Cape Flats[1]	11	273.0	1.3	0.08
Lowland Pella[2,3]	20	792.0	–	–
	8	451.0	–	–
Mountain fynbos Jonkershoek[4]	21	1 425.9	9.3	0.29
Mountain fynbos Zachariashoek[4]	12	181.5	0.9	0.03
Mountain fynbos Bakkerskloof[4]	12	55.4	0.3	0.01

[1] Low (1983)
[2] O'Callaghan (1981)
[3] Mitchell et al. (1986)
[4] Van Wilgen and Le Maitre (1981)

ever, from resource allocation studies on individual species, it is known that some fynbos species effectively withdraw both N and P prior to senescence and leaf fall (Jongens-Roberts and Mitchell 1986; Stock et al. 1987; Mitchell et al. 1986) and it is expected that the inputs would be very low.

Thus, low soil fertility appears to have an underlying effect on the quantities and rates of nutrient return via litterfall. Decomposition of the sclerophyllous litter material from fynbos is extremely low (Mitchell et al. 1986; Mitchell and Coley 1987) as would be expected from the unfavourable climatic conditions and poor litter quality — i.e. high lignin, high holocellulose, and high phenolic content — associated with low nutrient contents of most fynbos elements. Therefore, turnover times are prolonged and in excess of the 3–4 years reported for Australian heath and Californian chaparral (Specht 1981; Schlesinger and Hasey 1981). In fact, during the first 18 months of decomposition, N and P were immobilized rather than released from the litter of three fynbos growth forms (proteoid, restioid, and ericoid) (Mitchell et al. 1986). Decomposition is not the major means of nutrient release and this process is probably dependent upon the periodic fires so characteristic of fynbos.

Fire and its central role in fynbos nutrient cycling

Fire appears to be the major mineralizing agent in fynbos, returning mineral elements held in the above ground phytomass and litter to the soil (Van Wilgen and Le Maitre 1981; Brown and Mitchell 1986; Stock and Lewis 1986a). Fires are a regular disturbance initiating successional changes, during which patterns of resource availability alter. These changes reportedly follow definite patterns, with one school suggesting that the availability of all resources (water, light, and nutrients) is elevated at the soil surface shortly after the disturbance, and that the availability of these resources diminishes with successional age (Odum 1969; Vitousek and Reiners 1975). Tilman (1986), in contrast, has argued that changes in plant species composition during succession reflect a shift from high light/low nutrient conditions during early succession to low light/high nutrient conditions later in the successional sequence. The results of studies in the fynbos biome support both these hypotheses.

Firstly, post-fire nutrient flushes appear to be a feature of fynbos systems and increased availability of N, P, and cations has been reported (Stock and Lewis 1986a; Brown and Mitchell 1986; Musil and Midgley 1990). The availability of these elements appears to decrease rapidly (after nine months for N) as elements become incorporated into plant biomass as well as being immobilized by decomposer organisms. This would support the idea of reducing nutrient availability with successional age. Rice and Pancholy (1972) have suggested that in early successional stages mineralization of N, and more particularly the nitrification process, could proceed unhindered, while in climax ecosystems nitrification is inhibited by plant-produced allelochemicals. This mechanism would act to prevent losses of easily leached nitrate. This proposal has been tested in fynbos and no evidence to support the hypothesis was forthcoming (Stock et al. 1988). It was found that mineralization, including nitrification, proceeded at approximately the same rate through a 20-year-old secondary succession, and it was rather changes in soil total N content and micro-climate factors (soil water content and temperature) which determined quantities of nitrogen released (Stock et al. 1988). Like N, P availability during early post-fire successional stages is enhanced (Brown and Mitchell 1986) but also rapidly (within four months) returned to pre-fire levels. It is suggested that the evidence to support initially high nutrient levels which decrease during succession is equivocal as most of the effects are of an extremely short duration and may be due to:

- the absence of plant uptake immediately after fire;
- the deposition of nutrient-rich ash at the soil surface; and
- thermal disruption of the soil fauna and flora.

These short duration fire-induced effects have little to do with the actual successional process after the first post-fire year.

Evidence for the second suggestion, namely Tilman's (1986) resource ratio hypothesis, is also available and relates particularly to elements such as N and S. These elements are volatilized by fire and differences occur in the N/P ratios at the soil surface between the pre- and post-fire stages. High P/low N immediately post-fire favours certain species, most notably the N_2-fixing guild, which would then be replaced as fixed N is returned to the system either via atmospheric deposition or biological N_2-fixation. Atmospheric inputs of N are low, generally < 2–3 kg ha^{-1} y^{-1} (Stock and Lewis 1986b; Van Wyk et al. 1991) but N_2-fixation may be significant as free living N_2-fixing bacteria (J Loos personal communication) and N_2-fixing members of the Fabaceae often do dominate in early successional stages of many ecosystems (Hansen and Pate 1987). As succession proceeds, greater quantities of soil organic matter accumulate and the decomposition of this would add to the quantities of nutrient available. Nutrients such as N from the larger soil N pools may become increasingly available as has been shown by the mineralization studies of Stock et al. (1988). Other elements such as P and some cations might accumulate in the ecosystem but as they are immobilized in the soil flora or else locked up in the above ground phytomass, they are largely unavailable. Thus, resource availability effects on fynbos secondary successional processes can be seen to be complex with equivocal evidence available to support both Odum's (1969) and Tilman's (1986) hypotheses. Further critical examination of patterns of resource availability through fynbos successional sequences is required before the vegetation dynamics of such systems can be fully understood.

Impact of nutrient enrichment on fynbos vegetation

The suggested strong controlling influence of low soil fertility on fynbos ecosystem structure and function has been tested in a number of studies using nutrient addition experiments (Witkowski 1988; Lamb and Klausner 1988). These studies provide a basis for predicting the outcome of conservation threats such as human-induced nutrient enrichment of fynbos ecosystems which is a process expected to increase dramatically as industry and agriculture expand to meet the demands of a regionally burgeoning population (Deacon this volume).

Sources of nutrient pollution in the south-western Cape include gaseous emissions from industry, drift from aerial application of herbicides, insecticides and fungicides, application of fire retardants, runoff from fertilized agricultural or forestry land, and deliberate fertilizer application to natural vegetation.

Input of nutrients from gaseous pollution in the south-western Cape is probably limited at present as the prevailing winds (SE and NW) ensure that air masses over current industrial areas are dispersed out to sea. $S0_2$ and 0_3 damage to plants was not observed near a petroleum refinery and a fertilizer factory, while fluorine damage was localized (Botha et al. 1989). Ca, total N, NH_4, NO_3, Na, PO_4, and total P measured in rainfall and stream flow following fire in mountain fynbos show that there is a net import of these nutrients from aerial deposition between fires (Van Wyk et al. 1991). These nutrients are apparently of natural origin, mostly marine, although adjacent wild fires could possibly contribute.

When examining the potential effect of nutrient enrichment on fynbos, it is necessary to consider the nutrient elements involved as well as the intensity and frequency of the input. Soil characteristics, vegetation type, and age since the last fire are also considerations. Nutrient additions in other ecosystems of the world have been shown to affect growth rates, productivity, plant phenology, species composition, and the rate of succession (Heddle and Specht 1975; Gray and Schlesinger 1981). Ecosystem process, such as the amount and quality of litterfall, may also change and decomposition and nutrient cycling rates may accelcrate with associated changes in vegetation composition and structure.

Witkowski (1988) and Lamb and Klausner (1988) both studied the effects in fynbos of a single application of a low-level nutrient addition in the form of single nutrient elements and balanced nutrient applications. In Witkowski's studies N applications induced slight increases in shoot growth and biomass accumulation, whereas P additions decreased growth (Witkowski 1988; Witkowski and Mitchell 1989; Witkowski et al. 1990b). Lamb and Klausner (1988) reported the reverse, namely increased growth in response to P but decreases when N was added. In addition,

Witkowski (1989a) reported significant changes in foliage projective cover among the restioid, grass, and annual components of the vegetation. The equivocal responses observed by the above authors were, however, small and it appears that low concentration single event nutrient additions are unlikely to effect major changes in productivity or species composition of fynbos. Limited changes in morphological characteristics, phenology, and allocation to reproductive structures were also evident in the proteoid, ericoid, and restioid physiognomic groups (Lamb and Klausner 1988; Witkowski 1988). Litterfall (especially leaf fall) increased in response to nutrient additions and the N and P content of the litter was elevated (Witkowski 1989b). Nitrogen levels in the surface soil quickly returned to earlier levels but elevated P levels were detectable two years after the addition (Witkowski et al. 1990a).

Observation shows that species composition after different fires on any one site varies within a natural range determined by many factors, of which nutrient status of the soils is only one. Therefore small nutrient additions are unlikely to cause changes outside the natural amplitude of vegetation variation for a particular site.

Ecosystem nutrient cycling processes invariably change in response to nutrient additions. In particular the passage of nutrients through the plant is often accelerated which results in higher N and P levels in litter. Increases in litter mass and nutrient content following nutrient addition to fynbos have been observed (Witkowski 1989b) while the return of nutrients to the soil was slow because, as predicted from the results of Mitchell et al. (1986), decomposition appears to limit the rate of nutrient turnover in fynbos ecosystems. Despite increased nutrient accretion in the litter layer, the litter quality (especially C/N, C/P ratios) does not alter significantly and decomposition rates remain slow and limit nutrient return to the soil. Vitousek (1982) suggested that slow decomposition rates represent a self-perpetuating system which maintains low soil nutrient levels. Nutrients are only released in the short post-fire period during which N levels may be particularly low due to thermal volatilization and leaching while the extra P added to the system may be immobilized in the large unavailable soil pool (Chapman et al. 1989). Thus, after a single fire the effects on ecosystem process of small nutrient additions are eliminated (Figure 10.2) and may have little or no long-term impact on the stability and resilience of low nutrient, fire-prone ecosystems such as fynbos.

Only anecdotal reports exist as to the possible effects of larger or more frequent inputs of nutrients into plant communities. Observation of 'natural' vegetation adjacent to agricultural fields, roadside verges, and other areas of human activity involving nutrient additions indicates that succession may be retarded, nutrient demanding early successional species favoured, and species diversity reduced. However, physical disturbance often accompanies the nutrient additions and the effects are difficult to separate. Pot experiments indicate that high levels of nutrient additions can reduce growth rates or kill a variety of sclerophyllous fynbos species, while weed species tend to be more plastic in their response to nutrients. Hence these species may be able to outcompete the natural vegetation. Witkowski (1988) indicated trends that may occur when higher nutrient additions are introduced. These included a favouring of early successional species, combined with a more rapid completion of the life cycle of some shrubby species. The effects of high nutrient additions are unlikely to disappear as rapidly as those of low nutrient additions as nutrient cycling processes would take longer to reach a new equilibrium. While the effect of these additions may be delayed by slow decomposition rates, following the release of nutrients after a fire, the equilibrium community may be markedly different from the original fynbos. The new community is unlikely to have the same structure as communities on richer soils in the fynbos biome as other soil characteristics (e.g. texture) are important in delimiting these (e.g. renoster shrublands). The most important danger is that alien weedy vegetation, in particular pines and Australian acacias, will invade. These plants respond with increased growth to much higher levels of nutrient enrichment than fynbos plants and they will be able to outcompete these (Witkowski 1991a).

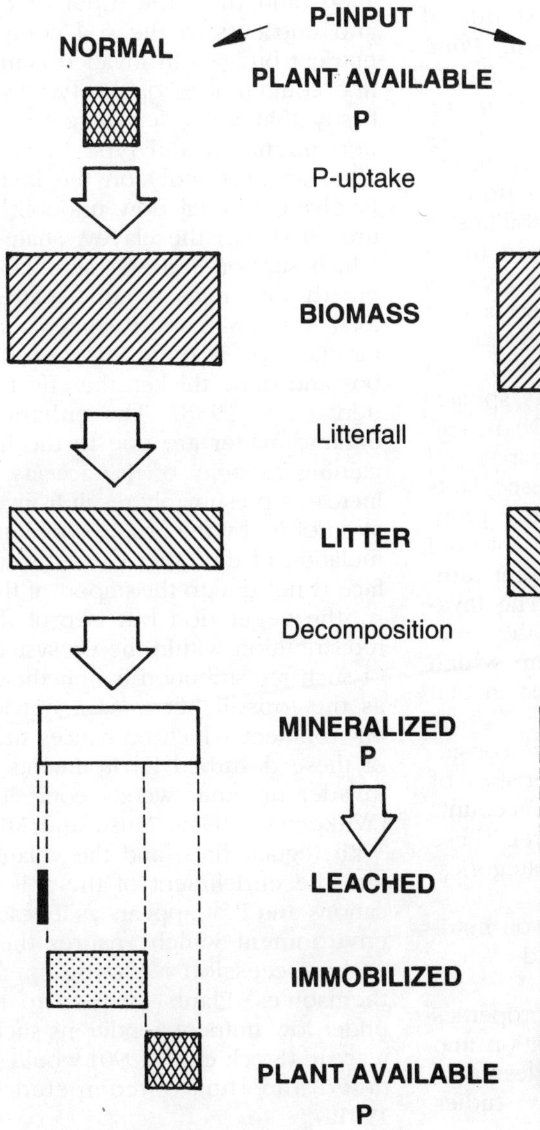

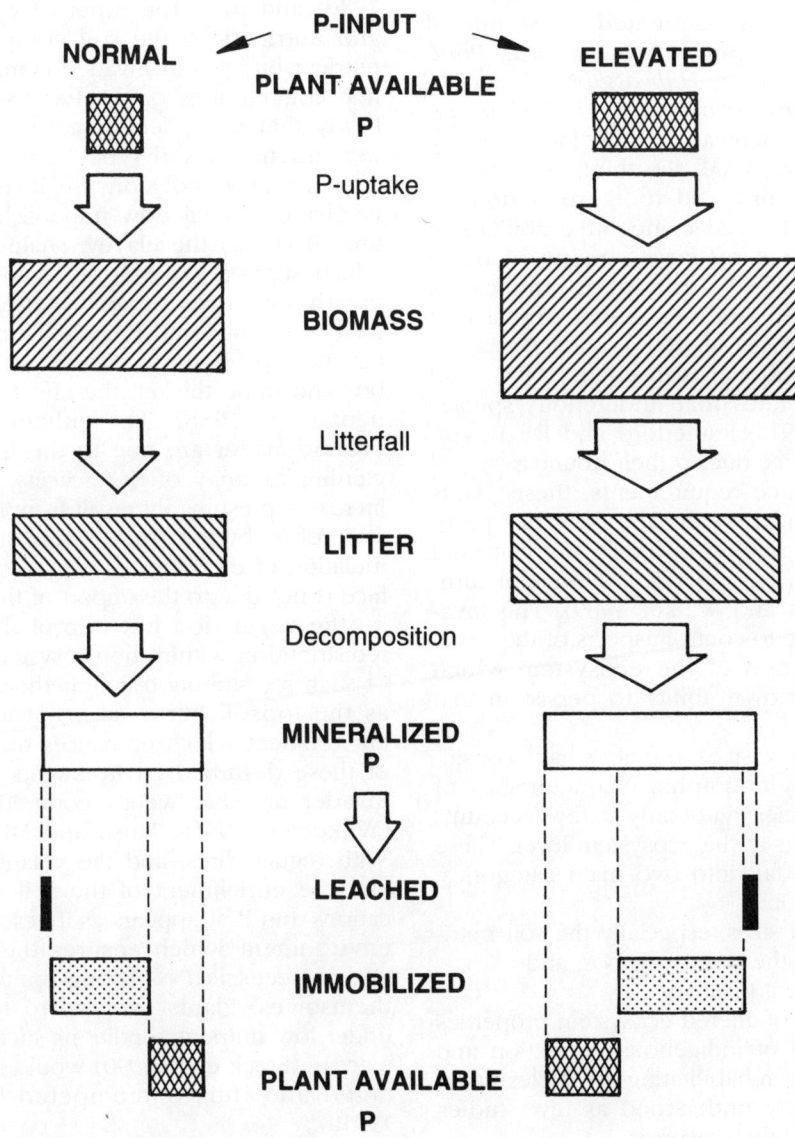

FIGURE 10.2 A model showing the effect of a phosphorus (P) addition on phosphorus pool sizes and the rates of nutrient cycling processes in a fynbos ecosystem (width of arrows indicates the rate of the processes; size of rectangles indicates the size of the pools).

Plant invasives and their impact on fynbos nutrient cycling

Woody alien invasive plant species have become a feature of fynbos landscapes (Richardson et al. this volume) with much of the fynbos biome being dominated by stands of *Acacia cyclops*, *A. saligna*, *A. mearnsii*, *Pinus radiata*, *P. pinaster*, *Hakea gibbosa*, and *H. sericea*. All these plants possess mechanisms to enhance nutrient acquisition including N_2-fixing symbionts, VAM, sheathing mycorrhizas, and extensive proteoid root production. In addition to enhanced competitive abilities in obtaining scarce resources, many of these species are able to dominate landscapes through elevated seed production (and lack of predators) (Dean et al. 1986), enhanced seedling survival (Richardson et al. 1987), and faster growth rates than indigenous species (Witkowski 1991a; Rutherford and De Bosenberg 1988). Thus, due to their abundance and differing resource requirements, these plants have the potential to alter the collective properties of an ecosystem including nutrient pool sizes, productivity, and rates of nutrient turnover (Vitousek and Walker 1989). The invasives may come to control aspects of the structure and function of the ecosystem which enhances their own ability to persist in that system.

Studies of invasives in fynbos have considered mainly demographic characteristics of individual species with only a few accounts detailing impacts at the ecosystem level. These latter accounts fall into two main categories, namely effects on:

- nutrient pool sizes, especially the soil nutrient reservoir in the case of fynbos; and
- rates of nutrient turnover.

The effects of altered ecosystem properties on the survival of indigenous vegetation and the feasibility of rehabilitating areas cleared of aliens are poorly understood as few studies have considered these aspects.

NUTRIENT POOL SIZES

The most notable feature of both pine and acacia invasives is that they alter the spatial distribution of nutrients within an ecosystem (Milton 1981; Versveld and Van Wilgen 1986; Witkowski and Mitchell 1987; Musil and Midgley 1990; Witkowski 1991b). With greater productivity both plant groups change fynbos ecosystems by producing considerably more above ground biomass. This in turn influences litterfall dynamics (704 g m^{-2} for acacias, Milton 1981; cf. 78 g m^{-2} for fynbos, Mitchell et al. 1986) and thus, the input of organic matter and nutrients to the soil compartment. Soil nutrient budgets in invaded compared to pristine communities of the fynbos biome show firstly that the effects are to a large extent dependent on soil type (Low 1988; Figure 10.3) and secondly on the invading species involved (H Engledow unpublished data; Figure 10.4). On the clayey, shale-derived soils which support renoster shrubland, *A. saligna* greatly enhances organic matter, as well as total N, Ca, Mg, and P contents in both the litter and topsoil; on sandy soils supporting fynbos and dune thicket, the effects are less evident (Low 1988). The enhanced levels of organic matter are due to the higher carbon gaining capacity of the acacias, while the N increases presumably result from increased fixation of N_2 by the acacia symbionts. The accumulation of the other elements at the soil surface is not due to the import of these elements by the vegetation but is probably due to a redistribution within the ecosystem. The effect of such redistribution is nonetheless important as the topsoil becomes a nutrient-enriched environment which en-hances the domination of these disturbed environments by the alien invader or other weedy competitive species (Witkowski 1991a; Musil and Midgley 1990). With regular fires, and the volatilization of N and the enrichment of the soil surface with cations and P, it appears as if acacias create an environment which ensures the survival of early succession N_2-fixing species such as themselves. Plants adapted to regeneration under low nutrient conditions such as the Proteaceae (Stock et al. 1990) would be disadvantaged and thus outcompeted (Witkowski 1991a).

Invading pines and hakeas do not fix nitrogen and their effects on ecosystems are more subtle. H Engeldow (unpublished data) studied the impacts of four different alien species at a site on a single soil type on the Cape Peninsula and found that invasive pines and hakea had less of an effect on soil properties (organic matter, pH, total N, and total P) than did the two acacia species (Figure 10.4).

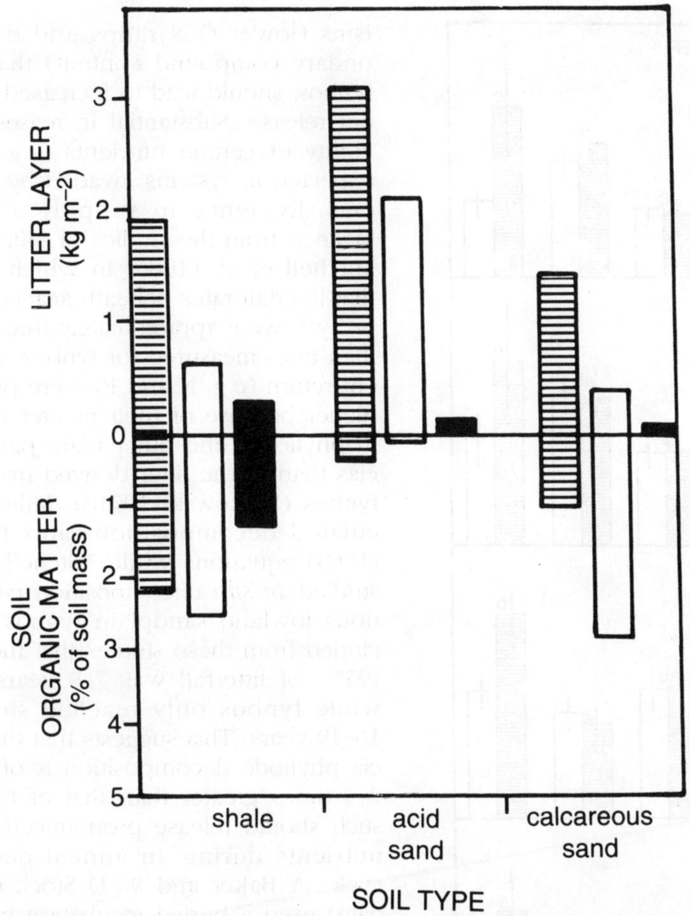

FIGURE 10.3 The litter layer mass and soil organic matter content (in surface 1 cm of soil) beneath stands of *Acacia saligna* ▤, closed ☐ and open ■ canopy of renoster shrubland (shale soil), fynbos (acid sands), and dune thicket (calcareous sands) communities in the south western Cape (data from Low 1988).

In addition the revegetation of post-fire soil environments in areas previously dominated by hakea and pines showed that the small physical changes induced by these species had little effect on the abundance and species composition of the post-burn vegetation (H Engeldow unpublished data). Only regrowth under *A. cyclops* and *A. saligna* differed significantly ($p < 0.05$) from the patterns of regrowth under the burnt-out canopies of indigenous *Leucospermum conocarpodendron* individuals. It is apparent, therefore, that each of the invasive woody species has its own particular effect on nutrient distribution within the ecosystem and those patterns of alteration are a result of the different resource acquisition strategies employed by the species. The effects that would occur as a result of the invasion of Mediterranean Basin grasses or other herbaceous elements rather than large perennial shrubs are unknown. However, it is not likely to be of major importance at the ecosystem level because of the shorter lifespans and lower phytomass of these growth forms.

NUTRIENT TURNOVER RATES

Mechanisms whereby invasive plants alter ecosystem level processes are diverse. Attention, however, is often concentrated on the process of nutrient release via decomposition as this is thought to be pivotal in controlling the fertility of terrestrial ecosystems. In invaded systems substantially increased litter and soil N pool sizes, coupled with a litter quality less resistant to decomposing organ-

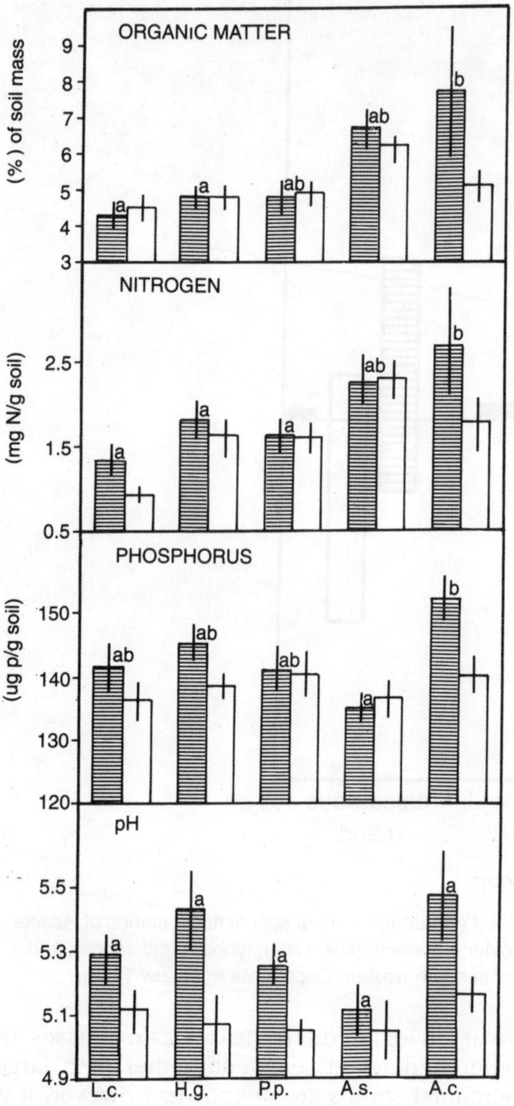

FIGURE 10.4 The organic matter, total nitrogen, total phosphorus, and pH values of soils beneath indigenous *Leucospermum conocarpodendron* (L.c.), alien *Hakea gibbosa* (H.g.), *Pinus pinaster* (P.p.), *Acacia saligna* (A.s.), and *A. cyclops* (A.c.) canopies ▤ and adjacent open areas □ sampled at Millers Point on the Cape Peninsula. Values are means ± standard deviations and different letters indicate significantdifferences ($p < 0.05$, One-way ANOVAS) between beneath-canopy sites of the five study species (H Engeldow unpublished data).

isms (lower C/N ratios and decreased secondary compound content) than indigenous fynbos, should lead to increased rates of nutrient release. Substantial increases in the availability of certain nutrients (e.g. N) could be expected in systems invaded by N_2-fixing acacias. Evidence in support of this can be gleaned from the studies of Milton (1981) and Mitchell et al. (1986) in which it was found that litterfall rates beneath acacia stands (700 g m^{-2} y^{-1}) were approximately three times higher than rates measured for fynbos. Rates of nutrient return (e.g. N and P) were proportionately higher because of their greater concentrations in phyllodes and other plant parts of the acacias than in the litter derived from indigenous fynbos (Witkowski 1991b). Milton (1981) calculated decomposition rates from Olson's (1963) equations while Mitchell et al. (1986) studied *in situ* decomposition rates of indigenous lowland sandplain fynbos. It was concluded from these studies that the steady-state (95%) of litterfall was 7–9 years for acacias, while fynbos only reached steady-state at 16–19 years. This suggests that the rate of acacia phyllode decomposition is of the order of 2–3 times greater than that of fynbos and as such should release pronounced quantities of nutrients during an annual decomposition cycle. A Baker and W D Stock (unpublished data) used a buried incubation bag technique to study differences in the rate of N and P mineralization between natural and acacia-invaded dune thicket and fynbos communities. Annual rates of mineral N and P release were not substantially greater in *Acacia saligna*-invaded fynbos, while in *Acacia cyclops*-invaded dune thicket N and P mineralization was demonstrably higher (particularly N) in the invaded sites (Table 10.4). It would appear, therefore, that the soil properties beneath fynbos (low pH, low nutrient state) and thicket (high pH, relatively high nutrient status) have a controlling influence upon decomposition rates and the recycling of nutrients. These results are in agreement with Low's (1988) observations (Figure 10.3) that acacias have little effect in enriching the organic matter content of acid sandy soils but that on calcareous sands substantial organic enrichment of the top 10 cm does occur. Large litter masses can accumulate in sandy acid environments which increase the fire suscept-

TABLE 10.4 The annual nitrogen and phosphorus release in soil (μg N or P g^{-1} soil y^{-1}) beneath uninvaded and Acacia-invaded fynbos and thicket ecosystems in the south-western Cape. From A Baker and W D Stock (unpublished data).

	Thicket on calcareous sands		Fynbos on acid sands	
	A. cyclops-invaded	Uninvaded	*A. saligna*-invaded	Uninvaded
Mineral N μg N g^{-1} soil y^{-1}	24.155	11.292	3.360	6.938
Mineral P μg N g^{-1} soil y^{-1}	3.960	2.730	0.812	0.580

ibility of these environments (Van Wilgen and Richardson 1985) and fire might therefore act as the predominant mineralizing agent as it does for the indigenous fynbos stands characteristic of these soils.

The assumption that organic enrichment of acacia-invaded sites is due to the efficiency with which these plants are able to fix atmospheric N_2 has not been rigorously tested in southern African ecosystems. In Greece, where *A. saligna* is a major species used for dune stabilization, acetylene reduction assays of pot-grown material have shown excised nodules to have a reducing activity of 0.445 + 0.053 nmoles C_2H_4 produced min^{-1} mg dry wt. $nodule^{-1}$ (Nakos 1977). This value is in the range typical of leguminous species. No *in situ* assays of N_2-fixation capabilities have been attempted either in *A. saligna*'s natural or planted habitats which makes inferences about the relevance of the fixation process in the field rather tenuous. W D Stock (unpublished data) has attempted to quantify the importance of the fixation process in altering invaded southern African ecosystems using the ^{15}N natural abundance technique (Shearer and Kohl 1986). Results to date have shown that $^{15}N/^{14}N$ isotope ratios of leaves, litter, and soil organic matter in acacia-invaded thicket and fynbos systems do not differ greatly from the uninvaded controls. This lack of difference may have two contrasting interpretations, namely that little fixation occurs to change $^{15}N/^{14}N$ ratios in the total plant, or secondly that these species selectively allocate ^{15}N to specific organs which were not sampled in field studies. From the unpublished data of J S Pate and W D Stock it is evident that glasshouse-grown, fully N_2-fixing acacia plants selectively accumulate ^{15}N in the nodules. This makes quantification of the contribution of N_2-fixing acacias to ecosystem nitrogen capital difficult as whole plants, including the roots of these perennial plants, are required before nitrogen fixation can be realistically quantified.

Therefore, it is extremely difficult to assess the importance of the fixation process in altering the nutrient cycling characteristics of fynbos ecosystems, even though the enhanced accretion of C and N to acacia-invaded systems is almost exclusively attributed to the N_2-fixation process (Milton 1981; Witkowski and Mitchell 1987; Musil and Midgley 1990). Further research is required to quantify why alien N_2-fixing plant species are so successful while the indigenous fixers, which include many species, are so sparse in most fynbos ecosystems.

Impact of flower harvesting on fynbos nutrient cycling

Flower harvesting has caused concern in low nutrient fynbos ecosystems as intense picking could lead to substantial nutrient losses and eventual depletion of the nutrient capital of the ecosystem. Replacement of nutrients removed in this manner could nonetheless be returned by fertilizer inputs. The desirability of such a practice is questionable, however, as nutrient enrichment experiments of natural veld (Witkowski 1988; Lamb and Klaussner 1988) have shown species and guilds to have unpredictable responses even in low nutrient addition experiments. Since few data are available concerning the range and quantities of flowers harvested from any one area, it is virtually impossible to quantify and determine the significance of individual elements lost. It is, however, feasible to estimate possible impacts by considering the allocation of elements to reproductive structures in ecosystem nutrient budgets.

The nitrogen budget of a *Protea repens*-dominated lowland fynbos stand (11 years old) shows that inflorescences of these shrubby species contain less than 0.5% of the N contained within the ecosystem (Table 10.5) and 14.1% of the N capital of the aerial biomass. These values are far higher than those obtained for *Protea compacta* and *P. obtusifolia* inflorescences from the Agulhas Plain where Esler et al. (1989) found that N and P constituted only 0.01% and 0.006% of the ecosystem nutrient capital for each element respectively (Table 10.5). Thus, if harvesting removes all blooms each year, the system could lose between 0.2 and 2.14 and 0.02–0.3 g N or P m^{-2} yr^{-1}, which at the higher levels is approximately 10 times the inputs received from atmospheric precipitation (Brown et al. 1984; Stock and Lewis 1986b). Esler et al. (1989) also found that calculating nutrient losses was complicated because the developmental stage of the harvested flower was also significant in terms of quantities of nutrients lost. Young inflorescences of *P. compacta* and *P. obtusifolia* had much lower N and P contents (< 50%) than fully matured blooms because of high nutrient storage in seeds of Proteaceae (Kuo et al. 1982). Thus, those species harvested in the bud to early opening stages (mostly blooms of *Protea*, *Leucospermum*, and *Erica* spp.) would suffer substantially lower nutrient removal rates than those plants harvested for their cones (*Leucadendron* spp.). Without further research to quantify the impacts of flower harvesting on ecosystem nutrient budgets, discussion of the relative importance of this process in exploited fynbos systems remains speculative. With demographic consequences due to progagule reduction being evident in intensively harvested areas (P Mustart and R M Cowling unpublished data) and fertilizers eliciting limited responses in most species present, methods of exploiting veld for wild flower production are still of the non-renewable 'resource mining' type. Urgent attention needs to be given to this aspect of fynbos exploitation so that both large numbers of the desired cut flower species and the overall species richness of an area can be maintained (see also Van Wilgen et al. this volume).

SUMMARY

Results presented in this chapter summarize a decade of fynbos ecosystem studies with the range of topics addressed in the various studies demonstrating the relevance of the process-functional approach in arriving at a predictive understanding of ecosystem function. All the ecosystem processes studied appear to be intimately associated with the low nutrient status of the soils. Low productivity, low decomposition rates, slow nutrient cycling, and the importance of fire as a mineralizing agent appear to be characteristic features of fynbos. Such inherent restrictions on ecosystem processes have imposed limitations on the resilience of fynbos to anthropogenic disturbances. Small-scale nutrient enrichments, the introduction of alien plant species, and nutrient removal via flower harvesting all affect ecosystem function. The quantification of the severity of these impacts is still rudimentary and in future conservation research programmes these aspects need further attention in collaboration with studies of the demographic consequences of such disturbances. For characteristic fynbos vegetation types to survive the pressures induced by burgeoning populations in the south-western Cape (Deacon this volume), research priorities must address problems from ecosystem process-functional level studies through to population studies at the species level.

TABLE 10.5 The potential nutrient losses (g m^{-2}) and proportion of total ecosystem nutrient capital (percentages shown in parentheses) lost as a result of flower harvesting of fynbos species from lowland sites in the south-western Cape.

Harvested species and site	N	P	Ca	Mg	K
	Nutrient elements (g m^{-2})				
Protea repens, acid sand, north of Cape Town[1]	1.25 (0.22)	0.1 (0.4)	2.08 (1.86)	0.12 (0.22)	0.97 (0.99)
Protea compacta, acid sand, Agulhas Plain[2]	0.2 (0.01)	0.02 (0.006)			
Protea obtusifolia, limestone, Agulhas Plain[2]	2.14 (0.04)	0.3 (0.032)			

[1] Low (1988)
[2] Esler et al. (1989)

ACKNOWLEDGEMENTS

We thank A Baker and H Engeldow for providing unpublished data. R Groves made useful comments on an earlier draft.

REFERENCES

ATKINS G A, PATE J S, SANFORD P J, DAKORA F D and MATTHEWS I (1989). Nitrogen nutrition of nodules in relation to 'N-hunger' in cowpea (*Vigna unguiculata* L. Walp). *Plant Physiology* **90**, 1 644–9.

BOTHA A T, VISSER J H and MOORE L D (1989). Evaluation of possible fluoride injury to vegetation in the vicinity of an industrial site near Cape Town. *South African Journal of Science* **85**, 741–5.

BROWN G and MITCHELL D T (1986). Influence of fire on the soil phosphorus status in Sand Plain Lowland Fynbos, south western Cape. *South African Journal of Botany* **52**, 67–72.

BROWN G, MITCHELL D T and STOCK W D (1984). Atmospheric deposition of phosphorus in a coastal fynbos ecosystem of the south western Cape, South Africa. *Journal of Ecology* **72**, 547–51.

BUWALDA J G and GOH K M (1982). Host-fungus competition for carbon as a cause of growth depression in vesicular-arbuscular mycorrhizal ryegrass. *Soil Biology and Biochemistry* **14**, 103–6.

CHAPMAN S B, ROSE R J and CLARKE R T (1989). A model of the phosphorus dynamics of *Calluna* heathland. *Journal of Ecology* **77**, 35–48.

DEAN S J, HOLMES P M and WEISS P W (1986). Seed biology of invasive alien plants in South Africa and Namibia. In *The ecology of biological invasions in southern Africa.* (eds Macdonald I A W, Kruger F J and Ferrar A A) Oxford University Press, Cape Town, 157–70.

ESLER K J, COWLING R M, WITKOWSKI E T F and MUSTART P L (1989). Reproductive traits and accumulation of nitrogen and phosphorus during the development of fruits of *Protea compacta* R. Br. (Calcifuge) and *Protea obtusifolia* Buek. ex Meisn. (Calcicole). *New Phytologist* **112**, 109–15.

FUGGLE R F (1981). Macro-climatic patterns within the fynbos biome National Programme for Environmental Science, Fynbos Biome Project, Final Report. CSIR, Pretoria.

GLYPHIS J P (1985). Herbivory and tannin polyphenols in mediterranean ecosystems. Ph.D. thesis, University of Cape Town.

GLYPHIS J P and PUTTICK G M (1988). Phenolics in some southern African mediterranean shrubland plants. *Phytochemistry* **27**, 743–51.

GRAY J T and SCHLESINGER W H (1981). Nutrient cycling in mediterranean-type ecosystems. In *Resource use in chaparral and matorral: a comparison of vegetation function in two mediterranean-type ecosystems.* (ed Miller P C) Springer-Verlag, Berlin, 259–89.

GROBBELAAR N, VAN ROOYEN M W and VAN ROOYEN N (1983). A qualitative study of the nodulating ability of legume species: List 6. *South African Journal of Botany* **2**, 329–32.

GROVES R H (1983). Nutrient cycling in Australian heath and South African fynbos. In *Mediterranean-type ecosystems: the role of nutrients.* (eds Kruger F J, Mitchell D T and Jarvis J U M) Springer-Verlag, Berlin, 179–91.

HANSEN A P and PATE J S (1987). Comparative growth and symbiotic performance of seedlings of *Acacia* spp. in defined pot culture or as natural understorey components of a eucalypt forest ecosystem in SW Australia. *Journal of Experimental Botany* **38**, 13–25.

HEDDLE E M and SPECHT R L (1975). Dark Island Heath (Ninety-Mile Plain, South Australia). VIII. The effect of fertilizers on composition and growth, 1950–1960. *Australian Journal of Botany* **23**, 151–64.

HIGGINS K B, LAMB A J and VAN WILGEN B W (1987). Root systems of selected plant species in mesic mountain fynbos in the Jonkershoek Valley, south western Cape Province. *South African Journal of Botany* **53**, 249–57.

JONGENS-ROBERTS S and MITCHELL D T (1986). The distribution of dry mass and phosphorus in an evergreen fynbos shrub species, *Leucospermum parile* (Salisb. ex J Knight) Sweet (Proteaceae), at different stages of development. *New Phytologist* **103**, 669–83.

JOUBERT J G V and STINDT H W (1979). The nutritive value of natural pastures in the district of Swellendam in the winter rainfall area of the Republic of South Africa. Department of Agricultural Technical Services, Technical Communication, 156.

KRUGER F J (1977). A preliminary account of aerial plant biomass in fynbos communities of the mediterranean-type climate zone of the Cape Province. *Bothalia* **12**, 301–7.

KUO J, HOCKING P J and PATE J S (1982). Nutrient reserves in seeds of selected Proteaceous species from south-western Australia. *Australian Journal of Botany* **30**, 231–49.

LAMB A J and KLAUSSNER E (1988). Response of the fynbos shrubs *Protea repens* and *Erica plukenetii* to low levels of nitrogen and phosphorus applications. *South African Journal of Botany* **59**, 558–64.

LAMONT B B (1981). Specialized roots of non-symbiotic origin in Heathlands. *Heathlands and Related Shrublands of the World B. Analytical Studies.* (ed Specht R L) Elsevier, Amsterdam, 83–195.

LOVELESS A R (1962). Further evidence to support a nutritional interpretation of sclerophylly. *Annals of Botany* **26**, 551–61.

LOW A B (1983). Phytomass and major nutrient pools in an 11-year post-fire coastal fynbos community. *South African Journal of Botany* **2**, 98–104.

LOW A B (1988). Do Australian Acacias alter natural soils in the Western Cape, South Africa? In *Time scales and water stress.* (eds Di Castri F, Moret C, Rambal S and Roy J). Proceedings of the 5th International Conference on Mediterranean Ecosystems. I.U.B.S., Paris, 637–42.

MAGGS J and PEARSON C H (1977). Litterfall and litter layer decay in coastal scrub at Sydney, Australia. *Oecologia* **31** 239–50.

MILLER P C, MILLER J M and MILLER P M (1983). Seasonal progression of plant water relations in fynbos in the Western Cape Province, South Africa. *Oecologia* **53**, 392–6.

MILTON S J (1981). Litterfall of the exotic acacias in the south western Cape. *Journal of South African Botany* **47**, 147–55.

MITCHELL D T, COLEY P G F, WEBB S and ALLSOPP N (1986). Litterfall and decomposition processes in the coastal fynbos vegetation, south-western Cape, South Africa. *Journal of Ecology* **74**, 977–93.

MITCHELL D T and COLEY P G F (1987). Litter production and decomposition of *Protea repens* growing in Sand Plain Lowland and Mountain Fynbos, south western Cape. *South African Journal of Botany*, **53**, 25–37.

MOONEY H A, KUMMEROW J, JOHNSON A W, PARSONS D J, KEELEY S, HOFFMAN A, HAYS R I, GILIBERTO J and CHU C (1977). The producers — their resources and adaptive responses. In *Convergent evolution in Chile and California: mediterranean climate ecosystems.* (ed Mooney H A) Dowden, Hutchinson and Ross, Strandsburg, Pennsylvania, 1–12.

MUSIL C F and MIDGLEY G F (1990). The relative impact of invasive Australian acacias, fire and season on the soil chemical status of a sand plain lowland fynbos community. *South African Journal of Botany* **56**, 417–19.

NAKOS G (1977). Acetylene reduction (N_2-fixation) by nodules of *Acacia cyanophylla. Soil Biology and Biochemistry* **9**, 131–3.

O'CALLAGHAN M (1981). Preburnt and postburnt fynbos standing crop at Pella, Cape. Honours thesis, Department of Botany, University of Cape Town.

O'NEILL R V, DeANGELIS D L, WAIDE J B and ALLEN T F H (1986). *A hierarchical concept of ecosystems.* Princeton University Press, Princeton, New Jersey.

ODUM E P (1969). The strategy of ecosystem development. *Science* **164**, 262–70.

ODUM E P (1986). Introductory review: perspective of ecosystem theory and application. In *Ecosystem theory and application.* (ed Polunin N) John Wiley and Sons, New York, 1–11.

OLSON J S (1963). Energy storage and the balance of producers and decomposers in ecological systems. *Ecology* **44**, 322–31.

PIERCE S M (1984). A synthesis of plant phenology in the Fynbos Biome. *South African National Scientific Programmes Report No.* **88**.

REID P C M and DE JAGER J M (1988). Geographical distribution of monthly mean daily global solar radiation over South Africa. *South African Journal of Plant and Soil* **6**, 46–9.

RICE E L and PANCHOLY S K (1972). Inhibition of nitrification by climax ecosystems. *American Journal of Botany* **59**, 1 033–40.

RICHARDSON D M, VAN WILGEN B W and MITCHELL D T (1987). Aspects of the reproductive ecology of four Australian *Hakea* species (Proteaceae) in South Africa. *Oecologia* **71**, 345–54.

RICHARDSON D M and KRUGER F J (1990). Water relations and photosynthetic characteristics of selected trees and shrubs of riparian and hillslope habitats in the south-western Cape Province, South Africa. *South African Journal of Botany* **56**, 214–26.

RUTHERFORD M C (1978). Karoo-fynbos biomass along an elevational gradient in the western Cape. *Bothalia* **12**, 555–60.

RUTHERFORD M C and DE BOSENBERG J W (1988). Some responses of indigenous Western Cape vegetation to the Australian invasive, *Acacia cyclops.* In *Time scales and water stress.* (eds Di Castri F, Floret O, Rambal S and Roy J) Proceedings of the 5th International Conference on Mediterranean Ecosystems. I.U.B.S., Paris, 631–6.

SCHLESINGER W H and HASEY M M (1981). Decomposition of chaparral shrub foliage: losses of organic and inorganic constituents from deciduous and evergreen leaves. *Ecology* **62**, 762–74.

SHEARER G and KOHL D H (1986). N_2 fixation in field settings: estimations based on natural ^{15}N abundance. *Australian Journal of Plant Physiology* **13**, 699–756.

SMALL E (1972). Photosynthetic rates in relation to nitrogen recycling as an adaptation to nutrient deficiency in peat bog plants. *Canadian Journal of Botany* **50**, 2 227–33.

SOMMERVILLE J E M (1983). Aspects of coastal fynbos phenology. M.Sc. thesis, University of Cape Town.

SPECHT R L (1969). A comparison of the sclerophyllous vegetation characteristics of mediterranean-type climates in France, California and Southern Australia. II. Dry matter, energy and nutrient accumulation. *Australian Journal of Botany* **17**, 293–308.

SPECHT R L (1981). Nutrient release from decomposing leaf litter of *Banksia ornata*, Dark Island heathland, South Australia. *Australian Journal of Ecology* **6**, 59–63.

STAPHORST J L and STRIJDOM B W (1975). Specificity in the rhizobium symbiosis of *Aspalathus linearis* (Burm. Fil.) R. Dahlgr. ssp. *linearis. Phytophylactica* **7**, 95–6.

STEINBERGER Y, SCHMIDA A and WHITFORD W G (1990). Decomposition along a rainfall gradient in the Judean desert, Israel. *Oecologia* **82**, 322–4.

STOCK W D (1985). An investigation of nitrogen cycling processes in a coastal fynbos ecosystem in the south western Cape Province, South Africa. Ph.D. thesis, University of Cape Town.

STOCK W D and LEWIS O A M (1986a). Soil nitrogen and the role of fire as a mineralizing agent in a South African coastal fynbos ecosystem. *Journal of Ecology* **74**, 317–28.

STOCK W D and LEWIS O A M (1986b). Atmospheric input of nitrogen to a coastal fynbos ecosystem of the south western Cape Province, South Africa. *South African Journal of Botany* **52**, 273–6.

STOCK W D, SOMMERVILLE J E M and LEWIS O A M (1987). Seasonal allocation of dry mass and nitrogen in a fynbos endemic Restionaceae species *Thamnochortus punctatus* Pill. *Oecologia* **72**, 315–20.

STOCK W D, LEWIS O A M and ALLSOPP N (1988). Soil nitro-

gen mineralization in a coastal fynbos succession. *Plant and Soil* **106**, 295–8.
STOCK W D, PATE J S and DELFS J (1990). Influence of seed size and quality on seedling development under low nutrient conditions in five Australian and South African members of the Proteaceae. *Journal of Ecology* **78**, 1005–20.
STRIBLEY D P and READ D J (1974). The biology of mycorrhiza in the Ericaceae III. Movement of carbon-14 from host to fungus. *New Phytologist* **73**, 731–41.
TILMAN D (1986). Evolution and differentiation in terrestrial plant communities: the importance of the soil resource: light gradient. In *Community Ecology.* (eds Case T and Diamond J) Harper and Row, New York, 359–80.
VAN DER HEYDEN F and LEWIS O A M (1989). Seasonal variation in photosynthetic capacity with respect to plant water status of five species of the mediterranean climate region of South Africa. *South African Journal of Botany* **55**, 509–15.
VAN WILGEN B W and LE MAITRE D (1981). Preliminary estimates of nutrient levels in fynbos vegetation and the role of fire in nutrient cycling. *South African Forestry Journal* **119**, 24–8.
VAN WILGEN B W and RICHARDSON D M (1985). The effects on alien shrub invasions on vegetation structure and fire behaviour in South African fynbos shrublands: a simulation study. *Journal of Applied Ecology* **22**, 955–66.
VAN WYK D B, LESCH W and STOCK W D (1991). The effect of fire on stream water quality and nutrient budgets of the Swartboskloof catchment. In *Swartboskloof: a fynbos catchment.* (eds Van Wilgen B, Kruger F J and Van Hensbergen H J) Springer-Verlag, Berlin (in press).
VERSVELD D B and VAN WILGEN B W (1986). Impact of woody aliens on ecosystem properties. In *The ecology and management of biological invasions in Southern Africa.* (eds Macdonald I A W, Kruger F J and Ferrar A A) Oxford University Press, Cape Town, 239–47.
VITOUSEK P (1982). Nutrient cycling and nutrient use efficiency. *American Naturalist* **119**, 553–72.
VITOUSEK P M and REINERS W A (1975). Ecosystem succession and nutrient retention: a hypothesis. *Bioscience* **25**, 376–81.
VITOUSEK P M and SANFORD R L Jr. (1986). Nutrient cycling in moist tropical forest. *Annual Review of Ecology and Systematics* **17**, 137–67.
VITOUSEK P M and WALKER L R (1989). Biological invasion by *Myrica faya* in Hawaii: plant demography, nitrogen fixation, ecosystem effects. *Ecological Monographs* **59**, 247–65.
VON WILLERT D J, HEPPICH M and MILLER J M (1989). Photosynthetic characteristics and leaf water relations of mountain fynbos vegetation in the Cedarberg area (South Africa). *South African Journal of Botany* **55**, 288–98.
WHITTAKER R H (1970). *Communities and ecosystems.* The Macmillan Company, Collier-Macmillan Limited, London.
WITKOWSKI E T F (1988). Response of a sand-plain lowland fynbos ecosystem to nutrient additions. Ph.D. thesis, University of Cape Town.
WITKOWSKI E T F (1989a). Response to nutrient additions by the plant growth forms of sand-plain lowland fynbos, South Africa. *Vegetatio* **79**, 89–97.
WITKOWSKI E T F (1989b). Effects of nutrient additions on litter production and nutrient return in a nutrient-poor Cape fynbos ecosystem. *Plant and Soil* **117**, 227–35.
WITKOWSKI E T F (1991b). Effects of invasive alien acacias on nutrient cycling in the coastal lowlands of the Cape fynbos. *Journal of Applied Ecology* **28**, 1–15.
WITKOWSKI E T F and MITCHELL D T (1987). Variations in soil phosphorus in the fynbos biome, South Africa. *Journal of Ecology* **75**, 1 159–71.
WITKOWSKI E T F and MITCHELL D T (1989). The effects of nutrient additions on above-ground phytomass and its phosphorus and nitrogen contents of sand-plain lowland fynbos. *South African Journal of Botany* **55**, 243–9.
WITKOWSKI E T F, MITCHELL D T and STOCK W D (1990a). Response of a Cape fynbos ecosystem to nutrient additions: nutrient dynamics in fertilised soils. *Acta Oecologica* **11**, 165–79.
WITKOWSKI E T F, MITCHELL D T and STOCK W D (1990b). Response of a Cape fynbos ecosystem to nutrient additions: shoot growth and nutrient contents of a proteoid (*Leucospermum parile*) and an ericoid (*Phylica cephalantha*) evergreen shrub. *Acta Oecologica* **11**, 311–26.
WITKOWSKI E T F (1991a). Growth and competition between seedlings of *Protea repens* and the alien invasive *Acacia saligna*, in relation to nutrient availability. *Functional Ecology* **5**, 101–10.

11

Human settlement

H J Deacon

INTRODUCTION

The human species is one of many that have successfully invaded the fynbos biome. Although other organisms may be more numerous, people have an importance that belies numbers because of the impact of their activities on the environment. The purpose of this chapter is to outline the archaeological and historical evidence for human settlement and to indicate some of the ways in which people have functioned as agents of ecological change.

The fynbos biome, with its mild climate and scenic setting, is a favoured human habitat. As populations have grown and urbanization has increased, the need to minimize the negative impacts and to maintain and develop the quality of life has become a priority. Lacking rich sources of exploitable minerals and fossil fuels, there has not been the same level of primary industrial development as is found in other growth areas of South Africa. This poses problems for sustaining economic prosperity. Socioeconomic factors that have a wider frame of reference than the region will dictate future developments but, much as in California, the attractiveness of the environment — natural and cultural — will act as a drawcard. A major consideration becomes how to balance the conservation of the environment and the inevitability of increasing demands for space, facilities, and services. Although rapid changes are taking place, they are evolutionary rather than revolutionary and the past is of 'vital importance' (Christopher 1976) to understanding the present.

INITIAL SETTLEMENT

Africa south of the Sahara is the biogeographic region where humans evolved (Tuttle 1988). The early evolutionary stages were largely confined to tropical latitudes but the range expanded to higher latitudes with the appearance of archaic forms of *Homo sapiens.* It is from about half a million years ago that dispersion included the southernmost part of the African continent that constitutes the fynbos biome.

The soils of the fynbos biome are normally too acid to preserve bone and for this reason human fossils other than those from recent times are rare. More durable markers of people are stone artefacts, and patches of flaked stone map the locations used by Stone Age people.

The oldest sets of artefacts that are found in the fynbos landscape are those of the Earlier Stone Age and are associated with the Acheulian industry. Similar artefacts are widespread in Africa outside forested zones and were made over a period of almost a million years, ending perhaps 200 000 years ago (Deacon 1975). The type locality where such artefacts were first studied in detail at the turn of the century was Stellenbosch and the artefacts were seen as significant in demonstrating the high antiquity of people living in South Africa (Peringuey and Corstophine 1900). The Stellenbosch finds along the Eerste River valley are typical of occurrences in the intermontane valleys of the Cape Fold Mountains. They are restricted to lower elevations and with one exception — Montagu Cave — to open sites. Settlement also included the whole of the coastal platform.

One of the more informative Acheulian sites is in dune sands on the farm Elandsfontein between the town of Hopefield and the Langebaan Lagoon. The vegetation is fynbos growing on an acid sand mantle that overlies a more calcareous sand in which a rich fossil mammal fauna and Acheulian artefacts are preserved (Singer and Wymer 1968). The concentration of fauna and artefacts is due to deflation in the past exposing the water table, thus creating interdune pools that attracted animals and people. Higher water tables and vegetation cover in subsequent times, inferred from soil profiles, suggest that fynbos has expanded over the site at the expense of thicket vegetation.

Among the fossils found is a human skull with very prominent brow ridges and a marked post-orbital constriction (Figure 11.1). It is archaic in these attributes and along with the Kabwe skull from Zambia represents the ancestral Middle Pleistocene (700 000–125 000 years) populations inhabiting this part of Africa (Rightmire 1988). In this context it is difficult to disentangle the activities of humans from those of carnivores and both would have contributed to the finds made there. The fossil fauna is typically that of the African Middle Pleistocene and a diverse large herbivore community is present. Giraffe, pigs, elephant, buffalo, and a variety of grazing and browsing antelope are represented by genera that may be extant but by species largely ancestral to modern forms (Klein 1983). The site provides the last recorded presence in the region of the short-necked giraffe (*Sivatherium*), a sabre-toothed cat (*Megantereon*), and a gelada baboon (*Theropithecus*) almost the size of a person (Hendey and Deacon 1977). Situated on the edge of the coastal sands, the fauna shows the animals that ranged between the parkland on the plains of the coastal platform inland and what was then a wider belt of thicket along the coast. The Middle Pleistocene environments may have been richer than in subsequent times because thereafter people played an increasingly important role in environmental change.

Before the beginning of the Late Pleistocene, 125 000 years ago, the Acheulian tool kit had been replaced by Middle Stone Age technology characterized by flake blades. Middle Stone Age artefact occurrences are more widespread than those of Acheulian times and occupations are recorded at open stations as well as in rock shelters, some at high altitudes in the mountains (Deacon 1989). The food resources that necessitated Middle Stone Age people 'mapping' themselves on to the landscape in this way would have been geophytes — plants with underground corms or bulbs like *Watsonia*. Much of the productivity in fynbos ecosystems is in plants with underground storage organs. Geophytes are most prominent in the early post-fire succession (Le Maitre and Midgley this volume) and patches of geophytes will have a sustained yield only if regularly burnt. The implication is that from at least the beginning of the Late Pleistocene, fire was being used as a tool to farm the fynbos. There are abundant hearths in human occupation sites of this time showing the ability to make fire at will. Corms of geophytes are rich in carbohydrates but low in fats and protein, however the latter were obtainable from other plants, shellfish, and meat.

The Klasies River main site on the Tsitsikamma coast provides an important 20 metre sequence through the earlier half of the Late Pleistocene between 120 000 and 60 000 years ago. This is a rare situation where a local source of alkaline ground water has permitted the preservation of bone and shell as well as other artefacts in this time range. The main site is notable because fragmentary human fossils that are anatomically modern (Singer and Wymer 1982; Deacon 1988) come mainly from two horizons dated to 90 000 and 120 000 years and are among the oldest known anywhere. The dating is important support for the hypothesis that modern people evolved in an African centre (Stringer and Andrews 1988). The human remains are mainly cranial parts and show cut and percussion marks that are consistent with cannibalism (White 1987). In these latitudes where there is no chronic shortage of protein, cannibalism seems anomalous and may have been linked to social disruption as was the episode of cannibalism in the nineteenth century in the Drakensberg area. In the organization of activities around domestic hearths, the patterned disposal of food waste, and the way in which artefacts were used as symbols, the behaviour of these Pleistocene Stone Age groups is different in detail but not in kind from their Holocene Bushman successors. They were anatomically and behaviourally modern and were involved in managing fynbos ecosystems.

A large proportion of Middle Stone Age occurrences date to the Last Interglacial *sensu lato* (130 000–75 000 years ago) and many excavated sequences show a discontinuity between these earlier occupations and those of the end of the Pleistocene. The discontinuities

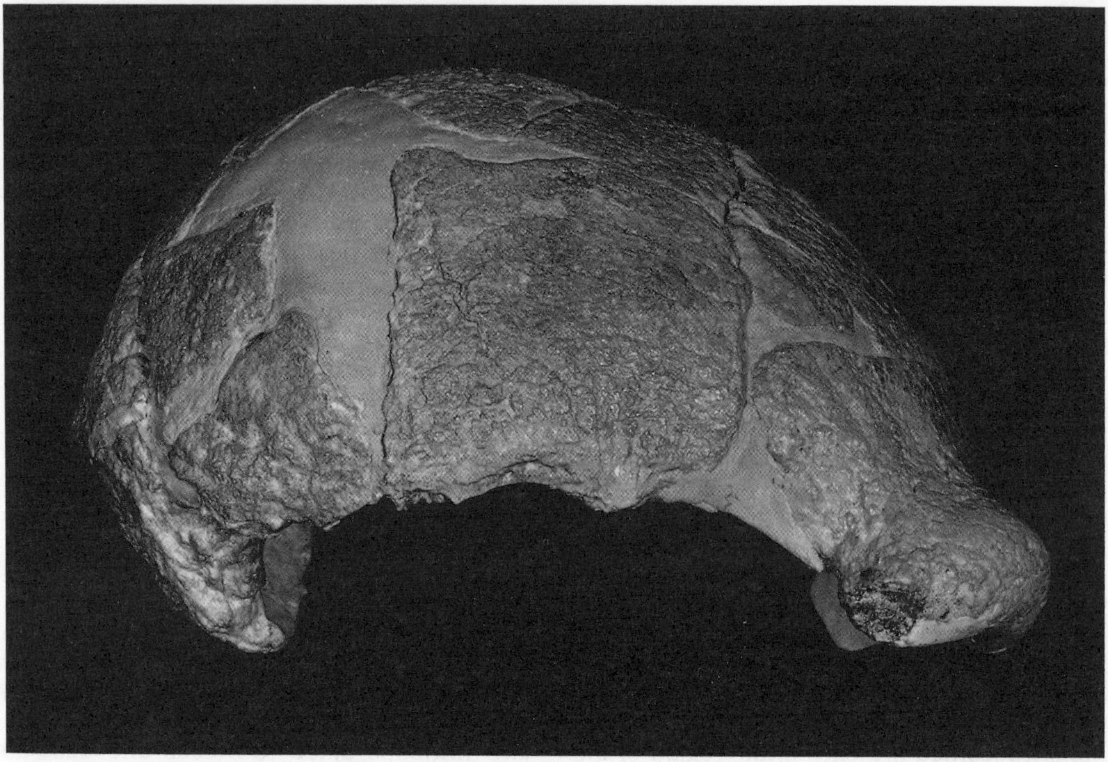

FIGURE 11.1 Saldanha skull: the oldest human fossil from the fynbos region (photograph of original Saldanha calvarium in the South African Museum Collections) (Photo: C Booth).

represent an extended period of reduced archaeological visibility and thus a reduction in the density of people inhabiting not only the fynbos biome but the whole subcontinent. The timing suggests climatic changes of the Late Pleistocene were a forcing factor (Deacon and Thackeray 1984). The implications are that after initial settlement, there was not a simple exponential increase in the number of people living in the region, and that in the recent geological history, some periods have been more favourable than others for the maintenance of human populations.

SETTLEMENT IN END PLEISTOCENE AND HOLOCENE TIMES

Hunter-gatherers of the Later Stone Age occupied the fynbos biome from 21 000 years ago. The stone tool technology shows some reduction in the size of flake blanks that may relate to the introduction of the bow, and new artefacts like polished bone tools and personal ornaments occur (Deacon 1984). There is striking similarity in items of material culture, like the use of ostrich eggshell ornaments and engraved water containers and tortoiseshell bowls, between Later Stone Age people living in the fynbos biome and extant hunter-gatherers in Botswana. This suggests a close correspondence in economic and social organization as well as ideology. The Later Stone Age inhabitants of the fynbos would have been part of a pan-Bushman lifestyle that showed little regional differentiation in southern Africa.

During the Last Glacial Maximum (18 000 years ago) — the most extreme climatic conditions of the recent past — people were able to live in the fynbos biome, although much of the arid interior of South Africa was effectively depopulated. In the intermontane valleys, on the coastal forelands, and on the continental shelf exposed by sea levels lowered by maximally 120 metres there were zebra, wildebeest, and eland and these were among the principal species hunted. The zebrine (Equus capensis), the hartebeest (Megalotragus priscus), the giant buffalo (Pelorovis antiquus) — all large herbivores that became extinct at the end of the Pleistocene (Klein 1983) — were components in this fauna. Their importance would have been as large mowers increasing productivity of the grazing (Brink 1987) and their extinction can be ascribed to environmental change and human activities like veld burning. With changes in the biota went changes in the lifestyle of the people.

The amelioration of climates after the Last Glacial Maximum started as early as 14 000 years ago but it is in the Holocene (10 000 years ago) that archaeological sites are more numerous. Evidence from these sites suggest territorially organized groups of hunter-gatherers functioning as dedicated patch foragers. They probably used fire-stick farming to encourage natural fields of geophytes, and snared small antelope in the surrounds of these patches. One proxy measure for population increase is the number of radiocarbon dates per millennium obtained for the region. Many more younger occupation events have been recorded and dated through archaeological research. This is consistent with a significant growth in population in the Holocene (Figure 11.2). In the richer habitats of the southern Cape, there are large numbers of burials (FitzSimons 1926; Hall and Binneman 1987) that may have served to legitimize access to resources. The picture that emerges towards the end of the Holocene is of a Stone Age population of hunter-gatherers close to the carrying capacity for the region.

Two thousand years ago herding became established as a competing way of life in the fynbos biome. The centre of domestication of sheep and cattle was in the north African littoral and Saharan uplands and it took almost 6 000 years for the transmission of stock and stock-keeping practices to travel the length of the continent (Deacon 1984). In southern Africa herding is related to the spread of people speaking Hottentot or Khoikhoi dialects related to the central Kalahari languages spoken by Bushmen. On linguistic grounds eastern Botswana-western Zimbabwe is indicated as the core area from which herding practices were transmitted (Westphal 1963) through southern Africa and initially herding appears to have followed a pathway along the Namibian and Cape coasts, with later movements along the Orange River.

Sheep farming may have preceded the introduction of cattle by several hundred years on evidence from the site of De Kelders near Gansbaai (Schweitzer 1979). Characteristic bag-shaped pottery with lugs for suspension on draught oxen occur at Boomplaas by 300 AD. The evidence from the latter site is of the early herding communities using a stock post system to disperse their flocks and culling upwards of 75% of lambs as in modern market-orientated sheep farming (Von den Driesch and Deacon 1985). Stock farming was practised on an intensive scale from an early time in areas of suitable grazing. Cattle as an important component are seen at Kasteelberg near Saldanha at least by 800 AD (Klein and Cruz-Uribe 1989) and in the contact period, from the 1590s when external trade became established, large herds are recorded. An example is an estimated 20 000 cattle as well as sheep which were driven into the Salt River area bordering Table Bay in the first half of December 1652. Herds of this size could not be kept long in one place and these animals were driven off again in the same month (Thom 1952, 1954). An estimated population of 50 000 Khoikhoi (Elphick and Malherbe 1989) in the south-western Cape in the mid-seventeenth century and ratios that ranged as high as 5–10 cattle and up to 20 sheep per individual (Thom 1952) give figures of thousands of head of stock being pastured mainly on the Cape forelands.

Right of access to diverse resource-rich patches was the objective of Later Stone Age foragers. Stock keeping constrained Khoikhoi settlement to areas of sufficient grazing and effectively to certain terrain types. Conflict between foragers, known to Van Riebeeck as the Sonqua (Thom 1952), and the Khoikhoi herders would have a greater time depth than that recorded at the beginnings of colonialism. Documenting the history of this interaction is an archaeological problem that is being actively researched (Parkington and Hall 1987). Of relevance here is the impact of 1 500 years of herder settlement that followed on and overlapped with a much greater time depth of hunter-gatherer settlement. It is too simplistic to equate significant anthropogenic

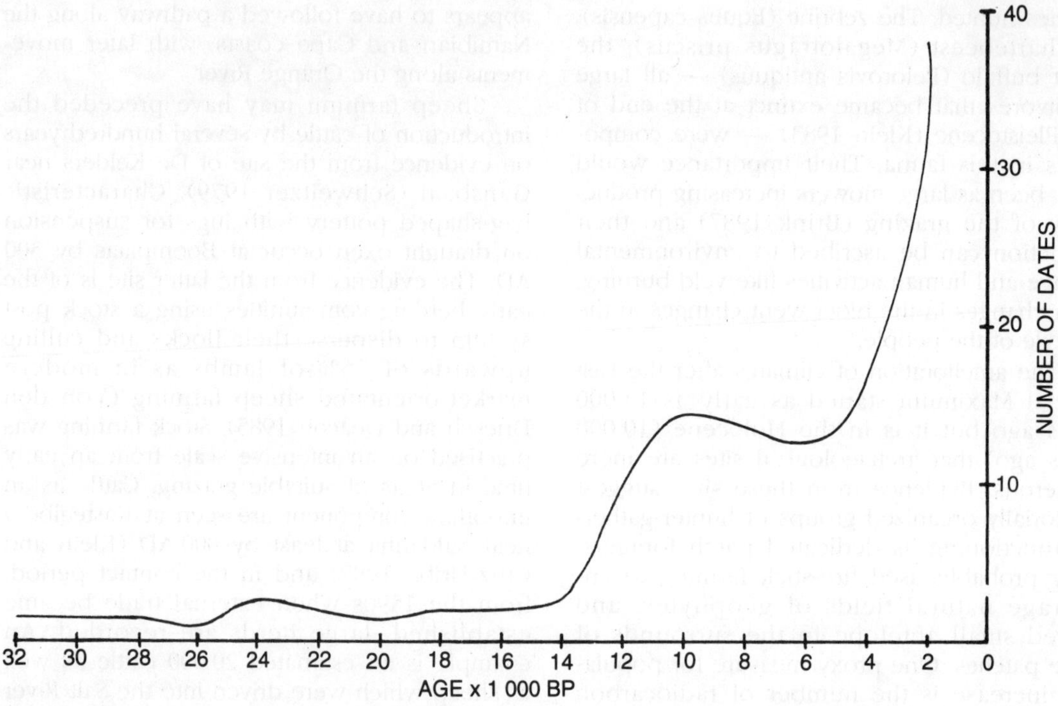

FIGURE 11.2 The number of radiocarbon-dated archaeological observations for the fynbos region is a proxy measure of the increase in pre-colonial human populations notably after 16 000 BP.

impacts on fynbos ecosystems exclusively with modern times. The impacts, although not quantified, involved decreased intervals between fires relative to the natural fire regime and increased soil erosion through impairment of the vegetation cover. Later impacts have been incremental.

COLONIAL SETTLEMENT

After Portuguese pilots had pioneered the route, ships of various nations plied the coastal waters and stopped over at the Cape (Raven-Hart 1967). Supplies of water, wood, and game, in addition to livestock bartered from the Khoikhoi, were available. It was to formalize and control this trade in the face of English competition that in 1652 the Dutch East India Company set up a supply station at the Cape (Figure 11.3). This decision had significant consequences for the Khoikhoi and Bushman inhabitants, for the future history of southern Africa, and for the ecology of the fynbos biome.

The influx of trade goods had increased competition and conflict between the Khoikhoi clans that was compounded by the intervention of the Company. The Khoikhoi were not united in their resistance to the occupation of their traditional lands and a rapid breakdown of the social and economic system followed after 1652 (Elphick 1977). The ravages of introduced epidemic diseases like smallpox further weakened the response and within decades the Khoikhoi ceased to be a political force and were progressively reduced to landless serfs (De Kiewiet 1957). Khoikhoi resistance to European contact that started in 1488 with stone throwing when Diaz came ashore at Mossel Bay (Axelson 1973) was manifest through the succeeding centuries and rigorously suppressed. In the new colonial order that emerged, however, elements of Khoikhoi plant and animal lore, nomadism,

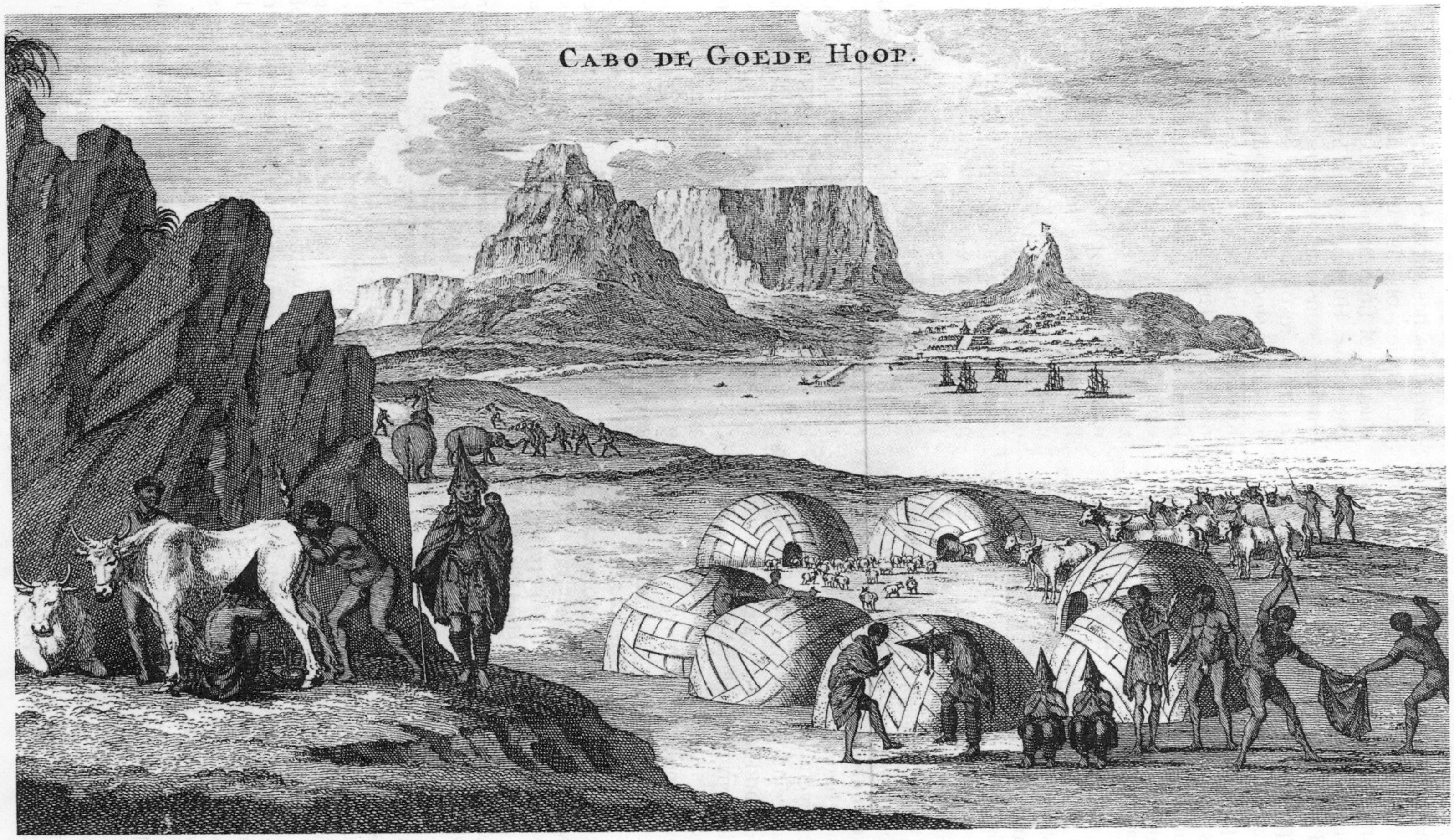

FIGURE 11.3 Cabo de Goede Hoop: Abraham Bogaert's illustration of the Cape in 1702 showing activities around a Khoikhoi kraal, an elephant hunt in the middle distance, and the settlement below Table Mountain (Raven-Hart 1970).

and veld management practices were integrated with and facilitated the expansion of the European farming frontier.

Through making land grants to immigrant settlers and employees, what was strictly a trading company became an agent of colonization. The Company was not geared to administer a settler population that increased not only in numbers but also in aspirations and in demands for resources, particularly land. In Company times populations were small. There were some 3 000 free people and slightly over 4 000 slaves recorded in the 1731 census (Armstrong and Worden 1989). In 1806 at the beginning of the second British occupation of the Cape, and within the boundaries of the time, the population consisted of almost equal numbers of Khoikhoi, slaves, and Europeans and was some 75 000 persons (Christopher 1976). Half this population was concentrated around Cape Town and the Boland. These numbers are minuscule when compared to comtemporary figures and settlement density remained relatively low prior to urbanization in the present century. However, as settlement was dispersed and aimed at extracting maximum benefit from the environment, the ecological impact was significant.

The movement of the farming frontier has attracted the interest of historical geographers (Christopher 1976, 1982; Guelke and Shell 1983; Guelke 1989). After initial colonization, expansion was rapid. The Company ceased to grant freehold land in 1717 but permits for loan farms continued to be issued freely. By 1760 the area farmed extended over the whole fynbos biome. Land use was extensive rather than intensive. Expansion of the farming frontier was driven by the availability of land at nominal cost and natural resources like game and timber that substituted for capital. Proximity to the market was a major factor in determining that the cultivation of cereals and viticulture was restricted to arable lands on the freehold farms in the south-western Cape. This corresponds to the area of mediterranean-type climate *sensu stricto* and is the area associated with predominantly base-rich substrates (Deacon et al. this volume). Stock raising became the main activity on the leased lands on the frontier.

Africa is the only continent on which something of the diversity of Pleistocene large mammals has survived into the present. The decimation of this fauna followed the introduction of firearms and the process started at the Cape. Hunting was a source of income through skins and ivory as well as a *raison d'être* to reconnoitre new settlements and to prosecute stock barter arrangements with the Khoikhoi, enforced by strength of arms if necessary. The survival of the large mammal fauna was not consistent with farming and as early as 1656 Van Riebeeck's journal records that bounty was paid for the extermination of carnivores (Thom 1954). The rhinoceros, hippopotamus, the red hartebeest, and the endemic sable-like blue antelope became extinct and other species like the elephant, the eland, the Cape buffalo, the bontebok, and the mountain zebra were reduced to relict herds. Timber was another natural resource that seemed plentiful in 1652 but six years later a proclamation was issued restricting the cutting of yellowwood (Podocarpus spp.) (Thom 1954). There was a consistent demand for larger building timbers in Cape Town where up to a third of the early colonial population was housed. Afromontane forest patches east of Cape Town, particularly in the Swellendam area, were heavily exploited in the next century. It is probably distance and the absence of any safe port that saved the main extent of the Afromontane forests in the Knysna area from the same degree of over-exploitation. The disturbance of the indigenous forest ecosystems was all the greater because the preferred species were not dominant.

The climate at the Cape was not suitable for the cultivation of the African cereals (Deacon 1986 for references) and this may explain why cereal cultivation was not established earlier. Eurasian cultigens and the plough were introduced by the Company. Wheat, rye, barley, and grapes for wine making became the main crops. Within the first decade of the eighteenth century attempts were being made to curb agricultural production on arable lands by halting immigration. This was because of the difficulty of finding markets for the produce. The arable farming frontier thus moved slowly.

Wine was the principal export of the colony from its beginnings until the middle of the last century. Viticulture was controlled by a slave-owning landed gentry (Guelke and Shell

1983). The social and economic importance of this influential group is symbolized in the homesteads that grew more sumptuous as the power of the Company declined. The gentry were able to derive initial benefit from marketing opportunities offered by British rule but in the nineteenth century they stood to lose most from the emancipation of slaves and competition with other wine exporting countries. At the end of the nineteenth century, deciduous fruit became an alternative crop, a hedge against the overproduction of wine. This led to a significant increase in the area of arable land under cultivation and the arable farming frontier expanded, notably into the intermontane valleys. Wheat growing also contributed further to the movement of the arable farming frontier as the population of the country increased. The progressive intensification of agriculture on arable lands has been aided by mechanization allowing the cultivation of larger fields and deeper ploughing to bring otherwise marginal lands into use. The diversity and extent of habitats necessary to support communities of indigenous plants and animals have been reduced as a consequence of the area disturbed by ploughing. The conservationist is confronted with the problem of determining what areas should be set aside for protection and how large such areas should be (Bond 1989; Rebelo this volume).

European settlement introduced a system of individual ownership that was new in this context. Although individual holdings were large in relation to contemporary farm units in Europe, they were smaller units of land than the communal holdings of the Khoikhoi. The potential for overstocking and overgrazing and mismanagement through too frequent burning was increased. The results are perhaps most evident in the renoster shrublands — the anthropogenically induced vegetation that replaced grassland and thicket on the base richer substrates over wide areas of the fynbos biome. Like the Khoikhoi before them, the Company and independent farmers kept many more sheep than cattle. In 1695 the Company had a herd of 1 005 cattle and 5 582 sheep and the colonists owned 47 447 sheep apart from cattle (Thom 1936). From 1700 the Company reduced its direct involvement in stock raising thereby encouraging private enterprise and the movement of the stock keeping frontier. In the next 15 years the number of small stock trebled and may have numbered over a million by the end of the century. While meat production remained the main source of income the native breeds, which fared better on unimproved pastures, were favoured. Replacement by exotic breeds began at the end of the Company rule and most significant was the introduction of wool sheep. Thom (1936) suggests it was the merino that saved the Cape colony from bankruptcy in the middle of the last century and wool production financed the opening up of the interior of the subcontinent to European settlement. The prosperity generated at this time can be linked to an improved transport network and the growth of country towns. The history of European expansion in southern Africa is intimately tied to the quest for title to land for raising stock (Christopher 1976). This can be seen as a means to avoid poverty rather than to generate material wealth and served ultimately to preserve class status in a society in which class and race divisions were reinforcing. The ideology linked to the pastoral frontier had ecological repercussions in that there was limited incentive to reduce stocking levels through investment in selective breeding.

The European colonization of the Cape was a horticultural event because a new and different southern flora became accessible to northern hemisphere gardeners. While tied to the Company rule and relatively isolated, the scale of introduction of alien species to the Cape, apart from ornamental shrubs and trees, was limited to cultivars and the weeds of cultivation. This changed when the Cape became part of a wide flung British trading empire and a staging post for the colonization of Australia and New Zealand. It is largely through such contacts in the last century that there was an interchange of plants, but no animals, between the two southern continents. Woody Australian species have proved effective colonizers (Richardson et al. this volume); the converse is true in Australia where it is the more herbaceous plants introduced from the Cape that have become invasive. Some North American and European pines introduced to provide for timber needs have also shown invasive vigour and few indigenous plants or animals have adapted to the pine plantation habitat.

The planting of crops or pastures creates

ecosystems that are less diverse than the natural ecosystems they replace. They need husbanding and nurturing to retain soil fertility and to control weeds and animal pests. As agriculture has intensified, it has become more reliant on chemicals for fertilizers and pesticides. The fruit industry is heavily dependent on chemicals to control fruit fly and the codling moth and provides an example of the difficulty of managing artificial ecosystems (Smit 1964). Given the current perturbed state of artificial ecosystems in the fynbos biome, chemical and biological controls are necessary to maintain production but pressure to make more use of biological controls is mounting.

Colonial settlement initiated many changes in the fynbos landscape. The Company, and the settlers it introduced, had different priorities which became a potential source of conflict. Through the high reliance on natural resources, however, the initial impact was on the natural pastures, timber, and game. The area of cultivation and therefore disturbance remained limited outside the south-western Cape until the present century. The expansion of the pastoral frontier was rapid and uncontrolled. With the strategic importance of the Cape its primary interest, the British administration was no more effective than the Company in containing this frontier within the boundaries of the Cape colony. Through access to markets its administration offered scope for the intensification of agriculture. This can be traced from the fledgling capitalism of the landed gentry on their wine estates to the company-operated large modern concerns and co-operatives.

PERSPECTIVE AND PROSPECTS

A concern of this chapter has been to show that the fynbos biome has been a locus for human activities for upwards of half a million years. No part of the region is pristine or 'natural'. The artificial character is most obvious in the presence of exotic species and in commensuals adapted to living in the built-environment, itself an artefact. This artificial character would take in even the remote mountain catchments because these too are managed by fire to control run-off (Van Wilgen et al. this volume). The human impact has been cumulative and is superimposed on the changes brought about through the natural turnover of species and climatic change. The impact is the result of the consequences of 100 000 years of fire-stick farming, 2 000 years of stock raising, and 300 years of cultivation. Added to this is the disturbance caused by roads, dams, and urban expansion.

It may have been impossible for the Bushman patch foragers to have anticipated the invasion of their domain by Khoikhoi with domestic animals. The Khoikhoi may have been in a better position to have anticipated a clash with the seafaring Europeans who for 150 years came to trade and left, but then in 1652 came and stayed. The 40 freeburgers who were granted farms along the Eerste River at Stellenbosch, if pressed, should have been able to predict the expansion of the farming frontier beyond the Hottentots Holland Mountains if not the continuous urban development across the Cape Flats. Authorities from the Company to modern central and local government have had an equally poor record of anticipating and planning for development. The situation has reached a stage where demands on the environment in terms of production and recreation are less sustainable, the population is growing through natural increase, and significant immigration and basic resources like water are proving to be in short supply.

The human impact will increase in proportion to population growth. It is the demand for services and labour that has fuelled a recent phase of immigration and urban growth. New peri-urban informal settlements in the Cape metropolitan area like Khayelitsha, established in the last decade, house a guestimated quarter to half a million people and surveys indicate current immigration to Khayelitsha alone is 100 persons per day (Seekings 1990). This immigration is not directly from underdeveloped areas like the Transkei but involves the relocation of people from rural and older established urban areas in the western Cape. Still, it is significant that only 11% of the inhabitants of Khayelitsha were born in the region. Projections for the increase in population over the next decade in peri-urban settlements alone are conservatively put at a million people. The growing numbers of new settlers make the total of 158 Huguenots pale into insignificance not the least because the latter had 300 years of unrivalled opportunity to establish themselves. The demographic profile of the fynbos biome has changed and is changing dramatically.

Sustaining economic growth and creating job opportunities for a population that may increase substantially in the next decade will be a problem if living standards are to be maintained and improved. With limited primary resources, it is the intangibles that have to do with the quality of life that become major assets. These can have worth in monetary terms but assessment depends on cultural — that is learnt — values. The preservation of the indigenous biota and the conservation of ecosystems are activities in the realm of intangibles.

The Fynbos Biome Project was launched some 15 years ago to provide a scientific basis for the management of the biome. In the conferences and publications that followed some of the conflicts in needs, concepts, and priorities between the scientist, manager, and public were aired. There was not and could not be a simple consensus within these groups on an agenda for research, conservation, and utilization. What is important, however, is that in the public discourse that was engendered, the fynbos became more than a subject of scientific interest: it became established as a cultural symbol, an intangible of value. It is not solely on the hard won knowledge of fynbos ecosystems that the future conservation status of the region rests. As in all human affairs ideology plays a significant role. It determines how we view the past, our relationship with the physical, biotic, and cultural environment in which we live, and in this context the value we place on the fynbos.

ACKNOWLEDGEMENTS

The research on which this chapter is based has been supported in its various facets by grants from the Centre for Science Development of the Human Sciences Research Council to study the Late Pleistocene archaeology and by the Foundation for Research Development through the Special Programme on Climate Change and the International Geosphere Biosphere Program to study palaeoecology. J Deacon and C Rademeyer steered my reading in useful directions and provided comments on the chapter which are gratefully acknowledged. W R Siegfried kindly made available a copy of an unpublished manuscript on the human impact in the fynbos and supplied some references. The reviewers of the chapter, N G Penn and J Sealy, are thanked for their comments.

REFERENCES

ARMSTRONG J C and WORDEN N A (1989). The slaves, 1652–1834. In *The shaping of South African Society, 1652–1840.* (eds Elphick R and Giliomee H) Maskew Miller Longman, Cape Town, 109–83.

AXELSON E (1973). *Cango to the Cape: early Portuguese explorers.* Faber and Faber, London.

BOND W J (1989). Describing and conserving biotic diversity. In *Biotic diversity in southern Africa: concepts and conservation.* (ed Huntley B J) Oxford University Press, Cape Town, 2–18.

BRINK J S (1987). The archaeozoology of Florisbad, Orange Free State. MA thesis, University of Stellenbosch, Stellenbosch.

CHRISTOPHER A J (1976). *Southern Africa.* Dawson, Folkestone.

CHRISTOPHER A J (1982). Towards a definition of the nineteenth century South African frontier. *South African Geographical Journal* **64**, 97–113.

DEACON H J (1975). Demography, subsistence and culture during the Acheulian in southern Africa. In *After the Australopithecines.* (eds Butzer K W and Isaac G L) Mouton, The Hague, 543–70.

DEACON H J (1988). The origins of anatomically modern people and the South African evidence. *Palaeoecology of Africa* **19**, 193–9.

DEACON H J (1989). Late Pleistocene palaeoecology and archaeology in the southern Cape, South Africa. In *The human revolution: behavioural and biological perspectives on the origins of modern humans.* (eds Mellars P and Stringer C) Edinburgh University Press, Edinburgh, 547–64.

DEACON H J and THACKERAY J F (1984). Late Pleistocene environmental changes and implications for the archaeological record in southern Africa. In *Late Cainozoic palaeoclimates of the Southern Hemisphere.* (ed Vogel J C) Balkema, Rotterdam, 375–90.

DEACON J (1984). Later Stone Age people and their descendants in southern Africa. In *Southern African prehistory and paleoenvironments.* (ed Klein R G) Balkema, Rotterdam, 221–328.

DEACON J (1986). Human settlement in South Africa and archaeological evidence for alien plants and animals. In *The ecology and management of biological invasions in southern Africa.* (eds Macdonald I A, Kruger F J and Ferrar A A) Oxford University Press, Cape Town, 3–19.

DE KIEWIET C W (1957). *A history of South Africa: social and economic.* Oxford University Press, London.

ELPHICK R (1977). *Kraal and castle: Khoikhoi and the founding of white South Africa.* Yale University Press, New Haven.

ELPHICK R and MALHERBE V C (1989). The Khoisan to 1828. In *The shaping of South African society, 1652–1840.* (eds Elphick R and Giliomee H) Maskew Miller Longman, Cape Town, 3–65.

FITZSIMONS F W (1926). Cliff dwellers of Tzitzikama: results of recent excavations. *South African Journal of Science* **23**, 813–17.

GUELKE L (1989). Freehold farmers and frontier settlers, 1657–1780. In *The shaping of South African society, 1652–1840.* (eds Elphick R and Giliomee H) Maskew Miller Longman, Cape Town, 6–108.

GUELKE L and SHELL R (1983). An early colonial land gentry, land and wealth in the Cape Colony 1682–1731. *Journal of Historical Geography* **9**, 265–83.

HALL S and BINNEMAN J (1987). Later Stone Age burial variability in the Cape: a social interpretation. *South African Archaeological Bulletin* **42**, 140–52.

HENDEY Q B and DEACON H J (1977). Studies in palaeontology and archaeology in the Saldanha region. *Transactions of the Royal Society of South Africa* **42**, 371–81.

KLEIN R G (1983). Palaeoenvironmental implications of Quaternary large mammals in the fynbos region. In *Fynbos palaeoecology: a preliminary synthesis.* (eds Deacon H J, Hendey Q B and Lambrechts J J N) *South African National Scientific Programmes Report* **75**, CSIR, Pretoria, 116–38.

KLEIN R G and CRUZ-URIBE (1989). Faunal evidence for prehistoric herder-forager activities at Kasteelberg, western Cape Province, South Africa. In *The South African Archaeological Bulletin* **44**, 82–97.

PARKINGTON J and HALL M (1987). *Papers in the prehistory of the western Cape, South Africa.* British Archaeological Reports, Oxford.

PERINGUEY L and CORSTOPHINE G S (1900). Stone implements from Bushman's Crossing. *South African Philosophical Society Proceedings* **11**, 24.

RAVEN-HART R (1967). *Before van Riebeeck: callers at South Africa from 1488 to 1652.* Struik, Cape Town.

RAVEN-HART R (1970). *Cape Good Hope 1652–1702: the first fifty years of Dutch colonization as seen by callers.* Balkema, Cape Town.

RIGHTMIRE G P (1988). *Homo erectus* and later Middle Pleistocene humans. *Annual Review of Anthropology* **17**, 239–59.

SCHWEITZER F R (1979). Excavations at Die Kelders, Cape Province, South Africa: the Holocene deposits. *Annals of the South African Museum* **78**, 101–233.

SEEKINGS J (1990). *Survey of residential and migration histories of residents of the shack areas of Khayelitsha.* Research unit for the sociology of development, University of Stellenbosch, Occasional Paper **15**, 1–66.

SMIT E (1964). *Insects in South Africa and how to control them.* Oxford University Press, Oxford.

SINGER R and WYMER J (1968). Archaeological investigations at the Saldanha skull site in South Africa. *South African Archaeological Bulletin* **25**, 63–74.

SINGER R and WYMER J (1982). *The archaeology of the Klasies River Mouth Caves.* Chicago University Press, Chicago.

STRINGER C B and ANDREWS P (1988). Genetic and fossil evidence for the origin of modern humans. *Science* **239**, 1 263–71.

THOM H B (1936). *Die geskiedenis van die skaapboerdery in Suid-Afrika.* N. V. Swets and Zeitlinger, Amsterdam.

THOM H B (1952). *Journal of Jan van Riebeeck* **1**, Balkema, Cape Town.

THOM H B (1954). *Journal of Jan van Riebeeck* **2**, Balkema, Cape Town.

TUTTLE R H (1988). What's new in African paleoanthropology. *Annual Review of Anthropology* **17**, 391–426.

VON DEN DRIESCH A and DEACON H J (1985). Sheep remains from Boomplaas Cave, South Africa. *The South African Archaeological Bulletin* **40**, 39–44.

WESTPHAL E O J (1963). The linguistic prehistory of southern Africa: Bush, Kwadi, Hottentot, and Bantu linguistic relationships. *Africa* **33**, 237–67.

12

Plant and animal invasions

D M Richardson, I A W Macdonald, P M Holmes and R M Cowling

INTRODUCTION

A striking feature of fynbos landscapes is the presence, if not dominance, of invasive alien trees and shrubs, especially those of the genera *Acacia*, *Hakea*, and *Pinus*. These taxa now dominate thousands of hectares of natural and semi-natural vegetation, significantly modifying communities, and threatening many indigenous taxa with extinction. The rapid spread of these plants into, and the transformation of, ecosystems unmodified by humans is unusual when compared with other southern African biomes (Macdonald 1984; Macdonald et al. 1986; Macdonald and Richardson 1986), other mediterranean climate regions of the world (Kruger et al. 1989), or indeed any other continental regions (Drake et al. 1989; Crawley 1989; Richardson and Bond 1991). Very few animal species have been successful in the fynbos, and widespread successes have usually been associated with human activities.

It is not clear whether, as suggested by Taylor (1969), fynbos communities are inherently highly invasible by plant species, or whether the current severity of plant invasions simply reflects the bias of early and frequent introductions and the subsequent influence of humans. Uncoupling a community's 'inherent invasibility' (or openness to invasion) from the cumulative effects of many extrinsic factors is difficult, but is essential for understanding and managing invasions. In this chapter we address both issues and attempt to link plant and animal invasions in fynbos with broader issues in biogeography and community ecology. The patterns of invasion of the biome by *Homo sapiens* and their role in other invasions are described by Deacon (this volume).

Patterns and processes of invasion are studied at different levels of organization, ranging from the biome (e.g. broad-scale distribution patterns), landscape (e.g. rates and patterns of invasion of a single species in a catchment), community (e.g. the relationship between non-equilibrium dynamics and invasibility), and organism (e.g. life history strategies and functional groups of invaders). First, we present a general overview of the status and ecological impacts of plant and animal invaders. In subsequent sections we focus on both the invaders (by considering the attributes that have permitted invasion), and the invaded systems (by determining community properties that regulate invasions). For plants, we combine the two approaches to define invasion windows (*sensu* Johnstone 1986) and functional groups among the invaders. We also seek new insights on the roles of various life history strategies by comparing the invasive success of congeners and other closely related taxa.

AN OVERVIEW OF ALIEN INVASIONS IN THE FYNBOS BIOME

The archaeological record shows that the fynbos biome was peopled long before 150 000 BP, at which stage people first began using fire and, therefore, increased enormously their ability to influence natural communities. However, it was only with the commencement of pastoralism in the biome between 1 700 and 2 000 years ago that the first alien species were introduced (Deacon 1986). The first alien plant species reliably recorded in the biome was the Eurasian burclover, *Medicago polymorpha*, which was associated with domestic sheep remains in deposits dated to *c.* 1 200 BP (Deacon 1986). There is no palaeobotanical evidence to show that natural communities were significantly modified by such early introductions (Coetzee et al. 1983).

The major impact of people on the biome undoubtedly began after European settlement at the Cape in 1652 AD. Unlike the areas further north in southern Africa, Iron Age cultures had not penetrated the fynbos biome by this time (Deacon 1986). Accordingly, modifications to natural ecosystems before this date were restricted to those caused by nomadic pastoralism, hunting, and gathering. The stage was thus set for a rather unusual experiment: the introduction of a wide range of alien organisms (Table 12.1) into relatively pristine ecosystems.

Ecological impact of plant invasions

Of all introduced organisms, the invasive trees and shrubs are thought to have exerted the greatest ecological effects on the biome (Macdonald and Richardson 1986; Macdonald et al. 1988). These effects include the alteration of coastal sediment movement patterns, the acceleration of river bank erosion, a reduction in stream flow, changes in fire regime, and the alteration of the composition of natural plant and animal communities (Macdonald and Richardson 1986; Versfeld and Van Wilgen 1986). Just how pervasive these influences have been is shown by the assessment that 36 of the 224 native birds recorded from the terrestrial ecosystems of the biome have invaded the biome in response to the new habitats created by woody alien plants (Macdonald and Richardson 1986).

Most of the early management programmes were prompted by the threat that tree and shrub invasions were considered to pose to water supplies in the region (Fenn 1980). More recently, regional (Hall et al. 1980; Hall and Ashton 1983; Hall and Veldhuis 1985; Hall 1987) and local (Richardson and Van Wilgen 1986a, b; Richardson et al. 1989) surveys have highlighted the significant threat that these invasions pose to the endemic flora. If the proportion of all the Red Data Book species in the biome (approximately 1 326) threatened by woody aliens is the same as in the three subsamples of threatened taxa for which threats have been analysed, it can be estimated that approximately 750 species are currently at risk (Hall 1987; Macdonald, Loope et al. 1989). Increasingly, the conservation of this flora is the major motivation for control programmes which are often extremely expensive (Moll and Campbell 1976; Taylor 1977; Macdonald et al. 1985; Macdonald, Clark and Taylor 1989; Van Wilgen et al. this volume). This concern is warranted: global comparisons fail to reveal any other continental plant invasions threatening species loss at this scale (Hall 1987; Macdonald et al. 1988; Macdonald, Loope et al. 1989).

Recent research has indicated that herbaceous plant invasions may also be reducing indigenous plant species richness in the biome (Campbell et al. 1980; Macdonald 1988; Vlok 1988). However, herbaceous invaders have not been identified as potential causes of the extinction of fynbos species (Goldblatt 1979; Hall 1987).

Ecological impact of animal invasions

Of the animal invaders, the Argentine ant, *Iridomyrmex humilis*, has received the most attention as regards its impact on natural communities: it has long been recognized as a threat to natural invertebrate communities (Skaife 1955). The implications of the Argentine ant's displacement of indigenous ant species for the long-term survival of the approximately 1 000 fynbos plant species dependent on native ants for seed dispersal have, however, only recently been noticed (Slingsby 1982; Bond and Slingsby 1983; Bond and Slingsby 1984; Bond and Breytenbach 1985; Giliomee 1986). Fortunately, the invasion of remaining areas of fynbos does not appear to be rapid (Prins et al. 1991), although established populations have been found in some relatively undisturbed areas (Donnelly and Giliomee 1985; De Kock and Giliomee 1989). No research has been done on the effects of any of the other invertebrate invaders of the biome.

Of the mammals, only the Himalayan thar, *Hemitragus jemlahicus*, currently poses a significant threat to natural communities in the biome (Brooke et al. 1986; Macdonald and Richardson 1986). Fortunately, the patches of heavy defoliation and associated soil erosion caused by these animals at the peak population levels reached in the early 1970s only covered about three square kilometres on the Table Mountain range (Lloyd 1975). A sustained management programme

TABLE 12.1 The identity, history, and extent of infestations of the most widespread alien species in each of the major taxonomic groups. Within each group species are ranked in decreasing order of importance where this has been determined.

Taxonomic group and species	Region of origin	Year of first introduction	Reason for introduction	Reason for dissemination	Extent of area invaded in the biome (% $^1/_4$°grids)	(% nature reserves)
Vascular plants						
Acacia saligna	Australia	1848	Botanic garden	Tannin/timber/ sand-stabilization	60	51 (n = 73)
Acacia longifolia	Australia	1827	Horticulture	Sand-stabilization/ unintentional	42	26 (n = 73)
Acacia cyclops	Australia	1857	Botanic garden	Timber/ sand-stabilization	65	52 (n = 73)
Pinus pinaster	Mediterranean	1680s	Timber	Timber	30	45 (n = 73)
Leptospermum laevigatum	Australia	1850	Sand-stabilization	Hedges	10	22 (n = 73)
Hakea sericea	Australia	1858	Botanic garden	Hedges/ unintentional	30	37 (n = 73)
Pinus radiata	California	1865	Timber	Timber	17	19 (n = 73)
Pinus halepensis	Mediterranean	1830	Timber	Timber	?	5 (n = 73)
Paraserianthus lophantha	Australia	1835	Botanic garden	Horticulture/ unintentional	14	12 (n = 73)
Acacia melanoxylon	Australia	1848	Botanic garden	Horticulture/ timber	26	27 (n = 73)
Acacia mearnsii	Australia	1858	Botanic garden	Tannin/timber	47	44 (n = 73)
Hakea gibbosa	Australia	1835	Botanic garden	Hedges/sand-stabilization	4	8 (n = 73)
Invertebrates						
Iridomyrmex humilis	Brazil	1901	Unintentional	Unintentional	?	69 (n = 16)
Vespula germanica	Europe	1970	Unintentional	Unintentional	3	All reserves on Cape Peninsula
Theba pisana	Mediterranean	1881	Unintentional	Unintentional	20	All coastal reserves

TABLE 12.1 continued.

Vertebrates						
Reptiles						
Ramphotyphlops braminus	South-east Asia	pre 1830	Unintentional	Unintentional	1	? none
Birds						
Numida meleagris	Eastern Cape	1890s	Sport hunting	Sport hunting	95	77 (n = 15)
Columba livia	Europe	1652	Agriculture	Agriculture	70	13 (n = 15)
Sturnus vulgaris	Europe	1897	Acclimatization	Natural expansion	94	93 (n = 15)
Fringilla coelebs	Europe	1890s	Acclimitization	Natural expansion	1	7 (n = 15)
Passer domesticus	India	1964	Natural expansion	Natural expansion	94	80 (n = 15)
Mammals						
Hemitragus jemlahicus	Asia	1935	Zoo	Natural expansion	0.5	7 (n = 15)
Sus scrofa	Asia	?1652	Agriculture	Biological control	5	7 (n = 15)
Sciurus carolinensis	North America	1890s	Acclimatization	Pets	7	29 (n = 14)
Rattus rattus	Asia	pre 1862	Unintentional	Unintentional	21	50 (n = 14)
Mus musculus	Asia	pre 1834	Unintentional	Unintentional	9	57 (n = 14)

Data sources: **Vascular plants:** Macdonald and Jarman (1984); Shaughnessy (1986); Macdonald, Jarman and Beeston (1985); Macdonald and Richardson (1986); I A W Macdonald (unpublished data). **Invertebrates:** Skaife (1962); De Kock and Giliomee (1989); Whitehead and Prins (1975); Cooke (1984); Van Bruggen (1964); Prins et al. (1991); W F Sirgel (unpublished data). **Vertebrates:** Brooke et al. (1986); McLachlan (1978); South African Bird Atlas Project, Harrison (1987); Lloyd (1975); Botha (1989); Millar (1980); Sclater (1901); Macdonald, Powrie and Siegfried (1986).

has held their numbers in check since this time.

No introduced bird species have invaded the relatively unmodified tracts of mountain fynbos. The ecological impacts of introduced birds are confined largely to the lowland areas which have already been extensively transformed (Moll and Bossi 1984). As seen elsewhere in the expanded range of the European starling, *Sturnus vulgaris* (Macdonald et al. 1988), this species competes for nest holes with native bird species (e.g. Van der Merwe 1984). There is no clear indication that this competition has led to the local extinction of any bird species. However, the disappearance of the native Pied starling, *Spreo bicolor*, from the Cape Peninsula might be a case in point (I A W Macdonald unpublished data; but see Brooke et al. 1986).

ANIMAL INVASIONS

Which animals have invaded?

Very few introduced animals have successfully invaded natural ecosystems in the fynbos biome. Those species that have established widespread feral populations have done so only in association with people, e.g. the European starling, *Sturnus vulgaris* (Winterbottom and Liversidge 1954; Liversidge 1962), the house sparrow, *Passer domesticus* (Vierke 1970), and the black rat, *Rattus rattus* (Smithers 1983) (Table 12.1). Even some of the less successful invaders are dependent upon artificial habitats: both the grey squirrel, *Sciurus carolinensis*, and the chaffinch, *Fringilla coelebs*, are totally dependent upon stands of introduced tree species for their survival in the biome (Millar 1980; Maclean 1985).

Similarly, the two most widespread invertebrate invaders, the Argentine ant and the white dune snail, *Theba pisana* (Van Bruggen 1964; Middlemiss 1975), show local distributions strongly linked to patterns of human disturbance: the ant is generally restricted to areas with vehicular access (De Kock and Giliomee 1989) and the snail to road verges (W F Sirgel personal communication). The introduced wasp, *Vespula germanica*, is one of the few invertebrate invaders that is now well established in undisturbed areas of natural vegetation in the biome. It is, however, confined mainly to areas of dense forest rather than open fynbos vegetation (Whitehead and Prins 1975).

Factors influencing the success of animal invasions

NUTRITIONAL LIMITATIONS

Invasions of three of the biome's alien mammals have been intensively investigated. For the grey squirrel, the study concluded that the alien had not penetrated natural fynbos communities because it was totally dependent on introduced tree species for food and shelter (Millar 1980). Similarly, typical fynbos habitats are apparently unable to support viable populations of feral pigs, *Sus scrofa*, which can be maintained only in plantations of introduced trees or extensive marshes (Botha 1989). The low nutritional content of fynbos herbage (Campbell 1986) is the probable reason for the failure of African ungulates alien to the biome to survive in fynbos nature reserves without dietary supplementation or other management intervention (e.g. Barnard and Van der Walt 1961; Millar 1970; Anonymous 1978). Introduced ungulates browse heavily on alien *Acacia* species (e.g. Middlemiss 1963; Millar 1970; Middlemiss 1975; Van der Merwe 1977) which have a higher nutritional content than fynbos plants (Witkowski 1991). The only introduced mammal that has invaded the fynbos successfully is the Himalayan thar. This species is apparently pre-adapted to the low quality diet provided by natural fynbos vegetation (Lloyd 1975).

Introduced birds are restricted mainly to transformed ecosystems in the biome. Where they do enter areas of natural vegetation, they feed on the fruits of invasive alien plants, e.g. European starlings on the funicles of *Acacia cyclops* (Glyphis et al. 1981).

Nutritional limitations to invasion are not confined to vertebrate herbivores. Almost all alien invertebrate herbivores that have successfully invaded the biome have been restricted in their diet to introduced plant species and, as such, have remained pests of agriculture, forestry, and horticulture. Examples are the numerous alien aphids (Potgieter and Durr 1961; Durr and Martin 1976; Durr 1983), and pests of eucalypts (Cillié 1983; Winstanley 1984a) and conifers (Bruzas 1981;

Winstanley 1984b; Tribe 1990). In a few cases, these invertebrates are associated with invasive alien species, e.g. the butterfly, *Zophopetes dysmephila*, with *Phoenix reclinata* palms on Table Mountain (Moll and Campbell 1976; De Villiers and McDowell 1982). The high levels of secondary chemicals in fynbos plant species (Glyphis and Puttick 1988) have probably prevented their utilization by introduced invertebrate herbivores. The only really successful invertebrate alien herbivore is *Theba pisana* (Van Bruggen 1964; Middlemiss 1975), a snail which favours the relatively nutrient-rich coastal areas. The notion that the vegetation places nutritional limits on invertebrates is further supported by the observation that several native herbivores have undergone massive population increases in response to the introduction of alien food plants, e.g. the pine tree emperor moth, *Nudaurelia cytherea*, which is now a major pest of introduced pines (Geertsema 1975). The same applies to the moths *Psycharium pellucens* and *Tolna limula* and various stick insects which now feed on pines (G D Tribe personal communication).

THE LIMITING ROLE OF GENERALIST PREDATORS

Another factor which has been implicated as limiting the success of alien animals in the biome is pressure from generalist predators. Predation by the leopard, *Panthera pardus*, has possibly contributed to the failure of feral pigs as invaders (Botha 1989). All the large carnivores had already been eliminated on Table Mountain before its invasion by Himalayan thars (Stuart et al. 1985). A low-key hunting campaign (Brooke et al. 1986) controlled the numbers of thars and prevented them from expanding their range on to the other mountains of the Cape Peninsula. This suggests that mortality from predation could limit or prevent such an invasion.

Predation by native ant species has been identified as a factor limiting the successful establishment of introduced invertebrate bio-control agents in southern Africa (Prins et al. 1991). However, in the fynbos biome, where many bio-control agents have been introduced into agricultural areas, their establishment has been limited mainly by predation by the introduced Argentine ant. Notwithstanding this, the point remains that, even for invertebrates, the pressure exerted by generalist predators can prevent the successful establishment of an introduced species.

IS FYNBOS MORE INVASIBLE THAN ADJACENT BIOMES?

There are indications that the fynbos biome is more susceptible to alien animal invasions than are the other southern African biomes (Macdonald, Powrie and Siegfried 1986). With the recent initiation of a standardized approach to the collection of atlas data on birds throughout the subcontinent (Harrison 1987), it is now possible to test this hypothesis on the alien birds of the biome.

All alien species showed significantly higher overall frequencies of occurrence in the fynbos biome (Table 12.2). This could be construed as indicating that the fynbos was more susceptible to invasion by alien birds than the adjacent biomes. However, the alien birds are all to some extent dependent on human-modified habitats in southern Africa (Brooke et al. 1986) and it could equally well be argued that, as human population densities tend to be higher in the fynbos biome than in the adjacent biomes (Zietsman and Van der Merwe 1986), these differences in reporting frequencies simply reflect differences in the extent of human disturbance. To test this hypothesis, those grid squares without records of a particular alien species were excluded from the analysis. Thus, the reporting frequencies were compared for only those squares in each biome category where one can assume the level of human disturbance was high enough to enable the alien bird involved to become established. When this approach was adopted, the reporting frequencies for the helmeted guineafowl, *Numida meleagris*, and the European starling remained higher in the fynbos, and, for all but the starling in the eastern Cape, the differences were significant. The reporting frequencies of the feral pigeon, *Columba livia*, and the house sparrow, however, tended to be higher in the other biomes. For the sparrow this difference was statistically significant.

The conclusion is that the fynbos biome is not distinctly different from the other biomes

TABLE 12.2 **The percentage frequency of the occurrence of alien bird species on field sighting sheets returned to the South African Bird Atlas Project for grid squares classified as falling in the fynbos biome or in 'other biomes' in the western and eastern Cape atlassing regions. Using the data sets for the western Cape and eastern Cape bird atlassing regions as at March 1990, the frequency of grid squares in different reporting frequency classes for the fynbos biome was compared with those for all the other biomes. In the case of the western Cape, the other biomes were the succulent and Nama-karoo. In the eastern Cape, the Nama-karoo was predominant but there were also grassland, savanna, and forest elements present in this 'other biome' category. Only the western Cape data set was used for the *Numida meleagris* as it is there that the species was definitely introduced.**

Species	Region	Overall % frequency of occurrence		Significance of Chi-squared test ($P =$)	% frequency in grid squares where species was recorded		Significance of Chi-squared test ($P =$)
		Fynbos	Other		Fynbos	Other	
Numida meleagris	Western	44.3	4.9	****	46.4	31.5	0.006
Columba livia	Eastern	12.2	5.0	****	15.0	14.5	0.27
	Western	7.8	1.6	****	11.9	17.0	0.26
	Combined	9.1	3.2	****	12.9	15.1	0.49
Sturnus vulgaris	Eastern	32.5	13.8	****	37.5	27.6	0.14
	Western	53.7	9.0	****	55.4	37.8	****
	Combined	47.7	11.3	****	50.7	31.1	****
Passer domesticus	Eastern	23.8	30.5	****	26.3	43.0	0.009
	Western	40.8	24.7	****	42.7	52.8	****
	Combined	36.0	27.5	****	38.2	47.1	****

**** = < 0.001

in respect to its susceptibility to avian invasions. The generally greater extent to which the biome's ecosystems have been modified and transformed (Macdonald 1989), has allowed alien birds to become more widespread here than in the adjacent biomes. However, it is the particular autecological characteristics of the alien species that determine the degree of success within and beyond the fynbos biome. The generally higher and more dispersed precipitation in the fynbos biome relative to the karoo biomes makes it more suitable for the European starling (Macdonald, Powrie and Siegfried 1986). Conversely, the southern African house sparrow (mainly derived from the *indicus* subspecies of *Passer domesticus*; Summers-Smith 1988) is better suited to arid and semi-arid areas (I A W Macdonald unpublished data). The limited distribution of helmeted guineafowls in the drier portions of the western Cape could be due, in part, to their restriction to areas within range of drinking points (Urban et al. 1986).

The rates of animal invasions

Adequate data on which to base estimates of rates of spread exist for only a few vertebrate species (Figure 12.1). The results are predictable: the more mobile birds show more rapid range expansions than do the mammals. The small size of the fynbos biome means that linear rates of the expansion of invasion fronts can be estimated only with difficulty. For example, although the invasion of the house sparrow *Passer domesticus* began after 1961 and was complete by 1969, the invasion front cannot be plotted accurately because simultaneous 'first arrivals' were being recorded from scattered localities throughout the length and breadth of the biome during this period (Vierke 1970; Brooke 1986). As has been emphasized previously, almost all of these successful invaders are associated with transformed habitats and the rates of range expansions are thus more a function of this modification than of any intrinsic characteristic of the fynbos biome.

That the colonization of the biome by the house sparrow was completed in a shorter period than was that of the European starling should not be taken as showing that the house sparrow was better suited to conditions present in the biome. On the contrary, the starling is now probably the most numerous bird species in the biome, whereas the sparrow is relatively uncommon (Table 12.1; Macdonald 1987; Hockey et al. 1989). The reasons for the sparrow being able to complete its colonization of the biome more rapidly were probably the following:

- it had already undergone the early phase of its exponential population growth elsewhere in southern Africa, and was widespread along the entire northern edge of the biome before it began its invasion; and
- by the 1960s the density of human habitation in the biome was many times higher than at the turn of the century when the starling began its invasion.

The reality is that the sparrow is actually poorly suited to conditions present in the biome (relative to those pertaining, for example, in the drier karoo biome to the north where its invasion was much more rapid) (Vierke 1970).

The rates of range expansion by introduced invertebrates also vary considerably. For example, the rate of invasion of the Argentine ant into natural fynbos communities is very slow. The ant has been present on the Cape Peninsula for 80 years, and yet there are still tracts of mountain fynbos immediately above infested suburban areas where the ant is not present (Prins et al. 1991; T B Oatley personal communication). Observations from elsewhere in the biome show that the ant can survive in comparable mountain fynbos ecosystems (Giliomee 1986; De Kock and Giliomee 1989). Its absence from the Peninsula sites is a reflection of its poor ability to disperse or to establish populations in fynbos.

The rates of expansion of other alien insects have been very rapid: thus, the eucalyptus tortoise beetle, *Trachymela tincticollis*, became widespread throughout the entire winter rainfall region of the biome within a few years of its discovery in 1982 on the Cape Peninsula (Cillié 1984). In 1983 its easternmost distributional limit was at Bredasdorp. In 1984 it was recorded in Port Elizabeth on the eastern edge of the biome. The latter record might represent human-assisted extension in range but the species is known to be dis-

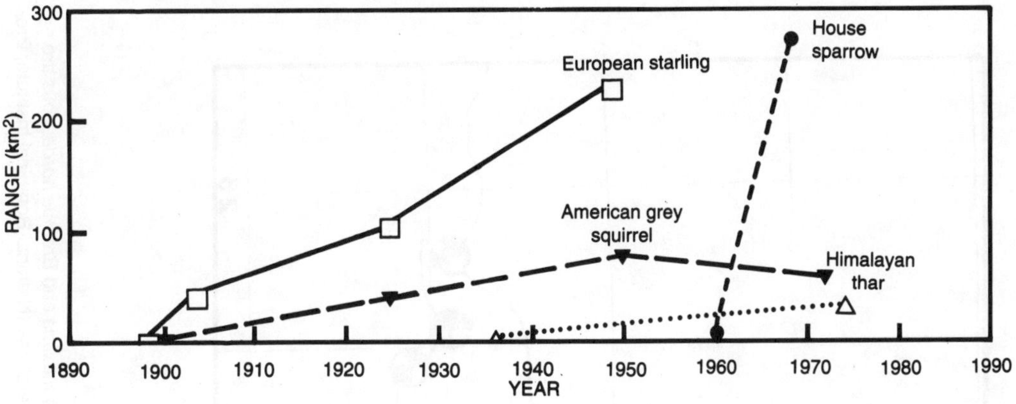

FIGURE 12.1 The range expansions (km^2) of vertebrate invaders in the fynbos biome (77 000 km^2 in extent).

persed by wind (Tribe and Cillié 1984). To control this beetle, the parasitic wasp *Enoggera reticulata* was released on the Cape Peninsula and at scattered localities along the False Bay coast in January 1986. It was recorded 500 km from the release sites 10 months later, and by February 1987 had reached Port Elizabeth, some 600 km from the release sites (Urban et al. 1987).

The linear rates of advance of invasion fronts can be estimated for a few alien animals in the fynbos (Balinsky 1962). Thus, the grey squirrel front moved at about 1.3 km yr^{-1} over its first 50 years, the European starling at about 13 km yr^{-1} over its first 56 years, while the wasp *Enoggera reticulata* spread at a rate of approximately 600 km yr^{-1} in its first year (accurately measured at 2 km day^{-1}; Urban et al. 1987).

It appears that the rates of invasion are regulated more by the dispersal characteristics of the invading animals than by any intrinsic characteristic of the biome. This generalization apparently holds for both vertebrates and invertebrates. However, rates of expansion are reduced where the natural communities present a barrier to the species involved e.g. the feral pig (Botha 1989), the grey squirrel (Millar 1980), and the Argentine ant (Prins et al. 1991). In such cases, the range expansion following the initial establishment of invasion nuclei has been much slower than the dispersal characteristics of the species would allow.

PLANT INVASIONS

Plant species were introduced, both intentionally and accidentally, in considerable numbers during the first 150 years after the establishment of the Cape colony (Shaughnessy 1986; Wells et al. 1986). However, very few of these species became invasive and only one, *Pinus pinaster*, is currently ranked among the 12 most important invaders of the biome (Table 12.1). These early introductions were virtually all of European origin. It was only after 1830, when tree and shrub species were intentionally imported from areas of similar climate (especially southern and western Australia) and were extensively propagated, that most of the important invaders of natural vegetation became established. Many of these species are now extremely widespread in the biome (Table 12.1; Figures 12.2, 12.3).

Wells et al. (1983) estimated that by 1982 there were 109 alien plant species invading the terrestrial habitats of the winter rainfall area. Of these, 62% were hemicryptophytes and 34% were trees and shrubs. Stream bank habitats were considered separately and for the whole of South Africa: 101 alien species were recorded invading this habitat, of which 43% were hemicryptophytes and 47% trees and shrubs (Wells et al. 1983). Several recent analyses have shown that herbaceous species generally outnumber tree and shrub species in local alien floras in the fynbos biome (e.g. Kruger et al. 1989). This numerical dominance

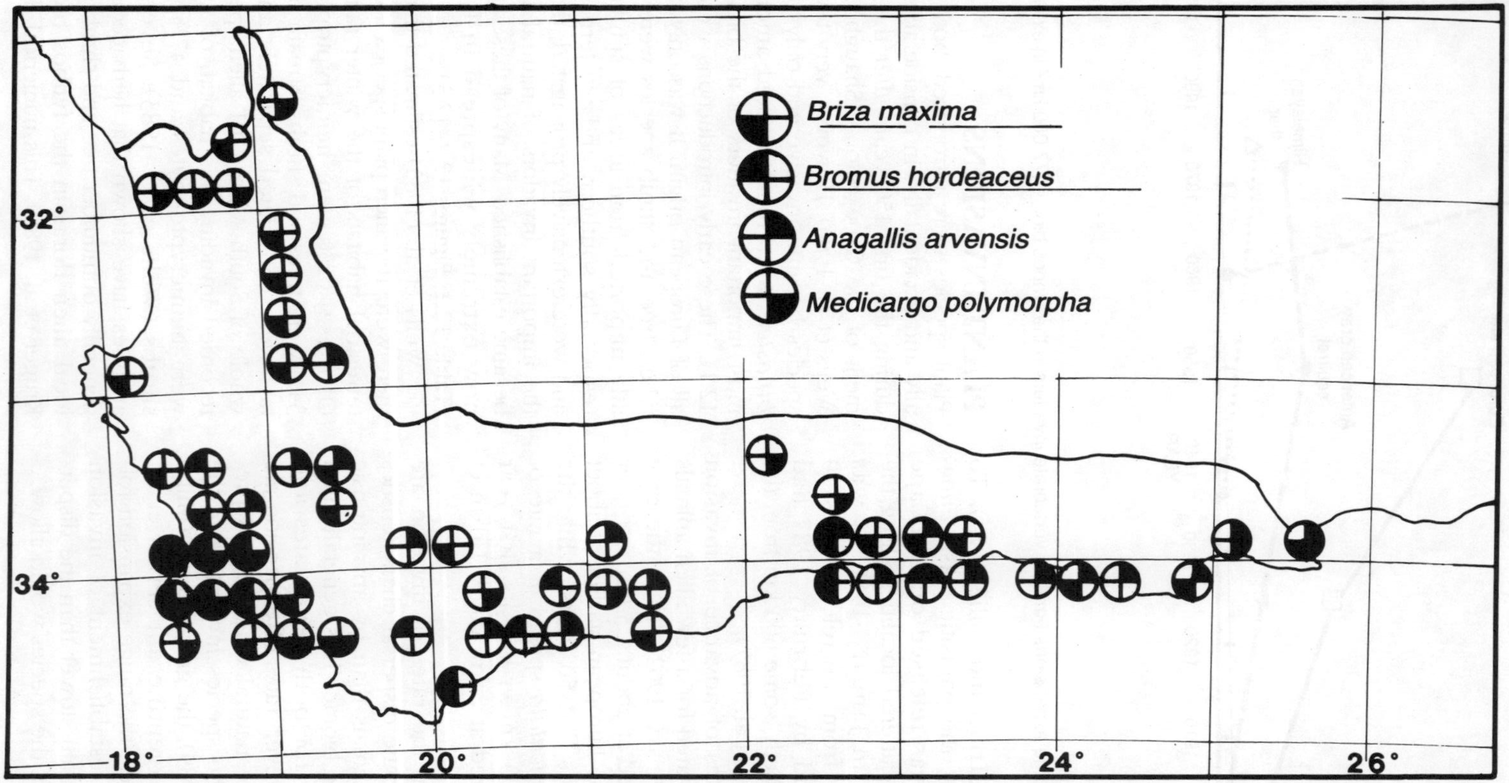

FIGURE 12.2 The distribution ranges of selected naturalized herbaceous species in the fynbos biome. The data for *Briza maxima* and *Bromus hordeaceus* are from H P Linder (unpublished data); *Anagallis arvensis* and *Medicago polymorpha* are from the PRECIS database (National Botanical Institute, Pretoria).

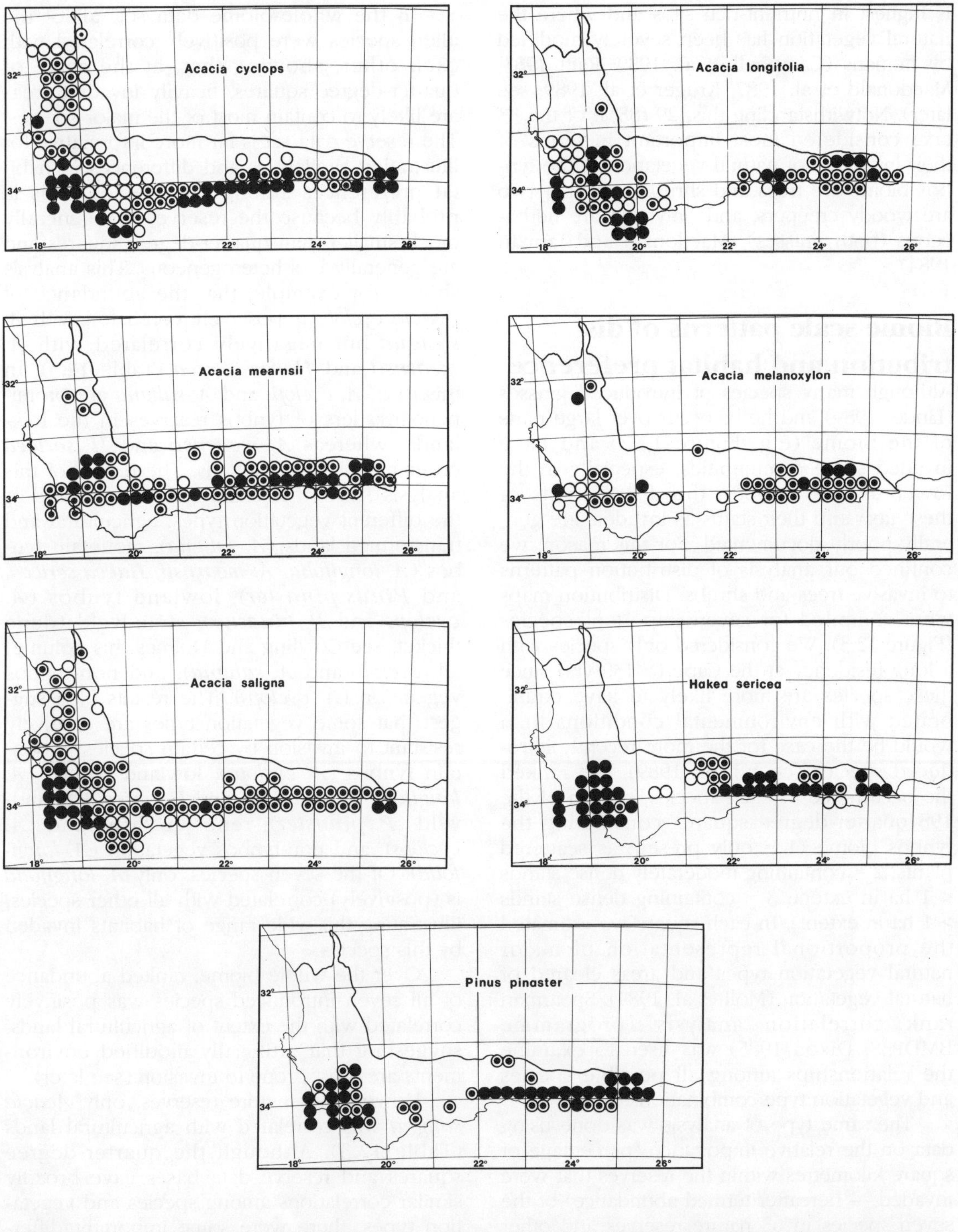

FIGURE 12.3 The distribution of important invasive plant species in the fynbos biome. Large solid dots = quarter-degree squares with dense stands > 1 ha in extent; small dots = quarter-degree squares with moderately dense stands < 1 ha in extent; open circles = quarter-degree squares with scattered plants only.

is highest in nutrient-rich sites and where the natural vegetation has been severely modified by humans (Campbell et al. 1980; Raitt 1983; Macdonald et al. 1987; Kruger et al. 1989; see later). Notwithstanding this, 29 (88%) of the 33 taxa considered most important in terms of their invasions of natural vegetation in the fynbos biome are trees and shrubs, a further two are woody creepers, and only two are herbaceous (both grasses) (Macdonald and Jarman 1984).

Biome-scale patterns of distribution and habitat preference

Although many species of introduced grasses (Linder 1989) and herbs occur over large parts of the biome (e.g. Figure 12.2) and have invaded some communities, especially in the lowlands (Vlok 1988), the distributions of these taxa and their status as invaders are generally poorly documented. For this reason, we confined our analysis of distribution patterns to invasive trees and shrubs. Distribution maps were compiled for seven important species (Figure 12.3). We considered only species with a long residency at the Cape (> 130 yrs) since these species are more likely to have equilibrated with environmental conditions than would be the case for the more recently introduced species (cf. Wilson 1989). We ranked the occurrence of each species in each of the 198 quarter-degree squares constituting the fynbos biome (1 = only present as scattered plants; 2 = containing moderately dense stands < 1 ha in extent; 3 = containing dense stands > 1 ha in extent). In each square we estimated the proportional representation of major natural vegetation types and areas cleared of natural vegetation (Moll et al. 1984). Spearman rank correlation analysis (programme BMDP3S; Dixon 1985) was used to examine the relationships among all possible species and vegetation type combinations.

The same type of analysis was done using data on the relative importance (percentage of square kilometres within the reserves that were invaded — hereafter termed abundance) of the seven species in 63 nature reserves and other protected areas in the biome (I A W Macdonald et al. unpublished data) together with the proportional representation of major vegetation types in those reserves (Moll et al. 1984).

In the whole-biome data set, all of the alien species were positively correlated with each other, showing that, at the scale of quarter-degree squares, heavily invaded areas are likely to contain most of the major species. The reserve data set is far more informative for illustrating similarities and differences in habitat preferences among the species. This is probably because the reserves are generally much smaller than quarter-degree squares and are generally less heterogeneous. This analysis shows, for example, that the abundance of *Acacia cyclops* is positively correlated with *A. saligna* but negatively correlated with *A. mearnsii* and *Hakea sericea* (Table 12.3). In this case, *A. cyclops* and *A. saligna* are prominent invaders of fynbos reserves in the lowlands, whereas *A. mearnsii* and *H. sericea* occur mainly in mountains. The results of this analysis show the 'characteristic' invaders of the different vegetation types: agricultural and transformed lands (*A. saligna*); mountain fynbos (*A. longifolia*, *A. mearnsii*, *Hakea sericea*, and *Pinus pinaster*); lowland fynbos (*A. cyclops* and *P. pinaster*); strandveld (dune thicket, see Cowling and Holmes this volume) (*A. cyclops* and *A. saligna*); and non-fynbos vegetation (*A. cyclops*). The results also suggest that some vegetation types are relatively resistant to invasion by certain species: mountain fynbos (*A. cyclops*); lowland fynbos (*A. longifolia, A. mearnsii,* and *H. sericea*); strandveld (*P. pinaster*); renoster shrubland (*A. cyclops*), and non-fynbos vegetation (*A. longifolia*). Of the seven species, only *A. longifolia* is (positively) correlated with all other species, illustrating the wide range of habitats invaded by this species.

Over the whole biome, ranked abundance of all seven introduced species was positively correlated with the extent of agricultural lands, suggesting that artificially modified environments are most prone to invasion (see later).

However, in nature reserves, only *Acacia saligna* was correlated with agricultural lands (Table 12.3). Although the quarter-degree squares and reserve data bases gave broadly similar correlations among species and vegetation types, there were some important differences. In the reserve data, these were probably due to bias against grassy fynbos and lowland vegetation types (including transformed land), and in the quarter-degree square

and shrubs (see text) and the extent of major vegetation types (Moll et al. 1984) in: (a) the 198 quarter-degree squares constituting the fynbos biome (the upper part of the matrix); and (b) 63 nature reserves and other protected areas. Other non-fynbos vegetation types include various thicket types, Afromontane forest, karroid shrublands, and various ecotonal communities (Moll et al. 1984).

a) Quarter-degree squares ($n = 198$) / b) Nature reserves ($n = 63$)	Agricultural/ transformed lands	Mountain fynbos	Grassy fynbos	Lowland fynbos	Strandveld (thicket)	Renoster shrubland	Other non-fynbos	*Acacia cyclops*	*A. longifolia*	*A. mearnsii*	*A. melanoxylon*	*A. saligna*	*Hakea sericea*	*Pinus pinaster*
Agricultural/transformed lands		N.S.	---	+++	+	N.S.	--	+++	+++	++	+++	+++	+	+
Mountain fynbos	-		--	---	-	++	---	N.S.	N.S.	+++	++	N.S.	+++	N.S.
Grassy fynbos	N.S.	N.S.		-	--	-	+++	-	+	+++	+++	-	N.S.	++
Lowland fynbos	N.S.	---	N.S.		+++	---	-	+++	N.S.	-	N.S.	+++	--	N.S.
Strandveld (thicket)	N.S.	-	N.S.	+		---	---	+++	N.S.	---	N.S.	+	-	N.S.
Renoster shrubland	N.S.	N.S.	N.S.	-	N.S.		--	---	N.S.	++	N.S.	N.S.	+	N.S.
Other non-fynbos	N.S.	--	N.S.	N.S.	N.S.	N.S.		+	++	+++	+	N.S.	N.S.	+++
Acacia cyclops	N.A.	---	N.S.	+++	++	-	+		+++	+++	+++	+++	+	+++
A. longifolia	N.S.	+++	N.S.	-	N.S.	N.S.	-	N.S.		+++	+++	+++	+++	+++
A. mearnsii	N.S.	++	N.S.	-	N.S.	N.S.	N.S.	--	+++		+++	+++	+++	+++
A. melanoxylon	N.S.	N.S.	N.S.	N.S.	N.S.	N.S.	N.S.	N.S.	+	+++		+++	+++	+++
A. saligna	++	N.S.	N.S.	N.S.	+	N.S.	N.S.	++	++	N.S.	N.S.		++	+++
Hakea sericea	N.S.	+++	N.S.	-	N.S.	N.S.	N.S.	-	+++	+++	++	N.S.		+++
Pinus pinaster	N.S.	+++	N.S.	++	-	N.S.	N.S.	N.S.	+++	++	+	N.S.	+++	

N.S. not significant
+/- $P < 0.05$
++/-- $P < 0.01$
+++/--- $P < 0.001$

database due to bias against records of invasive species in remoter parts of the biome. Some vegetation types were intercorrelated: for example, the extent of mountain fynbos in quarter-degree squares was positively correlated with renoster shrubland, and negatively correlated with grassy fynbos, lowland fynbos, strandveld, and non-fynbos vegetation (Table 12.3). These correlations reflect the geographical proximity of the major vegetation types and are important for understanding the current distribution of invasive species. For example, areas of extant lowland fynbos and strandveld are usually separated from mountain fynbos by large tracts of transformed land, thus precluding the ready exchange of invasive plants between these vegetation types.

The extent of mountain fynbos in quarter-degree squares over the whole biome was positively correlated with the ranked abundance of *A. mearnsii*, *A. melanoxylon*, and *H. sericea*. Subdividing mountain fynbos into wet, mesic, and dry types (*sensu* Moll et al. 1984) revealed that the abundance of *A. longifolia* and *Pinus pinaster* was positively correlated with the extent of wet and mesic mountain fynbos, but negatively correlated with the extent of the dry type. The reserves' database showed a strong correlation between the extent of mountain fynbos and the abundance of *A. longifolia, H. sericea,* and *P. pinaster.* Mountain fynbos occurs in 66% of the quarter-degree squares in the fynbos biome and is the dominant cover of the Cape Fold Belt mountains (Campbell 1985). *Hakea sericea* and *P. pinaster* are the most widespread invasive species in this vegetation type (Macdonald and Richardson 1986).

Hakea sericea, introduced to the biome in around 1858 and planted on a limited scale around Cape Town, Stellenbosch/Franschhoek, and George, spread rapidly eastwards from Cape Town and Franschhoek and in all directions from George to cover almost 4 800 km^2 or 14% of the area of mountain fynbos about 120 years after its introduction (Kluge and Richardson 1983; Macdonald 1984). This species occurs in 30% of quarter-degree squares in the biome and forms dense stands in 19% of squares (Figure 12.3). Its present distribution is positively associated with the occurrence of nutrient-poor soils derived from quartzite and sandstone, and its spread eastwards from Cape Town and northwards from George has been restricted to some extent by barriers of unsuitable nutrient-rich substrata (Fugler 1979; Richardson 1984). Hence, the occurrence of *H. sericea* over the whole biome is positively correlated with mountain fynbos (and spuriously with renoster shrubland — see above), and negatively correlated with lowland fynbos and strandveld. In the reserves' data set, *H. sericea* was positively correlated with mountain fynbos and negatively correlated with lowland fynbos (Table 12.3).

Pinus pinaster was introduced to the Cape in about 1680 and has been widely planted throughout the fynbos biome. Probably because of its long residency and wide dissemination, this species occurs in all vegetation types. It now occurs in 30% of quarter-degree squares in the biome, with dense stands in 19% of squares (Figure 12.3). A positive correlation exists between its ranked abundance and the extent of grassy fynbos and non-fynbos. In nature reserves, the abundance of this species is positively correlated with mountain fynbos and lowland fynbos and negatively correlated with strandveld (Table 12.3). This species (and other *Pinus* spp.) often forms mixed stands with *H. sericea*; the most recent estimate of the area invaded by these taxa throughout the biome is 7 592 km^2 (Macdonald et al. 1985).

Another important invader of mountain fynbos, *Acacia longifolia*, has recently shown rapid increases in both its distribution and density (McLachlan et al. 1980; Taylor et al. 1985) and is currently ranked the second most important alien plant invader in the biome (Macdonald and Jarman 1984). At present it occupies 42% of quarter-degree squares, but dense stands occur in only 12% of squares. This species differs from *Hakea* and *Pinus* species in that, besides invading remote mountain regions, it has also invaded forest fringes and riparian habitats (Pieterse 1986). Its occurrence over the whole biome is positively correlated with agricultural lands, grassy fynbos, and non-fynbos vegetation types. In reserves, the abundance of this species is positively correlated with mountain fynbos and negatively correlated with lowland fynbos and non-fynbos vegetation types (Table 12.3).

Acacia cyclops and *A. saligna* are the most important invasive tree species in lowland fynbos and strandveld (Macdonald and Richardson 1986), and are also the most widespread invasive trees in the biome, occurring in 65% and 60% of quarter-degree squares respectively (Figure 12.3). Dense stands of these species, occurring in 18% and 9% of squares respectively, are mainly confined to areas near planting sites centred on the Cape Flats (both species) and along the southern Cape coast between 21° and 26° east (*A. cyclops*). The occurrence of *A. cyclops* is positively correlated with agricultural lands, lowland fynbos, strandveld, and non-fynbos vegetation, but negatively correlated with grassy fynbos and renoster shrubland. The negative correlation between the abundance of *A. cyclops* and the extent of mountain fynbos in reserves is probably due to the geographical isolation of extant tracts of these vegetation types from sites where this species was planted (mainly in lowland fynbos and strandveld). However, where mountain fynbos adjoins coastal dunes, such as on the Cape Peninsula and between Gordon's Bay and Cape Agulhas, *A. cyclops* readily invades mountain fynbos (e.g. Taylor et al. 1985). The occurrence of *A. saligna* in the whole biome is also positively correlated with lowland fynbos, strandveld, and agricultural land. In the reserve data set, the abundance of this species is positively correlated only with agricultural lands and strandveld (Table 12.3).

Landscape-scale patterns

THE SPREAD OF *PINUS RADIATA* NEAR STELLENBOSCH

The several species of serotinous trees and shrubs that have invaded mountain fynbos show similar patterns of invasion. The spread of *Pinus radiata* from a plantation in the Jonkershoek Valley near Stellenbosch illustrates this pattern. Richardson and Brown (1986) analysed sequential aerial photographs, population age structures of self-sown stands, and fire records to reconstruct the history of invasion between 1938 and 1986. In 1938, the area adjoining the plantation contained no pine trees (Figure 12.4). By 1953, scattered stands of this species (< 21 plants/ha) were evident in 27%, and moderate stands (21–1 000 plants/ha) in 4% of 237 delimited one hectare blocks adjacent to the plantation. In 1967, *P. radiata* had invaded all but 5 blocks, and by 1977 only 15% of blocks had fewer than 21 plants (Figure 12.5). Invading trees established throughout the 3 km wide study area. This pattern shows that the invasion of fynbos by serotinous pines (*P. halepensis*, *P. pinaster*, and *P. radiata*) is a two-phase process:

- satellite foci establish when the winged seeds are dispersed by wind over several kilometres; and
- the satellite foci grow and coalesce to form dense even-aged stands (Kruger 1977a; Richardson and Brown 1986; Richardson 1988, 1989).

The other important serotinous invader of mountain fynbos, *Hakea sericea*, shows similar invasion patterns (Richardson 1989).

THE SPREAD OF *ACACIA CYCLOPS* ON THE CAPE PENINSULA

The results of surveys of the invasive plants in the Cape of Good Hope Nature Reserve in 1952, 1961, and 1977/8 show the pattern of spread of the bird-dispersed *Acacia cyclops* (Figure 12.6). Although control operations have been carried out in the reserve since 1941, these were unsystematic and totally unsuccessful until 1981 when a systematic plan of action was adopted (Macdonald et al. 1989). Despite the early control efforts, the area under dense stands of *A. cyclops* increased from less than 2% in 1952 to around 12% in 1977 (Figure 12.7). In general, dense stands mapped during the 1961 and 1977/8 surveys occurred progressively further from the dense stands mapped during 1952. The large increases in the area under dense stands between 1961 and 1977/8 are due to the rapid expansion of existing foci and the establishment of new stands well away from existing dense stands (Figure 12.6).

SPREAD OF LEGUMES AND OTHER TAXA ALONG RIVER COURSES

Studies of sequential air photographs have shown that certain *Acacia* species invade most rapidly along watercourses. The direction and rate of spread suggest that seed dispersal in water is the major cause of these patterns (Brownlie 1982). Riparian habitats throughout the biome are heavily invaded by numerous species e.g. *A. longifolia*, *A. saligna*, *A. mearn-*

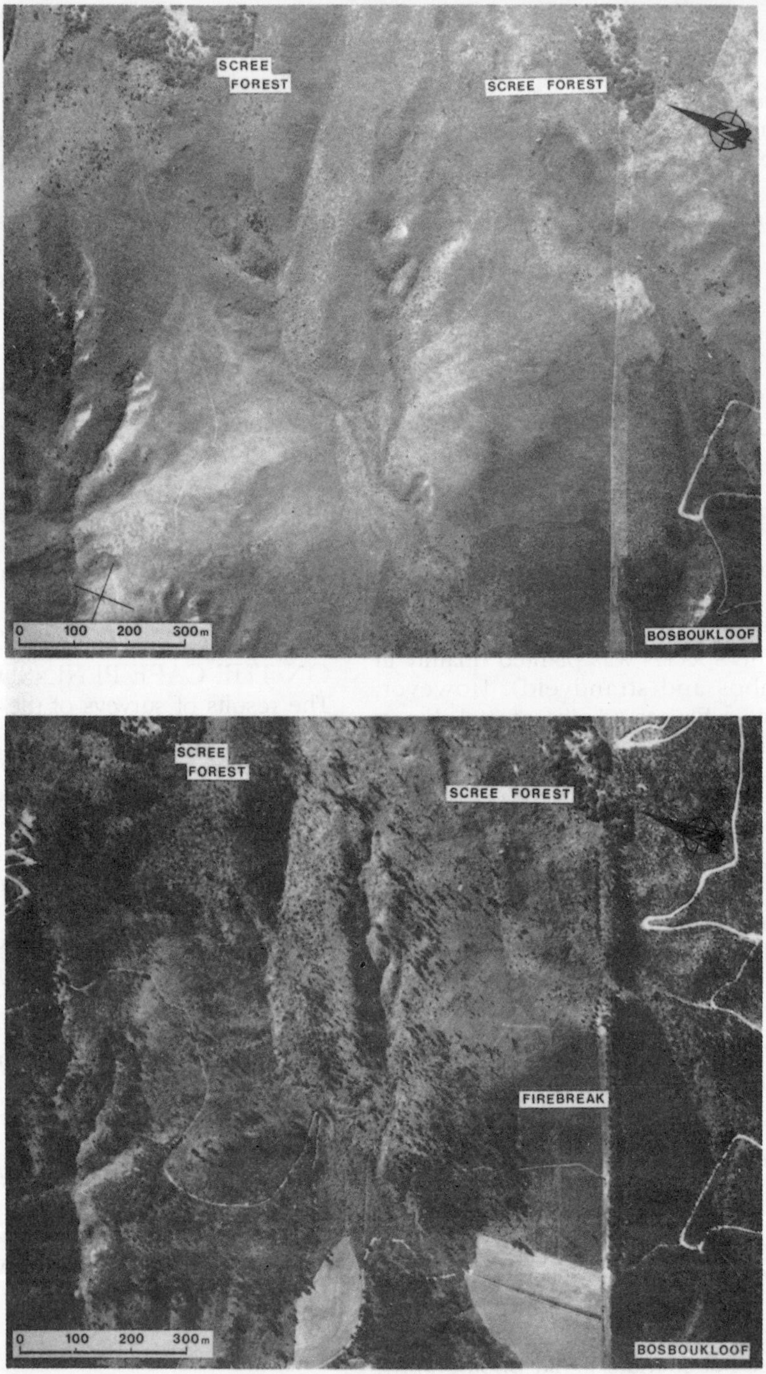

FIGURE 12.4 The invasion of mountain fynbos near Stellenbosch by pines (*Pinus halepensis*, *P. pinaster*, and *P. radiata*). The top photograph shows the uninvaded fynbos in 1938 before the establishment of the Bosboukloof plantation. *Pinus radiata* spread rapidly from the plantation, and *P. halepensis* and *P. pinaster* invaded from self-sown stands in other parts of the Jonkershoek Valley. The bottom photograph, taken in 1981, shows the significant change in vegetation structure caused by these invasions. Reproduced from Richardson and Brown (1986).

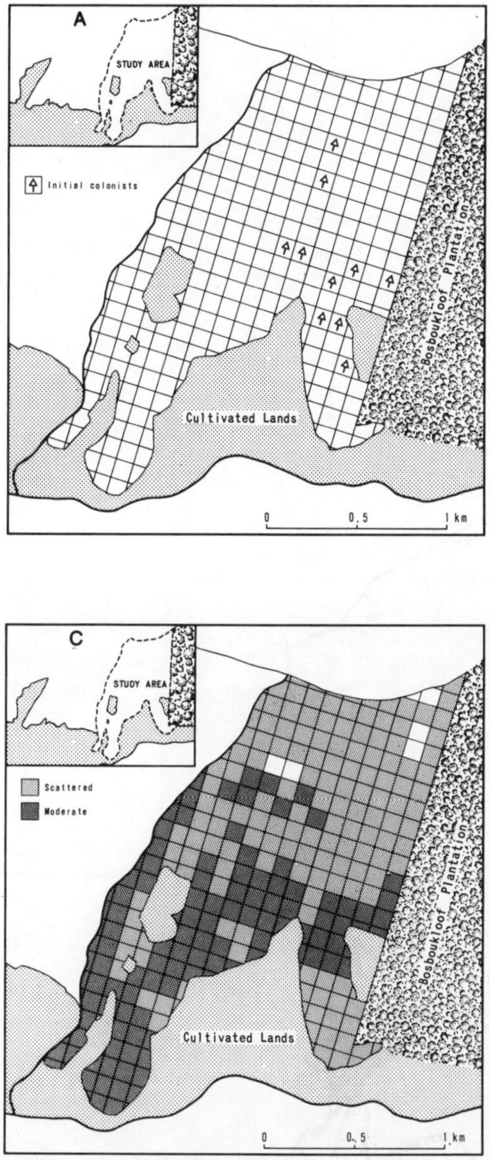

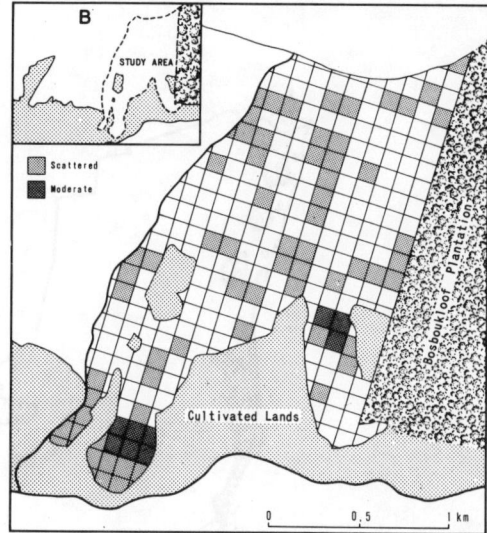

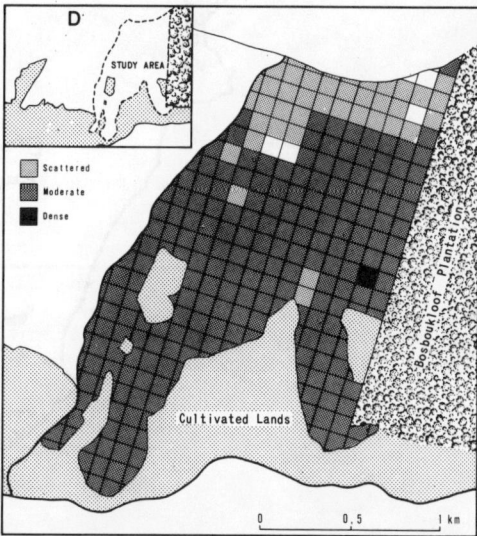

FIGURE 12.5 The pattern of invasion of *Pinus radiata* from a plantation in the Jonkershoek Valley near Stellenbosch. Figure A shows the position of initial colonists in 1953. Figures B, C, and D show the density of *P. radiata* trees in 1 ha blocks in 1953, 1967, and 1977 respectively. Reproduced from Richardson and Brown (1986).

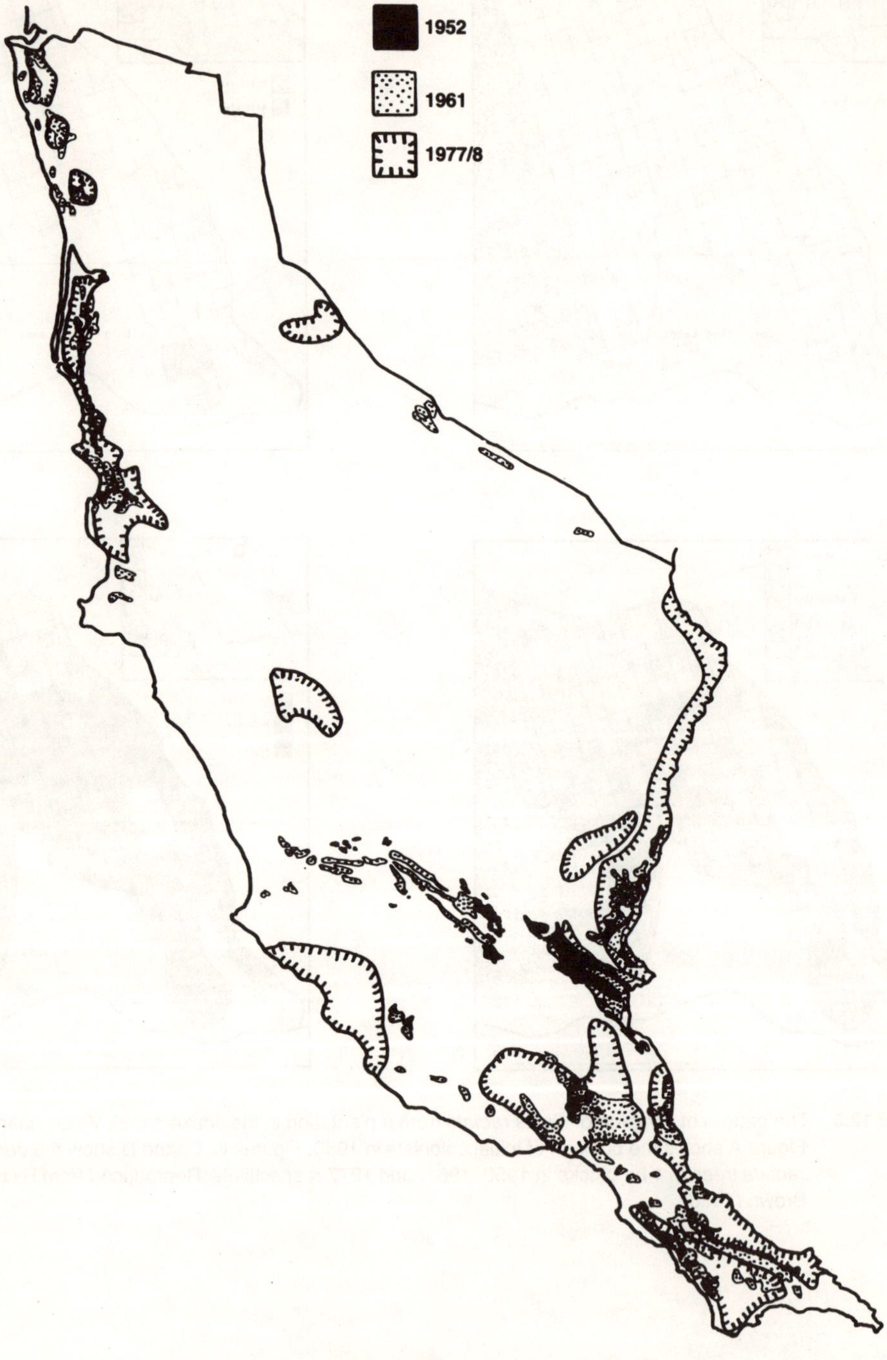

FIGURE 12.6 The distribution of dense stands of *Acacia cyclops* in the Cape of Good Hope Nature Reserve in 1952, 1961, and 1977/8. Redrawn from Macdonald, Clarke and Taylor (1989).

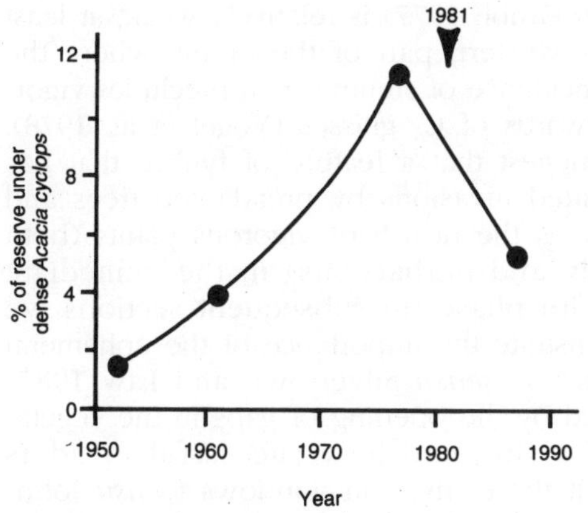

FIGURE 12.7 The increasing extent of dense *Acacia cyclops* stands in the Cape of Good Hope Nature Reserve between 1952 and 1987. The rapid increase in the area of the reserve under dense stands between 1952 and 1977 was despite the sporadic control operations initiated in 1941. Only following the implementation of systematic control measures in 1981 was the trend reversed. Calculated from data in Macdonald, Clarke and Taylor (1989).

sii, *Paraserianthus* (= *Albizia*) *lophantha* and *Sesbania punicea* (Macdonald and Jarman 1984). A helicopter survey of the lowlands in the western Cape in 1983 showed that *S. punicea* was present in 64% of the 532 km of river course surveyed (Bruwer 1983) — this only 17 years after the first warning of the potential of *S. punicea* as a weed in the biome (Hoffmann and Moran 1988).

DETERMINANTS OF PLANT INVASIBILITY

The many hypotheses that have been put forward to explain the success of invasive plants in fynbos address physiological and life history adaptations to the physical environment, dispersal mechanisms, species interactions, and other factors (see review in Richardson and Cowling in press). We support many of these arguments but believe that a comprehensive explanation of plant invasions requires further elucidation of the major determinants of community structure and of the interactions among these and the life history strategies of the invaders. For example, we need to know why fynbos can support dense stands of introduced trees when this life form is relatively uncommon in the indigenous flora, and why introduced herbs and grasses are confined to lowland fynbos and disturbed sites. In this section we examine the properties and processes (biotic and abiotic) that regulate the membership of fynbos communities by introduced plants.

Why can fynbos support dense stands of alien trees and shrubs?

Much of the fynbos biome is bioclimatically suited to tree growth (Miller et al. 1983) and yet indigenous trees are relatively uncommon in fynbos. Indigenous forests are generally restricted to riparian zones and sheltered ravines. Although deforestation and frequent burning have certainly reduced the extent of forests in parts of the biome (Hendey 1983; Manders et al. in press), the notion of large-scale displacement of forest by fynbos throughout the biome over the past 500 years as postulated by Acocks (1975) and others is untenable (Sugden 1989). Recent palaeoecological and biogeographic studies testify to the prevalence of fynbos over much of its present range during the late Pleistocene and Holocene (Linder et al. this volume). Despite this, the fynbos environment can sustain productive planted forests and is readily invaded by introduced trees. The ready incorporation of introduced trees into fynbos communities, and the higher steady-state biomass of invaded compared with non-invaded communities (see review in Richardson and Cowling in press) strongly suggest a 'surplus' of resources associated with a vacant niche in fynbos for this life form (Moll et al. 1980). Salient features that distinguish uninvaded fynbos from other mediterranean climate regions with equivalent rainfall and soil nutrient status are the low height and above-ground biomass of the vegetation, despite roughly equivalent photosynthetic capacities of dominant species, similar annual biomass increments in the first few years after fire, and analogous disturbance regimes (Richardson and Cowling in press). Fynbos shrubs are apparently constrained by the need to survive fire and to exploit the brief period after fire; they lack the capacity (either as individuals or as a community) to optimize resource use later in stand development (Richardson and Kruger 1990; Richardson and Cowling in press; Stock et al. this volume).

The tree flora of the Cape Province is of Afromontane origin, comprising mainly tropical species that migrated from the north-east (Cowling 1983; Linder et al. this volume). These taxa lack the mycorrhizal associates and other features required for success in low-nutrient fire-prone systems (Lamont 1983). Most have fleshy fruits (Knight 1988), a strategy not suited to the nutrient-deficient soils of the fynbos (Milewski and Bond 1982). These trees rarely invade fynbos communities, except in relatively nutrient-rich and mesic sites or where fires have been excluded for 40 years or more (Cowling et al. this volume). The indigenous trees that do regularly occur in fynbos (e.g. *Protea nitida*, *Maytenus oleoides*, and *Widdringtonia nodiflora*) lack mechanisms for rapid colonization of disturbed areas. Those pioneer trees that were present in the biome (e.g. *Casuarina*; Coetzee and Muller 1984) have become extinct, and the geographic isolation of the fynbos biome from a source of preadapted trees has prevented the natural immigration of this guild. For some reason that we cannot explain, a guild of pioneer trees has not evolved in the fynbos flora.

We suggest that the large discrepancy between the potential and actual steady-state biomasses for the fynbos due to the dearth of pioneer trees is an important determinant of invasibility (see also Richardson and Cowling in press). This notion has been expressed in various ways by several authors (Campbell et al. 1979; Moll et al. 1980; Deacon 1991). However appealing this hypothesis, it does not, on its own, account for the apparent openness of fynbos to invasion by trees. Many other treeless communities, notably grasslands, can also support trees. Unlike the fynbos, however, trees are usually excluded from these communities by fire, frost, or the resident biota (especially vigorous grasses and herbivores), or interactions among these factors (Richardson and Bond 1991). Frost is infrequent in the fynbos, and fire excludes only trees that take longer than about 8 years to mature. Herbivory levels in most fynbos areas are apparently low, and competition between tree seedlings and herbaceous plants in the regeneration niche (*sensu* Grubb 1977) is relatively weak, at least in the western part of the biome where the low incidence of summer rain precludes vigorous swards of C_4 grasses (Vogel et al. 1978). We suggest that a feature of fynbos that has permitted invasions by preadapted trees and shrubs is the dearth of vigorous plants (both woody and herbaceous) in the immediate post-fire phase. In subsequent sections we demonstrate the importance of the ephemeral resources (*sensu* Silvertown and Law 1987) created by the opening of gaps in the vegetation by fire, and how successful invaders exploit these invasion windows (*sensu* Johnstone 1986). Further, we illustrate how the relatively benign nature of environmental conditions (e.g. soil moisture) and of the resident biota of the fynbos region amplifies the invasibility of fynbos for preadapted trees and shrubs.

Determinants of distribution of introduced grasses and herbs

Many species of introduced grasses (Linder 1989) and herbs occur over large parts of the biome (e.g. Figure 12.2). In a survey of introduced annuals at five sites in lowland fynbos and one in mountain fynbos, Vlok (1988) found that the level of invasion (expressed as percentage cover) was negatively associated with the species richness and percentage cover of indigenous annuals and geophytes. All the study sites were disturbed by grazing, mowing, fertilizer application, mole rat activity, or a combination of these factors. Many of the alien species encountered at these sites are important weeds in other mediterranean climate regions (Kruger et al. 1989). Introduced herbs and grasses are conspicuously rare in untransformed mountain fynbos (e.g. McDonald and Morley 1988). Although 29 introduced annual herb species were recorded in the Cape of Good Hope Nature Reserve, all were confined to disturbed areas such as road verges (Macdonald et al. 1987).

Hobbs and Atkins (1988) have shown that disturbance and nutrient addition facilitated the establishment of introduced annuals in the western Australian wheatbelt. It is likely that nutrient enrichment (e.g. by pollution) of the oligotrophic soils of mountain fynbos would lead to increased invasion by annuals and other herbs. The one invaded mountain fynbos site surveyed by Vlok (1988) had been enriched by nutrients from adjacent cultivated lands.

The invasion of mountain fynbos by introduced herbs is also facilitated by afforestation or invasions by trees. Thus, the introduced annuals *Briza maxima* and *Hypochoeris radicata*, absent from the Biesievlei catchment near Stellenbosch before afforestation in 1945, were dominant species in the understorey 35 years later (Richardson and Van Wilgen 1986a).

The fynbos environment: the habitat barrier

Deacon et al. (this volume) provide a detailed account of the environmental factors that have shaped the biota of the fynbos biome. Very few studies have related invasive success or failure directly to environmental factors. Here, we discuss only a few salient abiotic features that appear to have influenced plant invasions in the fynbos since European colonization. We concentrate on factors that influence:

- the production and protection of viable seeds;
- dispersal; and
- germination.

THE PRODUCTION AND PROTECTION OF VIABLE SEEDS

The nutrient-poor, fire-prone conditions of the fynbos present several barriers for introduced plants. The tolerance of low nutrient availability is clearly a fundamental requirement for invasive success in the fynbos. Most of the successful invasive trees and shrubs produce relatively large, high quality seeds that permit rapid growth and survival of seedlings after fire (Dean et al. 1986). In the case of *Hakea sericea*, seedlings may grow for up to 125 days in sterile acid-washed sand before showing symptoms of mineral deficiency (Mitchell and Allsopp 1984). The nutrient-rich seeds of this species in the fynbos attest to the 'incredibly efficient mechanisms for mobilizing certain nutrient elements from leaves and fruits to developing seeds' evident in many species of the Proteaceae (Pate et al. 1986).

Legume seeds require a period of after-ripening to induce the hard seededness that causes dormancy (Tran and Cavanagh 1984). The typical mediterranean climate in most years favours the production of hard seeds in the summer-fruiting *A. cyclops*, *A. saligna*, and *A. longifolia*. Holmes (1989a) has ascribed the lack of dormancy in the 1986 seed crop of *Acacia saligna* at Pella to the atypically moist conditions which prevailed that year during the period between dehiscence and seed fall. The climate-induced after-ripening in *A. longifolia* and *A. saligna* accelerates markedly at the start of hot dry conditions. This is of great benefit to these species, since seeds are released in early summer, which allows time for seed burial by ants before the onset of the fire season (D Donnelley personal communication).

FACTORS INFLUENCING DISPERSAL

The abiotic dispersal of seeds and other propagules is an important factor in tree and shrub invasions, particularly in mountain fynbos (dispersal by wind) and along rivers (dispersal by water). The fynbos biome is subjected to high velocity winds throughout the year (Deacon et al. this volume). This factor has facilitated the rapid spread of introduced species with winged seeds. In those (western) parts of the biome with a true mediterranean-type climate, this factor is especially important since the strongest winds occur during the dry hot summers when most large fires occur and also when dehydration causes the partial opening of pine cones (at least in *P. pinaster* and *P. radiata*; *P. halepensis* seems to require fire; Richardson 1989). This means that most winged seeds are released (either when adult pines and hakeas are killed in fire or when seeds are released from dehydrated pine cones) when the strongest winds occur.

The riparian zones of the many streams that drain the rugged Cape Fold Belt mountains are another abiotic feature of the biome which has important implications for the dispersal of introduced trees and shrubs. These riparian zones are periodically scoured by spates that disperse and may scarify the hard-coated seeds of introduced legume species (Tran and Cavanagh 1984) and large buoyant fruits or seeds such as the acorns of *Quercus robur* (Knight 1985). The synchronous dispersal of propagules, the breaking of seed dormancy, and the preparation of safe sites (*sensu*

Harper et al. 1961) for establishment have facilitated the rapid spread of species such as *Acacia longifolia*, *A. saligna*, *Paraserianthus lophantha*, and *Sesbania punicea*.

FACTORS INFLUENCING GERMINATION

Late summer fires favour the recruitment of *Hakea* and *Pinus* species with canopy-stored seeds and hard-seeded legumes with soil-stored seeds by stimulating germination immediately before the onset of the winter rains. For canopy-stored seeds, the period of exposure to predators between dissemination and germination is minimized, and the period for seedling establishment, before the next unfavourable season, is maximized (see Bond 1984).

The seeds of *Hakea sericea* remain viable in a dried state after release but germinate quickly under moist conditions (Mitchell and Allsopp 1984). With frequent watering, the seeds of *H. gibbosa*, *H. sericea*, and *H. suaveolens* start germinating after 30 days, reaching maximum germination (50%, 98%, and 90% for the three species respectively) between 60 and 70 days (Richardson et al. 1987). Germination in *Pinus halepensis*, *P. pinaster*, and *P. radiata* seeds improves with exposure to low temperatures (stratification) (references in Dean et al. 1986), and thus is delayed until after the first cold period in autumn.

In contrast to the relatively thin-coated seeds of *Hakea* and *Pinus*, introduced *Acacia* species have thick, hard-coated seeds which may be water-impermeable and thus able to remain dormant in the soil (Cavanagh 1980). In many *Acacia* species, exposure to the heat of a fire renders the coat permeable. Because dormancy is controlled by the seed coat only (Cavanagh 1980), germination may then proceed as soon as conditions are suitable. Seeds in the litter layer may be killed by fire, whereas those buried in the soil may experience a temperature pulse which simply breaks dormancy (Holmes 1986b; Pieterse and Cairns 1986a). Both *A. longifolia* and *A. saligna* have a rapid, positive response to heat treatment (Pieterse and Cairns 1986b; Jeffery et al. 1988). By contrast, *A. cyclops* readily loses dormancy in the absence of fire (Holmes 1989a), does not respond to dry heat treatment, and may rather respond to fluctuating temperatures experienced on a bare soil surface (Jeffery et al. 1988). *Acacia cyclops* is bird-dispersed and seeds are scarified during passage through the bird's gut (Glyphis et al. 1981). *Acacia* seedling emergence from the upper 50 mm of soil was correlated positively with mean daily rainfall (Holmes 1989c) and negatively with mean daily net evaporation (Holmes 1989b). In both studies, seedlings emerged throughout the winter months, indicating that winter temperatures do not limit germination. *Acacia cyclops* seedlings emerge in the shade of a closed stand (Holmes 1989b) but may not survive in shade exceeding 75% (Milton 1982).

Disturbance regimes

Disturbance has long been cited as a major 'cause' of biological invasions (Elton 1958; Harper 1965; Baker 1986; Crawley 1987; Fox and Fox 1986; Orians 1986; Rejmanek 1989), and the fact that introduced trees and shrubs have invaded apparently 'undisturbed' fynbos communities has been considered to be unusual (Kruger 1977a, b; Mooney 1982; Crawley 1989; Drake et al. 1989; Hobbs 1989). But just what is understood by 'disturbance'? We distinguish between the 'natural' disturbance regime and the prevailing (human-modified) disturbance regime; it appears that what is perceived as 'disturbance' in most of the literature on invasions is neither of these, but rather the difference between the two ('perturbation' *sensu* White and Pickett 1985 or 'exogenous disturbance' *sensu* Fox and Fox 1986). If the extent of agricultural land in quarter-degree squares is taken as a crude index of the level of 'perturbation' in different parts of the biome, then perturbation is a significant factor regulating invasions. The extent of agricultural lands in quarter-degree squares is positively correlated with the occurrence and abundance of all the major invasive trees and shrubs (Table 12.3) (see later). The positive association between the level of perturbation and the extent of invasion is also shown by comparing the invasive floras in nature reserves in the biome. The most heavily 'perturbed' reserves in the fynbos biome (e.g. Bontebok National Park, Cape Flats Nature Reserve, Rondevlei, and Tygerberg) have the highest proportions of introduced species (Table 12.4). In these examples, the abun-

TABLE 12.4 The sizes of native and introduced floras in reserves in the fynbos biome, with an analysis of the composition of the introduced floras in terms of life forms. Trees and shrubs are most prominent in mountain fynbos reserves which have highly leached soils, whereas the more fertile soils of lowland fynbos favour herbaceous species (data from Kruger et al. 1989).

Vegetation type and reserve	Size of flora		Percentage of introduced species in given category		
	Native	Introduced	Trees and shrubs	Herbs	Other
Mountain fynbos					
Cape of Good Hope	1 050	71	42	56	1
Fernkloof	773	31	48	48	3
Jonkershoek	1 231	53	43	49	8
Table Mountain	1 362	78	50	45	5
Vogelgat	697	25	32	64	4
Zachariashoek	623	8	38	63	0
Lowland fynbos					
Cape Flats	165	34	6	91	3
Goukamma	356	15	27	67	7
Pella	379	18	28	72	0
Rondevlei	229	42	24	62	4
Renoster shrubland					
Bontebok National Park	446	46	46	50	4
Tygerberg	373	116	20	74	6
Strandveld (thicket)					
Rocherpan	207	12	8	83	9
Fynbos-karoo transition					
Nieuwoudtville	276	19	5	84	11

dance of important invasive trees and shrubs and the large alien floras in modified environments are probably largely attributable to high levels of intentional planting (and therefore the enhanced availability of propagules) and to the increased frequency of soil disturbance resulting in reduced plant cover which relieves competitive inhibition. Other contributing factors associated with increased soil disturbance are increased dispersal of propagules and modified fire regimes. In addition, these perturbed areas often carry abnormally high densities of large herbivores (e.g. Prŷs-Jones et al. 1985), resulting in further disruptions to natural disturbance regimes (Macdonald et al. 1987) and a relaxation of competition. On a smaller scale, Macdonald (1984) demonstrated that *Acacia cyclops* and *A. saligna* proliferated more rapidly on regularly disturbed portions of the Pella research site (on the lowlands of the western Cape). Although mountain fynbos reserves have relatively small invasive floras (Table 12.4), those few species have invaded extensively. Invasions in mountain fynbos are driven by the fire regime but can occur without any exogenous disturbance.

Whereas disturbance regimes in much of the extant lowland fynbos and strandveld have been noticeably altered by modern human habitation (Moll and Bossi 1984), most mountain fynbos communities remain in a relatively pristine condition (Aschmann 1973; Moll and Bossi 1984). Although the contemporary large mammal fauna of the biome differs considerably from that during the mid-Pleistocene when people first occupied the region

(Hendey 1983; Deacon this volume), the elimination of species such as elephant, black rhinoceros, hippopotamus, eland, and red hartebeest is unlikely to have influenced montane plant communities substantially since these species were concentrated in the lowlands (Skead 1979). The use of fire by pre-historic pastoralists to promote the growth of palatable plants probably reduced tree and shrub cover in parts of the region (Hendey 1983).

The natural disturbance regime in much of the fynbos biome is characterized by intense fires at intervals of between 6 and 30 years (Kruger 1977b; Kruger and Bigalke 1984). As is the case in other mediterranean climate regions, fire plays a fundamental role in structuring plant and animal communities in the fynbos, and therefore in regulating invasions. Fire may encourage invasion by activating mass seed release from serotinous cones, by stimulating germination of soil-stored seeds, and by creating safe sites for germination and seedling survival. Fires of moderate to high intensity may kill seeds stored close to the soil surface (to a depth of 40 mm or more; Holmes 1989b). Some invaders with soil-stored seeds may have an advantage over equivalent fynbos species in that their seeds can survive higher soil temperatures e.g. *Acacia saligna* versus *Podalyria calyptrata* (Jeffery et al. 1988). However, other species may be less adapted to fire e.g. *A. cyclops* (Jeffery et al. 1988). The dichotomy in response of the two invasive *Acacia* species is corroborated by evidence from the Pella research site where *A. saligna* was shown to be favoured, and *A. cyclops* reduced, by repeated burning (Macdonald and Knight 1985). Fire may also exclude a seed-regenerating invader directly if it occurs at intervals less than the invader's juvenile period or if, as in the case of *Hakea salicifolia*, the seeds are inadequately insulated against the heat of crown fires (Richardson et al. 1987). There have been no detailed studies of the effects of disturbances other than fire on invasions.

Biotic factors

Interactions between introduced species and the resident biota have both restricted and facilitated plant invasions in the fynbos. Many types of interactions are involved (e.g. competition, predation, herbivory, disease, parasitism, mutualisms) but few have been studied in any detail. The role of human beings in plant and animal invasions has already been discussed. Rather than catalogue all potentially significant factors, we have chosen only a few salient factors that are implicated in invasions of the most important species.

BIOTIC FACTORS THAT REDUCE INVASIBILITY

Competition from resident plant species often prevents invasions. The inhibiting effect of competition is most often appreciated when disturbance facilitates invasion through the removal of competition. The lack of vigorous competition from herbs and grasses in the early post-fire phase in fynbos is considered an important factor contributing to the ready invasion of fynbos by pines. The importance of competition in the regeneration niche (*sensu* Grubb 1977) in determining the distributional limits of pines throughout the world, and the apparent openness of fynbos to invasion by pines have been discussed elsewhere (Richardson 1990; Richardson and Bond 1991; Richardson and Cowling in press). The relative abundance of C_4 grasses in 'Waboomveld' (vegetation dominated by *Protea nitida*) (Linder 1989) suggests that these communities should be less open to invasion by introduced trees and shrubs, but this hypothesis remains untested.

Seed predation by birds and rodents is known to reduce recruitment in *Acacia* spp. (Holmes 1990), *Hakea sericea* (Richardson 1985), *Pinus* spp. (Fish 1989), and *Quercus robur* (Knight 1985). There is no evidence, however, that post-dispersal seed predation has *prevented* invasion. Introduced plants have few enemies in the fynbos (e.g. Hoffmann and Moran 1988 for *Sesbania punicea*). Indigenous insects feed on the seeds of *A. cyclops* (Holmes and Rebelo 1988) but the effect on recruitment and invasion is apparently small in most parts of the biome. The success of biocontrol programmes — especially those involving seed-attacking insects — underscores the importance of predator release in plant invasions in the fynbos. For example, Weiss and Milton (1984) report an average of 5.6 viable seeds m^{-2} in the soil under *A. longifolia* canopies in Australia, compared with

7 370 viable seeds m^{-2} in the Cape. Similarly, Richardson et al. (1987) counted up to 570 seeds per shrub on *Hakea gibbosa* in the Cape, nearly four times the maximum reported by Beadle (1940) for Australia. Neser (1984) has suggested that *H. sericea* 'may have evolved the capacity to produce a superabundance of seeds to be able to survive in the presence of the specialized seed-attacking insects' (see also Neser and Kluge 1985).

Pathogenic fungi on pines, although problematic in forest nurseries in the biome, have not influenced invasions to any extent. Similarly, the susceptibility of certain *Banksia* species to infection with the pathogenic fungus *Phytophthora cinnamomi* in cultivation in the Cape (Von Broembsen 1984) is not thought to reduce the invasive potential of some species (Richardson et al. 1990). Die-back in stands of *H. sericea* throughout its range in the fynbos has been attributed to infection by the fungus *Colletotrichum gloeosporioides* (Morris 1982). This die-back, first reported in 1966, causes large-scale mortality in some areas and has the potential to reduce substantially the invasiveness of *H. sericea* (Richardson and Manders 1985). The fungus is thought to be indigenous to the Cape (M J Morris personal communication).

BIOTIC FACTORS THAT INCREASE INVASIBILITY

One of the most striking mutualisms between introduced plants and resident biota involves seed dispersal of certain invasive trees. Seeds of the Australian *Acacia* species are dispersed by various indigenous and introduced birds (Broekhuyzen 1960; Middlemiss 1963; Winterbottom 1970; Glyphis et al. 1981; Pieterse 1986; Hofmeyr 1989; Knight and Macdonald 1991), mammals (Davidge 1978; David 1980; Taylor and Macdonald 1985), and ants (Pieterse and Cairns 1990; Holmes 1990). Holmes (1989c, 1990) suggests that indigenous ants may play a critical role in accumulating (burying) seeds in sparse stands, out of reach of resident granivores which have the potential to consume the entire seed crop. Some introduced plants have established opportunistic dispersal mutualisms with animals introduced from areas far removed from the natural range of the plant. The introduced European starling is an important disperser of Australian *Acacia* seeds (Glyphis et al. 1981). Similarly, the introduced squirrel *Sciurus carolinensis* is the principal disperser of *Pinus pinea* seeds (Millar 1980). Self-sown stands of *P. pinea* are almost always aggregated close to stands of another introduced tree, *Quercus robur*, which is the principal source of food for the squirrels (Millar 1980) but which is restricted to riparian zones (Knight 1985; Richardson 1989).

Mycorrhizae increase photosynthetic efficiency, improve nutrient uptake and drought resistance, and enhance resistance to pathogenic fungi (Donald 1979). The lack of mycorrhizal symbionts has prevented the establishment of trees such as pines in several parts of the world (references in Marais and Kotzé 1978). Pearson (1950) has suggested that 'it is not unlikely that the many trees that failed when transplanted to the soil of South Africa were unsuccessful because their mycorrhizae found conditions unfavourable'. The early introduction of mycorrhizal symbionts with wide host ranges, and the rapid spread of these symbionts, probably facilitated the successful establishment of pines (which seem to rely on introduced fungi for mycorrhizae) in the fynbos (see Pearson 1950). Hoffman and Mitchell (1986) have suggested that vesicular-arbuscular mycorrhizae on *A. saligna* have enabled this species to cope with the low nutrient conditions in the fynbos.

WHAT MAKES A PLANT A SUCCESSFUL INVADER OF FYNBOS?

Of the hundreds of plant species of all life forms introduced to the Cape, and of the 367 species considered to be naturalized (Kruger et al. 1989), fewer than 20 species (mostly trees and shrubs) have invaded natural (unmodified) fynbos. The fact that so few introduced species have been incorporated into fynbos communities indicates the stringent requirements for membership of these systems. (We exclude the many species that 'invade' in severely modified communities e.g. the roadside weeds and pests of agriculture that may occur transiently in some fynbos communities.) In this section we describe salient features of the biology and ecology of those species that have been successful, and explore reasons for the differences in invasive success among congeners. In an attempt to gain a predictive understanding of what makes a successful invader in fynbos, we explore the barriers and windows to invasion: we define templates of invasibility and examine interactions between life history strategies and invasiveness.

What species have invaded?

The most successful invasive trees and shrubs in the fynbos display a wide range of life history traits (Table 12.5). All species except *Pinus halepensis* have juvenile periods of five years or less. Seed mass ranges from 0.3 mg (*Leptospermum laevigatum*) to 82 mg (*Paraserianthus lophantha*). Seeds of the *Acacia* species and *P. lophantha* are dispersed mainly by water and animals and accumulate in the soil. The *Hakea* and *Pinus* species have canopy-stored seeds that are dispersed by wind. Four species resprout after fire, although this capacity is poorly developed in *A. cyclops*. Tall individuals of *P. pinaster* and *P. radiata* sometimes survive ground fires of moderate intensity. The matrix of life history traits defines two basic reproductive syndromes associated with fire resilience: one describes the hard-seeded legumes, the other characterizes the hakeas and pines. It is, however, more useful to compare the performance as invaders of closely related taxa (e.g. congeners) that have shown different degrees of success. It is frequently difficult to isolate inherent features of the invader that have facilitated success from the cumulative effects of extrinsic factors. Nevertheless, this approach does allow one to identify sets of life history traits (i.e. strategies) that have been successful. This has been useful in the delimitation of invasion windows (see later). In this section we contrast successful with less successful traits and strategies in the genera *Acacia*, *Hakea*, and *Pinus*.

ACACIA CYCLOPS, *A. LONGIFOLIA*, AND *A. SALIGNA*

Australian *Acacia* species have been remarkably successful as invaders of fynbos; seven of the 15 most important invasive plants in the biome belong to this genus (Macdonald and Jarman 1984). The most important invasive *Acacia* species have fairly similar life history traits (Table 12.5), but some differences have important implications for invasion in fynbos. *Acacia saligna* has large persistent seed banks in the soil (because of high percentage viability and water-impermeable dormancy). Seed banks accumulate rapidly and germination is cued by disturbance (a disturbance-coupled seed bank *sensu* Grubb 1988). The seeds, which are primarily ant-dispersed, are long-lived unless stimulated to germinate by fire (Holmes 1989c). *Acacia cyclops* also has large seed banks, but seeds have variable viability and dormancy, and most seeds survive less than a year (Holmes 1989c). Seed banks of *A. cyclops* are disturbance-uncoupled (see also Hall 1961). Both species have been successful invaders, but apparently by using different strategies. For *A. saligna*, the key processes are the extended seed longevity and rapid germination after fire. *Acacia cyclops*, however, has capitalized on the opportunistic dispersal mutualism with local animals and its ability to germinate without disturbance-related cues; the relatively small seed bank is a buffer against local extinction (Holmes 1989c). Whereas *A. saligna* releases most of its annual seed crop soon after maturation (Milton and Moll 1982), *A. cyclops* retains a large proportion of the seed crop in the canopy for several months: O'Dowd and Gill (1986) report over 80% retention after 180 days in Australia. These two species seem to typify the arillate (*A. cyclops*) and non-arillate (*A. saligna*) dispersal syndromes described by O'Dowd and Gill (1986). The salient differences in seed bank dynamics between the two species are shown in Figure 12.8. Another successful *Acacia* species in fynbos, *A. longifolia*, appears to be intermediate between these two species, but more similar to *A. saligna*. It shares with *A. saligna* the disturbance-coupled, large, viable seed bank with water-impermeable dormancy (Pieterse and Cairns 1987, 1988), but enjoys to a moderate extent the dispersal advantages of *A. cyclops* (Pieterse 1986). *Acacia longifolia* does not have the extended fruiting season of *A. cyclops*, nor does it 'advertise' its seeds to the same extent (Milton and Moll 1982), suggesting that it is not primarily adapted for seed dispersal by birds. Despite this, the dispersal of seeds by Red-winged starlings, *Onychognathus morio*, has been an important factor in the invasion of mountain fynbos by *A. longifolia* (Pieterse 1986; D M Richardson unpublished data).

TABLE 12.5 The life history traits of major invasive plants in the fynbos biome. Species are listed in order of importance (ranked according to the current magnitude of invasion, the extent of potential habitat, and the potential rate of spread; Macdonald and Jarman 1984). Data are from Richardson and Cowling (in press). Information on the history of introduction and current extent of invasion for each species is given in Table 12.1.

Species	Life form	Juvenile period (years)	Fresh seed mass (mg)	Seed storage strategy[1]	Dispersal agent[2]	Fire tolerance (adults)[3]	Sprouting ability[4]
Acacia saligna	Tree/shrub	3	16.4	S	Wa/A	0	2
Acacia longifolia	Tree/shrub	3	10.8	S	Wa/B/A	0	0
Acacia cyclops	Tree/shrub	3	35.0	S	B/Wa	0	1
Pinus pinaster	Tree	6	45.0	C1	Wi	1	0
Leptospermum laevigatum	Tree/shrub	5	0.3	C1	Wa/Wi	0	0
Hakea sericea	Tree/shrub	2	29.7	C2	Wi	0	0
Pinus radiata	Tree	5	34.0	C2	Wi	1	0
Pinus halepensis	Tree	15	16.0	C2	Wi	0	0
Paraserianthus lophantha	Tree	3	82.0	S	Wa	0	0
Acacia melanoxylon	Tree	3	16.7	S	Wa	0	2
Acacia mearnsii	Tree	3	12.8	S	Wa	0	2
Hakea gibbosa	Tree/shrub	2	37.1	C2	Wi	0	0

1 S = Soil-stored; C1 = canopy-stored (weakly serotinous); C2 = canopy-stored (strongly serotinous).
2 A = ants; B = birds; Wa = water; Wi = wind (listed in order of importance).
3 0 = poor (thin bark); 1 = moderate (adult plants occasionally survive fires); 2 = good (adult plants often survive fires).
4 0 = does not sprout; 1 = occasionally sprouts, but sprouting does not ensure persistence after fire; 2 = good sprouting ability.

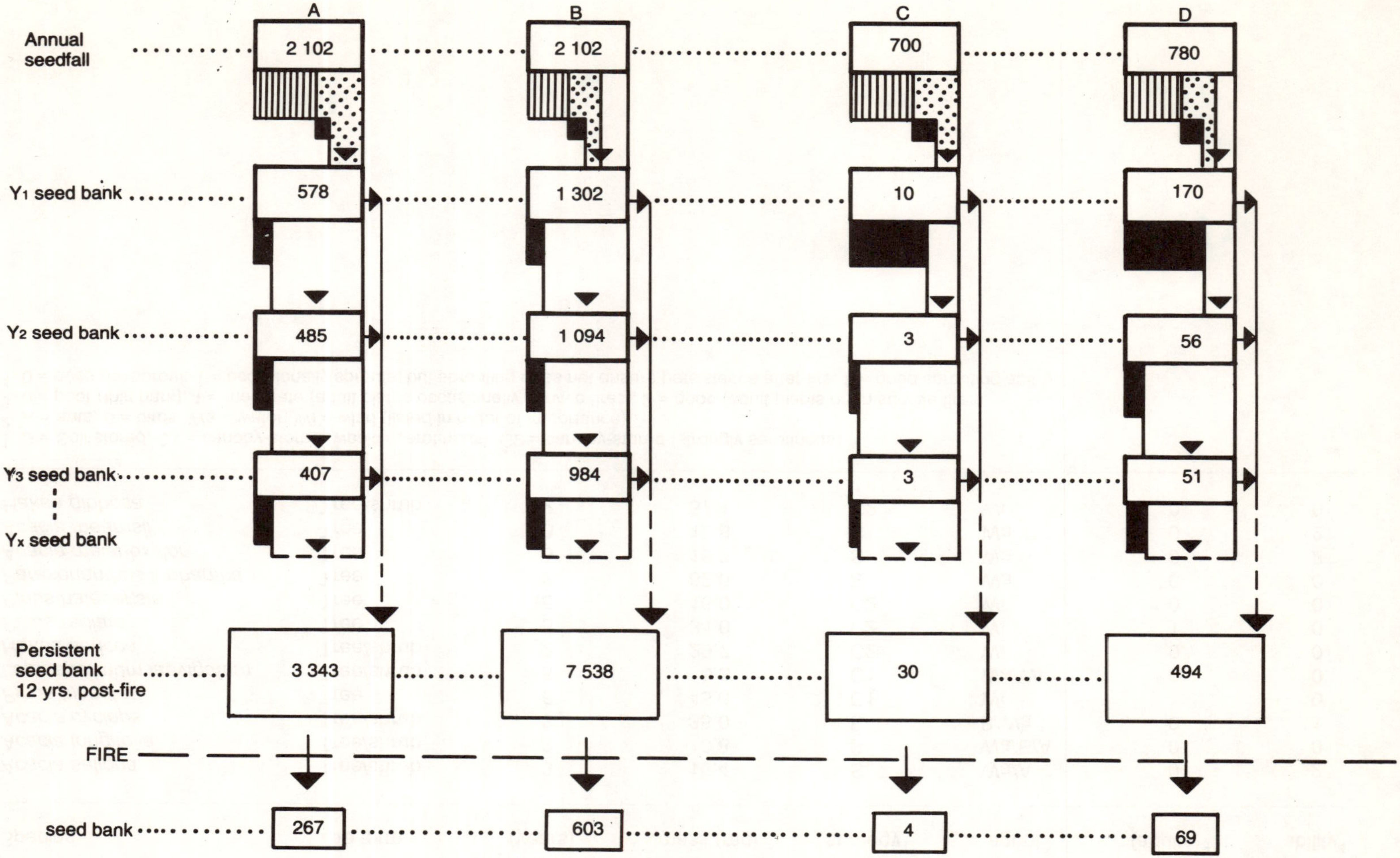

FIGURE 12.8 Seed bank dynamics and regeneration after fire in nascent foci (A and C) and established foci (B and D) of *Acacia saligna* (A and B) and *A. cyclops* (C and D). Arrow breadths are proportional to transition probabilities: open arrow = seed survival; striped arrow = predation; dotted arrow = ant dispersal; solid arrow = loss through germination and decay. Numbers are seeds m^{-2}. Both species have large soil-stored seed banks. However, whereas the seed bank in *A. saligna* is persistent and disturbance-coupled, the seeds of *A. cyclops* have variable viability and dormancy, germinate independent of disturbance, and usually survive less than a year. Based on information in Holmes (1989c).

FOUR *HAKEA* SPECIES

Four shrub species of the Australian Proteaceae (*Hakea sericea, H. gibbosa, H. suaveolens,* and *H. salicifolia*) were introduced to the Cape between 1840 and 1860. *Hakea sericea* is highly invasive, *H. gibbosa* and *H. suaveolens* are moderately invasive, and *H. salicifolia* has not invaded natural vegetation. The most important trait which separates *H. sericea* from the other species is its large seed bank in its adopted environment; the seed production of *H. sericea* is four times that of *H. gibbosa* and more than 16 times that of *H. suaveolens* for shrubs of 8 kg above-ground dry mass. These three species all form dense thickets in fynbos, but only *H. sericea* colonizes distant areas. The limited areal expansion of *H. gibbosa* and *H. suaveolens* was attributed to low seed numbers. *Hakea salicifolia* produces abundant seeds in the Cape, but these are unable to survive fynbos fires due to inadequate insulation by the small follicles (Richardson et al. 1987).

PINUS SPECIES

More than 50 *Pinus* species have been planted in the Cape (Poynton 1979). Although many of these are naturalized (i.e. reproduce naturally), only five species are listed among the 33 most important plant invaders in the fynbos biome (Macdonald and Jarman 1984). The most successful invasive pines in mountain fynbos (*P. halepensis, P. pinaster,* and *P. radiata*) are fire-resilient species characterized by relatively small seed size, low seed/wing loadings, short juvenile periods, moderate to high degrees of serotiny, and poor fire tolerance as adults (Richardson et al. 1990). These species and several others form a discrete group in the genus that are preadapted for colonization and growth in nutrient-poor fire-prone environments (McCune 1988; Richardson et al. 1990).

Comparing the success of congeners, and relating this to life history strategies provide a useful perspective of critical phases in the invasion process in fynbos. A problem with this approach is that these are natural experiments with no controls. No two introduced species have the same history of introduction and dissemination. The outcome of an introduction depends on factors inherent to both the organism and the environment, but extrinsic factors can override these considerations (Richardson 1989). To factor out the role of extrinsic factors, it is necessary to ascertain how the life history of the invader interacts with emergent properties of the target community. In the next section, we define the invasion windows that exist in the fynbos.

Defining plant invasion windows

A characteristic feature of many successions seems to be the dependence of invading species on just the right window in the successional sequence to enter the community and proliferate.

(Van Hulst 1987)

We recognize three basic categories of invasion (or succession *sensu* Van Hulst 1987) among the major tree and shrub invaders in fynbos. One characterizes the invasion and proliferation of serotinous, wind-dispersed trees and shrubs such as pines and hakeas in mountain fynbos. A second category of invasions involves trees and shrubs with soil-stored seeds that are dispersed by animals, water, wind, or a combination of these (e.g. Australian *Acacia* species). The third type includes invasions of hard-seeded legumes and other taxa along river courses. All three types of invasion are regulated by environmental factors, dynamic processes in the target community, life history strategies of the invaders, and interactions among these factors. In this section, we use information detailed in previous sections to develop models for these categories of invasion. We then compare the life history strategies of invaders in the first two categories.

INVASIONS BY SEROTINOUS WIND-DISPERSED SPECIES

The incorporation of introduced serotinous trees and shrubs into mountain fynbos communities depends on fire-induced changes in community structure. Fires cause large spatial and temporal fluctuations in the population sizes of component species, especially the non-sprouting overstorey shrubs, and therefore alter competitive hierarchies. The recurrent fires thus create transient invasion windows by simultaneously activating seed release from serotinous cones and introducing fluctuations in community connectance (*sensu* Pimm 1989) when the non-selective botanical barrier (*sensu* Johnstone 1986) is removed. Local extinction and invasion (or recolonization) of certain native components are normal and regular processes (Cowling and Gxaba 1990), but only in the immediate post-fire phase (*sensu* Kruger and Bigalke 1984). Mountain fynbos communities appear to be resistant to invasion by indigenous species in subsequent phases of succession (Kruger 1984). Invasions by introduced species are also largely confined to the immediate post-fire phase (Kruger 1977a; Richardson and Brown 1986; Richardson 1988; Fish 1989) and can proceed with no exogenous disturbance. We propose that the success of invaders such as *Pinus pinaster* (Kruger 1977a; Richardson 1989) or *P. radiata* (Richardson and Brown 1986) can be attributed to the way they exploit this invasion window. The two fundamental features that characterize the life history strategies of these successful invaders are their inherent dispersal ability and their superior fire resilience when compared with the native species. The first feature improves mobility and facilitates invasion. The second buffers established populations against local extinction, thus disrupting the prevailing non-equilibrium system (Figure 12.9). A more comprehensive account of this hypothesis is given by Richardson and Cowling (in press).

INVASIONS BY HARD-SEEDED ANIMAL-DISPERSED ACACIAS

Invasion commences with the deposition of seed under perching sites, typically in clumps associated with indigenous fruit-bearing plants (Glyphis et al. 1981). As the initial colonists mature, these attract vertebrate dispersers that deposit seeds of various species and disperse seeds outwards. As the complexity (species richness of fruiting species) of the clump increases, so long-distance dispersal becomes more effective (greater numbers of more species of dispersers are attracted and more seeds are dispersed). Also, as seed production increases, so the level of short-distance dispersal by vertebrates, ants (D Donnelley personal communication), and wind (Pressinger 1985) increases. Dispersal by vertebrates is especially important in creating new satellite populations, whereas dispersal by ants and wind leads to lateral expansion of foci. Both types of dispersal lead to the coalescence of foci to form continuous stands. For resprouting species (*A. saligna*), resilience to fire is achieved as soon as colonists are established. For non-sprouting species (*A. cyclops* and *A. longifolia*) resilience (immunity of populations to local extinction through the effects of disturbances such as fire) is achieved only once seed reserves have accumulated in the soil.

A COMPARISON BETWEEN THE TWO CATEGORIES OF INVASIONS

Both types of invaders spread as a ragged edge of isolated colonists rather than as an advancing front. The first stage of invasion is therefore effectively similar for these two groups despite the different life history strategies of the invaders. Both categories of invaders have enjoyed wide dissemination by humans (i.e. the creation of artificial nascent foci with large contact areas between seed sources and fynbos). This has to some extent blurred the inherent differences in population growth rates and dispersal. For serotinous invaders, propagules are introduced to the community by wind dispersal mainly immediately after fire, and are deposited relatively evenly over a range of microsites. Propagules of animal-dispersed invaders are introduced especially to older vegetation, and are clumped around fruit-bearing plants and other perch sites.

The processes involved in thicket development (the expansion and coalescence of scattered foci) are quite different for the two groups, because of the fundamental differences in seed bank dynamics and the responses of seeds to fire. These differences have important implications for explaining the differences in distribution of different species, and for planning control operations. The accumulation of seed banks is critical for establishing the level of population resilience shown by invaders such as *Acacia longifolia* (Pieterse and Cairns 1988) and *Pinus halepensis* (Richardson 1988). Fire is of fundamental importance for both types of invaders. Disturbance-coupled recruitment is critical for most species in both groups (*Acacia cyclops* is

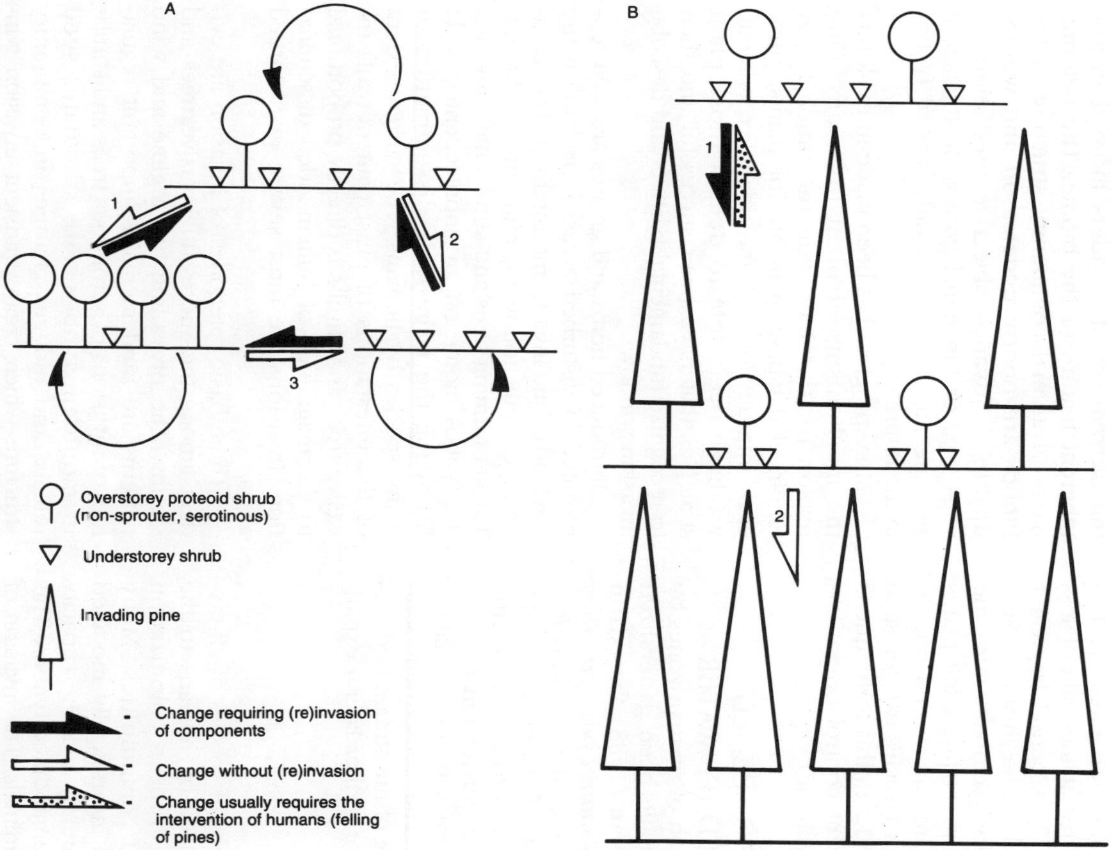

FIGURE 12.9 A schematic representation of fire-induced changes in the structure of mountain fynbos shrublands: A = in the absence of invasive pines (note the cyclical replacement processes); B = invaded by pines (modified from Richardson and Cowling in press). In the absence of invasive pines, the recruitment of overstorey shrubs is highly variable. They may proliferate and suppress understorey species (A1). Adverse conditions during or after a fire may result in the poor recruitment or local extinction of overstorey shrubs (A2). The restoration of either condition requires recolonization by components. Once pines are established in fynbos communities, their superior fire resilience disrupts the prevailing non-equilibrium system. As the invaders proliferate (B2), competition with fynbos elements is intensified, eventually leading to their local extinction as residual seed stores are depleted. There is no cyclical replacement without abnormal conditions (e.g. the felling of pines by humans), and a depauperate steady-state results.

apparently an exception, Holmes 1989c). Once thickets have developed, populations are usually self-sustaining and the only way to restore the natural non-equilibrium condition is to initiate an abnormal sequence of disturbances. Fire is the only major disturbance in fynbos, and fuel accumulation rates are too slow to allow intense fires at intervals of less than about eight years. This period is greater than the juvenile period of the invaders. Successive fires can therefore seldom return the system to non-equilibrium. Felling adult plants and then burning the area after seed release from serotinous cones or follicles can reinstate non-equilibrium (Figure 12.9); this technique is the foundation of current control operations (Van Wilgen et al. this volume).

INVASIONS BY HARD-SEEDED LEGUMES DISPERSED BY WATER

The processes of invasion of riparian zones by hard-seeded legumes have been generalized by Hoffmann and Moran (1988), based on their observations of *Sesbania punicea.* They suggest that the presence of dense stands in a river obstructs the flow of water, especially during flooding. This leads to erosion of the watercourses and to the conversion of well-defined rivers into diffuse systems of shallow streamlets and trickles. The consequent sedimentation and widening of the stream bed create an ideal substrate for further lateral expansion of the stand.

CONCLUSIONS

Very few of the hundreds of species of all taxonomic groups introduced by humans to the fynbos biome since European colonization have invaded natural communities. Many species are established in artificially modified environments but apparently lack the capacity to enter and persist in unmodified fynbos. This is particularly true for animals. The invasion of natural communities by certain plant species is unusual. The few plant species that have invaded have had exceptionally severe impacts. No animals have had similarly devastating effects.

All the successful vertebrate invaders in fynbos are well-known invaders in other parts of the world; they are essentially commensals of humans. In all cases, some exogenous disturbance to the receiving community is a prerequisite for faunal invasions. In particular, it appears that factors which reduce pressure from native predators facilitate these invasions. Dispersal characteristics are important and apparently regulate the rates of invasion by the *current* set of invaders more than any inherent feature of the biome. The determinants of animal community structure in the fynbos are poorly understood and we are unable to advance objective predictions of what type or functional groups of animals, if any, are most likely to invade natural fynbos in the future.

More progress has been made in exploring the mechanisms of plant invasions. One reason for this has been the much larger empirical database available for plants. Since each introduction is a transplant experiment, we have been able to delimit those plant strategies that have been successful and then to recognize fundamental barriers and describe invasion windows.

Introduced herbs and grasses are well represented in disturbed areas, especially in the lowlands, but invade mountain fynbos communities only following nutrient enrichment. Thicket-forming trees and shrubs have invaded the fynbos biome on an unprecedented scale following the large-scale dissemination of many species by humans. The current extent of the infestations of most (but not all) invasive trees and shrubs is directly proportional to the magnitude of human-aided dissemination. Invasions are most severe in disturbed areas.

Three basic categories of invasion are evident among the successful invasive trees and shrubs. The prevailing fire regime and wind patterns are fundamentally important regulators of invasions of serotinous trees and shrubs in mountain fynbos. Fire stimulates seed release and induces changes in community structure which create transient invasion windows. Fire is also an important factor in the invasion of lowland communities. Human-aided dissemination and the opportunistic establishment of other dispersal mutualisms with indigenous or introduced biota have been important for the major invaders of the lowlands. Once established in fynbos communities, the superior fire resilience of the invaders disrupts the prevailing non-equilibrium system.

As the invaders proliferate, competition with fynbos elements is intensified, eventually leading to their local extinction as residual seed stores are depleted. The rapid spread of several species along watercourses in the biome can be attributed to the efficient dispersal of seeds in water and the scouring of stream banks during floods which provides safe sites for establishment.

Although some of the major tree invaders of fynbos, notably the pines, also invade in other parts of the world (Richardson and Bond 1991), there can be little doubt that fynbos communities are exceptionally open to invasion by this life form. These invasions proceed at a similar rate to the spread of pines across deglaciated landscapes of the northern hemisphere in the early Holocene i.e. several kilometres per generation (cf. Delcourt and Delcourt (1987) for *Pinus contorta* in North America and Birks (1989) for *P. sylvestris* in the British Isles). Available evidence suggests that a major determinant of these invasions is the profound imbalance in ecosystem-level resources which is probably due to the geographic isolation of the biome which has prevented immigration, and historical events that have regulated speciation and extinction in the fynbos biota.

ACKNOWLEDGEMENTS

The work of D M Richardson formed part of the conservation forestry research programme of the South African Forestry Research Institute (now the Division of Forest Science and Technology, CSIR). I A W Macdonald's research was funded by the Nature Conservation Research Committee of the Foundation for Research Development (FRD). P M Holmes and R M Cowling acknowledge financial assistance from the Core Programme of the FRD. We thank the South African Bird Atlas Project and the National Botanical Institute for allowing the use of unpublished data. L Henderson, H P Linder, P J Pieterse, M C Rutherford, and W F Sirgel kindly provided unpublished distribution data. R K Brooke, A J Prins, H G Robertson, G D Tribe, and V B Whitehead are thanked for commenting on the sections on alien animals. R Hengeveld, D C le Maitre, and M Williamson reviewed the whole chapter and provided many useful comments.

REFERENCES

ACOCKS J P H (1975). Veld types of South Africa. *Memoirs of the Botanical Survey of South Africa* **40**, 1–128.

ANONYMOUS (1978). Report of the Director of Nature and Environmental Conservation. Report No. 34, 1977/8. Provincial Administration of the Cape of Good Hope, Cape Town.

ASCHMANN H (1973). Man's impact on the several regions with mediterranean climate. In *Mediterranean-type ecosystems: origin and structure.* (eds Di Castri F and Mooney H A) Ecological Studies 7, Springer-Verlag, Berlin, 363–71.

BAKER H G (1986). Patterns of plant invasion in North America. In *Ecology of biological invasions of North America and Hawaii.* (eds Mooney H A and Drake J A) Ecological Studies 58, Springer-Verlag, New York, 44–57.

BALINSKY B I (1962). Patterns of animal distribution on the African Continent. *Annals of the Cape Provincial Museum* **2**, 299–310.

BARNARD P J and VAN DER WALT K (1961). Translocation of the Bontebok (*Damaliscus pygargus*) from Bredasdorp to Swellendam. *Koedoe* **4**, 105–9.

BEADLE N C W (1940). Soil temperatures during forest fires and their effect on the survival of vegetation. *Journal of Ecology* **28**, 180–92.

BIRKS H J B (1989). Holocene isochrone maps and patterns of tree-spreading in the British Isles. *Journal of Biogeography* **16**, 503–40.

BOND W J (1984). Fire survival of Cape Proteaceae — influence of fire season and seed predators. *Vegetatio* **56**, 65–74.

BOND W J and BREYTENBACH G J (1985). Ants, rodents and seed predation in Proteaceae. *South African Journal of Zoology* **20**, 150–4.

BOND W J and SLINGSBY P (1983). Seed dispersal by ants in shrublands of the Cape Province and its evolutionary implications. *South African Journal of Science* **79**, 231–3.

BOND W J and SLINGSBY P (1984). Collapse of an ant plant mutualism — the Argentine ant, *Iridomyrmex humilis* and myrmecochorous Proteaceae. *Ecology* **65**, 1 031–7.

BOTHA S A (1989). Feral pigs in the Western Cape Province: Failure of a potentially invasive species. *South African Forestry Journal* **151**, 17–25.

BROEKHUYZEN G J (1960). Larger Stripe-breasted Swallow *Cecropis cucullata* feeding on vegetable matter. *Ostrich* **31**, 26.

BROOKE R K (1986). Bibliography of alien birds in southern and south-central Africa. *Occasional Report* **14**, Ecosystem Programmes, CSIR, Pretoria.

BROOKE R K, LLOYD P H and DE VILLIERS A L (1986). Alien and translocated terrestrial vertebrates in South Africa. In *The ecology and management of biological invasions in southern Africa.* (eds Macdonald I A W, Kruger F J and Ferrar A A) Oxford University Press, Cape Town, 63–74.

BROWNLIE S F (1982). The effects of recent landuse on a fynbos site. Unpublished report. School of Environmental Studies, University of Cape Town, Cape Town.

BRUWER J P (1983). Besmetting van *Sesbania punicea* en ander onkruide in die lope van sekere riviere in Wes-Kaap. Unpublished report. Department of Agriculture, Elsenburg.

BRUZAS A (1981). The pine woolly aphid, *Pineus pini* (Macquart) Adelgidae. Pamphlet 273, Directorate of Forestry, Pretoria.

CAMPBELL B M (1985). A classification of the mountain vegetation of the fynbos biome. *Memoirs of the Botanical Survey of South Africa* **50**, 1–121.

CAMPBELL B M (1986). Plant spinescence and herbivory in a nutrient-poor ecosystem. *Oikos* **47**, 168–72.

CAMPBELL B M, GUBB A and MOLL E J (1980). The vegetation of the Edith Stevens Cape Flats Nature Reserve. *Journal of South African Botany* **46**, 435–44.

CAMPBELL B M, MCKENZIE B and MOLL E J (1979). Should there be more tree vegetation in the mediterranean climate region of South Africa. *Journal of South African Botany* **45**, 543–57.

CAVANAGH A K (1980). A review of some aspects of the germination of acacias. *Proceedings of the Royal Society of Victo-*

ria **91**, 161–80.
CILLIE J J (1983). Australian tortoise beetle — a new threat to South African eucalypts. *Forestry News* 1/83, 18–19.
CILLIE J J (1984). The Eucalyptus Tortoise Beetle *Trachymela tincticollis* Blackburn (Chrysomelidae: Coleoptera). Pamphlet 273, Directorate of Forestry, Pretoria.
COETZEE J A and MULLER J (1984). The phytogeographic significance of some extinct pollen types from the Tertiary of the Southwestern Cape (South Africa). *Annals of the Missouri Botanical Gardens* **71**, 1 088–99.
COETZEE J A, SCHOLTZ A and DEACON H J (1983). Palynological studies and the vegetation history of the fynbos. In *Fynbos palaeoecology: a preliminary synthesis.* (eds Deacon H J, Hendey Q B and Lambrechts J J N). *South African National Scientific Programmes Report* **75**, CSIR, Pretoria, 156–73.
COOKE M J (1984). The European Wasp — a new alien invader in South Africa. *African Wildlife* **38**, 219.
COWLING R M (1983). The occurrence of C_3 and C_4 grasses in fynbos and allied shrublands in the South Eastern Cape South Africa. *Oecologia* **58**, 121–7.
COWLING R M and GXABA T (1990). Effects of a fynbos overstorey shrub on understorey community structure: implications for the maintenance of community-wide species richness. *South African Journal of Ecology* **1**, 1–7.
CRAWLEY M J (1987). What makes a community invasible? In *Colonization, succession and stability.* (eds Gray A J, Crawley M J and Edwards P J) Blackwell, Oxford, 429–53.
CRAWLEY M J (1989). Invaders. *Plants Today* **2**, 152–6.
DAVID J H M (1980). Demography and population dynamics of the striped fieldmouse, *Rhabdomys pumilio*, in alien *Acacia* vegetation on the Cape Flats, Cape Peninsula, South Africa, Ph.D. thesis, University of Cape Town.
DAVIDGE C (1978). Activity of chacma baboons (*Papio ursinus*) at Cape Point. *Zoologica Africana* **13**, 143–55.
DEACON H J (1991). Historical background of invasions in the southern African mediterranean region. In *Biogeography of Mediterranean invasions.* (eds Groves R H and Di Castri F) Cambridge University Press (in press).
DEACON J (1986). Human settlement in South Africa and archaeological evidence for alien plants and animals. In *The ecology and management of biological invasions in southern Africa* (eds Macdonald I A W, Kruger F J and Ferrar A A) Oxford University Press, Cape Town, 3–19.
DEAN S J, HOLMES P M and WEISS P W (1986). Seed biology of invasive alien plants in South Africa and South West Africa/Namibia. In *The Ecology and management of biological invasions in southern Africa.* (eds Macdonald I A W, Kruger F J and Ferrar A A) Oxford University Press, Cape Town, 157–70.
DE KOCK A E and GILIOMEE J H (1989). A survey of the Argentine ant, *Iridomyrmex humilis*, (Hymenoptera: Formicidae) in South African fynbos. *Journal of the Entomological Society of southern Africa* **52**, 157–64.
DELCOURT P A and DELCOURT H R (1987). Long-term forest dyna-mics of the temperate zone. A case study of Late-Quaternary forests in eastern North America. Springer-Verlag, New York.
DE VILLIERS A L and MCDOWELL C R (1982). The indigenous exotic vegetation of Table Mountain. *African Wildlife* **36**, 120.
DIXON W J (1985). BMDP statistical software. University of California Press, Berkeley.
DONALD D G M (1979). Nursery and establishment techniques as factors in productivity of man-made forests in Southern Africa. *South African Forestry Journal* **109**, 19–25.
DONNELLY D and GILIOMEE J H (1985). Community structure of epigaeic ants (Hymenoptera: Formicidae) in fynbos vegetation in the Jonkershoek Valley. *Journal of the Entomological Society of southern Africa* **48**, 247–57.
DRAKE J A,DI CASTRI F, GROVES R H, KRUGER F J, MOONEY H A, REJMANEK M and WILLIAMSON M (eds) (1989). *Biological invasions: a global perspective*, SCOPE **37**, Wiley, Chichester.
DURR H J R (1983). A list of additional host plants of aphids (Hemiptera: Aphidoidea) in South Africa. *Phytophylactica* **15**, 67–9.
DURR H J R and MARTIN R (1976). A list of previously unrecorded host plants of aphids (Hemiptera: Aphididae) in South Africa. *Phytophylactica* **8**, 79–82.
ELTON C (1958). *The ecology of invasions by plants and animals.* London, Methuen.
FENN J A (1980). Control of hakea in the Western Cape. In *Proceedings of the third National Weeds Conference.* (eds Neser S and Cairns A L P), Balkema, Cape Town, 167–73.
FISH K (1989). Biotic resistance to the invasion of *Pinus radiata* into mountain fynbos. Unpublished B.Sc. (Hons) report, Botany Department, University of Cape Town.
FOX M D and FOX B J (1986). The susceptibility of natural communities to invasion. In *Ecology of biological invasions: an Australian perspective.* (eds Groves R H and Burdon J J) Australian Academy of Science, Canberra, 57–66.
FUGLER S R (1979). Some aspects of the autecology of three *Hakea* species in the Cape Province, South Africa. M.Sc. thesis, University of Cape Town.
GEERTSEMA H (1975). Studies on the biology, ecology and control of the pine emperor moth *Nudaurelia cytherea cytherea* (Fabr.) (Lepidoptera: Saturniidae). *Annale Universiteit van Stellenbosch* **50** Series A(1), 1–170.
GILIOMEE J H (1986). Seed dispersal by ants in the Cape Flora threatened by *Iridomyrmex humilis* (Hymenoptera: Formicidae). *Entomological Generalis* **11**, 217–19.
GLYPHIS J P, MILTON S J and SIEGFRIED W R (1981). Dispersal of *Acacia cyclops* by birds. *Oecologia* **48**, 138–41.
GLYPHIS J P and PUTTICK G M (1988). Phenolics in some southern African mediterranean shrubland plants. *Phytochemistry* **27**, 743–51.
GOLDBLATT P (1979). An analysis of the flora of southern Africa: its characteristics, relationships, and origins. *Annals of the Missouri Botanical Garden* **65**, 369–436.
GRUBB P J (1977). The maintenance of species-richness in plant communities: the importance of the regeneration niche. *Biological Review* **52**, 107–45.
GRUBB P J (1988). The uncoupling of disturbance and recruitment, two kinds of seed bank, and persistence of plant populations at the regional and local scales. *Ann. Zool. Fennici* **25**, 23–36.
HALL A V (1961). Distribution studies of introduced trees and shrubs in the Cape Peninsula. *Journal of South African Botany* **27**, 101–10.
HALL A V (1987). Threatened plants in the fynbos and karoo biomes, South Africa. *Biological Conservation* **40**, 11–28.
HALL A V and ASHTON E R (1983). Threatened plants on the Cape Peninsula. Threatened Plants Research Group, University of Cape Town, Cape Town.
HALL A V, DE WINTER M, DE WINTER B and VON OOSTERHOUT S A M (1980). Threatened plants of southern Africa. *South African National Scientific Programmes Report* **45**, CSIR, Pretoria, 1–224.
HALL A V and VELDHUIS H A (1985). South African Red Data Book: Plants — Fynbos and Karoo Biomes. *South African National Scientific Programmes Report* **117**, CSIR, Pretoria, 1–157.
HARPER J L (1965). Establishment, aggression, and cohabitation in weedy species. In *The genetics of colonizing species.* (eds Baker H G and Stebbins G L) Academic Press, New York, 243–65.
HARPER J L, CLATWORTHY J N, MC NAUGHTON I H and

SAGAR G R (1961). The evolution and ecology of closely related species living in the same area. *Evolution* **15**, 209–27.
HARRISON J A J (1987). The Southern African Bird Atlas Project. *South African Journal of Science* **83**, 400–1.
HENDEY Q B (1983). Palaeontology and palaeoecology of the fynbos region: an introduction. In *Fynbos palaeoecology: a preliminary synthesis.* (eds Deacon H J, Hendey Q B and Lambrechts J J N) *South African National Scientific Programmes Report* **75**, CSIR, Pretoria, 87–99.
HOBBS R J (1989). The nature and effects of disturbance relative to invasions. In *Ecology of biological invasions: a global perspective.* (eds Drake J A, Di Castri F, Groves R H, Kruger F J, Mooney H A, Rejmanek M and Williamson M H) Wiley, New York, 389–405.
HOBBS R J and ATKINS L (1988). Effect of disturbance and nutrient addition on native and introduced annuals in plant communities in the Western Australian Wheatbelt. *Australian Journal of Ecology* **13**, 171–9.
HOCKEY P A R, UNDERHILL L G, NEATHERWAY M and RYAN P G (1989). *Atlas of the birds of the southwestern Cape.* Cape Bird Club, Cape Town.
HOFFMAN M T and MITCHELL D T (1986). The root morphology of some legume spp. in the south-western Cape and the relationship of vesicular-arbuscular mycorrhizas with dry mass and phosphorus content of *Acacia saligna* seedlings. *South African Journal of Botany* **52**, 316–20.
HOFFMANN J H and MORAN V C (1988). The invasive weed *Sesbania punicea* in South Africa and prospects for its biological control. *South African Journal of Science* **84**, 740–2.
HOFMEYR J (1989). European Swallows feeding on rooikrans. *Promerops* **187**, 15–17.
HOLMES P M (1989a). Decay rates in buried alien *Acacia* seed populations of different densities. *South African Journal of Botany* **55**, 299–303.
HOLMES P M (1989b). Effects of different clearing treatments on the seed-bank dynamics of an invasive Australian shrub, *Acacia cyclops*, in the Southwestern Cape, South Africa. *Forest Ecology and Management* **28**, 33–46.
HOLMES P M (1989c) A comparative study of the seed bank dynamics of two congeneric alien invasive species. Ph.D. thesis, University of Cape Town.
HOLMES P M (1990). Dispersal and predation in alien *Acacia. Oecologia* **83**, 288–90.
HOLMES P M and REBELO A G (1988). The occurrence of seed-feeding *Zulubius acaciaphagus* (Hemiptera, Alydidae) and its effects on *Acacia cyclops* seed germination and seed banks in South Africa. *South African Journal of Botany* **54**, 319–24.
JEFFERY D J, HOLMES P M and REBELO A G (1988). Effects of dry heat on seed germination in selected indigenous and alien legume species in South Africa. *South African Journal of Botany* **54**, 28–34.
JOHNSTONE I M (1986). Plant invasion windows: a time-based classification of invasive potential. *Biological Review* **61**, 369–94.
KLUGE R L and RICHARDSON D M (1983). Progress in the fight against hakea. *Veld & Flora* **69**, 136–8.
KNIGHT R S (1985). A model of episodic, abiotic dispersal for oaks (*Quercus robur*). *South African Journal of Botany* **51**, 265–9.
KNIGHT R S (1988). Aspects of plant dispersal in the south-western Cape with particular reference to the roles of birds as dispersal agents. Ph.D. thesis, University of Cape Town.
KNIGHT R S and MACDONALD I A W (1991). Acacias and Korhaans: an artificially assembled seed dispersal system. *South African Journal of Botany* **57**, 220–5
KRUGER F J (1977a). Invasive woody plants in the Cape fynbos with special reference to the biology and control of *Pinus pinaster.* In *Proceedings of the second National Weeds Conference of South Africa*, 57–74, Balkema, Cape Town.
KRUGER F J (1977b). Ecology of Cape fynbos in relation to fire. In *Proceedings of the symposium on the environmental consequences of fire and fuel management in mediterranean ecosystems.* (eds Mooney H A and Conrad C E), USDA Forest Service General Technical Report WO-3 57–74.
KRUGER F J (1984). Effects of fire on vegetation structure and dynamics. In *Ecological effects of fire in South African ecosystems.* (eds Booysen P de V and Tainton N M) Springer-Verlag, Berlin, 219–43.
KRUGER F J and BIGALKE R C (1984). Fire in fynbos. In *Ecological effects of fire in South African Ecosystems.* (eds Booysen P de V and Tainton N M) Springer-Verlag, Berlin, 67–114.
KRUGER F J, BREYTENBACH G J, MACDONALD I A W and RICHARDSON D M (1989). The characteristics of invaded mediterranean-climate regions. In *Biological invasions: a global perspective.* (eds Drake J A, Di Castri F, Groves R H, Kruger F J, Mooney H A, Rejmanek M and Williamson M H) SCOPE **37**, Wiley, Chichester, 181–213.
LAMONT B B (1983). Strategies for maximizing nutrient uptake in two mediterranean ecosystems of low nutrient status. In *Mediterranean-type ecosystems: the role of nutrients.* (eds Kruger F J, Mitchell D T and Jarvis J U M) *Ecological Studies* **43**, Springer-Verlag, Berlin, 246–73.
LINDER H P (1989). Grasses in the Cape Floristic Region: phytogeographical implications. *South African Journal of Science* **85**, 502–5.
LIVERSIDGE R (1962). The spread of the European Starling in the Eastern Cape. *Ostrich* **33**, 13–16.
LLOYD P H (1975). A study of the Himalayan Thar (*Hemitragus jemlahicus*) and its potential effects on the ecology of the Table Mountain Range. Unpublished Report. Cape Department of Nature and Environmental Conservation.
MACDONALD I A W (1984). Is the fynbos biome especially susceptible to invasion by alien plants? A re-analysis of available data. *South African Journal of Science* **80**, 369–77.
MACDONALD I A W (1987). Advances in our understanding of alien invasions of the fynbos biome: 1980–1985. *Occasional report* **19**, Ecosystems Programme, CSIR, Pretoria.
MACDONALD I A W (1988). Invasive alien plants and their control in southern African nature reserves. In *Proceedings of the conference on Science in the National Parks.* (ed Thomas L K) Fort Collins, July 1986, 63–79.
MACDONALD I A W (1989). Man's role in the changing face of southern Africa. In *Biotic diversity in southern Africa: concepts and conservation* (ed Huntley B J) Oxford University Press, Cape Town, 51–77.
MACDONALD I A W, CLARK D L and TAYLOR H C (1987). The alien flora of the Cape of Good Hope Nature Reserve. *South African Journal of Botany* **53**, 398–404.
MACDONALD I A W, CLARK D L and TAYLOR H C (1989). The history and effects of alien plant control in the Cape of Good Hope Nature Reserve, 1941–1987. *South African Journal of Botany* **55**, 56–75.
MACDONALD I A W and JARMAN M L (1984). Invasive alien organisms in the terrestrial ecosystems of the fynbos biome. *South African National Scientific Programmes Report* **85**, CSIR, Pretoria, 1–72.
MACDONALD I A W, JARMAN M L and BEESTON P (eds) (1985). Management of invasive alien plants in the fynbos biome. *South African National Scientific Programmes Report* **111**, CSIR, Pretoria.
MACDONALD I A W, GRABER D M, DEBENEDETTI S, GROVES R H and FUENTES E R (1988). Introduced species in nature reserves in mediterranean-type climate regions of the world. *Biological Conservation* **44**, 37–66.
MACDONALD I A W and KNIGHT R S (1985). What influences

Acacia invasions at Pella? Poster presented at the 7th annual research meeting of the Fynbos Biome Project, Stellenbosch, 29–30 July 1985.

MACDONALD I A W, KRUGER F J and FERRAR A A (eds) (1986). *The ecology and control of biological invasions in southern Africa.* Oxford University Press, Cape Town.

MACDONALD I A W, LOOPE L L, USHER M B and HAMMAN O (1989). Wildlife conservation and the invasion of nature reserves by introduced species: a global perspective. In *Biological invasions: a global perspective.* (eds Drake J A, Di Castri F, Groves R H, Kruger F J, Mooney H A, Rejmanek M and Williamson M H) SCOPE **37**, Wiley, Chichester, 215–55.

MACDONALD I A W, POWRIE F J and SIEGFRIED W R (1986). The differential invasion of southern Africa's biomes and ecosystems by alien plants and animals. In *The ecology and management of biological invasions in southern Africa.* (eds Macdonald I A W, Kruger F J and Ferrar A A) Oxford University Press, Cape Town, 209–25.

MACDONALD I A W and RICHARDSON D M (1986). Alien species in terrestrial ecosystems of the fynbos biome. In *The ecology and management of biological invasions in southern Africa.* (eds Macdonald I A W, Kruger F J and Ferrar A A) Oxford University Press, Cape Town, 77–91.

MACDONALD I A W, WISSEL C, KNIGHT R S and HOLMES P M (1987). Optimizing the control of alien woody plants in the fynbos. Poster presented at the 9th annual research meeting of the Fynbos Biome Project, Saasveld Forestry Research Centre, George, 23–5 June 1987.

MACLEAN G L (1985). *Roberts' Birds of Southern Africa.* John Voelcker Bird Book Fund, Cape Town.

MANDERS P T, RICHARDSON D M and MASSON P H (in press). Is fynbos a stage in succession to forest? Analysis of the perceived ecological distinction between two communities. In *Fire in mountain fynbos* . (eds Van Wilgen B W, Richardson D M, Kruger F J and Van Hensbergen H J), *Ecological Studies* **93**, Springer-Verlag, Berlin (in press).

MARAIS L J and KOTZE J M (1978). Effect of mycorrhizae on the growth of *Pinus patula* Schlecht et Cham. seedlings in South Africa. *South African Forestry Journal* **107**, 12–14.

MC CUNE B (1988). Ecological diversity in North American pines. *American Journal of Botany* **75**, 353–68.

MCDONALD D J and MORLEY M (1988). A checklist of the flowering plants and ferns of Swartboskloof, Jonkershoek, Cape Province. *Bothalia* **18**, 233–60.

MCLACHLAN G R (1978). A population of *Typhlops braminus* (Daudin) on the Cape Peninsula. *Zoologica Africana* **13**, 353–4.

MCLACHLAN D, MOLL E J and HALL A V (1980). Resurvey of the alien vegetation in the Cape Peninsula. *Journal of South African Botany* **46**, 127–46.

MIDDLEMISS E (1963). The distribution of *Acacia cyclops* in the Cape Peninsula by birds and other animals. *South African Journal of Science* **59**, 419–20.

MIDDLEMISS E (1975). The rondevlei Bird Sanctuary 1952–1974: a record of an environment. Divisional Council of the Cape, Cape Town.

MILEWSKI A V and BOND W J (1982). Convergence of myrmecochory in mediterranean Australia and South Africa. In *Ant-plant interactions in Australia.* (ed Buckley R C) W. Junk, The Hague, 89–98.

MILLAR J C G (1970). The Cape of Good Hope Nature Reserve, a report and management plan. Unpublished Report, Department of Nature and Environmental Conservation, Cape Town.

MILLAR J C G (1980). Aspects of the ecology of the American Grey Squirrel *Sciurus carolinensis* Gmelin in South Africa. M.Sc. thesis, University of Stellenbosch.

MILLER P C, MILLER J M and MILLER P M (1983). Seasonal progression of plant water relation in fynbos in the Western Cape Province, South Africa. *Oecologia* **56**, 392–6.

MILTON S J (1982). Effects of shading on nursery grown *Acacia* seedlings. *Journal of South African Botany* **48**, 245–72.

MILTON S J and MOLL E J (1982). Phenology of Australian acacias in the south-western Cape, South Africa. *Botanical Journal of the Linnean Society* **84**, 295–327.

MITCHELL D T and ALLSOPP N (1984). Changes in the phosphorus composition of seeds of *Hakea sericea* (Proteaceae) during germination under low phosphorus conditions. *New Phytologist* **96**, 239–47.

MOLL E J and BOSSI L (1984). A current assessment of the extent of the natural vegetation of the fynbos biome. *South African Journal of Science* **80**, 355–8.

MOLL E J and CAMPBELL B M (1976). *The ecological status of Table Mountain.* Department of Botany, University of Cape Town, Cape Town.

MOLL E J, CAMPBELL B M, COWLING R M, BOSSI L, JARMAN M L and BOUCHER C (1984). A description of vegetation categories in and adjacent to the fynbos biome. *South African National Scientific Programmes Report* **83**, CSIR, Pretoria.

MOLL E J, MCKENZIE B and MCLACHLAN D (1980). A possible explanation for the lack of trees in the fynbos, Cape Province, South Africa. *Biological Conservation* **17**, 221–8.

MOONEY H A (1982). Mediterranean-type ecosystems — research progress and opportunities. *South African Journal of Science* **78**, 5–7.

MORRIS M J (1982). Gummosis and die-back of *Hakea sericea* in South Africa. In *Proceedings of the fourth National Weeds Conference of South Africa.* (eds Van der Venter H and Mason M) Balkema, Cape Town, 51–4.

NESER S (1984). Insect enemies of *Hakea sericea* in Australia. In *Proceedings of the 4th International Conference on Mediterranean Ecosystems.* (ed Dell B) University of Western Australia, Nedlands, 126–7.

NESER S and KLUGE R L (1985). A seed-feeding insect showing promise in the control of a woody invasive plant: the weevil *Erytenna consputa* Pascoe on *Hakea sericea* (Proteaceae) in South Africa. In *Proceedings of the Sixth International Symposium of the Biological Control of Weeds.* (ed Delfosse E S) Agricultura Canada, Vancouver, 805–9.

O'DOWD D J and GILL A M (1986). Seed dispersal syndromes in Australian *Acacia.* In *Seed Dispersal.* (ed Murray D R) Academic Press Australia, Sydney, 87–121.

ORIANS G H (1986). Site characteristics favoring invasion. In *Ecology of biological invasions of North America and Hawaii.* (eds Mooney H A and Drake J A) Springer-Verlag, New York, 133–48.

PATE J S, RASINS E, RULLO J and KUO J (1986). Seed nutrient reserves of Proteaceae with special reference to protein bodies and their inclusions. *Annals of Botany* **57**, 747–70.

PEARSON A A (1950). Cape Agarics and Boleti. *Transactions of the British Mycological Society* **33**, 276–316.

PIETERSE P J (1986). Aspekte van die demografie van *Acacia longifolia* (Andr.) Willd. in die Banhoekvallei in die Suidwes-Kaap. M.Sc. thesis, University of Stellenbosch.

PIETERSE P J and CAIRNS A L P (1986a). The effect of fire on an *Acacia longifolia* seed bank in the south-western Cape. *South African Journal of Botany* **52**, 233–6.

PIETERSE P J and CAIRNS A L P (1986b). An effective technique for breaking the seed dormancy of *Acacia longifolia. South African Journal of Plant and Soil* **3**, 85–7.

PIETERSE P J and CAIRNS A L P (1987). The effect of fire on an *Acacia longifolia* seed bank and the growth, mortality and reproduction of seedlings establishing after a fire in the South West Cape. *Applied Plant Science* **1**, 34–8.

PIETERSE P J and CAIRNS A L P (1988). The population dynamics of the weed *Acacia longifolia* (Fabaceae) in the

absence and presence of fire. *South African Forestry Journal* **145**, 25–7.
PIETERSE P J and CAIRNS A L P (1990). Investigations on the removal by animals of *Acacia longifolia* (Fabaceae) seed from the soil surface at Banhoek in the SW Cape. *South African Journal of Plant and Soil* **7**, 155–7.
PIMM S L (1989). Theories of predicting success and impact of introduced species. In *Biological invasions: a global perspective.* (eds Drake J A, Di Castri F, Groves R H, Kruger F J, Mooney H A, Rejmanek M and Williamson M H) SCOPE **37**, Wiley, Chichester, 351–67.
POTGIETER J T and DURR H J R (1961). A host plant index of South African plant lice (Aphididae) with a list of species found on each plant recorded. *Annale van die Universiteit van Stellenbosch* **36**, 217–39.
POYNTON R J (1979). *Tree planting in southern Africa. Vol 1. The pines.* Department of Forestry, Pretoria.
PRESSINGER F (1985). The influence of wind on the dispersal pattern of *Acacia saligna* seeds. Poster presented at the 7th annual research meeting of the Fynbos Biome Project, Stellenbosch, 29–30 July 1985.
PRINS A J, ROBERTSON H G and PRINS A (1991). Pest ants in urban and agricultural areas of southern Africa. In *Applied myrmecology: a world perspective.* (ed Van der Meer R K) Westview Press, Boulder, Colorado (in press).
PRŶS-JONES R, CLARK D and FOSTER K (1985). Stocking density of large herbivores in fynbos of the Cape of Good Hope Nature Reserve, in relation to veld age. Poster presented at the 7th annual research meeting of the Fynbos Biome Project, Stellenbosch, 29–30 July 1985.
RAITT L M (1983). An analysis of the flora and floral phenology of a disturbed area of the Cape Flats. Unpublished paper, Fifth Annual Weeds Conference of Southern Africa, Stellenbosch.
REJMANEK M (1989). Invasibility of plant communities. In *Ecology of biological invasions: a global perspective.* (eds Drake J A, Di Castri F, Groves R H, Kruger F J, Mooney H A, Rejmanek M and Williamson M H) SCOPE **37**, Wiley, Chichester, 369–88.
RICHARDSON D M (1984). A cartographic analysis of physiographic factors influencing the distribution of *Hakea* spp. in the South-Western Cape. *South African Forestry Journal* **128**, 36–40.
RICHARDSON D M (1985). Studies on aspects of the integrated control of *Hakea sericea* in the south-western Cape Province, South Africa. M.Sc. thesis, University of Cape Town.
RICHARDSON D M (1988). Age structure and regeneration after fire in a self-sown *Pinus halepensis* forest on the Cape Peninsula, South Africa. *South African Journal of Botany* **54**, 140–4.
RICHARDSON D M (1989). The ecology of invasions by *Pinus* (Pinaceae) and *Hakea* (Proteaceae) species, with special emphasis on patterns, processes and consequences of invasion in mountain fynbos of the southwestern Cape Province, South Africa. Ph.D. thesis, University of Cape Town.
RICHARDSON D M (1990). Pines on the move. Invasions in perspective. *African Wildlife* **44**, 162–7.
RICHARDSON D M and BOND W J (1991). Determinants of plant distribution: evidence from pine invasions. *American Naturalist* **137**, 639–68.
RICHARDSON D M and BROWN P J (1986). Invasion of mesic mountain fynbos by *Pinus radiata. South African Journal of Botany* **52**, 529–36.
RICHARDSON D M and COWLING R M (in press). Why is mountain fynbos invasible and which species invade? In *Fire in South Africanmountain fynbos ecosystem.* (eds Van Wilgen B W, Richardson D M, Kruger F J and Van Hensbergen H J) *Ecological Studies* **93**, Springer-Verlag, Berlin (in press).
RICHARDSON D M, COWLING R M and LE MAITRE D C (1990). Assessing the risk of invasive success in *Pinus* and *Banksia* in South African mountain fynbos. *Journal of Vegetation Science* **1**, 629–42.
RICHARDSON D M and KRUGER F J (1990). Water relations and photosynthetic characteristics of selected trees and shrubs of riparian and hillslope habitats in the southwestern Cape Province, South Africa. *South African Journal of Botany* **56**, 214–25.
RICHARDSON D M and MANDERS P T (1985). Predicting pathogen-induced mortality in *Hakea sericea*, an aggressive alien plant invader in South Africa. *Annals of Applied Biology* **106**, 243–54.
RICHARDSON D M, MACDONALD I A W and FORSYTH G G (1989). Reductions in plant species richness under stands of alien trees and shrubs in the fynbos biome. *South African Forestry Journal* **149**, 1–8.
RICHARDSON D M and VAN WILGEN B W (1986a). Effects of thirty-five years of afforestation with *Pinus radiata* on the composition of Mesic Mountain Fynbos near Stellenbosch. *South African Journal of Botany* **52**, 309–15.
RICHARDSON D M and VAN WILGEN B W (1986b). The effects of fire in felled *Hakea sericea* and natural fynbos and implications for weed control in mountain catchments. *South African Forestry Journal* **139**, 4–14.
RICHARDSON D M, VAN WILGEN B W and MITCHELL D T (1987). Aspects of the reproductive ecology of four Australian *Hakea* species (Proteaceae) in South Africa. *Oecologia* **71**, 345–54.
SCLATER W L (1901). *The mammals of South Africa. Volume 2. Rodentia, Chiroptera, Insectivora, Cetacea and Edentata.* R H Porter, London.
SHAUGHNESSY G L (1986). A case study of some woody plant introductions to the Cape Town area. In *The ecology and management of biological invasions in southern Africa.* (eds Macdonald I A W, Kruger F J and Ferrar A A) Oxford University Press, Cape Town, 37–43.
SILVERTOWN J and LAW R (1987). Do plants need niches? Some recent developments in plant community ecology. *Trends in Ecology and Evolution* **2**, 24–6.
SKAIFE S H (1955). The Argentine Ant. *Transactions of the Royal Society of South Africa* **34**, 355–77.
SKAIFE S H (1962). The distribution of the Argentine Ant *Iridomyrmex humilis* Mayr. *Annals of the Cape Provincial Museum* **2**, 297–8.
SKEAD C J (1979). *Historical mammal incidence in the Cape Province. Vol 1. The Western and Northern Cape.* Cape Department of Nature and Environmental Conservation, Cape Town.
SLINGSBY P (1982). The Argentine ant — how much of a threat? *Veld & Flora* **68**, 102–4.
SMITHERS R H N (1983). *The mammals of the Southern African Subregion.* University of Pretoria, Pretoria.
STIRTON C H (ed) (1978). *Plant invaders — beautiful but dangerous.* Department of Nature and Environmental Conservation of the Cape Provincial Administration, Cape Town.
STUART C T, MACDONALD I A W and MILLS M G L (1985). History, current status and conservation of large mammalian predators in the Cape Province, Republic of South Africa. *Biological Conservation* **31**, 9–17.
SUGDEN J M (1989). Late Quaternary palaeoecology of the central and marginal uplands of the Karoo, South Africa. Ph.D. thesis, University of Cape Town.
SUMMERS-SMITH J D (1988). *The sparrows.* T & A D Poyser, Calton, UK.
TAYLOR H C (1969). Pest-plants and nature conservation in the winter rainfall region. *Journal of the Botanical Society of South Africa* **55**, 32–8.
TAYLOR H C (1977). Aspects of the ecology of the Cape of

Good Hope Nature Reserve in relation to fire and conservation. In *Proceedings of the symposium on environmental consequences of fire and fuel management in Mediterranean ecosystems,* USDA Forest Service General Technical Report WO-3 483–7.

TAYLOR H C and MACDONALD S A (1985). Invasive alien woody plants in the Cape of Good Hope Nature Reserve. I. Results of a first survey in 1966. *South African Journal of Botany* **51**, 14–20.

TAYLOR H C, MACDONALD S A and MACDONALD I A W (1985). Invasive alien woody plants in the Cape of Good Hope Nature Reserve. II. Results of a second survey from 1976 to 1980. *South African Journal of Botany* **51**, 21–9.

TRAN V N and CAVANAGH A K (1984). Structural aspects of dormancy. In *Seed physiology. Volume 2. Germination and reserve mobilization.* (ed Murray D R) Academic Press, Sydney, 1–44.

TRIBE G D (1990). Phenology of *Pinus radiata* log colonization by the pine bark beetle, *Hylastes angustatus* (Herbst) (Coleoptera: Scolytidae). *Journal of the Entomological Society of southern Africa* **53**, 93–100.

TRIBE G D and CILLIE J J (1984). Eucalyptus tortoise beetles on the move? *Forestry News* 2/84, 15.

URBAN A J, TRIBE G D and CILLIE J J (1987). Dispersal of the introduced parasitoid, *Enoggera polita* (Hymenoptera: Pteromalidae), and its initial colonization of Eucalyptus Tortoise Beetle, *Trachymela tincticollis* (Coleoptera: Chrysomelidae), in the southern Cape Province. *Proceedings of the 6th Congress of the Entomological Society of southern Africa.* Stellenbosch, 77–8.

URBAN E K, FRY C H and KEITH S (eds) (1986). *The birds of Africa. Volume 2.* Academic Press, London.

VAN BRUGGEN A C (1964). The distribution of introduced mollusc species in Southern Africa. *Beaufortia* **11**, 161–9.

VAN DER MERWE C V (1977). 'n Plantegroei opname van die De Hoop Natuurreservaat. *Bontebok (Old series)* **1** (2), 1–29.

VAN DER MERWE F (1984). Europese spreeus verdring gryskopspegte. *African Wildlife* **38**, 152–7.

VAN HULST R (1987). Invasion models of vegetation dynamics. *Vegetatio* **69**, 123–31.

VERSFELD D B and VAN WILGEN B W (1986). Impact of woody aliens on ecosystem properties. In *The ecology and management of biological invasions in southern Africa.* (eds Macdonald I A W, Kruger F J and Ferrar A A) Oxford University Press, Cape Town, 239–46.

VIERKE J (1970). Die Besiedlung Suedafrikas durch den Haussperling (*Passer domesticus*). *Journal fur Ornithologie* **111**, 94–103.

VLOK J H J (1988). Alpha diversity of lowland fynbos herbs at various levels of infestation by alien annuals. *South African Journal of Botany* **54**, 623–7.

VOGEL J, FULS A and ELLIS R (1978). The geographic distribution of kranz grasses in South Africa. *South African Journal of Science* **74**, 209–15.

VON BROEMBSEN S (1984). Occurrence of *Phytophtora cinnamom* on indigenous and exotic hosts in South Africa, with special reference to the south western Cape Province. *Phytophylactica* **16**, 227–9.

WEISS P W and MILTON S J (1984). *Chrysanthemoides monilifera* and *Acacia longifolia* in Australia and South Africa. In *Proceedings of the 4th International Conference on mediterranean ecosystems.* (ed Dell B) University of Western Australia, Nedlands, 159–60.

WELLS M J, ENGELBRECHT V M, BALSINHAS A A and STIRTON C H (1983). Weed flora of South Africa 3: more power shifts in the veld. *Bothalia* **14**, 967–70.

WELLS M J, POYNTON R J, BALSINHAS A A, MUSIL K J, JOFFE H, VAN HOEPEN E and ABBOTT S K (1986). The history of introduction of invasive alien plants to southern Africa. In *The ecology and management of biological invasions in southern Africa.* (eds Macdonald I A W, Kruger F J and Ferrar A A) Oxford University Press, Cape Town, 21–35.

WHITE P S and PICKETT S T A (1985). Natural disturbance and patch dynamics: an introduction. In *The ecology of natural disturbance and patch dynamics.* (eds Pickett S T A and White P S) Academic Press, Orlando, 3–13.

WHITEHEAD V B and PRINS A J (1975). The European Wasp *Vespula germanica* (F) in the Cape Province. *Journal of the Entomological Society of Southern Africa* **38**, 39–42.

WILSON J B (1989). Relations between native and exotic plant guilds in the Upper Clutha, New Zealand. *Journal of Ecology* **77**, 223–35.

WINSTANLEY J K (1984a). The eucalyptus snout beetle, *Gonipterus scutellatus* Gyllenhal (Curculionidae). Pamphlet 273, Directorate of Forestry, Pretoria.

WINSTANLEY J K (1984b). The Italian beetle, *Hylotrupes bajulus* (Cerambiycidae: Coleoptera). Pamphlet 273, Directorate of Forestry, Pretoria.

WINTERBOTTOM J M (1970). The birds of the alien acacia thickets of the south western Cape. *Zoologica Africana* **5**, 49–57.

WINTERBOTTOM J M and LIVERSIDGE R (1954). The European starling in the south west Cape. *Ostrich* **25**, 89–96.

WITKOWSKI E T F (1991). Growth and competition between seedlings of *Protea repens (L.) L.* and the alien invasive, *Acacia saligna, (Labill.) Wendl.* in relation to nutrient availability. *Functional Ecology* **5**, 101–10.

ZIETSMAN H L and VAN DER MERWE I J (1986). *Population census atlas of South Africa.* Institute for cartographic analysis Publication 15/1986. University of Stellenbosch.

13

Preservation of biotic diversity

A G Rebelo

The Cape Floristic Region (Figure 13.1) is an area of pronounced plant species richness and endemism (Bond and Goldblatt 1984; Cowling et al. this volume). The region includes more than 8 550 species of which about 68% are endemic. The region includes five largely unrelated vegetation types, namely fynbos, renoster shrubland, thicket, karroid shrubland, and Afromontane forest. These are described in Cowling and Holmes (this volume).

The Cape Floristic Region has been, and is being, extensively transformed by pastoral, agricultural, and urban development (Deacon and Siegfried this volume), and alien plant encroachment (Richardson et al. this volume). This alteration is most evident in renoster shrubland and lowland fynbos, and least in mountain fynbos (Moll and Bossi 1984; Jarman 1986). In the lowlands of the western Cape only six per cent of renoster shrubland and 14% of fynbos are currently untransformed (Boucher 1981a).

The majority of species of larger (> 50 kg) mammals and birds, most especially the carnivores and scavengers, were exterminated during the 1800s (Skead 1980; Brooke 1984; Smithers 1986; Rookmaaker 1989), but no regional Red Data Book, other than that for plants (Hall and Veldhuis 1985), exists. The 1 320 Red Data Book plant species for the Cape Floristic Region (excluding karroid shrubland) comprise 56% of the southern African total, although the Cape Floristic Region comprises less than four per cent of the total area (Hall et al. 1980; Hall and Veldhuis 1985).

The above situation compares with those of the most beleaguered regions of the world (Hall 1987a). However, a framework of preserved areas exists and an efficient conservation strategy could be implemented using remaining untransformed areas. Progress must be speedy since urban growth and alien plant encroachment are rapidly reducing viable options.

This chapter provides an overview of the preservation of biotic diversity in the Cape Floristic Region. It summarizes what is currently preserved and compares this to the optimal reserve configuration, based on preserving maximum plant species diversity in a minimal area of fynbos, in order to outline the deficiencies in the existing reserve network. However, conservation priorities must also focus on unique areas most threatened by imminent transformation. Areas containing the most Red Data Book species and extensively transformed vegetation types are obvious candidates for high priority conservation measures. Having established the priorities and location of reserves, the chapter investigates arguments for reserve sizes, using island biogeographic and minimum viable population size principles, primarily in response to Kruger's (1977) assertion that the minimum reserve size in fynbos vegetation in the Cape Floristic Region is 10 000 ha. The chapter ends with a short speculation on the preservation of alien species, human use, and global climatic change.

HISTORY AND STATUS OF PRESERVATION

Background

Although the South African nature reserve system did not develop according to any preconceived strategy for maximizing the preservation of biological diversity (Siegfried 1989), conservation strategies in the Cape Floristic Region have emphasized the preservation mainly of plant species richness (Kruger 1977; Bond et al. 1988). This has three origins. Firstly, the large mammal and large bird species, which form

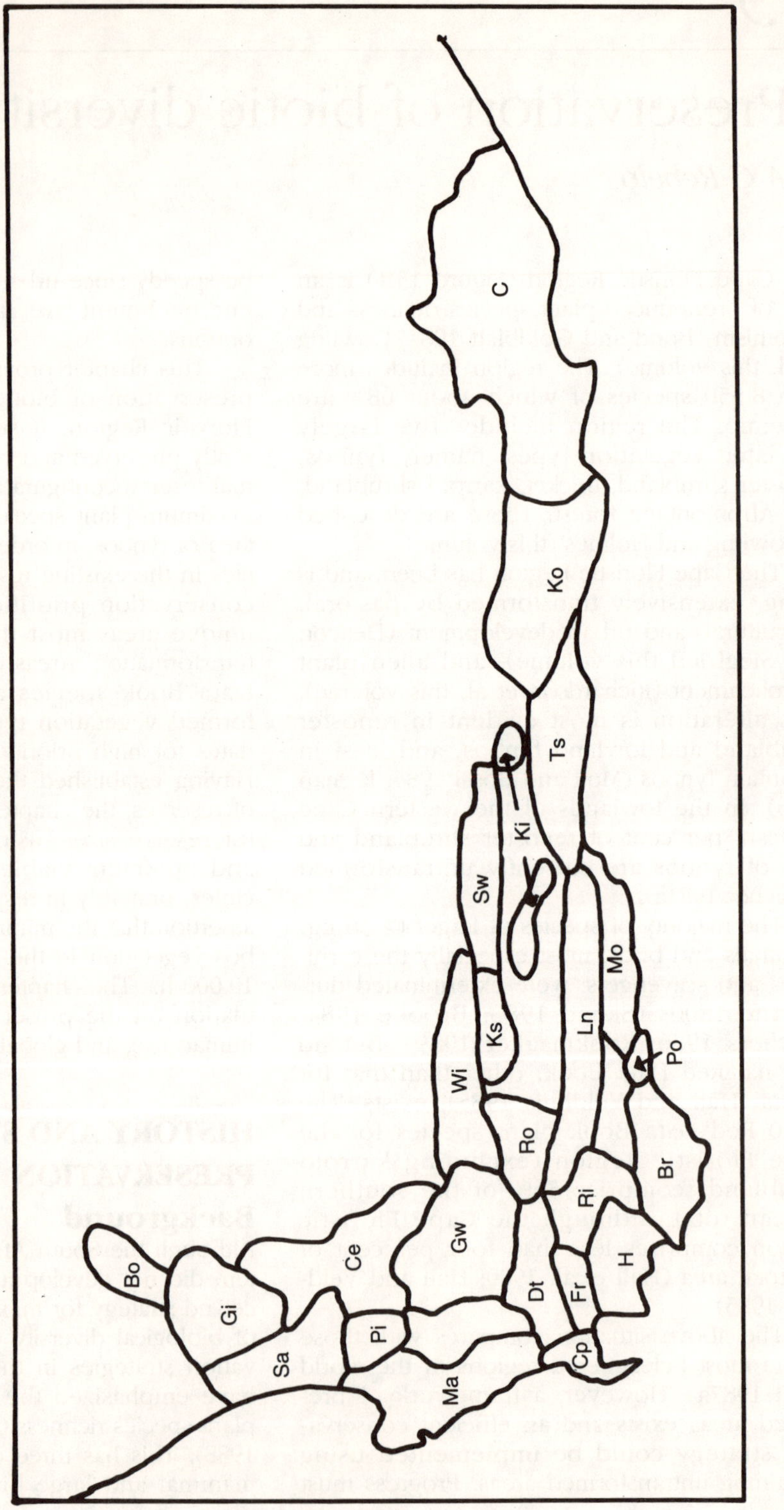

FIGURE 13.1 The centres of endemism for Proteaceae in the Cape Floristic Region (Rebelo and Siegfried 1990). The names of centres are given in Table 13.3.

the focus of species-orientated preservation programmes elsewhere in southern Africa, were exterminated in the Cape Floristic Region long before current conservation movements came into being (Skead 1980; Rookmaaker 1989). Secondly, the high plant species richness and endemism of the Cape Floristic Region (Cowling et al. this volume) far outshadow that of any vertebrate taxon. Although the invertebrate fauna is probably as diverse as the flora (Rebelo 1987b), with the exception of butterflies little is known about richness in these taxa. Thirdly, by far the largest area preserved in the Cape Floristic Region was proclaimed in order to protect mountain water catchment areas (Greyling and Huntley 1984; Grove 1987) which fortuitously coincided with the areas of high plant species richness.

The preservation of species richness is probably appropriate for fynbos, although the establishment and management of some large reserves in fynbos have been based largely on single species (e.g. Cederberg Wilderness Area based on the Clanwilliam Cedar, *Widdringtonia cedarbergensis*, Manders 1986; Kogelberg State Forest based on the Marsh Rose, *Orothamnus zeyheri*, Boucher 1981b; Luckhoff 1982). It appears that the large mammals which were exterminated from the Cape Floristic Region never played a major role in the dynamics of nutrient-poor fynbos communities (Skead 1980; Morrow et al. 1983). Attempts by provincial and local conservation authorities to introduce large mammals ('big game') into fynbos reserves have failed, owing to lack of grazing and deficiency diseases associated with the nutrient-poor soils (Van Rensberg 1975; Zumpt and Heine 1977).

Large mammal species were apparently confined to non-fynbos (renoster shrubland, thicket, and forest) (Skead 1980). Much of the renoster shrubland was apparently derived from a grassland, possibly dominated by *Themeda triandra* (Muir 1929; Skead 1980; Cowling 1984; Cowling et al. 1986; Cowling and Holmes this volume). Although European stock farming was initiated only in 1703 (Le Cordeur 1986), Skead (1980) suggests that this transformation was complete in the western Cape by 1750. In the southern Cape the transformation appears to have occurred between Sparrman's 1775 travels and the travels of Barrow in 1797 and Burchell in 1814 (Muir 1929). This period probably coincides with the elimination of the large mammals in the region and the start of settled farming (Skead 1980). It certainly precedes the large-scale wool farming that was initiated in the 1820s (Le Cordeur 1986). Since the alleged transformation was completed before the travels of the early naturalists, it is possible that many plant species may have become extinct before the region was explored botanically (Hall and Veldhuis 1985).

The development of the reserve system

By far the largest proportion of preserved land in the Cape Floristic Region (878 000 ha) was proclaimed by the Forestry Directorate of the Department of Environment Affairs as State Forests (Table 13.1). The Forestry Directorate was the first to establish reserves in the Cape Floristic Region, primarily in response to the destruction of Afromontane forest (Grove 1987). Although established as forestry reserves, which were *de facto* preserved as water catchment areas, these were without comprehensive protection (Huntley 1978). The importance of preserving water catchment areas was highlighted by research in which annual runoff from mature forest plantations was found to be 50–100% less than that of natural fynbos vegetation (reviewed in Versveld and Van Wilgen 1986). This resulted in the rapid acquisition of land during the 1960s (Taylor 1978) and strict controls on afforestation, under the Mountain Catchment Areas Act of 1970, to protect South Africa's limited water supply (Greyling and Huntley 1984; Macdonald 1989). Rapid advances in the conservation policies of the Forestry Directorate resulted in the establishment of several wilderness areas and forest nature reserves between 1971 and 1982 (Figure 13.2; Kruger 1977). These were established for scientific research in natural ecosystems, the aesthetic values they engendered, and the physical and spiritual opportunities they afforded (Ackerman 1972). It was planned to transfer one-third of State Forests to wilderness areas, the remainder available for future developments and protected as *de facto* nature reserves in the interim (Ackerman 1972).

As a result of the government's decentral-

TABLE 13.1 The ownership and controlling bodies of reserves in the Cape Floristic Region. Data from the Percy FitzPatrick Institute of African Ornithology database (Siegfried 1989) and Cowan (1987).

Authority	Area preserved (X 10^3 ha)	Proportion of total (%)
CDNEC[1]:		
Provincial nature reserves	82	4.6
Former State Forests	878	48.9
Private and local authority nature reserves:		
Subsidized by CDNEC[1]	53	2.9
Non-subsidized[2]	22	1.2
State owned and controlled reserves[3]	69	3.8
Contractual nature reserves[4]	693	38.6

[1] CDNEC = Chief Directorate of Nature and Environmental Conservation of the Cape Provincial Administration.
[2] Private nature reserves and natural heritage sites.
[3] National Botanical Institute, National Parks Board, and South African Defence Force.
[4] Contractual mountain catchment areas and contractual national parks.

ization policy, the majority of unafforested State Forests were transferred to the Chief Directorate of Nature and Environmental Conservation of the Cape Provincial Administration during 1987–1988 (Hilton-Taylor and Le Roux 1989). It is uncertain whether the Chief Directorate of Nature and Environmental Conservation will be able to maintain the former State Forests as efficiently as the Forestry Directorate, as the costs of alien removal, fire-break maintenance, and fire-control were cross-subsidized by the afforestation activities of the Forestry Directorate (Van Wilgen et al. this volume).

The Chief Directorate of Nature and Environmental Conservation has the smallest area preserved of all the provincial authorities (Cowan 1987), despite the Cape Province being the largest province in South Africa. Only 82 000 ha (0.9%) of the Cape Floristic Region area was preserved in provincial nature reserves in 1986. This probably reflects the department's mission to co-ordinate the interests of inland fisheries, conservation, pest control, and museum services. Thus the department viewed conservation alone as inadequate: 'Wildlife must be regarded as a by-product of wise usage of the land' (Anon 1952). Consequently, management and research were directed towards developing a programme of wildlife management suited to farming conditions (Anon 1952), apparently at the expense of obtaining land for provincial nature reserves. The majority of early provincial nature reserves were acquired for breeding fish and mammals to restock reserves and to sell to private landowners (Scott 1986). During the 1970s emphasis shifted to the preservation of representative vegetation types (Scott 1986). Recently, some provincial nature reserves under the control of the Chief Directorate of Nature and Environmental Conservation have been rezoned or deproclaimed (McDowell et al. 1991; Wood 1991).

Other state-controlled reserves in the Cape Floristic Region include the National Parks Board with five reserves totalling 30 000 ha; the National Botanical Institute which maintains portions of its three national botanical gardens as nature reserves, in addition to two nature reserves, which together total 800 ha; and the South African Defence Force which has 13 training and military areas (37 000 ha) managed as nature reserves since 1978. Although the state-controlled land outside the former State Forests amounts to only 1.7% of the area of the Cape Floristic Region (Table 13.1), these reserves are mainly in the lowlands and are thus critical in terms of species and habitats preserved.

Private nature reserves and natural heritage sites comprise the smallest area of con-

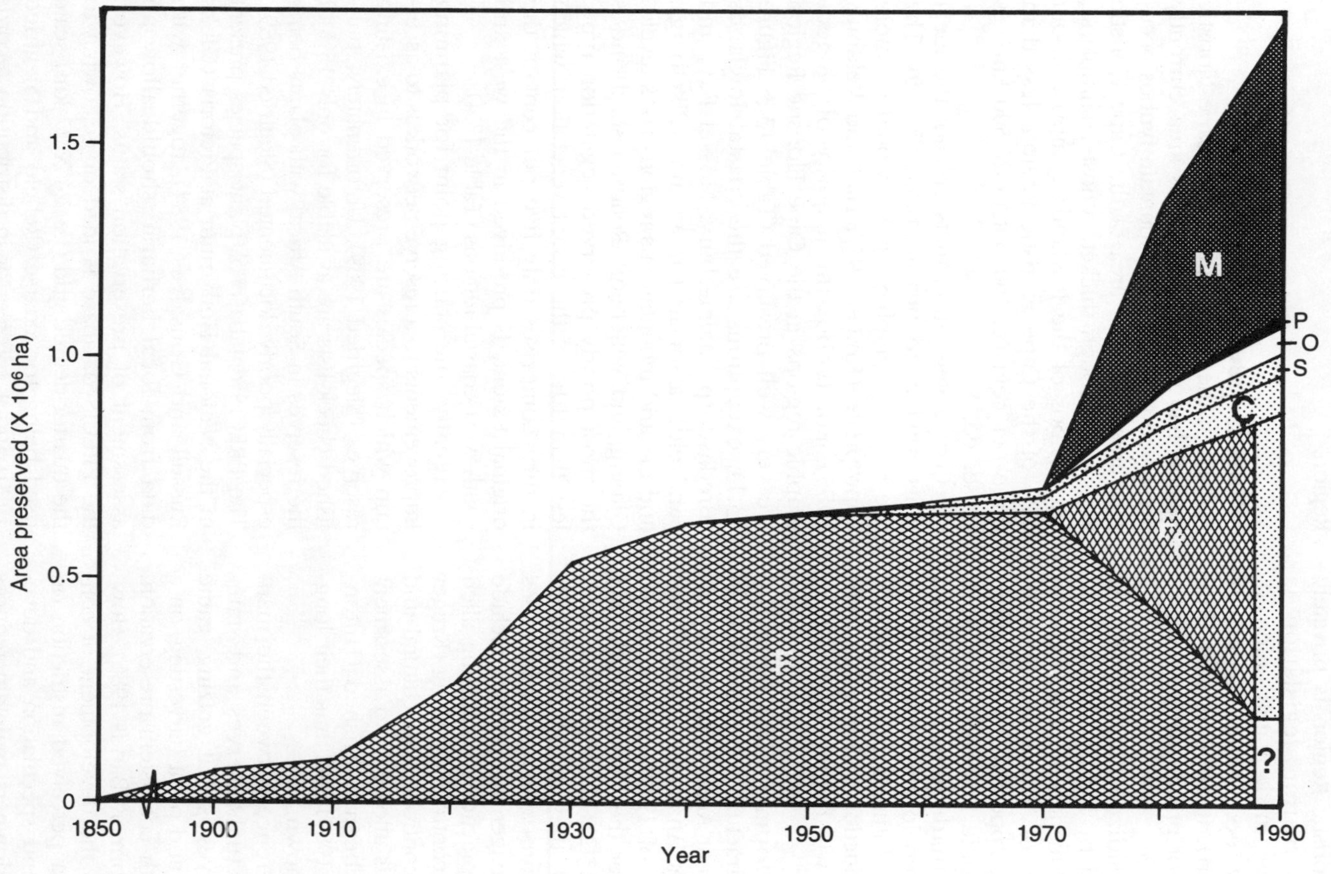

FIGURE 13.2 The historical allocation of land to the reserve system in the Cape Floristic Region. The ownership and management categories are: C = Chief Directorate of Nature and Environmental Conservation (provincial nature reserves); F = former State Forests; S = privately owned, subsidized by Chief Directorate of Nature and Environmental Conservation (subsidized nature reserves); P = privately owned and controlled (private nature reserves and natural heritage sites); O = state owned and controlled (National Botanical Institute, National Parks Board, South African Defence Force); M = contractual nature reserves (mountain catchment areas and national parks). Note that State Forests (F) only received formal protection between 1970 and 1987 (Ff), after which they were transferred to the Chief Directorate of Nature and Environmental Conservation. A proportion of unafforested State Forests has not been transferred to the Chief Directorate of Nature and Environmental Conservation and will probably be used for future afforestation. Data from the Percy FitzPatrick Institute of African Ornithology database (Siegfried 1989) and Cowan (1987).

served land (Table 13.1). These have no secure long-term conservation status (Hilton Taylor and Le Roux 1989). Since about 80% of land in the Cape Floristic Region is privately owned, and conservation legislation is emphatically restrictive, inadequately enforced, and provides few economic incentives (McDowell 1986a, b), it appears that little increase in the area of private nature reserves can be expected.

Neither do subsidized nature reserves, which are controlled by local authorities and obtain a subsidy and management advice from the Chief Directorate of Nature and Environmental Conservation, appear to have a secure long-term future. Some reserves have been deproclaimed or inadequately managed (McDowell 1986a; Rebelo and Holmes 1988), setting precedents for future deproclamation of other reserves. A reduction and possible withdrawal of subsidies will further threaten their status.

A recent development has been the proclamation of contractual preservation areas as Mountain Catchment Areas (via the Department of Environment Affairs under the Water Conservation Act 63 of 1970) and contractual national parks (under the National Parks Act 57 of 1976) (Figure 13.2). Although these areas are privately owned, they are managed integrally with core nature reserves for purposes of conservation. Management policies include prescribed burning practices, combating alien invasive plants, and controlled grazing (Kruger 1982). Limited and controlled agricultural and urban development is allowed. No assessment of these areas has been published (Hilton-Taylor and Le Roux 1989) and thus their long-term viability is unknown.

The major difference between the management of contractual reserves and state-owned nature reserves is that grazing, game farming, and flower and plant harvesting, on a scale compatible with ecosystem preservation, are allowed in the former (Bands 1985). However, an obstacle to the development of contractual reserves is a perceived restriction on farming practices, land speculation and development, which landowners maintain reduces the resale value of their land (McDowell 1986a). Nevertheless, contractual nature reserves appear to be the only method by which very large tracts of land can currently be acquired for preservation (Figure 13.2), and already they comprise a large proportion of the total area preserved in the Cape Floristic Region.

What is preserved?

Fynbos in the mountains of the Cape Floristic Region has more than half its area currently preserved (Table 13.2). Mountain fynbos occupies about half the area of the Cape Floristic Region. Although thicket, renoster shrubland, and fynbos of the lowlands comprise about 40% of the Cape Floristic Region, less than three per cent (*c.* 760 out of 28 500 km^2) is preserved.

Coarse vegetation units are not the ideal units in which to measure biotic diversity. The use of biogeographic regions would be more appropriate (Kruger 1977; Hall and Veldhuis 1985). For fynbos the majority of biogeographic regions in the Cape Floristic Region are very well preserved (Table 13.3; Figure 13.1); exceptions are the coastal lowlands (Bredasdorp, Malmesbury, Mossel Bay, and Sandveld), associated mountains (Piketberg), and the arid mountainous regions (Bokkeveld, Gifberg, and Witteberg). Renoster shrubland is the most poorly preserved vegetation type: less than 10% of the extant vegetation, which in turn comprises only five per cent of the original extent, is preserved in the west and south-west coastal regions (Table 13.4).

Ideally, the starting point for planning improvements to a reserve network is to ascertain which species are preserved in existing reserves (Siegfried 1989). Unfortunately, published check-lists are available for only 15% of the reserves in South Africa, with plants being especially poorly documented (Siegfried 1989). The PRECIS database of plant species present in the National Herbarium at Pretoria (Gibbs Russell and Gonsalves 1984), together with data from local herbaria, should allow an assessment of preservation status. However, the PRECIS database is inadequate, owing to the quarter-degree grid (i.e. 24 X 27 km) scale used being too coarse (Rebelo and Siegfried 1990) and a ±30% error in distribution records in terms of species (Rebelo and Cowling 1991). Based on the Proteaceae, Rebelo and Siegfried (1990) estimate that 93% of mountain fynbos plant species are protected in the exist-

TABLE 13.2 The changes in the conservation status of vegetation types (based on Acocks 1975) within the Cape Floristic Region. Vegetation categories in parenthesis refer to types described by Cowling and Holmes (this volume).

Acocks veld type	Total area (X 10^3 ha)[1]	Per cent of total area protected						Per cent of natural area remaining[1]
		1974[2]	1983[3]	1987[4]		1989[5]	Herein[6]	
Knysna forest (Afromontane)	384.4	47.9	3.5	4.3		—	—	76
Strandveld (dune thicket)	445.3[7]	1.1	0.5	0.5		0.7	6.8	76
Renosterveld:								
mountain (renoster shrubland)	475.4	0.0	12.1	1.5		2.3	1.3	73
coastal (renoster shrubland)	1528.5	0.9	0.6	0.5		2.5	0.3	15
Macchia:								
coastal (fynbos)	877.0	2.1	2.4	1.8		4.7	3.3	53
mountain (fynbos)	1834.5	15.3	33.6*	54.0*	(28.3)	53.1	52.6	89
false (fynbos, grassy fynbos)	1896.5	21.1	2.2*	2.0*	(26.9)	19.6	47.6	97

1 Moll and Bossi (1984): estimates for Acocks (1975) veld types.
2 Edwards (1974): nature reserves and unafforested State Forest land.
3 Scheepers (1983): as above.
4 Cowan (1987): includes private and state mountain catchment areas.
5 Hilton-Taylor and Le Roux (1989): includes private and state mountain catchment areas.
6 Data from Tables 13.3 and 13.4 (Siegfried 1989). The vegetation types do not correspond exactly. Data as for Cowan (1987), but include private nature reserves and unafforested State Forest land in the east of the region (i.e. 'false fynbos').
7 Excludes portion beyond the Cape Floristic Region.
* Data for the fynbos types from 1983 to 1985 are incorrect partly due to differences in the vegetation type classifications used. Recomputed values (from Cowan 1987) are presented in parentheses.

TABLE 13.3 The areas currently reserved and proposed for the preservation of fynbos vegetation in the major phytogeographic zones of the Cape Floristic Region. The codes refer to the areas depicted in Figure 13.1.

	Code	Total extant area[1] of fynbos (ha)	Area conserved[2] (ha)	Proportion conserved (%)	Area proposed for conservation[2] (ha)	Proportion proposed for conservation[3]	Number of reserves Existing total	Number of reserves >100 km²	Number of reserves Proposed
North-western province									
Cederberg District	Ce	256 900	221 601	86	0	0	3	3	0
Great Winterhoek District	Gw	220 600	162 092	73	1 750	1	7	4	1
Piketberg District	Pi	49 360	0	0	23 200	47	0	0	2
Sandveld District									
Sandveld Zone	Sa	230 600	0	0	27 500	12	0	0	2
Bokkeveld Zone	Bo	85 000	5 070	6	0	0	1	0	0
Gifberg Zone	Gi	237 000	0	0	0	0	0	0	0
South-western province									
Malmesbury District	Ma	97 200	3 409	4	57 339	59	13	0	17
Peninsula District	Cp	27 800	27 755	100	0	0	10	1	0
Riviersonderend District	Ri	91 200	73 157	80	0	0	4	2	0
Franschhoek District	Fr	59 600	58 519	98	1 100	2	7	4	1
DuToitskloof District	Dt	136 900	122 212	89	16 300	12	8	2	11
Houwhoek District	H	99 100	64 416	65	41 325	42	14	2	6
Bredasdorp District	Br	231 700	17 699	8	68 450	30	16	1	7
Potberg District	Po	13 000	2 500	19	500	4	1	0	1
Mosselbay District	Mo	154 900	2 705	2	680	0.4	5	0	1
Coastal mountain province									
Koo Langeberg District	Kl	72 200	59 936	83	0	0	4	1	0
Langeberg District	Ln	185 300	85 023	46	10 950	6	5	4	4
Outeniqua District	Ou	161 900	156 043	96	19 001	12	33	5	8
Kouga District	Ko	665 800	178 213	27	36 632	6	18	4	6
South-eastern province									
Cockscomb District	C	260 100	150 288	58	105 250	40	23	5	8
Inland mountain province									
Swartberg District									
Swartberg Zone	Sw	121 600	121 600	100	0	—	3	1	—
Klein Swartberg Zone	Ks	57 400	57 000	99	0	0	5	2	0
Karoo Island Zone	Ki	168 500	55 250	33	0	0	4	3	0
Witteberg District	Wi	16 100	0	0	0	0	0	0	0
Total		3 699 760	1 624 688	43.9%	409 925	11%	184	44	75
Total original area of fynbos		4 608 000		35.3%		9%			

1 Calculated from Moll and Bossi (1984).
2 Data from the Percy FitzPatrick Institute of African Ornithology database (Siegfried 1989).

TABLE 13.4 The areas currently preserved and proposed for conservation in the major non-fynbos vegetation types in the Cape Floristic Region.

	Total extant area of vegetation[1] (ha)	Area conserved[2] (ha)	Proportion conserved (%)	Area proposed for conservation[3] (ha)	Proportion proposed for conservation (%)	Area transformed by agriculture[4] (ha)	Proportion transformed (% total)
Renoster shrubland							
West coast	27 294	2 484	9.1	12 198	44.7	637 497	95.9
South-west coast	19 984	660	3.3	8 505	42.6	469 582	95.9
South coast	356 994	2 935	0.8	40 320	11.3	414 633	53.7
Central mountain	581 336	8 865	1.5	6 350	1.1	71 443	10.9
Total	985 608	14 944	1.5	67 373	6.8	1 593 155	61.8
Thicket[5]							
West coast	192 049	27 335	14.2	53 287	27.7	213 611	52.7
South coast	130 628	12 590	9.6	17 724	13.6	6 076	4.4
Total	322 677	36 713	11.4	71 011	22.0	219 687	40.5

1 Moll and Bossi (1984).
2 Jarman (1986) and Cowan (1987).
3 Jarman (1986).
4 Estimated from Moll and Bossi (1984).
5 Mosaic of dune thicket and fynbos.

ing reserve network. No estimates are available for renoster shrubland or lowland vegetation types.

Status of nature reserves

Although it has become standard to consider most preserved areas as 'nature reserves' in the Cape Floristic Region, these reserves encompass widely diverging management objectives. Thus, emphasis in the acquisition of provincial nature reserves has been the preservation of bontebok (*Damaliscus dorcas dorcas*), mountain zebra (*Equus zebra zebra*), geometric tortoise (*Psammobates geometricus*), feeding and roosting areas for wading birds, and fish breeding. Many subsidized nature reserves are managed as botanical gardens, game farms, reclaimed agricultural lands, and other activities incompatible with classical conservation concepts. Management practices geared towards maintaining large mammals in fynbos (e.g. Cape of Good Hope Nature Reserve, De Hoop Nature Reserve, Bontebok National Park) may result in the degradation of natural vegetation by too frequent burning, bush cutting, the input of fertilizers and trace elements (e.g. from salt-licks), the provision of drinking troughs, and the planting of pasture grasses (Van Rensburg 1975; Zumpt and Heine 1977; Novelli 1986; Scott 1986; Van Wilgen et al. this volume; D Clark personal communication). That the environment is degraded by these 'habitat improvement' activities suggests that resident, large mammals are not an integral part of the ecology of fynbos. By contrast, the breeding biology of the geometric tortoise requires a fire regime compatible with flora preservation, so that management practices aimed at maintaining tortoises effectively preserve the natural habitat (Greig 1981; Van Wilgen et al. this volume).

Although a classification of nature reserves on the basis of management goals is urgently required, data are not available for the majority of subsidized and private nature reserves. In addition, the management policies for the State Forests acquired by the Chief Directorate of Nature and Environmental Conservation are still being formulated. A thorough review of the management strategies of reserves in the Cape Floristic Region is long overdue.

CONFIGURATION OF RESERVES AND THE RESERVE NETWORK

Assuming that the goal of a conservation strategy is to preserve maximum biotic diversity in a minimal area, then three aspects must be considered: the number of reserves, their location, and their size. It is the location of individual reserves and their relative contribution towards preserving total species richness which determine the number of reserves required (Rebelo and Siegfried in press). The optimal location of reserves is determined primarily by centres of endemism, with species turnover and relative species richness contributing to the total number of reserves required (Rebelo and Siegfried in press). Because of practical considerations, the size of reserves is usually determined by factors other than preserving biotic diversity *per se* and will be considered later.

Three major prescriptions have been made over the past half century for the preservation of fynbos. Wicht (1945) sought to preserve it for its aesthetic and general scientific value, with the primary objective of conserving the native vegetation in reserves comprising 'well selected, representative, relatively large regions, which should be maintained with painstaking care'. Five reserves were designated as the core of this reserve configuration (Table 13.5), with numerous additional, unspecified, local reserves mooted. Kruger (1977) considered the optimum strategy to place reserves so as to represent different fynbos vegetation types. In the absence of a classification with the desired resolution, he used major fynbos vegetation types within biogeographic regions as the basis for designating 19 zones requiring preservation (Table 13.5). Based on the distribution of Proteaceae species, Rebelo and Siegfried (1990) used transects comprising 12 X 13 km grid squares through areas of high species richness to determine the location of reserves required to preserve each species in at least one reserve.

The dynamics of reserve allocation in the Cape Floristic Region were investigated by Rebelo and Siegfried (in press), using an iterative approach (Pressey and Nicholls 1989a, b) and the distribution of Proteaceae species on a 12 X 13 km square grid (Figure 13.3a). All Pro-

TABLE 13.5 A summary of the history of proposed nature reserve networks for fynbos vegetation in the Cape Floristic Region. Reserves are numbered in order of decreasing importance (P = primary reserves, S = secondary reserves).

Vegetation type[1]	Biogeographic centre (Weimarck 1941)	Wicht (1945)	Kruger (1977)	Rebelo and Siegfried (1990)
Mountain fynbos	North-western Centre			
	Cederberg subcentre	1. Cederberg	7. Cederberg	P6, S2
	Great Winterhoek subcentre		8. Groot Winterhoek	P7, P13, S3
	South-western Centre			
	French Hoek subcentre	2. Drakenstein-Kogelberg	10. Southern	P1, P12, S1, S6, S7
			11. Riviersonderend	P3, S9
	Hottentot Holland subcentre		9. Northern	P11
	Peninsula subcentre			P5
	Karoo mountain Centre	3. Swartberg	12. Swartberg	P8
			13. Little Karoo Islands	S10, S11, S12
	Langeberg Centre	4. Lemoenshoek	14. Lemoenshoek	P4
	(Potberg 'island')			P9
	Knysna Centre	5a. Outeniqua	15. Outeniquas	P14
	South-eastern Centre			
	Zitzikamma subcentre	5b. Tsitsikamma	17. Tsitsikamma	—
	Cockscomb subcentre		16. Kouga river drainage	P15
			18. Winterhoek Mountains	S5
	Zuurberg subcentre		19. Zuurberg	—
Coastal fynbos	South-western Centre			
	Malmesbury subcentre		1. West coast	P10, S4
	Bredasdorp subcentre		2. S coast — Elim flats	P2
	Langeberg Centre		3. S coast — limestone	S8
Arid fynbos	North-western Centre		4. Witteberg mountains	S12
	Karoo mountain Centre		5. East	S11
			6. West	S10

[1] *Sensu* Kruger (1977) and Taylor (1978), based on Moll et al. (1984). Arid fynbos refers primarily to asteraceous and restioid fynbos from the dry north and north-west of the region, and in the rain-shadows of the Langeberg and Cederberg mountains.

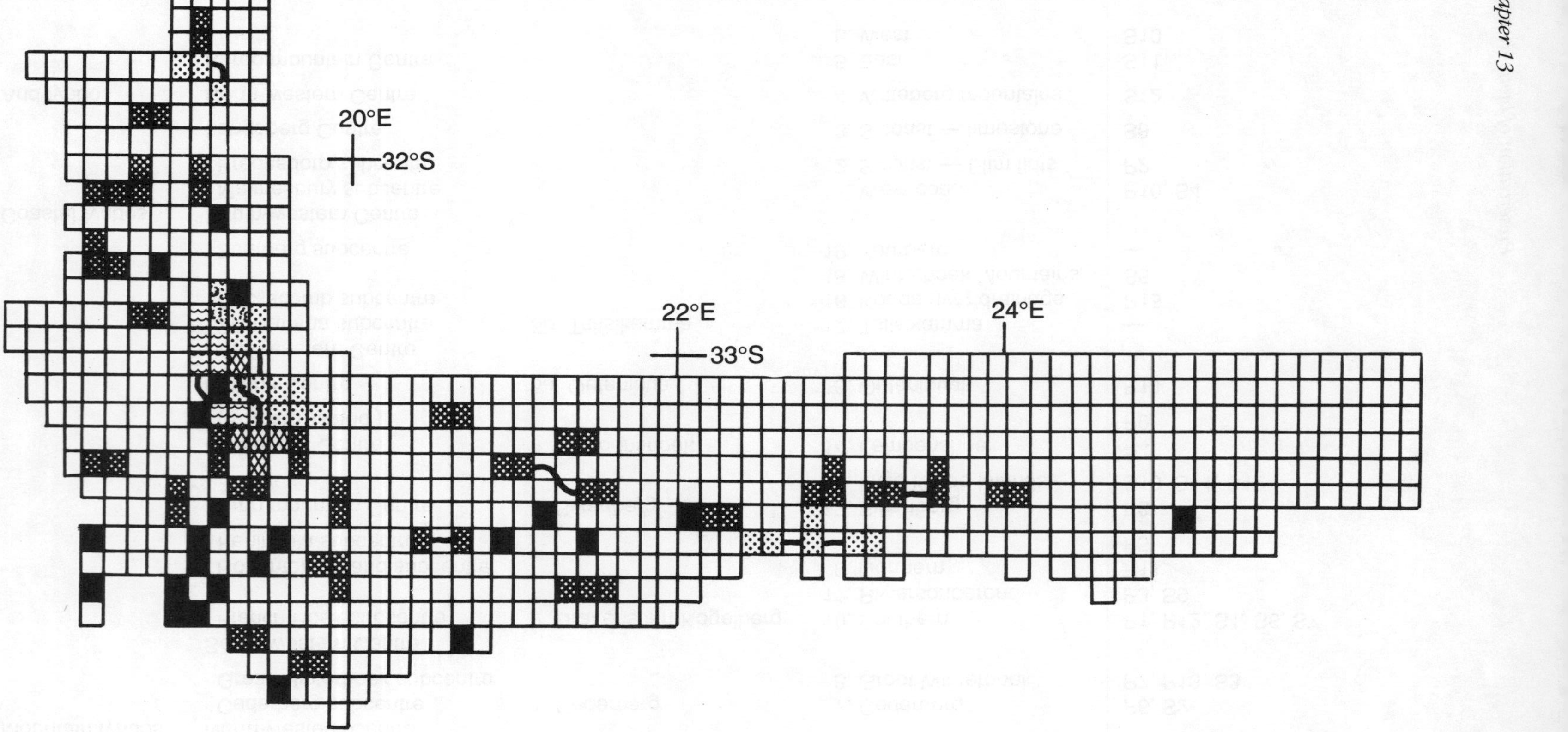

FIGURE 13.3a The spatial configuration of an optimal nature reserve network for fynbos vegetation in the Cape Floristic Region (Rebelo and Siegfried in press), based on an iterative model using the distribution of Proteaceae species in an eighth-degree grid (12 X 13 km). A reserve was selected, at each iteration, as the grid square containing the highest sum of rarity scores for all species not adequately preserved, where the rarity score equals the inverse of the proportion of total grid squares occupied by the species. Solid blocks are squares invariably selected as reserves. Shaded blocks indicate reserves for which several grid squares are equally suitable and solid lines connect squares which are not orthogonally adjacent. Different shading separates different adjacent reserves.

teaceae species can be preserved at least once in a reserve system of 53 grid squares (Figure 13.3a) or 17% of the total area of fynbos (although the area of 20 reserves (= grid squares) protecting one or two species could perhaps be less than 156 km^2). To preserve each species in at least two reserves requires slightly less than double the area above, whereas preserving each species in five reserves requires four times the area, and in 10 reserves requires six times the area. The configuration of these reserves validates Kruger's (1977) approach of placing a reserve in each biogeographic zone, but indicates that more reserves are required in the more species-rich areas.

The use of the Proteaceae as representative of the total taxa for the Cape Floristic Region is considered valid, since the species richness of Proteaceae is significantly correlated to total species richness, as well as to the species richness in all the major fynbos families — Ericaceae, Restionaceae, Bruniaceae, Rutaceae: Diosmae, Penaeaceae — and some larger genera — *Aspalathus* (Fabaceae), *Muraltia* (Polygalaceae) — at a quarter-degree (24 X 27 km) grid square scale (Rebelo and Siegfried 1990). This suggests that similar factors influenced speciation and dispersion in the major fynbos families. It is not known over what range of grid scales the relationship holds.

Assuming that all grid squares containing reserves which comprise greater than 50% of their area adequately preserve all species present, then the existing reserve configuration (Figure 13.3b) preserves 80% of fynbos species. The major gaps identified in the reserve network agree with the conclusions derived from an analysis of area preserved in biogeographical zones (Table 13.3); namely the lowlands and arid areas are not adequately preserved (Rebelo and Siegfried 1990).

Note that the above network considers only the preservation of maximum species richness in a minimal area. Nothing is known about the importance of corridors, sink-source areas in fynbos, or the seasonal importance of other vegetation types in supplying species for pollination, seed dispersal, or predation in determining the spatial configuration of the reserve network.

CONSERVATION PRIORITIES

A variety of approaches exist for assessing areas for preservation (Margules and Usher 1981). To date, only a single attempt at prioritizing areas for preservation in the Cape Floristic Region has been undertaken (Jarman 1986), and this was confined to an appraisal of the lowlands. The results of this survey formed the basis for the NAKOR National Plan for Nature Conservation (Burgers et al. 1987). Sites of existing natural vegetation were ranked according to the rarity of the vegetation type, habitat diversity, species and rare species richness, the size and shape of the area, and its proximity to neighbouring sites. A major omission from the analysis was an assessment of the degree of threat. Consequently, it has favoured the acquisition of large reserves, far removed from any threat, at a fraction of the cost per unit area of small potential reserves, under imminent threat of agricultural and urban transformation.

Suitable databases for assessing priority areas include Red Data Books and analyses of major threats to species in the Cape Floristic Region. Red Data Books for all vertebrate orders (Brooke 1984; Smithers 1986; Skelton 1987; Branch 1988), plants (Hall et al. 1980; Hall and Veldhuis 1985), and butterflies (Henning and Henning 1989) are available for South Africa. These list threatened and naturally rare species and summarize existing autecological information (Ferrar 1989). With the exception of birds and large mammals, many of which are still common in South Africa but extinct or threatened in the Cape Floristic Region, few threatened species are shared between the Cape Floristic Region and the rest of South Africa (Rebelo in press). A regional compilation of extinct, threatened, and rare mammal and bird species (Rebelo in press) was undertaken for the Cape Floristic Region.

What are the threats?

Agriculture has had the largest impact on vegetation types in the Cape Floristic Region (Table 13.6). Its effect has been largely confined to the lowlands and is most prominent in renoster shrubland, with its conversion to primarily wheat lands, vineyards, and pastures (Boucher 1981a; Hall 1981). Lowland fynbos

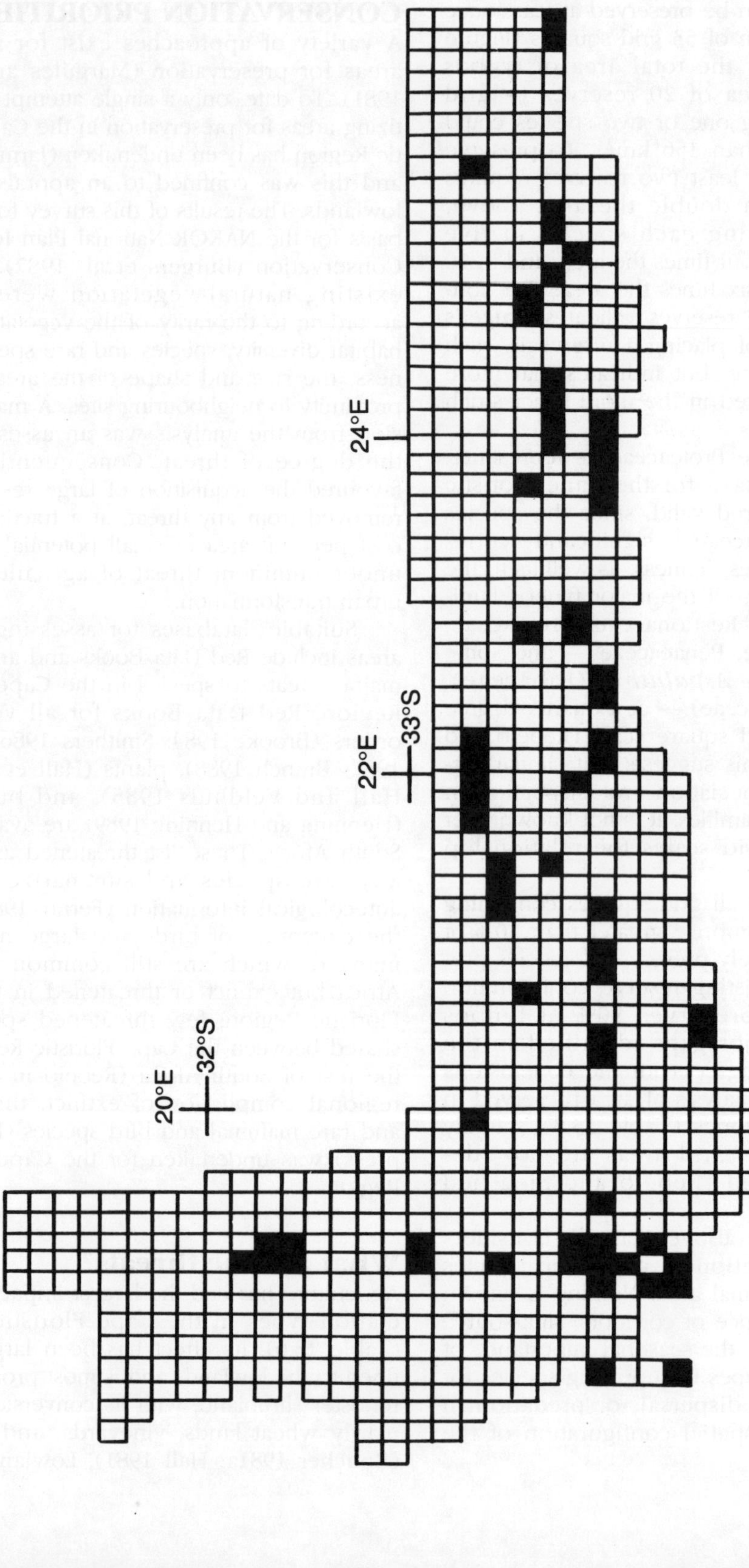

FIGURE 13.3b The distribution of state-owned reserves in the Cape Floristic Region. Solid blocks have more than 55% (8 000 ha) of the total area (14 400 ha) preserved (Rebelo and Siegfried 1990).

TABLE 13.6 **Threats to vegetation types within the Cape Floristic Region. Threats are categorized as high (H), medium (M), low (L), or no threats known (0). Where data are available, the area transformed by the threat is given as a percentage of the total area of the vegetation type.**

Threat	Fynbos mountain	Fynbos lowland	Renoster shrubland mountain	Renoster shrubland lowland	Thicket	Afromontane forest	Karroid shrubland
Agriculture and afforestation[1,2]	7	49	11	79	41	24	L
Alien invasives: Fabaceae[3]	10	36	0	7	43	M	L
Alien invasives: *Hakea* and *Pinus*[3]	26	0	0	4	7	0	0
Urbanization[2]	L	M	L	L	H	L	L
Dam building[2]	M	0	M	L	0	L	0
Water extraction	0	0	L	0	H	0	?

1 Moll and Bossi (1984).
2 Macdonald et al. (1985).
3 Macdonald (1989).

and thicket have been largely converted to pasture (Rebelo et al. 1991). Alien invasive plant species, together with agriculture, account for the bulk of habitat destruction in the Cape Floristic Region. Unlike agriculture, the removal of aliens may result in the recovery of vegetation (Richardson et al. this volume). Although urbanization accounts for less than one per cent of the area of the Cape Floristic Region, it is largely concentrated in the Greater Cape Town metropolitan area (Macdonald 1989). In a survey of the 484 ha of lowland fynbos remaining in the *c.* 30 000 ha metropolitan area, McDowell et al. (1991) found 74 Red Data Book Plant taxa, a ratio of 15.3 species per km^2. Much of the coastal area immediately adjacent to the sea (mainly thicket) is under threat of resort development (Cowling and Pierce 1985; Jeffery and Moll 1987). National legislation to restrict such development within one kilometre of the coastline was repealed by provincial authorities. Dams cover an insignificant area of the Cape Floristic Region (*c.* 1.3%, including farm dams) (Macdonald 1989) but became a major conservation issue following a proposal to dam the lower Palmiet River in the Kogelberg State Forest (Roberts 1982). Water is extracted from underground aquifers along the west coast for urban consumption, and is apparently drying up perennial streams in the area (F W Duckitt personal communication).

A similar pattern is evident for overall threats to specific Red Data Book species (Table 13.7). Agriculture and alien invasive plants contribute significantly to species decline in the Cape Floristic Region. However, threats vary considerably among groups (Table 13.7). Thus, birds and large mammals have largely been hunted and poisoned (Brooke 1984; Smithers 1986), whereas reptiles are especially susceptible to collection for the pet trade (Branch 1988), fish to predation by introduced alien fish (Gaigher et al. 1980; Skelton 1987), and butterflies to coastal development (Henning and Henning 1989). Whereas the majority of threats involve an obvious human impact, the importance of fire emphasizes the need to maintain natural disturbance regimes (Van Wilgen et al. this volume). Interestingly, fire has been identified as a threat to frogs, but not to fish, possibly reflecting differing perceptions to fire (independent of its effects on water discharge) by compilers of Red Data Books. Although hybridization has been recorded only as a threat to frogs and small mammals, the transfer of genes between populations and species of plants through the establishment of wild flower gardens probably occurs more frequently among widespread species than is realized.

Although 68% of the plants in the Cape Floristic Region are endemic, only 15% are listed in the Red Data Book, and only three per cent are threatened (Table 13.8). Of these, typical fynbos families (Proteaceae, Ericaceae, and Rutaceae) appear most threatened. Mammals are the most seriously threatened (14% of species) of taxa, but this is largely due to the extinction, by hunting, of large carnivores and ungulates with widespread distributions. Some 60% of mammals over 50 kg have been eliminated from the Cape Floristic Region (Rebelo in press), although many species have been reintroduced. Amphibians, with 13% threatened, have a quarter of their endemic species threatened. Overall threats to amphibians and reptiles do not appear to differ among families, although the pet trade is the greatest threat to girdled lizards (*Cordylidae*) and tortoises (*Chelonii*) (Branch 1988). Among birds, larger species from the scavenging, predatory, and plant-invertebrate feeding guilds are most threatened, mainly by hunting and poisoning (Brooke 1984). The Lycaenidae account for the vast majority of Red Data Book butterflies, and account for 82% of the species endemic to the Cape Floristic Region. Lycaenidae account for all of the threatened species in the Cape Floristic Region (Table 13.7), probably because of their association with both specific food plants and ant-host species (Henning and Henning 1989). Urbanization and coastal development are the greatest threat to butterfly species in the region (Table 13.7).

In summary, although different taxa are threatened by different factors, agriculture and alien plant invasions are the greatest overall threats to plants, animals, and ecosystems in the Cape Floristic Region.

Priority areas: where are the threats greatest?

Areas richest in Red Data Book species are shown in Figure 13.4. Mammals and birds are

TABLE 13.7 Threats to plant and animal taxa in the Cape Floristic Region listed as endangered or vulnerable in the Red Data Books. For each taxon, the number of species in the largest threat category has been scaled to equal 10. Data from Rebelo (in press).

Threat	Score Plant Elim[1]	Total[2]	Butterfly	Fish	Amphibian	Reptile	Bird	Mammal	Rank of threat
Agriculture	4.4	5.3	8.9	9	10	10	1.3	1.5	46
Alien invasive plants	10	10	4.4	6	0	4	0	0	24.4
Hunting/poisoning	—	—	0	0	0	0	10	10	20
Fire (frequency and season)	2.3	3.7	5.6	0	2.5	2	2.5	0	16.3
Urbanization/industrialization	5.6	3.8	10	1	0	0	0	0	14.8
Commercial collecting	1.6	1.3	?	0	0	10	0	0	11.3
Dams/weirs/roads	1.7	0.9	2.2	4	2.5	0	1.3	0	10.9
Alien predators	—	—	0	10	0	0	0	0	10
Pollution[3]	0.2	0.4	0	4	5	0	0	0	9.4
Afforestation	1	0.9	2.2	0	2.5	2	1.3	0	8.9
Mining/quarrying	0.8	1.2	0	1	0	4	0	0.5	6.7
Grazing/browsing	0.8	2.9	0	0	0	0	3.8	0	6.7
Hybridization	?	?	0	0	2.5	0	0	0.5	3
Intolerance of human presence	—	—	0	0	0	0	2.5	0	2.5
Casual flower picking	1.5	0.2	—	—	—	—	—	—	0.2
Genetic decline[4]	4.5	1.1	0	0	0	0	0	0	1.1
Mowing/human trampling	0.8	1	—	—	—	—	—	—	1
Number of spp. in the largest threat category	49	84	9	10	4	5	8	20	

1 Data for the plant species of the Agulhas Region (Hall and Veldhuis 1985). Typically for fynbos, little agricultural transformation has occurred: consequently this threat is under-represented relative to the entire Cape Floristic Region. This value does not contribute to the total 'rank of threat'.
2 Data for 232 species for which threats are listed in Hall and Veldhuis (1985).
3 Including: fertilizers, pesticides, salinization, and acid rain.
4 Speculative assessment due to inbreeding depression and stochastic processes inherent to small populations.

excluded, but these would almost certainly have been most abundant in non-fynbos vegetation (Skead 1980). The bontebok and extinct bluebuck (*Hippotragus leucophaeus*) were apparently endemic to renoster shrubland of the south-western coast (Smithers 1983), and buffalo and elephant were most frequently encountered along the south and east coast (Skead 1980). It cannot be ascertained from historical records whether quagga occurred in large numbers in the lowlands or were largely confined to karroid and renoster shrubland of the interior mountains (Bateman 1961; Skead 1980; Smithers 1983, 1986; Rookmaaker 1989).

Red Data Book amphibians, reptiles, butterflies, and plants are concentrated in the Greater Cape Town metropolitan area, encompassing the Cape Peninsula and the adjacent Cape Flats (Figure 13.4). Of the 82 Red Data Book plant species in the lowlands of the metropolitan area, 74 occur in fynbos and eight in thicket vegetation (McDowell et al. 1991). Similarly, most amphibians occur in acid waters associated with fynbos vegetation. A minor node of species richness in the Van Stadensberg area near Port Elizabeth is shared by Red Data Book reptiles, amphibians, butterflies, and plants (Figure 13.4).

Only fish have a divergent pattern from the above. The majority of endemic fish appear to have arisen from the confluence of the Orange and Olifants Rivers between the Palaeogene and Miocene (Dingle and Hendey 1984). Subsequent river capture of the Orange to the north resulted in the isolation of the Olifants River System and the evolution of eight extant endemic species (Skelton 1987). Agriculture and predatory sport fish have restricted many of these endemics to the upstream portions of their previous distribution range (Gaigher et al. 1980; Scott and Hamman 1984; Skelton 1987). The species in the river systems of the south and west coast tend to be generalists, but are usually absent from areas adjacent to agriculture (Scott and Hamman 1984; Skelton 1987).

Unfortunately, detailed data at a quarter-degree grid scale on the distribution of common species are not available, so that patterns of species richness are not available for taxa other than plants. For plants, the richness of Red Data Book species is strongly correlated with that of total species richness (Rebelo and Tansley submitted). Furthermore, for the Proteaceae, the distribution of non-Red Data Book species is also strongly correlated with total species richness. Thus, rare (Red Data Book) species tend to occur in areas which are also rich in non-Red Data Book species (Rebelo and Tansley submitted). Analyses of threatened Red Data Book species will therefore be biased by species richness, and should be corrected for species richness before geographical patterns of threats can be determined.

The geographical dispersion of grid squares with significantly more threatened (extinct, endangered, and vulnerable) Proteaceae species, relative to that predicted by regression analysis from the total number of Proteaceae species, is largely confined to the lowlands, with the Cape Peninsula and Cape Flats being the highest ranked area (Figure 13.5; Rebelo and Tansley submitted). This is primarily due to the urban expansion of the Greater Cape Town metropolitan area over much of one of the richer centres of endemism within the Cape Floristic Region (Figure 13.1).

The most urgent strategic requirement for the preservation of biotic diversity in the Cape Floristic Region is thus within the Greater Cape Town metropolitan area. This should involve the acquisition of nature reserves and corridors with emphasis on lowland fynbos communities, specifically the large areas to the immediate north, as proposed by Jarman (1986).

The areas with high numbers of threatened species are quite distinct from those containing more naturally rare (i.e. rare or critically rare) Red Data Book species than predicted from total Proteaceae species richness. Areas of high naturally rare species are concentrated in the high mountains of the south-west, with several minor, isolated outliers (Figure 13.5). Provided that the State Forests in these centres continue to be preserved, large-scale extinction of naturally rare species in the Cape Floristic Region could be avoided.

Note that these results relegate renoster shrubland to a far lower priority than suggested by Jarman (1986). This may be because renoster shrubland is poor in locally endemic species (Cowling et al. this volume). More data are urgently required on the distribution and abundance of renoster shrubland species.

TABLE 13.8 The endemic and threatened (extinct, endangered, or vulnerable) species in families containing the most Red Data Book species in the Cape Floristic Region (CFR). The values in parentheses are the percentage of the total number of species in the CFR accounted for by the subtotal. Data from Rebelo (in press).

	Number of species							
	Total		Endemic		Red Data Book		Threatened	
Plants								
Proteaceae	320		306		131		65	
Iridaceae	612		485		242		51	
Ericaceae	688		666		138		31	
Rutaceae	259		242		103		22	
Asteraceae	986		608		166		13	
Fabaceae	644		525		110		10	
Subtotal	3 509	(41)	2 834	(48)	890	(67)	192	(68)
Total for CFR	8 600		5 865		1 326		281	
Butterflies								
Lycaenidae	143	(61)	59	(82)	52	(96)	6	(100)
Total for CFR	234		72		54		6	
Fish								
Cyprinidae	15	(54)	11	(73)	10	(83)	5	(71)
Total for CFR	28		15		12		7	
Amphibians								
Heliophrynidae	4		4		2		2	
Ranidae	13		4		2		1	
Subtotal	17	(45)	8	(42)	4	(57)	3	(60)
Total for CFR	38		19		7		5	
Reptiles								
Sauria: Cordylidae	14		4		3		1	
Serpentes: Colubridae	25		0		3		1	
Sauria: Gekkonidae	18		3		3		0	
Subtotal	57	(52)	7	(36)	9	(53)	2	(60)
Total for CFR	109		19		17		5	
Birds								
Falconiiformes	22		0		6		5	
Gruiformes	15		0		7		4	
Subtotal	37	(13)	0	(0)	13	(62)	9	(75)
Total for CFR	288		6		21		12	
Mammals								
Carnivora	27		0		11		7	
Artiodacyta	20		2		10		8	
Perissodactyla	5		0		3		3	
Subtotal	52	(41)	2	(22)	24	(65)	18	(86)
Total for CFR	127		9		37		21	

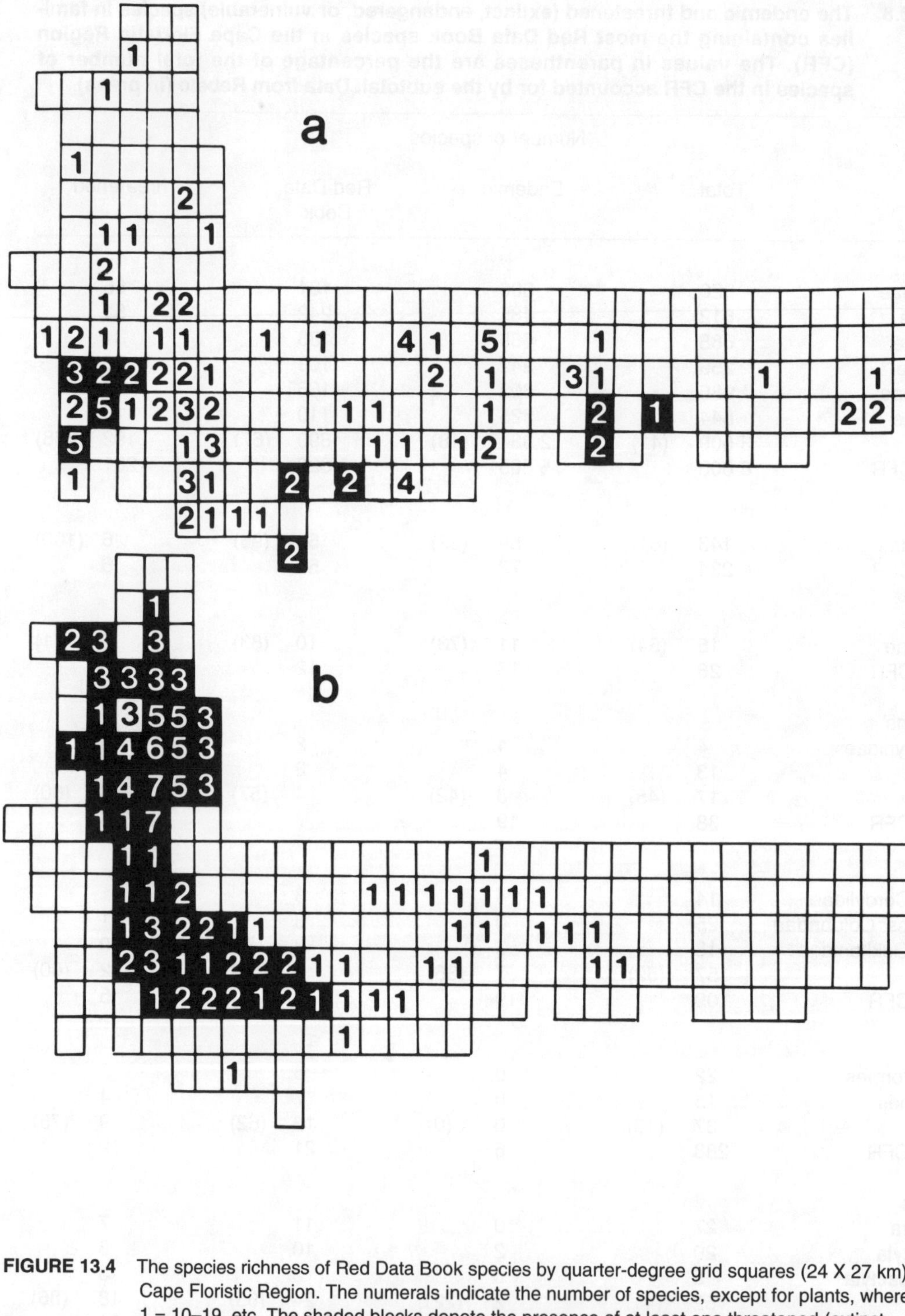

FIGURE 13.4 The species richness of Red Data Book species by quarter-degree grid squares (24 X 27 km) in the Cape Floristic Region. The numerals indicate the number of species, except for plants, where 0 = 1–9, 1 = 10–19, etc. The shaded blocks denote the presence of at least one threatened (extinct, endangered, and vulnerable) species, except for plants where they denote more than 15 threatened species. a = butterflies; b = fish; c = amphibians; d = reptiles (light shading refers to a species threatened by commercial pet collecting); e = plants. Data from Rebelo (in press).

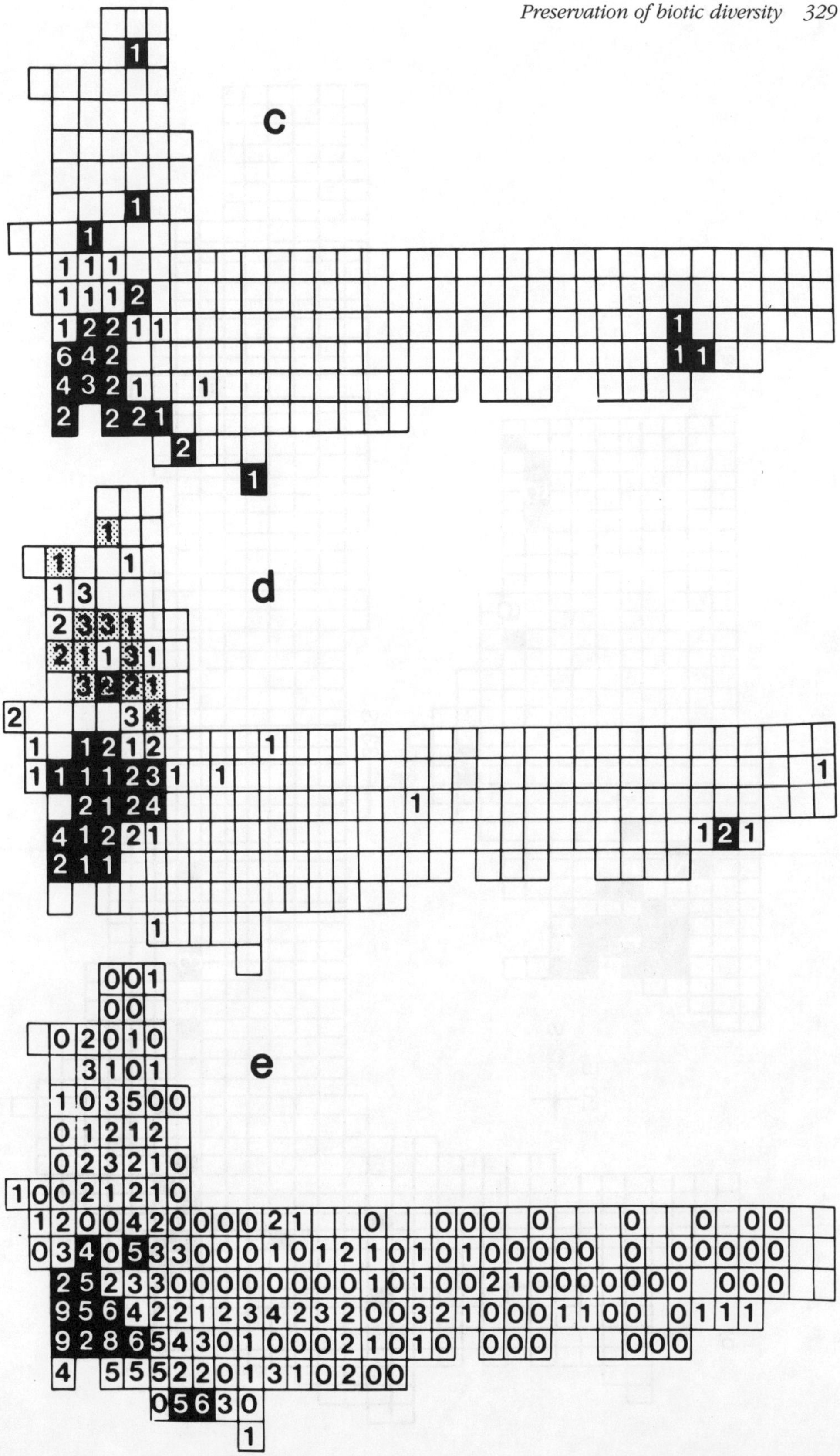
c
d
e

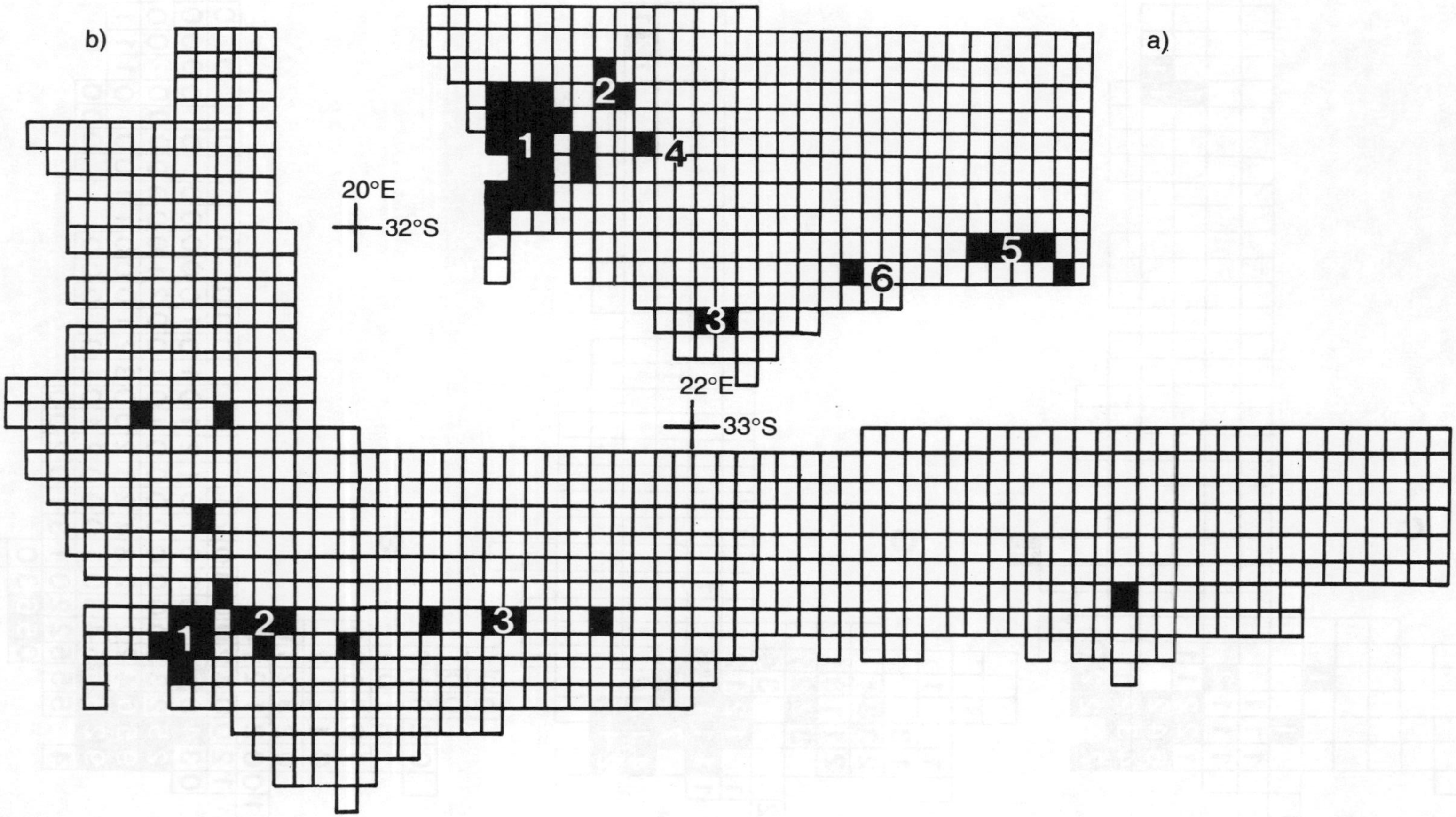

FIGURE 13.5 Priority areas for conservation in the Cape Floristic Region based on the distribution of (a) threatened (extinct, endangered, and vulnerable) and (b) naturally rare (rare) Proteaceae species. The shaded blocks (12 X 13 km) lie outside the 95% prediction limits of the regression of threatened or naturally rare species against total Proteaceae species richness (Rebelo and Tansley submitted).

EFFECTIVE RESERVE SIZE

Sizes of existing reserves

Forty-nine (20%) of the 244 reserves in the Cape Floristic Region are larger than 10 000 ha (Figure 13.6). More than half of these were formerly State Forests, 17 are contractual reserves (accounting for 94% of the contractual reserves), and six reserves are controlled by other authorities. The large majority of these reserves are in the mountains and preserve fynbos. There is only one fynbos reserve in the lowlands greater than 10 000 ha. No renoster shrubland reserves are larger than 10 000 ha. Many of these reserves are contiguous, particularly contractual reserves which usually abut on state-owned reserves, so that effective reserve sizes in the mountains are often larger than 100 000 ha. Similarly, some State Forests consist of isolated mountain peaks, but these are usually surrounded by contractual reserves.

Some 105 reserves (43%) are smaller than 500 ha (Figure 13.6). These comprise the majority of private and subsidized reserves. Only four renoster shrubland reserves exceed 500 ha in size, whereas other vegetation types contain many reserves larger than 500 ha. Forty-two reserves (17%) are less than 50 ha in size.

How do these reserve size classes compare with theoretical considerations for reserve size based on the autecology of species and ecosystem processes in the Cape Floristic Region?

Biogeographical considerations

An empirical study on the effects of island size on plant species richness has been undertaken in the relatively species-poor fynbos of the southern Cape Floristic Region (Bond et al. 1988). Fynbos 'islands' surrounded by Afromontane forest supported significantly fewer species than mainland sites of the same area, with up to 75% extinction on the smaller islands. This island effect disappeared at about 600 ha. Species were apparently lost from islands due to a change in the disturbance regime, with smaller islands burning less frequently. This hypothesis was supported by the lack of resprouting and lower, shorter-lived species on islands. There was no evidence for a collapse of mutualisms, whether dispersal or pollinator, nor were dioecious species especially prone to extinction. There is, however, no basis for extrapolating these results to more species-rich areas of the Cape Floristic Region, due to differences in species-area relations (Cowling et al. this volume) and the frequency of natural ignition events (Le Maitre and Midgley this volume).

A study of fynbos on limestone islands within acid sand fynbos compared with equivalent sizes of fynbos on extensive limestone deposits, yielded minimum reserve sizes of only 4–15 ha (Cowling and Bond in press). Only local limestone endemics and limestone specialist shrubs (calcicoles), those arguably most in need of preservation, were vulnerable to extinction. The frequencies of dispersal and pollination syndromes, growth forms, and height classes were unrelated to island size. Since the limestone islands are exposed to the same disturbance regimes as, and share pollinators and seed dispersers with, acid sand fynbos (Cowling and Bond 1991), these minimum areas are only applicable to reserves in which disturbance (specifically fire) regimes are maintained, and which have access to pollinators and seed dispersal agents.

There appears to be a case for preserving very small patches (< 4 ha) of vegetation, provided that the disturbance regimes are maintained. It is only in much larger reserves (*c.* 500 ha in the mountains of the southern Cape) that disturbance regimes are maintained naturally, so that all reserves less than 500 ha will invariably require management to maintain fire regimes and, thus, species richness.

Critical population size and habitat area: plants

The theory of minimum viable population size has emphasized the concept of a critical range of population sizes below which demographic, genetic, and environmental stochasticity strongly reduces a population's chances of long-term survival (Gilpin and Soule 1986). The mean population size of several fynbos plant species is below 50 (Kruger 1977; seven species in the Proteaceae — Tansley 1988), with a few species only ever recorded as occurring in populations of fewer than 10 individuals (e.g. *Sorocephalus imbricatus*, *S. palustris*, and *Mimetes stokoei* in the Proteaceae). Although seed bank sizes of these species are unknown, so that the species are not necessarily susceptible to inbreeding depression or

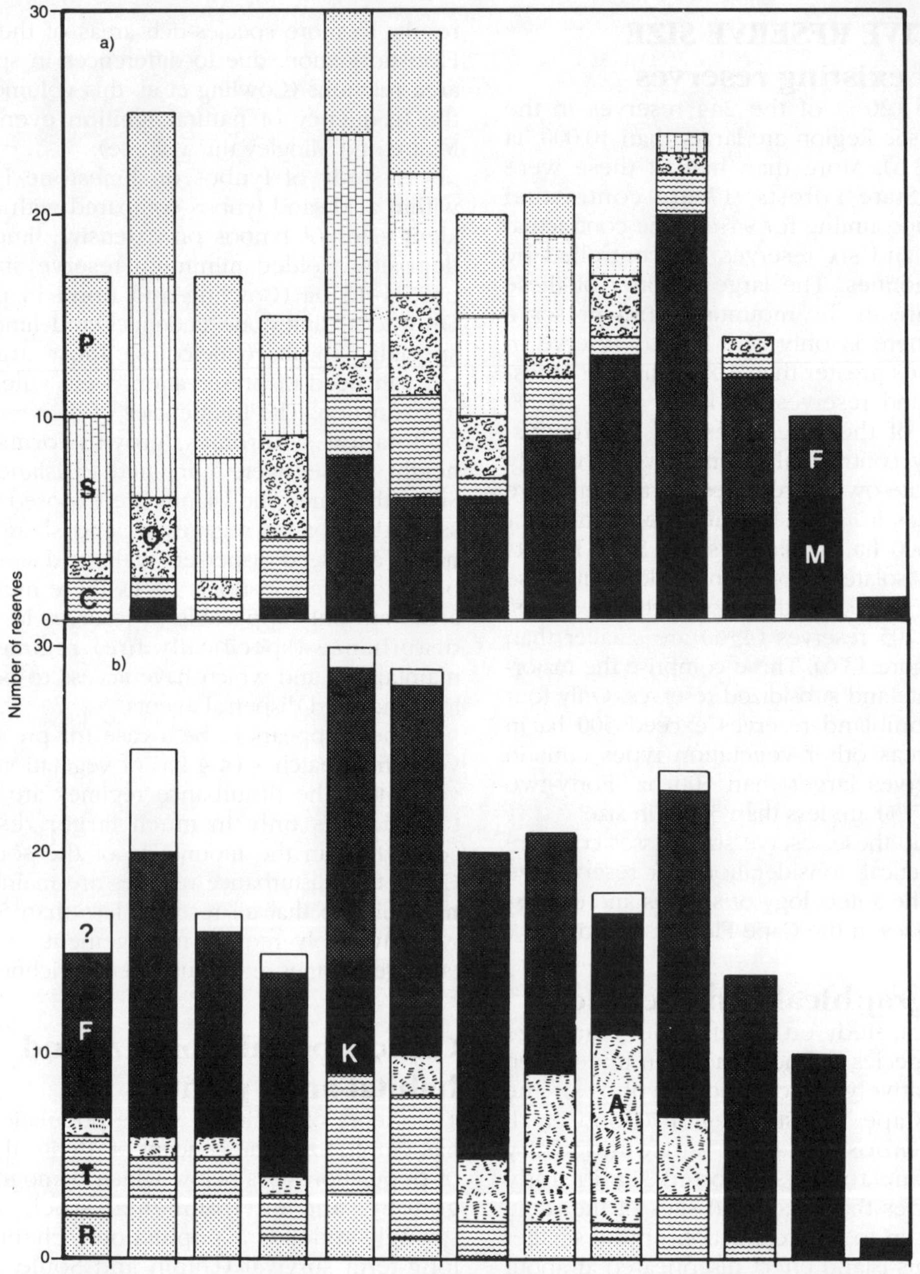

FIGURE 13.6 The frequency of reserve size classes in the Cape Floristic Region. (a) Categorized by owner: C = provincial nature reserves; F = State Forests; M = contractual reserves; O = other state reserves (National Botanical Institute, National Parks Board, South African Defence Force); P = private nature reserves and natural heritage sites; S = subsidized nature reserves. (b) Categorized by the dominant vegetation type within the reserve: A = Afromontane forest; F = fynbos; K = karroid shrubland; R = renoster shrubland; T = thicket; ? = no data. The abscissa is a logarithmic scale: white numerals refer to the exponent. Data from the Percy FitzPatrick Institute of African Ornithology database (Siegfried 1989) and Cowan (1987).

loss of heterozygosity, the areas occupied by these species are consistent with the notion that preserving seed banks requires a far smaller area than for adult plants. Thus, the preservation of rare fynbos plant species could be undertaken, with appropriate (intensive and expensive) management, in several thousand pocket-handkerchief reserves each several hectares in size. Such reserves will probably result in the local extinction of more common species which generally occur in extensive populations (Bond 1989), and sparse species, but should be adequate for preserving target rare species. However, the rarer species may be in the process of losing heterozygosity, hence data on the genetic status of rare species are required before any conclusions can be drawn.

Kruger's (1977) statement '(where) populations are so small it would in principle be best to include the total populations in reserves where possible' is a debatable viewpoint of .little use for determining minimum reserve size. Even where rare species tend to co-occur in rare habitats, these habitats often comprise small patches in the total landscape. Thus small, patch-sized reserves may be feasible, provided that ecological processes shared between patches, such as fire regimes and water table dynamics, are maintained (Cowling and Bond 1991).

Preserving higher trophic levels

Arguments based on the size of reserves to maintain minimum viable populations of large herbivores and carnivores have usually focused on single reserves. Seldom have reserve networks been considered. Thus, Kruger's (1977) arguments assume that each reserve must maintain minimum viable populations, and consequently his minimum viable reserve sizes are very large. If the entire reserve network is taken as the unit, then the majority of reserves need only contain the minimum viable family unit, provided that adequate corridors exist to allow movement of non-territorial and immature individuals between nature reserves. Where no corridors exist, each reserve designated for a particular species must contain a minimum viable population of those species.

The requirement for large reserve sizes does not mean, as argued by Kruger (1977), that the Protea Seedeater *Serinus leucopteris* would require a nature reserve of 1 000 ha as a minimum area for its survival (assuming a density of one pair of birds per 2 ha in its favoured habitat, under optimal fire management, and a minimum viable population of 150 breeding pairs, Kruger 1977). Presumably, 500 reserves each 2 ha in size would equally suffice, provided that the configuration of reserves could allow movements of Protea Seedeaters between reserves. Distances moved by Protea Seedeaters are unknown, but the species does not occur on the Cape Peninsula (Hockey et al. 1989), a 28 000 ha mountain island of apparently suitable fynbos habitat, separated by 40 km of lowland from the nearest mountain. Since the Peninsula was connected by mountains to the 'mainland' 20 000 years ago (Dingle and Rogers 1972), it could perhaps be countered that Kruger's estimates are too low. The Sentinel Rock Thrush (*Monticola explorator*) and Victorin's Warbler (*Bradypterus victorini*) are similarly absent from the Cape Peninsula despite the apparent abundance of suitable habitat (Hockey et al. 1989).

Because fynbos occurs on nutrient-poor soils, its carrying capacity for sustaining large herbivores is low (Cody et al. 1983). Thus, ecologically balanced reserves (in which large herbivores and top predators typical of the vegetation can be preserved indefinitely without intervention) in fynbos should be very large. Fortunately the mountainous topography of the region which favoured the establishment of contractual reserves (for water production) allows movement of large mammals between core statutory reserves.

Mountain fynbos is considered so unsuitable for stock farming that official estimates of carrying capacity have not been made, although the carrying capacity of lowland fynbos of the west coast is 7.5 ha per small livestock unit (C J Pienaar personal communication). Thus, preservation of the Cape mountain zebra (= 8 small livestock units, L Viljoen personal communication) would require a minimum of 30 000 ha of fynbos vegetation to support a viable population of about 500 animals (i.e. in 100 single stallion herds with a mean of *c.* 5 animals, Smithers 1983). However, the minimum reserve size (to support a minimum viable family unit) would be about 300 ha, provided individuals could move between reserves. It is impossible to establish whether the mountain zebra did, in fact, occur in fynbos historically. Two possibilities exist: either the quagga (*Equus quagga*) occupied the renoster shrubland with the mountain zebra in fynbos; or alternatively, both species could have occurred in renoster shrubland, with the mountain zebra being more efficient at retreating to the mountains when persecuted (Skead 1980). Since renoster shrubland in the mountains has double the carrying capacity (34.2 ha per zebra) of fynbos (C J Pienaar personal communication), preservation of the Cape mountain zebra should rather be emphasized in reserves containing this vegetation type. However, only *c.* 9 000 ha of mountain renoster shrubland in the mountains are currently preserved (Cowan 1987). At present Cape mountain zebra preservation is centred in the Karoo mountains (Smithers 1986), with their higher carrying capacity of 10.2–34.2 ha per zebra (L Viljoen personal communication).

Data for chacma baboon (*Papio ursinus*) suggest an area relationship per troop in fynbos on the southern Cape Peninsula of A = 40I + 80, where A is area (in ha) and I is the number of individuals per troop (Davidge 1976). Recorded troop sizes in fynbos range from 6–85 (Davidge 1976), requiring from 320 to 3 500 ha per troop. Although these troops are largely confined to fynbos, they regularly consume shellfish from the seashore and large quantities of seeds from the alien *Acacia cyclops* (Hall 1963; Davidge 1976). This suggests that far larger ranges would be required for animals confined to pristine fynbos and that most troops probably occupy portions of other vegetation types.

Fynbos nature reserves aimed at maintaining viable populations of large mammals have been unsuccessful. The siting of the original Bontebok National Park in fynbos, at Bredasdorp, culminated in a decline in the numbers of bontebok primarily due to nutrient deficiencies and concomitant disease (Barnard and Van Der Walt 1961; Van Rensburg 1975). Numbers only increased after the animals were translocated to a reserve containing renoster shrubland (Barnard and Van Der Walt 1961). Similarly, the 'big-game'-orientated management policy of the Cape of Good Hope Nature Reserve and Bontebok National Park necessitated the provision of salt-licks and other practices incompatible with flora conservation, such as frequent burning to convert fynbos to grassland, bush cutting, the cultivation of alien pasture grasses, and the broadcast dispersal of copper to overcome the major mammal trace element deficiency (Millar 1970; Van Rensburg 1975; Greyling and Huntley 1984; D Clarke personal communication). Despite these efforts, certain large mammal species have been unable to establish themselves or have become extinct in the reserves (Millar 1970; De Graaf et al. 1976; Van Der Walt et al. 1976a, b). Similarly, the lack of sufficient renoster shrubland (16% of the area of the reserve, Cowan 1987) in the Gamka Nature Reserve may be a contributory cause of the decline of the Cape mountain zebra from 25 in 1986 to 18 in 1990 (Smithers 1986; Percy FitzPatrick Institute of African Ornithology).

Grazing and browsing mammals tend to concentrate their foraging in burned areas of fynbos (Cody et al. 1983) with a disproportionately high consumption of flowers and other reproductive parts (Rebelo 1987b). Whereas the densities of large mammals might perhaps be regulated by judicious culling and forced 'migration' in nature reserves, the same is not true of smaller herbivores, such as rock hyraxes (*Procavia capensis*). This animal has been implicated in the lack of reproduction of plants in fynbos (Macdonald 1989). The only effective management of such animals is by preserving ecosystems which are large enough to support the top predators thought to be indigenous to fynbos, namely, leopards (*Panthera pardalis*), caracals (*Felix caracal*), and black eagles (*Aquila verreauxii*).

The home ranges of leopard in the Cape Floristic Region have been investigated for only two regions. In the Franschhoek/Dutoitskloof district, home ranges were 38 800 ha for a male and 48 700 ha for a female, with a total range of 44 400 ha and 67 000 ha, respectively (Norton and Lawson 1984). These are among the largest home ranges recorded for leopards outside arid regions. By contrast, the home ranges of three male leopards in fynbos in the Cederberg district were 4 950 ± 1 500 ha (mean ± standard deviation, total range = 8 480 ± 3 840 ha) (Norton and Henley 1987). Assuming the degree of territorial overlap is similar in the two districts, a rough estimate of leopard density is currently one per 1 100–8 000 ha, giving a total of between 450 and 1 000 leopards in fynbos in the Cape Floristic Region. Whether differences in density reflect carrying capacity or human interference is uncertain. Norton and Henley (1987) suggest that leopard densities are higher in the Cederberg, where rock hyraxes are the major component of the diet (Norton et al. 1986), although a large proportion of legally killed leopards were from the south-western Cape (Stuart et al. 1985). A 'safe zone' or 'open sanctuary' for leopards of 450 000 ha was proposed as early as 1977 in the mountains of the Outeniqua and Kouga districts (Stuart et al. 1985), and some leopard conservation areas have been established by the Chief Directorate of Nature and Environmental Conservation (Hilton-Taylor and Le Roux 1989). However, the Outeniqua State Forests are heavily afforested and have not been ceded to the Chief Directorate of Nature and Environmental Conservation, so that the scheme may need to be reassessed. Perhaps the optimal preservation of leopards should envisage fynbos as a sink area, with Cape Floristic Region reserves containing karroid and renoster shrubland, together with karroid shrubland on the escarpment to the north of the Cape Floristic Region, as suitable source areas.

Only a single caracal's home range, 6 500 ha for a young male in the Franschhoek/Houwhoek district, has been determined (Norton and Lawson 1984). The data suggest that although territories do include fynbos, most activity occurs in neighbouring vegetation types. Excursions into fynbos were rare and associated with dispersal behaviour, during which the individual covered a total range of 89 500 ha in 18 months (Norton and Lawson 1984).

Estimates of black eagle densities are not available for fynbos. Vernon (in Macdonald and Gargett 1984) estimates their density at 1–3 birds per 10 000 ha for the entire Cape Province. Within the Cape Floristic Region, densities are estimated at between 9–28 breeding pairs per 1 000 000 ha, with a total of about 230 pairs (Boshoff and Vernon 1980). Since birds are persecuted as stock thieves (Siegfried 1963), carrying capacity in the Cape Floristic Region is almost certainly higher than 16 000 ha per pair (assuming that birds foraged only in fynbos in the mountains), and can presumably be preserved in a much smaller area within the Cape Floristic Region if their persecution in neighbouring, more productive, stockland is terminated.

Assuming that the total reserve system could maintain a minimum viable population with individuals moving between reserves, then a minimum fynbos reserve size to maintain a single pair of top predators would be 16 000 ha in the south-western and 2 200 ha in the north-western Cape Floristic Region. These figures could presumably be reduced if distances between reserves were sufficiently small to allow territories to be divided between reserves, or if areas of renoster shrubland with its higher carrying capacity were included in the reserves.

Alternatively, should corridors not be suitable for leopard or zebra movements, a reserve of between 100 000 and 1 000 000 ha would be required to preserve a minimum viable population of 50 pairs.

Fire management

Fynbos is a fire-maintained ecosystem (Van Wilgen et al. this volume). Historically, fire management has changed from the extensive use of fire to obtain pasturage, to total protection in the middle of this century (Kruger 1979). Thereafter, the occurrence of large uncontrollable fires and research indicating that certain species required fire, lead to the use of fire as a conservation and management tool (Kruger 1979; Boucher 1981b). Subsequent research has focused on season of burn and fire frequency (Van Wilgen 1987; Van Wilgen et al. this volume). Recently, coupled with the devolution of authority for state lands to local authorities, block burning was curtailed pending the outcome of certain court cases in which private property was damaged during 'controlled' burns.

The Directorate of Forestry divided reserves into management units (compartments), and each was burned about every 12 years (Kruger 1977). Since managers consider 500–1 500 ha as optimal-sized fire compartments, Kruger (1977) regarded 5 000 ha as the minimum reserve size required to 'maintain an even distribution of habitat age-classes'. The reasons for maintaining an even distribution of habitat age-classes are not given, but are probably in order to maintain resident populations of large mammals, nectarivores, and seed-eating mammals and birds that might be confined to either only very young or mature vegetation. Historically, fires have occurred over large areas, often exceeding 10 000 ha in area (Kruger 1979). Large fires would have forced pollinators and predators to emigrate from the burned area, only recolonizing once flowering started. A guild of plants comprising about 150 species, mostly species of the Proteaceae, Bruniaceae, and Asteraceae which may dominate fynbos, store seeds in the canopy (serotiny) until after a fire (Le Maitre 1987; Le Maitre and Midgley this volume). This has been proposed as a predator satiation strategy (Bond et al. 1984). Under Kruger's scheme, where a compartment within the reserve is burned every year, high levels of seed and seedling predators may be maintained locally in the reserve, perhaps reducing recruitment of serotinous plant species. Because burns temporarily increase the carrying capacity of fynbos vegetation, the burning of one compartment per year is often used by reserve managers to maintain large mammals in fynbos, at the expense of plant species susceptible to grazing or short rotation burns (Rebelo 1987a).

In short, optimal fire management practices do not reflect the large areas burned historically, and presumably prehistorically, and should not be used to speculate on reserve sizes. Equally, it can be argued that all reserves below a certain arbitrary size are treated as a single fire compartment. Since the length of compartment perimeters is where money is spent in prescribed burning operations (Kruger 1977), reserves of all sizes can be 'optimally' managed as a single compartment to decrease costs. To date, no research on the size of fires on the patch dynamics of fynbos landscapes has been undertaken (Van Wilgen 1987). If large fires are essential for reducing seed predator numbers, it might be fortunate that managers are unable to suppress the extensive fires which occur periodically, despite current fire-control techniques (Kruger 1979; Van Wilgen 1987).

The role of corridors

The above discussion on optimal reserve sizes assumes that adequate corridors exist between reserves. However, virtually no data exist on the importance of corridors in the Cape Floristic Region.

Movement patterns of nectarivores are poorly known, but suggest that corridors may not be essential for nectar-feeding birds (Rebelo 1987c). For instance, Cape Sugarbirds (*Promerops cafer*) and Orangebreasted Sunbirds (*Nectarinia violacea*) occur in large numbers in lowland proteoid fynbos during spring, but are largely absent during other periods (personal observation). Presumably they overwinter in the neighbouring Outeniqua Mountains which entails a movement of 20–40 km over nectar-barren habitat. Mist-netting of sugarbirds in the south-western Cape indicates movements of over 30 km, both over nectariferous and nectar-barren habitats (Fraser et al. 1989). Corridors do, however, appear necessary for the insect visitors to protea inflorescences. For example, insects are relatively scarce at 'Protea Heights' where the nearest natural protea stands are seven kilometres away (J H Coetzee personal communication).

A large proportion of fynbos plant species have ant-dispersed propagules which seldom move more than a few metres from parent plants (Bond 1989; Cowling and Bond 1991), so the dispersal rates of these plant species along corridors is likely to be negligible. No data exist on the importance of corridors for plant population movements. Such data are essential as the only available land for the preservation of fynbos in the Greater Cape Town metropolitan area consists of corridors under power line servitudes and in road reserves (McDowell et al. 1991).

The use of road verges as corridors, or even nature reserves, has proved viable in Australia (Saunders et al. 1987; Van der Breggen and Dawson 1989). Some 23 200 ha of the Cape Floristic Region is under tar macadam or gravel roads, with 50 800 ha of road reserve potentially available for preservation (0.6% of the Cape Floristic Region) (Macdonald 1989). Road reserves form a large component of the potential area available for the preservation of lowland renoster shrubland. Unfortunately, there seems to be resistance to the idea in the Cape Floristic Region, with authorities favouring the maintenance of road verges as mowed grass parks (Department of Transport 1985; B Dawson personal communication). Within the Greater Cape Town metropolitan area, the area of road reserve alongside national roads (excluding the 8 ha N7–N1 interchange priority conservation site) alone amounts to more than 45 ha: a potential 10% increase in the area available for reserves (McDowell et al. 1991).

Preservation without reserves

Do we need additional reserves to preserve species diversity in the Cape Floristic Region? Two possible alternatives to a reserve network are preservation in botanical gardens and zoos and preservation within multiple land-use systems such as agricultural land.

EX SITU PRESERVATION

Part of the National Botanical Institute's mission is to provide the facilities, knowledge, and expertise necessary to ensure the conservation of the flora (National Botanical Institute 1990). Can the preservation of plants be adequately undertaken within its botanical gardens and seed storage banks?

For many species of plants it is possible to store far larger quantities of seed than total number of adult plants alive in the wild. However, in natural populations plant gene banks comprise both adults and seed banks. Long-lived seed banks can store genes for several generations: fortunately, the dormant seeds of these species usually store well *ex situ* (Ashton 1987). Some fynbos and renoster shrubland plants have long-lived seed banks (see Boucher 1981b), but this is not always the case for both serotinous (Le Maitre 1987) and non-serotinous species (Pierce and Cowling 1991). In reality, it is seldom possible to acquire large numbers of seeds between the identification of the threat and the elimination of species.

An advantage of preserving species in *ex situ* seed banks is the small spatial requirements relative to that of preserving adult plants. However, seeds cannot be stored indefinitely. Unless the seeds can be returned to a natural area before they lose viability, artificial propagation must be undertaken to maintain the seed bank (Ashton 1987).

Furthermore, problems of cultivation, including seed germination, seedling establishment, watering, transplanting, flower induction, pollination, and breeding systems, must be solved for each species before it can be adequately preserved. Although much research has been done in local botanical gardens, very little information has been published (National Botanical Institute 1990). The loss of genetic diversity can occur via the unintended selection of phenotypes for protracted seed longevity, rapid germination, fast growth, resistance to horticultural pests, and the selection of large flowered or 'perfect' forms (Given 1987). This may result in populations with seed germination cues, pathogen resistance, predator evasion mechanisms, and flowering and growth phenologies maladapted to field conditions. Unfortunately, merely cultivating a species may result in skewed genetic representation by the differential survival of individual plants (Given 1987). These maladaptations may be sufficient to prevent the species from establishing. By contrast, the selection of strains that are purged of deleterious recessive genes may be an essential first step to preserving species prone to inbreeding depression (Vrijenhoek 1989). In addition, there is the problem of hybridization, not only with different species, but also with phenotypes adapted to different environments. Although asexual propagation can be used where hybridization is a problem, it can result in the accelerated loss of genetic diversity if the selection of easily rooting or fast growing phenotypes occurs. Although *ex situ* seed banks and propagation are viable short-term options, their long-term prospects are poor. We do not even know how to preserve important crop species adequately (Conway 1989). As with animals (Gilpin and Soule 1986; Ralls et al. 1986) 'captive' propagation should be a stop-gap measure, aimed at rescuing species by rapid propagation and re-establishment in a natural habitat (Ledig 1986; Conway 1989). Not only is prolonged cultivation expensive, but even short duration failures may negate years of preservation (Ashton 1987).

PRESERVATION IN NON-RESERVE MULTIPLE-USE AREAS

About 80% of land in the Cape Floristic Region is privately owned (McDowell 1986a). The ideal preservation system is one in which human activities are compatible with the long-term preservation of biotic diversity.

The wild flower industry has championed

its contribution to the preservation of fynbos by its commercial utilization of this resource (Davis 1990; Van Wilgen et al. this volume). This has been advocated as an alternative to reserves, with productive utilization and land custodianship mooted as a strong force for resource conservation (Davis 1990). However, harsh harvesting techniques which drastically reduce plant seed banks (Greig 1984; Rebelo 1987b; Rebelo and Holmes 1988), coupled with lack of law enforcement, suggest that species may not be safely utilized, even in proclaimed mountain catchment areas and nature reserves. Recurrent pleas by the industry to invest in monospecific plantations of specific horticultural varieties to increase the standard of material produced, especially for the highly competitive overseas market (Anon 1990), suggest that there is no long-term conservation future for commercially exploited fynbos (Davis 1990).

Advances in modern technology are resulting in the cultivation of agriculturally marginal lands which once harboured natural vegetation. Current legislation favours the conversion of marginal lands to agriculture as a tax evasion strategy, and legislation to prevent the cultivation of virgin lands appears to be seldom enforced (McDowell 1986a).

The frequent burning of fynbos, at a cycle too short to allow the regeneration of most plant species, promotes grass cover and is often utilized to provide extra grazing. This practice can be legally restricted only in contractual nature reserves and mountain catchment areas.

The preservation of the Cape Floristic Region outside reserves is thus uncertain. As economic incentives increase, so natural vegetation will be replaced by planted pastures in renoster shrubland and orchards of alien and indigenous plants in fynbos. Long-term preservation thus appears to require the legislation of contractual reserves as a minimum option.

Synthesis: options for preserving fynbos

Although most mountain districts have at least a single statutory reserve of sufficient size (> 10 000 ha) to maintain top predators (Table 13.3; Figure 13.6a), only one sufficiently large fynbos reserve (De Hoop) exists in the lowlands. Since corridors for predators are unlikely on the lowlands it is vital, therefore, that the two proposed reserve networks (Dassenberg and Agulhas) in the Malmesbury and Bredasdorp districts (Jarman 1986) be acquired *in toto* for preservation within these districts to be ecologically viable.

With the above in mind, it has been expedient to emphasize that very small reserves are adequate for the preservation of specific plant species and, provided that corridors exist between nature reserves, very small reserves are also ecologically viable. This argument contrasts with historical considerations (Kruger 1977) maintaining that reserves should be as large as possible. Such arguments have led to a perception by current conservation planners that smaller reserves are not only too expensive to run, but also cannot achieve conservation objectives.

Relative costs per species or area do increase with smaller units of preservation. This is not merely a function of traditionally perceived management costs (e.g. fencing, transport, fire-control, and policing), but also the costs of maintaining ecological processes and the research required to accomplish this. Thus, very large reserves may have very low management requirements. Small reserves need to be managed to maintain populations and ecosystem processes. However, this can be done *ad hoc* as problems with individual species manifest themselves. *Ex situ* cultivation requires that seeds must be obtained, stored, and germinated, and plants must be grown and propagated. More importantly, research into possible problems must be undertaken before any preservation can be effective. The costs of a species preservation programme increases by between 10- to 10 000-fold at each of these three levels of intervention (Woodruff 1989). As small reserves are likely to be very much cheaper to manage than *ex situ* cultivation programmes, they have an integral role in the preservation of the flora, especially in the short- to medium-term. This applies especially to the preservation of extremely localized plant species in agricultural areas.

Conservation research priorities in the Cape Floristic Region should be centred on the effects of fragmentation, the role of corridors, and the maintenance of natural disturbance regimes. Soule and Kohm (1989) emphasize the need for a co-ordinated programme of comparative research on populations, communities, and ecosystems in relatively undisturbed and secure situations. Two such fynbos sites, (Pella and Jonkershoek) where preliminary baseline information has been collected, exist. Research and monitoring in these areas should continue, and suitable sites in other vegetation types should be identified. Although local botanical gardens are increasingly becoming involved in the preservation of plant species, the high numbers of threatened species and costs of *ex situ* cultivation require innovative approaches. A key area for research should be the use of devegetated roadside verges for establishing large '*ex situ* populations' of locally endangered species. With an emphasis on multi-species stands, road verges, in addition to a corridor function, may also serve an educative and tourist role. Experience gained while revegetating road verges could lay the foundations for the potential use and acquisition of agricultural lands as future nature reserves.

LONG-TERM PROSPECTS

Aliens

Much fynbos is invaded by *Hakea sericea* (Proteaceae), *Pinus pinaster* (Pinaceae), and *Acacia longifolia* (Fabaceae) in the mountains, and *A. cyclops* and *A. saligna* in the lowlands. Some 7 592 and 8 962 km^2 of mountain and lowland vegetation, respectively, are invaded by aliens, often forming monospecific stands (Macdonald and Richardson 1986; Richardson et al. this volume). *Hakea* and *Pinus* can be effectively controlled by felling all seed-producing plants a year before burning (Macdonald and Richardson 1986). In the past, the Forestry Directorate of the Department of Environment Affairs was responsible for clearing infestations in the mountains. With the conservation section of that department now subsumed under the Chief Directorate of Nature and Environmental Conservation, future control will not be subsidized by the newly privatized afforested areas (Van Wilgen et al. this volume).

The introduction of a weevil (*Erytenna consputa*) for the biological control of *Hakea* has greatly reduced seed production, thus reducing post-fire population sizes (Moran et al. 1986). Similarly, the introduction of *Trichilogaster acaciaelongifoliae* should lower the seed output and vegetative growth of *Acacia longifolia* (Moran et al. 1986), and the fungus *Uromycladium tepperianum* should do the same for *A. saligna* (Kluge et al. 1986). Indigenous agents have also begun impacting on aliens e.g. *Zulubius acaciaphagus*, a bug, on the seeds of *A. cyclops* (Holmes and Rebelo 1988); and *Colletotrichum gloeosporioides*, a fungus, on the stems of *H. sericea* (Kluge et al. 1986). However, changes in management attitudes suggest that while integrated mechanical and biological control may be feasible in plantations and agricultural land, long-term control must be totally biological.

Population growth and urbanization

The Greater Cape Town metropolitan area currently contains about 2.2 million people (Anon 1986). Since the relaxation of legislation controlling the movement of black people, there has been a large influx of impoverished and poorly educated people from rural areas outside the Cape Floristic Region. This will swell the population to 3.5 million people by the year 2000 and to 6.2 million people by 2020 (Anon 1986, 1991). The rapidity of these changes is highlighted by the proposals for a False Bay Coastal Park in 1986 (Jarman 1986) having to be scrapped in 1987 as the region had been zoned and partly developed as a black residential area in the interim (Burgers et al. 1987).

The conservation implications of the increase in population size has not yet been addressed, nor have strategies been proposed to protect species diversity within the burgeoning metropolitan area.

With the increasing population natural areas, including nature reserves, have become popular for recreation. The deterioration of footpaths, hiking trails, and recreation sites due to increased use has been identified as a major problem (Moll and Campbell 1976). Research in high-altitude, nutrient-poor heathlands has shown them to be susceptible to trampling, taking many years to recover and leaving them open to invasion by weedy species (Liddle 1975; Bayfield 1979). Good guidelines, based on scientific principles for the establishment and maintenance of trails are needed (Moll and Campbell 1976; Moll et al. 1978).

Global change

Any strategy for the long-term preservation of nature should encompass evolutionary and biogeographical considerations. These are strongly influenced by climatic change. One 'worst-case scenario' speculates that as a result of the 'greenhouse' phenomenon, the Cape Floristic Region could become warmer and drier in the west and warmer and moister in the east (Rebelo and Siegfried 1990). Moreover, since the cyclone belt should move southwards, the entire region would receive rain in summer, rather than in winter as is the case at present.

Many typical fynbos species are sensitive to the season of rainfall. Under summer rainfall, grasses invade oligotrophic soils and displace Restionaceae species. Restionaceae can be considered as slow maturing, perennial herbs (Steiner 1988), so that their replacement by faster growing grasses would facilitate short rotation fires, instead of the much longer fire intervals under winter rainfall conditions (Van Wilgen et al. this volume). Fynbos can readily be converted to grassland in the eastern Cape by repeated burning (Trollope 1973; Gibbs Russell and Robinson 1981), suggesting that grasslands may expand westwards and replace fynbos taxa if the greenhouse phenomenon is fully realized.

Fynbos has probably been displaced in certain areas by grassland, and vice versa, several times during the last two million years (Avery 1991). This would have involved the westward dispersal of many fynbos taxa along the Langeberg-Outeniqua and Swartberg mountain ranges to high altitude refugia under conditions of predominantly summer rainfall. This might explain the higher richness of fynbos plant taxa in the mountains to the west of the region, assuming that the Cape Floristic Region was never subject to a summer rainfall regime over its entire area (Cowling and Holmes this volume). However, there are three major differences between the historical and the envisaged greenhouse climate changes: the future rate of change is projected to be far faster than in the past, and many plant species may not be able to adapt to the new environmental pressures or disperse to potential refugia; the Cape Floristic Region is not completely preserved and past dispersal routes may no longer exist; and the rapid changes in temperature on the land may be more or less uncoupled from those in the sea, so that coastal climate changes may be unlike those of the past.

Noda of naturally rare species in mountain fynbos (Figure 13.5), which may be Pleistocene refugia or speciation centres, are well preserved in nature reserves. The challenge to conservation biologists is to develop a reserve network for the lowlands of the Cape Floristic Region which can accommodate alterations of habitat and movements of species brought about by climatic change.

SYNTHESIS

Although 19% of the Cape Floristic Region is preserved in statutory, private, and contractual nature reserves, by far the majority of this area (96%) comprises fynbos in the mountains. Only 1.5% of extant renoster shrubland, amounting to 0.6% of the original extent of the vegetation type, is preserved. Similarly, only three per cent of the original extent of fynbos on the lowlands is preserved. Despite recognition of the need for preserving lowland areas since the 1970s, the prioritization of available lowland areas in 1986, and the high ranking of the lowlands of the Cape Floristic Region on a national basis, little lowland has been added to the reserve network over the past decade. This neglect has been attributed to the lack of large mammal diversity (Greyling and Huntley 1984), but the cost of land with high agricultural potential is also a major obstacle (McDowell 1986a). Contractual national parks currently appear to be the most successful strategy in acquiring preserves for vegetation types on nutrient-poor soils. The problem with renoster shrubland appears intractable — even appealing schemes for reintroducing large mammals (e.g. rhino) would not overcome the high costs of acquiring suitable land.

The minimum reserve size for preserving minimum viable populations of higher trophic level herbivores and carnivores, appears to be about 10^5–10^6 ha. Reserves of this size class occur in most mountain centres of endemism, but few occur in the lowlands. Contrary to current management perceptions, reserves as small as 5 ha would be able to preserve species of plants provided that ecosystem processes and alien plants were adequately managed. Owing to the high numbers of threatened species, *ex situ* preservation is only feasible for a small proportion of species. Seed banking may offer a better solution, but large numbers of seeds may not be available from threatened populations. A strategy employing the cultivation of threatened species in suitable road verges to obtain plants and seeds for distribution and storage may be a possible answer.

The existing reserve network appears to be comprehensive for fynbos of the mountains, although centres of endemism to the north-west and arid interior are poorly preserved. By contrast the lowlands are very poorly preserved. With 92 Red Data Book plant species in the lowlands of the Greater Cape Town metropolitan area and a further 171 Red Data Book plant species on Table Mountain in the heart of Cape Town, this area must rank as the top priority area for conservation action on the subcontinent. With current rates of urbanization and projected population growth, the future of threatened species within the region appears bleak.

Long-term prospects for the preservation of lowland vegetation in the Cape Floristic Region will be determined primarily by how the rapid human population growth rate in the Greater Cape Town region is managed. Continued funding for the control of alien plant species is also required in the short-term.

ACKNOWLEDGEMENTS

I wish to thank W J Bond, R M Cowling, P M Holmes, W R Siegfried, M Usher, and an anonymous referee for comments, and J C Holmes for financial assistance to complete this study. Funding was also obtained from a University of Cape Town postgraduate bursary.

REFERENCES

ACKERMAN D P (1972). The proclamation of wilderness areas by the Department of Forestry. *South African Forestry Journal* **82**, 19–21.

ACOCKS J P H (1975). Veld types of South Africa. *Memoirs Botanical Survey South Africa* **28**.

ANON (1952). Department of Nature Conservation: General. *Department of Nature Conservation Report* **9**, 7–8.

ANON (1986). *Projections of the population of the Cape Town land use/ transport area 1980–2000.* Metropolitan Transport Planning Branch, City Engineer's Department, Cape Town.

ANON (1990). Kollig op uitvoerproteas. *Sappex News* **69**, 20–1.

ANON (1991). Environmental Focus. *Foundation for Research Development Report Series* **1**.

ASHTON P S (1987). Biological considerations in in situ vs *ex situ* plant conservation. In *Botanic Gardens and the world conservation strategy.* (eds Bramwell D, Hamann O, Heywood V and Synge H) Academic Press, London, 117–30.

AVERY D M (1991). Late Quaternary environmental change in southern Africa based on micromammalian evidence: a brief review. *Palaeoecology of Africa* (in press).

BANDS D P (1985). The influence of mountain catchment area control measures on land management in the Groot Winterhoek area of the western Cape: ecological, economic and social implications. M.Sc. thesis, Faculty of Forestry, University of Stellenbosch.

BARNARD P J and VAN DER WALT K (1961). Translocation of the Bontebok (*Damaliscus pygargus*) from Bredasdorp to Swellendam. *Koedoe* **4**, 105–9.

BATEMAN J A (1961). The mammals occurring in the Bredasdorp and Swellendam districts, Cape Province, since European settlement. *Koedoe* **4**, 78–100.

BAYFIELD N G (1979). Recovery of four montane heath communities on Cairngorm, Scotland, from disturbance by trampling. *Biological Conservation* **15**, 165–79.

BOND P and GOLDBLATT P (1984). Plants of the Cape Flora: a descriptive catalogue. *Journal South African Botany Supplement* **13**.

BOND W J (1989). The dynamic nature of biotic diversity. In *Biotic diversity in southern Africa. Concepts and conservation.* (ed Huntley B J) Oxford University Press, Cape Town, 2–18.

BOND W J, VLOK J and VIVIERS M (1984). Variation in seedling recruitment of Cape Proteaceae after fire. *Journal of Ecology* **72**, 209–21.

BOND W J, MIDGLEY J J and VLOK J (1988). When is an island not an island? Insular effects and their causes in fynbos shrublands. *Oecologia* **77**, 515–21.

BOSHOFF A F and VERNON C J (1980). The distribution and status of some eagles in the Cape Province. *Annals Cape Provincial Museum* **13**, 107–32.

BOUCHER C (1981a). Floristic and structural features of the coastal foreland vegetation south of the Berg river, western Cape Province, South Africa. In *Proceedings of a symposium on coastal lowlands of the western Cape.* (ed Moll E) University Western Cape, Bellville, 21–6.

BOUCHER C (1981b). Autecological and population studies of Orothamnus zeyheri in the Cape of South Africa. In *The biological aspects of rare plant conservation* (ed Synge H) Wiley, Chichester, 343–53.

BRANCH W R (ed) (1988). South African red data book — reptiles and amphibians. *South African National Scientific Programmes Report* **151**, CSIR, Pretoria.

BROOKE R K (1984). South African red data book — birds. *South African National Science Programmes Report* **97**, CSIR, Pretoria.

BURGERS C J, NEL J G and POOL R (1987). Background document for a meeting of the NAKOR national plan working group for region A1 on 30 April 1987 to assess proposals for conservation areas in the lowlands of the south-western Cape. Unpublished MS.

CODY M L, BREYTENBACH G J, FOX B, NEWSOME A E, QUINN R D and SIEGFRIED W R (1983). Animal communities: diversity, density and dynamics. In *Mineral nutrients in Mediterranean ecosystems.* (ed Day J A) *South African National Scientific Programmes Report* **71**, 91–110.

CONWAY W G (1989). The prospects for sustaining species and their evolution. In *Conservation for the twenty-first cen-*

tury. (ed Western D and Pearl M) Oxford University Press, New York, 199–209.

COWAN G I (1987). South African plan for nature conservation annual report (April 1986–March 1987). *Report of the Department of Environmental Affairs*, Pretoria.

COWLING R M (1984). A syntaxonomical and synecological study in the Humansdorp region of the Fynbos Biome. *Bothalia* **15**, 175–227.

COWLING R M and BOND W J (in press). How small can reserves be? An empirical approach in Cape Fynbos. Submitted to *Biological Conservation* (in press).

COWLING R M and PIERCE S M (1985). Southern Cape coastal dunes: an ecosystem lost? *Veld and Flora* **71**, 99–103.

COWLING R M, PIERCE S M and MOLL E J (1986). Conservation and utilization of south coast renoster shrubland, an endangered South African vegetation type. *Biological Conservation* **37**, 363–77.

DAVIDGE C (1976). Activity patterns of baboons (*Papio ursinus*) at Cape Point. M.Sc. thesis, University of Cape Town.

DAVIS G W (1990). Resource management, research, and the legal context. A report on a seminar/workshop held at Kirstenbosch, Cape Town, March 1990. South African Protea Producers and Exporters Association, Botrivier.

DE GRAAF G, VAN DER WALT P T and VAN ZYL L J (1976). Levensloop van 'n elandbevolking Taurotragus oryx in die Bontebok Nasionale Park. *Koedoe* **19**, 185–8.

DEPARTMENT OF TRANSPORT (1985). Handleiding vir die sny van gras en instandhouding van boom-en struik-aanplantings in nasionale padreserwes. Department of Transport, Pretoria.

DINGLE H V and HENDEY Q B (1984). Late Mesozoic and Tertiary sediment supply to the eastern Cape Basin (SE Atlantic) and palaeo-drainage systems in southwestern Africa. *Marine Geology* **56**, 13–26.

DINGLE H V and ROGERS J (1972). Pleistocene palaeogeography of the Agulhas Bank. *Transactions Royal Society South Africa* **40**, 155–65.

EDWARDS D (1974). Survey to determine the adequacy of existing conserved areas in relation to vegetation types: a preliminary report. *Koedoe* **17**, 3–38.

FERRAR A A (1989). The role of red data books in conserving biodiversity. In *Biotic diversity in southern Africa: concepts and conservation.* (ed Huntley B J) Oxford University Press, Cape Town, 136–47.

FRASER M W, MACMAHON L, UNDERHILL L G and REBELO A G (1989). Nectarivore ringing in the southwestern Cape. *Safring News* **18**, 3–18.

GAIGHER I G, HAMMAN K C D and THORNE SC (1980). The distribution, conservation status and factors affecting the survival of indigenous freshwater fishes in Cape Province. *Koedoe* **23**, 57–88.

GIBBS RUSSELL G E and GONSALVES P (1984). PRECIS — a curatorial and biogeographic system. In *Databases in systematics.* (eds Allkin R and Bisby F A) Academic Press, London, 137–53.

GIBBS RUSSELL G E and ROBINSON E R (1981). Phytogeography and speciation in the vegetation of the eastern Cape. *Bothalia* **14**, 467–72.

GILPIN M E and SOULE M E (1986). Minimum viable populations: processes of species extinction. In *Conservation Biology: the science of scarcity and diversity.* (ed Soule M E) Sinauer, Massachusetts, 19–34.

GIVEN D R (1987). What the conservationist requires of *ex situ* collections. In *Botanic Gardens and the world conservation strategy.* (ed Bramwell D, Hamann O, Heywood V, and Synge H) Academic Press, London, 103–16.

GREIG J C (1981). The geometric tortoise — symptom of a dying ecosystem. *Veld and Flora* **68**, 106–8.

GREIG J C (1984). The law of the jungle — Protea holosericea of Saw-edge Peak. *African Wildlife* **38**, 135–9.

GREYLING T and HUNTLEY B J (1984). Directory of Southern African conservation areas. *South African National Scientific Programmes Report* **98**, CSIR, Pretoria.

GROVE R (1987). Early themes in African conservation: the Cape in the nineteenth century. In *Conservation in Africa: people, policies and practice.* (eds Anderson D and Grove R) Cambridge University Press, Cambridge, 21–39.

HALL A V (1981). Conservation status of the vegetation of the western Cape coastal lowlands. In *Proceedings of a symposium on coastal lowlands of the western Cape.* (ed Moll E) University Western Cape, Bellville, 57–62.

HALL A V (1987a). Threatened plants in the fynbos and karoo biomes, South Africa. *Biological Conservation* **40**, 29–52.

HALL A V, DE WINTER M, DE WINTER B and VAN OOSTERHOUT S A M (1980). Threatened plants of Southern Africa. *South African National Scientific Programmes Report* **45**, CSIR, Pretoria.

HALL A V and VELDHUIS H A (1985). South African red data book: plants — fynbos and karoo biomes. *South African National Scientific Programmes Report* **117**, CSIR, Pretoria.

HALL K R L (1963). Variations in the ecology of the Chacma Baboon, *Papio ursinus. Symposium Zoological Society London* **10**, 1–28.

HENNING S F and HENNING G A (1989). South African red data book — butterflies. *South African National Scientific Programmes Report* **158**, CSIR, Pretoria.

HILTON-TAYLOR C and LE ROUX A (1989). Conservation status of the fynbos and karoo biomes. In *Biotic diversity in southern Africa. Concepts and conservation.* (ed Huntley B J) Oxford University Press, Cape Town, 202–23.

HOCKEY P A R, UNDERHILL L G, NEATHERWAY M and RYAN P G (1989). *Atlas of the birds of the southwestern Cape.* Cape Bird Club, Cape Town.

HOLMES P M and REBELO A G (1988). The occurrence of seed-feeding *Zulubius acaciaphagus* (Hemiptera, Alydidae) and its effects on *Acacia cyclops* seed germination and seed banks in South Africa. *South African Journal Botany* **54**, 319–24.

HUNTLEY B J (1978). Ecosystem conservation in southern Africa. In *Biogeography and ecology of southern Africa.* (ed Werger M J A) Junk, The Hague, 1 333–84.

JARMAN M L (1986). Conservation priorities in lowland regions of the fynbos biome. *South African National Scientific Programmes Report* **87**, CSIR, Pretoria.

JEFFERY D and MOLL E J (1987). *Langebaan to Agulhas a coastal survey.* Flora Conservation Committee of the Botanical Society of South Africa, Cape Town.

KLUGE R L, ZIMMERMANN H G, CILLERS C J and HARDING G B (1986). Integrated control for alien invasive weeds. In *The ecology and management of biological invasions in southern Africa.* (eds Macdonald I A W, Kruger F J and Ferrar A A) Oxford University Press, Cape Town, 295–303.

KRUGER F J (1977). Ecological reserves in the Cape Fynbos: toward a strategy for conservation. *South African Journal Science* **73**, 81–5.

KRUGER F J (1979). In *Fynbos ecology: a preliminary synthesis.* (eds Day J, Siegfried W R, Louw G N and Jarman M L) *South African National Scientific Programmes Report* **40**, CSIR, Pretoria, 43–57.

KRUGER F J (1982). Use and management of mediterranean ecosystems in South Africa — current problems. In *Medecos: dynamics and management of mediterranean-type ecosystems.* (eds Conrad C E and Oechel W C) USDA Forest Service General Technical Report PSW-58. Berkeley, CA, 67–114.

LE CORDEUR B A (1986). The occupations of the Cape, 1795–1854. In *An illustrated history of South Africa.* (ed Cameron T) Jonathan Ball, Johannesburg, 75–93.

LE MAITRE D C (1987). Dynamics of canopy-stored seed in relation to fire. In *Disturbance and the dynamics of fynbos biome communities* (eds Cowling R M, Le Maitre D C, Mc-

Kenzie B, Prys-Jones R P and Van Wilgen B W) *South African National Scientific Programmes Report* **135**, CSIR, Pretoria, 24–45.

LEDIG F T (1986). Heterozygosity, heterosis, and fitness in outbreeding plants. In *Conservation Biology: the science of scarcity and diversity.* (ed Soule M E) Sinauer, Massachusetts, 77–104.

LIDDLE M J (1975). A selective review of the ecological effects of human trampling on natural ecosystems. *Biological Conservation* **7**, 17–36.

LUCKHOFF H A (1982). Early history of the Kogelberg and Cape Hangklip areas and management of the State Forest. *Veld and Flora* **68**, 12–13.

MACDONALD I A W (1989). Man's role in changing the face of southern Africa. In *Biotic diversity in Southern Africa. Concepts and conservation.* (ed Huntley B J) Oxford University Press, Cape Town, 51–78.

MACDONALD I A W and GARGETT V (1984). Raptor density and diversity in the Matopos, Zimbabwe. *Proceedings Fifth Pan African Ornithological Congress,* 251–76.

MACDONALD I A W, JARMAN M L and BEESTON P (eds) (1985). Management of invasive alien plants in the Fynbos Biome. *South African National Science Programmes Report* **111**, CSIR, Pretoria, 140.

MACDONALD I A W and RICHARDSON D M (1986). Alien species in terrestrial ecosystems of the fynbos biome. In *The ecology and management of biological invasions in southern Africa.* (eds Macdonald I A W, Kruger F J and Ferrar A A) Oxford University Press, Cape Town, 77–91.

MANDERS P T (1986). An assessment of the current status of the Clanwilliam cedar (*Widdringtonia cedarbergensis*) and the reasons for its decline. *South African Forestry Journal* **139**, 48–53.

MARGULES C and USHER M B (1981). Criteria used in assessing wildlife conservation potential: a review. *Biological Conservation* **21**, 79–109.

MCDOWELL C (1986a). Legal strategies to optimise conservation of natural ecosystems by private landowners — economic incentives. *Comparative International Law Journal Southern Africa* **14**, 460–73.

MCDOWELL C (1986b). Legal strategies to optimise conservation of natural ecosystems by private landowners — restrictive legislation. *Comparative International Law Journal Southern Africa* **14**, 450–9.

MCDOWELL C R, LOW A B and MCKENZIE B (1991). Natural remnants and corridors in greater Cape Town: their role in threatened plant conservation. In *Nature conservation 2: the role of corridors.* (eds Saunders D A and Hobbs R J) Surrey Beatty and Sons Pty Limited, 27–39.

MILLAR J C C (1970). The Cape of Good Hope Nature Reserve: a report and management plan. Unpublished MS, Cape Divisional Council.

MOLL E J and BOSSI L (1984). Assessment of the extent of the natural vegetation of the Fynbos Biome of South Africa. *South African Journal of Science* **80**, 355–8.

MOLL E J and CAMPBELL B M (1976). *The ecological status of Table Mountain.* Department of Botany, University of Cape Town.

MOLL E J, MCKENZIE B, MCLACHLAN D and CAMPBELL B M (1978). A mountain in a city — the need to plan the human usage of the Table Mountain National Monument, South Africa. *Biological Conservation* **13**, 117–31.

MOLL E J, CAMPBELL B M, COWLING R M, BOSSI L, JARMAN M L and BOUCHER C (1984). A description of major vegetation categories in and adjacent to the fynbos biome. *South African National Science Programmes Report* **83**, 1–29, CSIR, Pretoria.

MORAN V C, NESER S and HOFFMANN J H (1986). The potential of insect herbivores for the biological control of invasive plants in South Africa. In *The ecology and management of biological invasions in southern Africa.* (eds Macdonald I A W, Kruger F J and Ferrar A A) Oxford University Press, Cape Town, 261–8.

MORROW P A, DAY J A, FOX M D, FROST P G H, JARVIS J U M, MILEWSKI A V and NORTON P M (1983). Interaction between plants and animals. In *Mineral nutrients in mediterranean ecosystems.* (ed Day J H) *South African National Science Programmes Report* **71**, CSIR, Pretoria, 111–24.

MUIR J (1929). The vegetation of the Riversdale area, Cape Province. *Memoirs of the Botanical Survey of South Africa* **13**.

NATIONAL BOTANICAL INSTITUTE (1990). Corporate strategic plan 1990–1991. Internal report, National Botanical Institute.

NORTON P M and HENLEY S R (1987). Home range and movements of male leopards in the Cedarberg Wilderness Area, Cape Province. *South African Journal Wildlife Research* **17**, 41–8.

NORTON P M and LAWSON A B (1984). Radio tracking of leopards and caracals in the Stellenbosch area, Cape Province. *South African Journal Wildlife Research* **15**, 17–24.

NORTON P M, LAWSON A B, HENLEY S R and AVERY G (1986). Prey of leopards in four mountainous areas of the Southern Cape Province. *South African Journal Wildlife Research* **16**, 47–52.

NOVELLI P (1986). Relationship between rainfall, population density and the size of the Bontebok lamb crop in the Bontebok National Park. *South African Journal Wildlife Research* **16**, 39–46.

PIERCE S M and COWLING R M (in press). Dynamics of soil-stored seed banks of six shrubs in fire-prone dune fynbos. *Journal of Ecology* .

PRESSEY R L and NICHOLLS O A (1989a). Efficiency in conservation evaluation: scoring versus iterative approaches. *Biological Conservation* **50**, 199–218.

PRESSEY R L and NICHOLLS O A (1989b). Application of a numerical algorithm to the selection of reserves in semi-arid New South Wales. *Biological Conservation* **50**, 263–78.

RALLS K, HARVEY P H and LYLES A M (1986). Inbreeding in natural populations of birds and mammals. In *Conservation Biology: the science of scarcity and diversity.* (ed Soule M E) Sinauer, Massachusetts, 35–56.

REBELO A G (1987a). Introduction. In *A preliminary synthesis of pollination biology in the Cape flora.* (ed Rebelo A G) *South African National Science Programmes Report* **141**, CSIR, Pretoria, 1–5.

REBELO A G (1987b). Management implications. In *A preliminary synthesis of pollination biology in the Cape flora.* (ed Rebelo A G) *South African National Science Programmes Report* **141**, CSIR, Pretoria, 193–211.

REBELO A G (1987c). Bird pollination in the Cape Flora. In *A preliminary synthesis of pollination biology in the Cape flora.* (ed Rebelo A G) *South African National Science Programmes Report* **141**, CSIR, Pretoria, 83–108.

REBELO A G (in press). The distribution of Red Data Book taxa in the Cape Floristic Region. *Transactions of the South African Royal Society.*

REBELO A G and COWLING R M (1991). The preservation of plant species in the Cape Floral Region: problems with the available data bases for the Riversdale Magisterial District. *South African Journal of Botany* **57**, 186–90.

REBELO A G, COWLING R M, CAMPBELL B M and MEADOWS M (1991). Plant communities of the Riversdale Plain. *South African Journal of Botany* **57**, 10–28.

REBELO A G and HOLMES P M (1988). Commercial exploitation of *Brunia albiflora* (Bruniaceae) in South Africa. *Biological Conservation* **43**, 195–207.

REBELO A G and SIEGFRIED W R (1990). Protection of fynbos vegetation: ideal and real-world options. *Biological Conserva-*

tion **54**, 17–34.

REBELO A G and SIEGFRIED W R (in press). Where should nature reserves be located in the Cape Floral Region, South Africa? Null models for spatial configurations based on iterative approaches for selecting a reserve network. *Conservation Biology.*

REBELO A G and TANSLEY S A (Submitted). Using rare plant species to select priority conservation areas in the Cape Floral Kingdom: the need to correct for total species richness? *South African Journal of Science.*

ROBERTS C P R (1982). Environmental implications of the proposed Palmiet River water and power development projects. *Veld and Flora* **68**, 4–6.

ROOKMAAKER L C (1989). *The zoological exploration of Southern Africa 1650–1790.* Balkema, Rotterdam.

SAUNDERS D A, ARNOLD G W, BURBIDGE A A and HOPKINS A J M (eds) (1987). *Nature conservation — the role of remnants.* Surrey Beatty, Chipping Norton.

SCHEEPERS J C (1983). The present status of vegetation conservation in South Africa. *Bothalia* **14**, 991–5.

SCOTT H A (1986). De Hoop Nature Reserve. *Cape Conservation Series* **7**.

SCOTT H A and HAMMAN K C D (1984). Freshwater fishes of the Cape. *Cape Conservation Series* **5**.

SIEGFRIED W R (1963). A preliminary report on black and martial eagles in the Laingsburg and Philipstown divisions. *Cape Dept Nature Conservation Investigational Report* **5**, 1–15.

SIEGFRIED W R (1989). Preservation of species in southern African nature reserves. In *Biotic diversity in southern Africa: concepts and conservation.* (ed Huntley B J) Oxford University Press, Cape Town, 186–201.

SKEAD C J (1980). *Historical mammal incidence in the Cape Province. Vol 1. The western and northern Cape.* Cape Department Nature Environmental Conservation, Cape Town.

SKELTON P H (1987). South African red data book — fishes. *South African National Science Programmes Report* **137**, CSIR, Pretoria.

SMITHERS R H N (1983). *The mammals of the southern African subregion.* University of Pretoria, Pretoria.

SMITHERS R H N (1986). South African red data book — terrestrial mammals. *South African National Science Programmes Report* **125**, CSIR, Pretoria.

SOULE M E and KOHM K A (eds) (1989). *Research priorities for conservation biology.* Island Press, Washington.

STEINER K E (1988). Dioecism and its correlates in the Cape Flora of Southern Africa. *American Journal of Botany* **75**, 1 742–54.

STUART C T, MACDONALD I A W and MILLS M G L (1985). History, current status and conservation of large mammalian predators in Cape Province, Republic of South Africa. *Biological Conservation* **31**, 7–19.

TANSLEY S A (1988). The status of threatened Proteaceae in the Cape Flora, South Africa, and the implications for their conservation. *Biological Conservation* **43**, 227–39.

TAYLOR H C (1978). Capensis. In *Biogeography and ecology of southern Africa.* (ed Werger M J A) Junk, The Hague, 171–229.

TROLLOPE W S W (1973). Fire as a method of controlling macchia (fynbos) vegetation on the Amatole mountains of the eastern Cape. *Proceedings Grassland Society southern Africa* **8**, 35–41.

VAN DER BREGGEN J P and DAWSON B L (1989). Report on a visit to western Australia. Unpublished MS, Department of Transport.

VAN DER WALT P T, VAN ZYL L J and DE GRAAF G (1976a). Levensloop van 'n Kaapse buffelbevolking Syncerus cafer in die Bontebok Nasionale Park. *Koedoe* **19**, 185–8.

VAN DER WALT P T, VAN ZYL L J and DE GRAAF G (1976b). Levensloop van 'n Kaapse rooiahartebeesbevolking Alcelaphus buselaphus caama in die Bontebok Nasionale Park. *Koedoe* **19**, 181–4.

VAN RENSBURG A P J (1975). Die geskiedenis van die Nationale Bontebokpark, Swellendam. *Koedoe* **18**, 165–90.

VAN WILGEN (1987). Fire regimes in the Fynbos Biome. In *Disturbance and the dynamics of Fynbos Biome communities.* (eds Cowling R M, Le Maitre D C, McKenzie B, Prys-Jones R P and Van Wilgen B W) *South African National Science Programmes Report* **135**, 6–14, CSIR, Pretoria.

VERSVELD D B and VAN WILGEN B W (1986). Impacts of woody plants on ecosystem processes. In *The ecology and management of biological invasions in southern Africa.* (eds Macdonald I A W, Kruger F J and Ferrar A A) Oxford University Press, Cape Town, 239–46.

VRIJENHOEK R C (1989). Population genetics and conservation. In *Conservation for the twenty-first century.* (ed Western D and Pearl M) Oxford University Press, New York, 199–209.

WEIMARCK H (1941). Phytogeographical groups, centres and intervals within the Cape Flora. *Lund. Univ. Arsskrift* **37**. Nr 5.

WICHT C L (1945). Preservation of the vegetation of the South-western Cape. *Special Publication Royal Society South Africa.*

WOOD J (1991). Romansriver and Haartebeesriver Provincial Nature Reserves. Unpublished report. Flora Conservation Committee, Botanical Society of South Africa, Cape Town.

WOODRUFF D S (1989). The problems of conserving genes and species. In *Conservation for the twenty-first century.* (ed Western D and Pearl M) Oxford University Press, New York, 76–88.

ZUMPT I F and HEINE E W P (1977). Some veterinary aspects of bontebok in the Cape of Good Hope Nature Reserve. *South African Journal Wildlife Research* **8**, 131–4.

14

Ecosystem management

B W van Wilgen, W J Bond and D M Richardson

INTRODUCTION

The Cape fynbos is one of the world's richest temperate floras crammed into 4.4% of South Africa's land surface. How much do we need to know to manage such a diverse biota, and will we ever know enough? There are substantial difficulties. Too many species exist in fynbos for all their management requirements to be known. Even the management of a single, well-studied species, (e.g. *Pinus radiata*), is fraught with uncertainties in management. Despite its small area, fynbos is renowned for the rapid species turnover from one habitat to the next and from one geographic region to the next (Cowling et al. this volume). We do not understand the reasons for this turnover. Thus, even though we do have a few well-studied examples, there are inadequate guidelines on how widely to generalize results from studies in one habitat and one region to the rest of the fynbos biome.

Despite the apparent immensity of the problem, substantial progress has been made in ecological research relevant to management. This chapter reviews the scientific basis for the current management of fynbos biome vegetation, and examines these in the light of scenarios of likely changes within the biome. We do not attempt to prescribe management methods, since management objectives may change according to land ownership and changing public perceptions. Instead, we discuss current principles, the evidence on which they are based, and their generality in different regions and scales of management. There are no absolutes in fynbos — we hope this chapter will be read not as a recipe but as a review which will indicate our understanding and expose our ignorance.

Researchers often have highly inflated notions on the ability of 'managers and decision makers' to take note of the results of their research. Regardless of the quality of technical knowledge, management is constrained by the size of the area managed and the available finance. In practice, a simple hierarchy of management intensity for fynbos landscapes can be recognized:

- no management — more remote or arid areas;
- 'natural burning zones' — area encircled by fire-breaks but, besides alien removal, with no management intervention. Mostly isolated mountain ranges with relatively low fire hazard from adjacent properties;
- 'block burning' — active management involving the control of invasive plants and prescribed burning at predetermined frequencies — many farms, nature reserves, and catchment areas. The scale can vary from blocks of 500–1 500 ha in catchments to patches of 2–3 ha in farms or nature reserves; and
- 'intensive management' including seed additions, translocations, seed harvesting, pest management, and fire.

Management for the most important goals (Table 14.1) centres largely on the application (or exclusion) of fire, and the control of alien invasive plants. Hence, much research has concerned the response of plants to the fire regime (see below). Most ecological research has been directed at level 3, some at level 2 and very little at level 4. Thus, the flower farmer and the manager of the small nature reserve have been least well served by research. Substantial horticultural research has been conducted on fynbos plants, especially those cultivated for the flower trade or by gardening enthusiasts. There is still much to be done to integrate horticultural insights into veld management at the more extensive scale, but we do not attempt to do so here. Instead, we review how a knowledge of the biology of plants and animals can

TABLE 14.1 Important management objectives and practices in the fynbos biome.

Sphere	Major objectives	Management practices
Nature conservation	Maintenance of species diversity	Prescribed burning to allow seed regeneration; control of invasive organisms
Water conservation	Maintain sustained flow of high quality water	Prescribed burning to increase water yield; control of alien woody plants
Fire management	Reduce fire hazard	Prescribed burning to reduce fuel loads; maintenance of fire breaks; control alien woody plants
Flower harvesting	Sustained yields for profit	Harvesting; prescribed burning to allow seed regeneration; re-seeding depleted areas; pest control
Grazing	Provide additional pasture for domestic livestock	Frequent burning to eliminate shrubs and encourage grasses
Recreation and tourism	Provide facilities for hiking and other outdoor activities	Establishment of trails and other infrastructure

be used to arrive at rational management decisions. In addition, we examine scenarios of habitat fragmentation, funding levels, and climate change for the biome. Managers will need to plan for, or at least understand, the consequences of such changes.

MANAGEMENT PRACTICES

The use of fire in fynbos biome ecosystems

The application of fire is the major management practice in fynbos biome ecosystems. Research into the effects of fire on fynbos has followed a demographic approach (Gill 1981). Most management prescriptions are based on an understanding of the population biology of important species, or even of a single species. Studies at the systems level (e.g. the effects of fire on nutrient cycles) have not been instrumental in deciding appropriate fire regimes for fynbos ecosystems. Burning operations are prescribed in terms of the fire regime (the frequency, season, and intensity at which fires occur), and are based on a knowledge of the effects of these three components on the vegetation.

FIRE FREQUENCY

There are physical limits to the frequencies of fire that can be applied in fynbos. Some fynbos communities accumulate enough fuel to sustain a fire under suitable conditions at a post-fire age of four years, and may burn at three years under extreme hot and dry weather conditions (Kruger 1977a). Fire cycles of less than four years are seldom possible. On the other hand, most attempts to exclude fire for longer than 45 years have met with failure, indicating that the exclusion of fire for longer than this should not be considered feasible.

Although fires in fynbos can burn at any time between four and 45 years, a narrower range of fire frequencies would be needed to ensure that many component species survive repeated fire. The only 'rule of thumb' for determining minimum fire frequencies in fynbos is given by Kruger and Lamb (1978). They suggest that prescribed burns should only take place once 50% of the population of the slowest maturing species in a given area has flowered for at least three successive seasons. This rule is usually applied to shrubs in the family Proteaceae, since these are generally the slowest to mature in fynbos. The time to reach maturity may vary geographically; for example a species may mature quickly in mesic areas, but take much longer to do so in drier areas. Minimum prescribed fire frequencies should take account of this. At the other end of the scale, several arguments have been put forward as important in determining maximum fire intervals. These include senescence in the Proteaceae (see below); lower vegetative cover and greater patchiness in regeneration following the inevitable fire in old fynbos, leading to increased erosion (Bond 1980); and the suppression of understorey species with long intervals between fire (Campbell and Van der Meulen 1980; Esler and Cowling 1990; Cowling et al. this volume). Fire frequencies greater than 25–30 years are seldom recommended due to these considerations.

A useful way to generalize the response of plant communities to different fire frequencies is the vital attributes' scheme proposed by Noble and Slatyer (1980). Although data for most fynbos species are lacking, several indicator species can be examined in this context. Serotinous Proteaceae, dominant shrubs in many fynbos areas, fall into Noble and Slatyer's class CI. These species are killed by fire and regenerate from seed stored in the canopy. CI species are 'intolerant' and recruitment is virtually confined to the period immediately after fire. The seed is short-lived after release and no effective seed reserves persist after the death of the parent and the opening of the cone. Populations of CI species may become locally extinct either as a result of fire before they are reproductively mature (no seed reserves) or if the interval between fires exceeds the lifespan of the individuals in the even-aged population. Frequent fires are a well-known cause of local extinction of slow maturing Proteaceae populations (Kruger 1977a; Van Wilgen and Kruger 1981). Old stands of fynbos are rare but Bond (1980), who has reported poor regeneration of CI Proteaceae after long (40–50 year) intervals between fires, suggested that such poor regeneration may be due to the depletion of seed reserves with senescence of the parents. For species of *Protea* at least, fire intervals of between 8 and 30 years are needed to ensure survival.

Populations of shrubs of the genus *Mimetes* (Proteaceae), found in many fynbos areas, are classed as GI species. GI species build up considerable seed stores in the soil, with longevities exceeding the lifespan of adult plants (> 12 years in the case of *Mimetes splendidus*, A J Lamb unpublished data). *M. splendidus* is an intolerant species and establishes only after fire, at which time the seed pool is exhausted. Data on the longevity of the soil-stored seed pool would be necessary to determine maximum periods between fires. Minimum periods are set by the age at which the species attain maturity and shed seed (about 5–6 years for *M. splendidus*). The seed banks of some non-Proteaceous, non-sprouting GI species (*Agathosma* and *Muraltia*) are small and largely dependent on the current year's input for maintenance (Pierce and Cowling 1991). Seeds of these species do lose viability in the soil (Pierce 1991), indicating that the seed bank is not long-lived. Furthermore, individual plant longevity of these species seldom exceeds 30 years in south-eastern Cape dune fynbos (Pierce and Cowling 1991). Like non-sprouting serotinous Proteaceae, these species would be vulnerable to drastic population reduction or even local extinction under certain fire frequencies.

Most species in fynbos are able to sprout (Noble and Slatyer's groups V, U, and W; see Van der Merwe 1966; Kruger 1987). Such species are less affected by fire frequency as they do not have to reach maturity to survive fires; they are also generally longer lived than non-sprouters. As stated above, non-sprouters are far more sensitive to changes in fire frequency and such plants therefore dictate the fire frequencies required to maintain species diversity.

The approaches described above would indicate that fire frequencies should be between 10 and 25 years to maintain species richness, and this is the most common approach adopted by managers (Van Wilgen et al. 1990a). However, in some cases the aims of management call for short fire frequencies. Fire-breaks, for example, are burnt frequently to reduce biomass (Van Wilgen 1982). Another example concerns areas managed for grazing. Some of the earliest research into the effects of fire on fynbos was directed at the eradication of Afromontane fynbos (see Cowling and Holmes this volume) in the grazing areas of the eastern Cape mountains (Trollope 1971,

1973). In order to eradicate shrubs such as *Erica brownleeae*, *Cliffortia linearifolia*, and *C. paucistaminea*, fires at frequencies of between two and four years were recommended. This effectively destroyed seedlings before they matured and resulted in a 'complete recovery' of the grass sward (Trollope 1973). Similar principles are applied in the management of the Bontebok National Park (Novellie 1984) and could be applied to reduce shrub densities of *Elytropappus rhinocerotis* and *Metalasia muricata*, and increase the cover of grasses in renoster shrublands (Cowling et al. 1986).

Paradoxically, it is sometimes argued that short fire frequencies may be appropriate for the management of fire-sensitive species. The Clanwilliam cedar (*Widdringtonia cedarbergensis*) is a case in point. Cedars are killed by fire and regenerate from seed only. Due to over-exploitation, the species has declined markedly (Manders 1986). The mortality of adult trees is high following intense summer fires, but seedling recruitment is good. Despite this, managers feel that adult mortality after long inter-fire periods (with resultant fires in old, dense vegetation) is unacceptably high. Some areas are therefore deliberately subjected to a regime of short frequency (four year), low intensity winter burns to reduce adult mortality. The drawback of this approach is that seedling recruitment following winter fires is poor (Manders 1987). In addition, the surrounding fynbos is subjected to the most unsuitable regime, both in terms of frequency, and of season (see below).

FIRE SEASON

The time of year at which fires occur is determined by climate and seasonal variations in the fuel properties of the vegetation, given that ignition sources are always available. Seasonal curing of the vegetation is not a feature of fynbos vegetation (Van Wilgen 1984) and fire season therefore depends largely on climatic factors. Owing to the mediterranean-type climate over much of the fynbos biome, fires occur mainly in the summer, but can occur in all months under suitable weather conditions (Van Wilgen 1987).

Fires in different seasons can have marked effects on elements of the vegetation and these can be nearly as pronounced as frequency effects. Shrubs (serotinous Proteaceae) and trees (the Clanwilliam cedar) that are killed by fire show maximum seedling recruitment after late summer and early autumn fires (Jordaan 1949, 1965, 1981; Bond et al. 1984; Manders 1985; Van Wilgen and Viviers 1985; Botha 1989; Midgley 1989). This has been attributed to seed predation by small mammals (see Johnson this volume). Regular prescribed burning outside the late summer-early autumn period could result in the local extinction of species (Bond et al. 1984) and is not usually applied where species conservation is an objective.

There is, however, much variation in the response of plants to fire season which is poorly understood. Autumn fires do not always guarantee good recruitment, just as spring fires do not always ensure poor recruitment (Figure 14.1). Other factors may play a role. Serotinous Proteaceae, for example, show a more marked sensitivity to season of burn on north than on south slopes in the Swartberg (Bond et al. 1984). There has been substantial research on the causes of Proteaceae sensitivity to fire season (Bond 1984; Coetzee and Giliomee 1987; Midgley and Vlok 1986; Midgley 1989; Johnson this volume). An understanding of the causes of this variation would be of considerable value in refining fire prescriptions in different areas, and may allow more flexibility in the approach to fire season in fynbos.

The response of other species in the fynbos biome to variations in fire season supports the contention that the late summer-early autumn period is the best time to burn (Figure 14.2). Seedling regeneration of the fynbos geophyte *Watsonia borbonica* is absent following fires between April and October, and occurs only after summer and autumn burns (Le Maitre 1984). The maximum flowering activity of most fynbos plants occurs in late winter and spring (Kruger 1981) which implies that the maximum seed loads will be available after fires in late summer or early autumn. Proteaceae with soil-stored seed show greatest recruitment following autumn burns, after the current seed crop has matured and been released (Le Maitre 1988). In the case of the shrubs *Paranomus bracteolaris* and *Leucadendron pubescens* growing in the Cederberg, this would mean that burning should take place after early December (Le Maitre 1988). Even

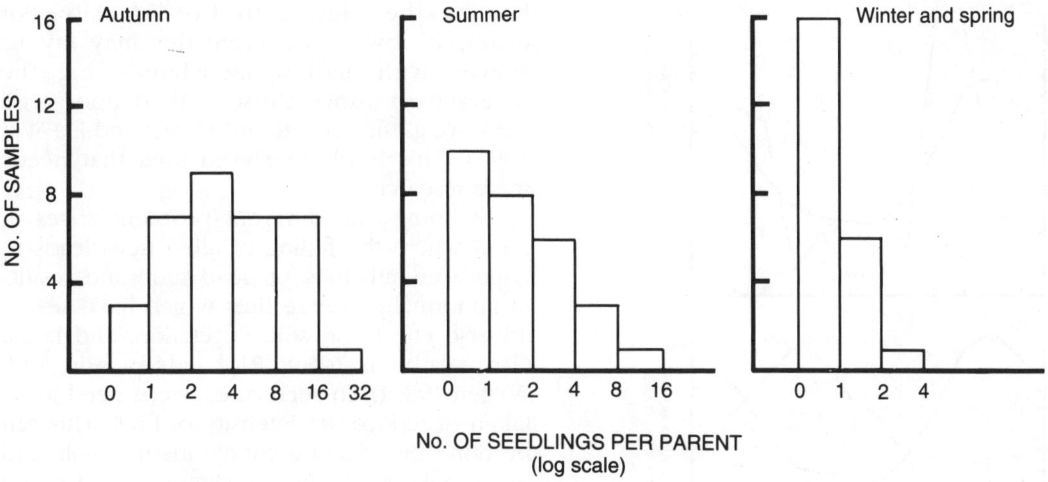

FIGURE 14.1 Regeneration success (number of seedlings per parent) censused after different fire seasons in the southern Cape mountains. Seedling numbers and pre-burn densities were censused after 30 fires with up to four replicate grids each of 65 1x1 m quadrats in each burn. Better recruitment follows autumn and summer fires, but much variation occurs, with poor recruitment (< 1 seedling per parent) after fires in all seasons. Reprinted from Bond et al. (1984) with the permission of Blackwell Scientific Publications.

some animals, such as the geometric tortoise (see below), show traits that reveal a susceptibility to winter or spring burns.

Cowling et al. (1986) have argued for autumn burns (February to April) in south coast renoster shrubland to promote grasses at the expense of shrubs. This recommendation was based on a knowledge of the life histories of the grass *Themeda triandra* and the shrubs *Metalasia muricata* and *Elytropappus rhinocerotis*. At this time of the year, stored carbohydrates in the roots of *Themeda triandra* are high; in addition, germination is promoted by autumn fires. Under frequent burns, to which the grasses are very resistant, the shrubs could be eliminated.

Although fires in summer and autumn are beneficial for most fynbos plant species, fires in autumn and winter may promote certain species over others. For example, *Stoebe plumosa* and *Metalasia* species (shrubs in the family Asteraceae) may benefit from winter burns (M Viviers personal communication). Their shallow soil-stored seed banks are apparently destroyed by intense summer fires, but are favoured by cool winter fires. These species also germinate before most others (Musil and De Witt 1990), giving them an advantage following winter burns. Such variation may not always be viewed as favourable. For example, frequent winter fires, applied to protect cedars (see above) may in fact endanger them, as *Stoebe plumosa* provides a fast-growing fuel bed that will increase, rather than decrease, the intensity of subsequent fires.

FIRE INTENSITY

Fire intensity depends on fuel loads, and the rate at which they burn. Fuel loads in fynbos vary from < 500 gm^{-2} to > 4 000 gm^{-2} (Van Wilgen 1982). The intensity at which a given fuel load will burn can vary by several orders of magnitude (Van Wilgen et al. 1985). Fire intensity can be manipulated either by reducing fuel loads (for example by burning more often) or by selecting the conditions that will lead to the desired type of fire (see below).

It is known that high intensity fires can have adverse effects on sprouting species (Van Wilgen 1982). However, very few studies have been done on the biological effects of intensity. The late summer-early autumn period usually has weather associated with high intensity fires, and indications are that at least some elements of the fynbos biota require high intensity fires for survival. The shrubs *Mimetes fimbriifolius* and *Leucospermum conocarpodendron* regenerate well after high inten-

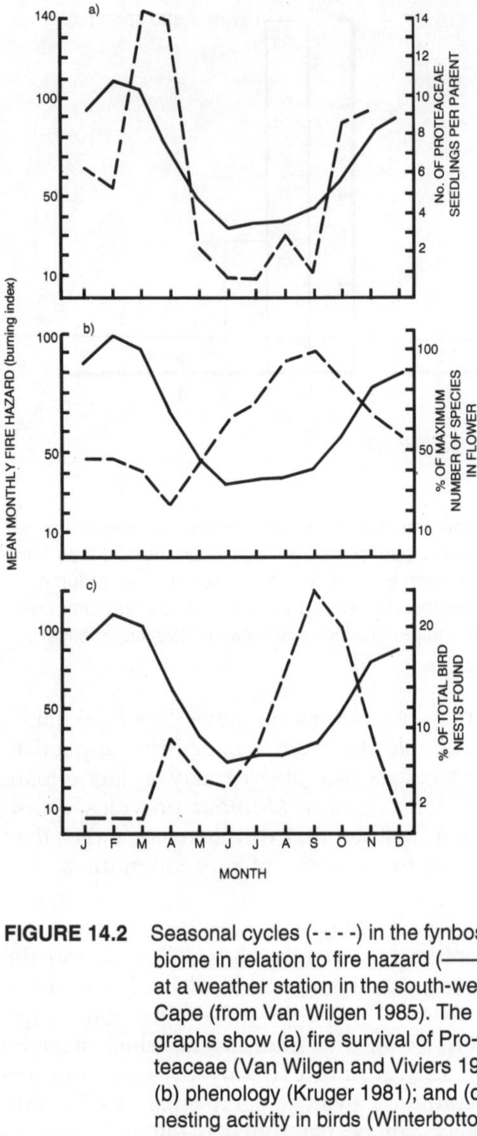

FIGURE 14.2 Seasonal cycles (- - - -) in the fynbos biome in relation to fire hazard (———) at a weather station in the south-western Cape (from Van Wilgen 1985). The graphs show (a) fire survival of Proteaceae (Van Wilgen and Viviers 1985); (b) phenology (Kruger 1981); and (c) nesting activity in birds (Winterbottom 1968). The most likely time for fires (late summer to early autumn) coincides with the period best suited to each cycle.

sity fires, but poorly after low intensity fires (Bond et al. 1990; Figure 14.3). Similar patterns have been observed in *Mimetes splendidus* (A J Lamb personal communication). Such findings present problems for managers, as the desired biological response requires high intensity fires, while safety factors require the opposite.

The relative abundance of species that regenerate from soil-stored seed banks after fire may be affected by fire intensity. For example, lower fire intensities may favour species with shallow seed banks (e.g. the Asteraceae) above those with deeper seed banks (e.g. the Proteaceae) (Musil and De Witt 1990). This is obviously an area that needs more research.

A somewhat different problem arises in areas where the felling of alien trees leads to large accumulations of dead fuel, and results in abnormally intense fires which have severe adverse effects on soil, vegetation, and fauna (Breytenbach 1989; Richardson and Van Wilgen 1986a). In such cases, steps need to be taken to reduce the intensity of fires. This can be done by selecting conditions that will lead to acceptable intensities (Figure 14.4) or by dispersing or physically removing fuel loads.

THE PRACTICAL APPLICATION OF APPROPRIATE FIRE REGIMES

Prescribed burning in the summer months (November to February) is seldom attempted in much of the biome, since the risk of runaway fires is too great. Burning is usually only feasible in March and April; but only about 12 days on average in these two months have suitable weather for prescribed burning in the western Cape. The fire danger during the remaining days is too high to meet the prescribed conditions, or too low to ensure that fires will burn (Van Wilgen and Richardson 1985a). Managers often opt for burning after mid-April to complete burning programmes, but should take into account the consequences in terms of adverse ecological effects. The problems faced by managers who have to complete burning programmes under rarely suitable conditions could be alleviated by the use of systems designed to select conditions that will lead to acceptable fire behaviour.

The adoption of fire danger rating systems, and fire behaviour prediction models, has received considerable attention in the fynbos biome (Van Wilgen 1984; Van Wilgen and Burgan 1984; Van Wilgen and Richardson 1985b; Juhnke and Fuggle 1987; Van Wilgen et al. 1990b). Fire behaviour prediction models developed in the United States of America (Rothermel 1972; Burgan and Rothermel 1984) have been adapted for use in fynbos, using appropriate fuel and weather inputs (Van Wilgen et al. 1985). With this system, it is possible to select environmental conditions that will lead to the desired type of fire. Examples are provided in Table 14.2 and in Figure 14.4. However, the practical application of these techniques has not been successful (Van Wilgen and Manders 1990). In cases where managers have used the technology, a considerable improvement in the ability to make appropriate and informed decisions has been achieved (R H Andrag, G Ruddock personal communication), and the approach holds promise for the future.

Expert systems offer considerable potential for applications where expertise is needed either quickly (as in disaster management), or across a wide range of ecological disciplines (such as in land management decisions, Noble 1987). Little information has been made available on how to apply expert systems to solve problems in ecology (Mentis and Walker 1989), but the approach deserves attention from land managers.

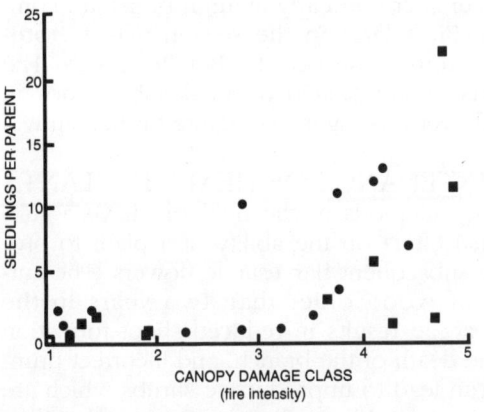

FIGURE 14.3 The relationship between fire intensity (represented by canopy damage classes) and seedling regeneration in two fynbos species with soil-stored seeds. Canopy damage class 1 represents low intensity fires, class 5 represents high intensity fires. The species are *Mimetes fimbriifolius* (solid circles) and *Leucospermum conocarpodendron* (solid squares). High success in germination only occurs after fires have reached an apparent threshold intensity. From Bond et al. (1990).

Flower harvesting

Fynbos plants are harvested for the cut flower trade along the southern lowlands, southern mountain foothills, and as far as the Langkloof in the east. In 1989 the total trade turnover was estimated at R29 million (Greyling and Davis 1989). Between 10 000 and 15 000 people derive their income from the wild flower industry in what are otherwise very poor agricultural conditions. The wild flower trade has thus added value to otherwise marginal agricultural land. A simple measure of this is the recent flurry of claims for compensation, running into millions of rand, for lost income after runaway fires (I Bell personal communication; B W van Wilgen and W J Bond personal observation).

Most of the income from the wild flower trade is derived from export. The market is based on Proteaceae which are now cultivated in several other parts of the world. The market has become highly competitive and the response of local growers and horticultural researchers has been to call for improved quality and new cultivars (Brits et al. 1983). These can best be obtained under cultivation. Almost all of the research on the wild flower trade has dealt with horticultural aspects of cultivating Proteaceae and breeding new varieties. Very few studies have been made on veld harvesting, apparently because this is seen as detrimental to the development of a competitive export market (e.g. Brits et al. 1983). Nevertheless, more than 75% of flowers are still harvested from the veld, mainly in a small area between Cape Hangklip and Mossel Bay (Cowling 1990). Managing veld for cut flowers requires the manipulation of fynbos succession for regeneration, the maintenance of healthy plants for flower production, the maintenance of adequate seed reserves to regenerate populations despite flower removal, and pest control to improve flower quality. Each of these is discussed separately.

MANIPULATION OF SUCCESSION

It is not possible to sustain flower harvesting without fire, despite a contrary belief among many farmers. Many species require fire as a direct or indirect cue for seed germination (e.g. Blommaert 1972; Brits 1986, 1987; Jeffery et al. 1988; Pierce 1990) and even the longer lived Proteaceae senesce and die after 30 to 50 years without fire (Bond 1980; Van Wilgen 1981). Because of life history differences between species, marketable flowers change in abundance after fire. Some species appear only after fire, persist a few years, and then die. The most important commercial examples are the 'everlastings' such as *Helichrysum vestitum*. Others flower within a year or two after fire (e.g. *Erica perspicua*) and continue to flower for many years. Finally, some species are relatively slow to mature and only produce marketable flowers some 5–6 years after fire (e.g. *Protea* and *Leucadendron* species). Thus, the farmer will have a shifting availability of different flower species as the veld ages. Depending on veld composition and the market, farmers may choose to burn on short rotations (e.g. for *Helichrysum*), or in blocks for a mix of different ages, thus providing a cross-section of species from all successional stages.

The implications of different fire regimes have been discussed earlier. In practice, most veld is burnt after 12 to 15 years when the vigour of *Protea* shrubs declines and stem lengths become too short for the trade. The favourite fire season varies from region to region and appears to be determined mostly by safety factors. On the south-western and southern Cape lowlands (Gansbaai to Stilbaai), late summer and early autumn burns are common. Fires later in the season may be preferred further west in the Bot River area. The effects of fire season on marketable crops in the flower picking areas require further study.

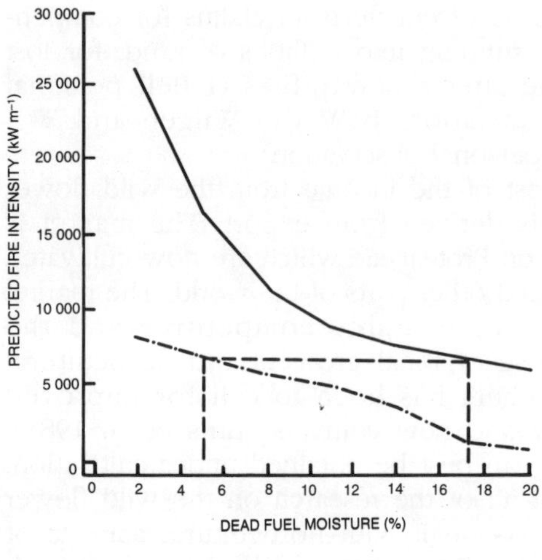

FIGURE 14.4 The effect of dead fuel moisture on predicted fire intensity (using Rothermel's fire model) for two fuel complexes. The upper curve (———) is for felled hakea, and the lower curve (•–•–•–) is for fynbos. The dashed line (– – –) indicates the moisture contents required to achieve similar fire intensities for the two fuel complexes. Reprinted from Richardson and Van Wilgen (1986a) with the permission of The Southern African Institute of Forestry.

MAINTENANCE OF HEALTHY PLANTS

Cutting methods in the field can have a substantial effect on the ability of a plant to produce subsequent harvestable flowers. The cutting of wood older than two years in the Proteaceae results in reduced shoot formation or the death of the branch, and incorrect pruning can lead to unproductive shrubs which are more susceptible to disease (Brits et al. 1986). Long stems are, however, required in the trade, and poor management of harvesting operations can lead to a decline in flower production.

Many species in fynbos have long stems unprotected by leaves. If stems are cut in these older sections below the leaves, the whole plant can die (Rebelo 1987). Poor harvesting procedures in the veld can thus lead to serious dieback of sensitive species including *Brunia albiflora*, *Erica pinea*, *Protea compacta*, and *Paranomus capitatus* (Rebelo 1987; Rebelo and Holmes 1988). There has been no research on determining the best methods for harvesting non-Proteaceous stems in the veld.

ADEQUATE SEED RESERVES

Clearly, the removal of flower heads or fruiting structures for the dried flower trade can influence the regeneration capacity of the veld. Indiscriminate harvesting of everlastings (*Helichrysum* and *Helipterum* spp.) has probably lead to their demise in the veld (Malan 1988). Many farmers are aware of the problem and harvest seeds from the veld, storing them for later broadcasting. However, the longevity of fynbos seed in storage may often be little more than a year (Eloff and Liede 1987; Pierce 1990). In addition, some species require burial by ants, germination cues can be complex and may require fire stimuli, while others are readily eaten by rodents or other seed predators. In general, seeds could be disseminated in the veld during the first autumn after a fire to ensure rapid germination at a time when vertebrate seed predators are scarce (Bond 1984; Midgley and Clayton 1990). However, the efficacy of seed harvesting and subsequent broadcasting in maintaining populations clearly need investigation.

For veld-harvested Proteaceae, a common alternative has been to limit harvesting to some fraction of the available flowers (Van Wilgen and Lamb 1986). For serotinous Proteaceae, state management agencies restricted picking to not more than half of the blooms or foliage each year with no picking at all for a full year before a prescribed fire. In fact, different species will have different sensitivities to picking intensity. Harvesting at levels in excess of 75% of blooms of the non-sprouting shrub *Phylica ericoides* diminishes the subsequent year's flower crop significantly compared to non-harvested controls; a harvesting intensity of 25% or less had little effect (D J Killian and R M Cowling unpublished data). Sprouting species, such as *Protea cynaroides*, are known to be resilient to heavy utilization (Vogts 1982). Non-sprouting species vary in their seed retention patterns. Some species hold seeds in cones for little more than a year, others retain a substantial fraction of seeds for three or four years (Bond 1985; Van Wilgen and Lamb 1986; Le Maitre 1990). Species with short-lived seed banks are more vulnerable and should be lightly harvested. Mustart and Cowling (submitted) suggest that seed bank sizes should not be decreased by more than 50% in seroti-

TABLE 14.2 Predictions of flame length (m) for different wind speeds and moisture contents of the dead fraction of the vegetation, using Rothermel's (1972) fire model and a fynbos fuel model (Van Wilgen 1984). Conditions that will lead to a fire with flame lengths of between 2 and 5 m are blocked off, to illustrate how the predictions can be used to select combinations of wind speed and fuel moisture that will result in a fire with these characteristics.

Moisture content of fine dead fuel (%)	Wind speed (km/hr) 0	5	10	15	20	25	30
2	1.1	3.4	5.3	7.0	8.4	9.8	11.1
5	1.0	3.0	4.8	6.3	7.6	8.8	10.0
8	0.9	2.7	4.2	5.6	6.8	7.9	8.9
11	0.7	2.2	3.5	4.6	5.5	6.4	7.3
14	0.4	1.2	1.8	2.4	2.9	3.4	3.8
17	0.3	1.1	1.7	2.2	2.7	3.1	3.5
20	0.3	1.0	1.6	2.1	2.6	3.0	3.4

nous Cape Proteaceae. This approximates the maximum degree (22 to 51%) by which unharvested seed banks would be reduced in the event of an unseasonable fire in late spring or early summer, before the current seed has matured. The harvesting of not more than 50% of current inflorescences or cones is recommended. Since there is lowered cone production in the post-harvest year, stem harvesting should only take place on alternate years. This would allow some measure of vegetative and subsequent reproductive restoration (Mustart and Cowling submitted).

There is, of course, little use in controlling harvesting without having full control over other aspects of management such as the timing of fires (see above). The rule regarding no harvesting before prescribed fires, for example, is difficult to apply because uncontrolled fires are still very common (Van Wilgen and Lamb 1986). One way of balancing this particular risk would be to reduce harvesting levels in older veld, since it has a greater probability of burning.

All work on the effects of harvesting on regeneration potential assume a strong relationship between pre-burn seed reserves and the eventual size of post-burn mature populations. This assumption has not been adequately tested. In lowland fynbos, for example, conditions for seedling survival after fire may be more important than seed bank size since seedling numbers are grouped in clumps vastly in excess of adult numbers (W J Bond and R M Cowling unpublished data).

PEST CONTROL

Pre-harvest pest control appears to be very limited in veld harvesting. However, there have been reports of the poisoning of sunbirds and sugarbirds which are said to damage inflorescences (Rebelo 1987). There seems little justification for this practice and it is hardly consistent with suggestions that veld harvesting will help conserve the fynbos ecosystem (Cowling 1990).

HARVESTING OF THATCHING REED

The fynbos restioid, *Thamnochortus insignis* ('dekriet'), is harvested locally in the sandveld of the southern Cape (Albertinia area) for use as thatching material. Linder (1990) provides an account of the biology and utilization of the species. Dekriet is much in demand because of its hard-wearing qualities and is sold as far afield as the Transvaal. The annual value of the crop is between R5–10 million (Linder 1990). Dekriet does not require fire for regeneration. Management is based on cutting good quality culms at a frequency (2–5 years) and season (early winter) which will ensure continued productivity of the plants. Competing shrubs are often removed leaving a relatively pure stand of dekriet. Should the market expand, there seems great potential for the cultivation of this plant.

The control of alien plants and animals

Invasive plants are a major problem in the fynbos biome (Richardson et al. this volume). The control of invasive trees and shrubs is the largest single task facing managers of most natural areas in the biome. In this section, we address the various control options briefly. These may be placed in three groups:

- biological control (hereafter bio-control);
- mechanical control; and
- chemical control.

We include all methods that involve the use of mechanical tools and/or fire under mechanical control: cut-and-burn, cut-and-leave, and burn standing.

CONTROL OPTIONS FOR PLANTS

A number of bio-control agents have been established in the biome (Table 14.3) and several other control options are available (Table 14.4). Clearly, some options are suitable for certain species but inappropriate for others. The effectiveness of any control option depends primarily on the life history attributes of the target species. In many cases, a combination of one or more control options (integrated control) produces the best results (Kluge et al. 1986). The situation is complicated when more than one weed species is present at a site, especially if the species differ markedly in their life history attributes. Other important constraints include managing control options within the confines of prescribed burning schedules. Wildfires often disrupt control operations. An expert system approach may offer the best way of selecting the best control option, but no such system has yet been developed.

Target species	Principal bio-control agents	Mode of operation	References
Acacia cyclops	Indigenous seed-feeding insects *Zulubius acaciaphagus* and *Remiptevans* sp. (Alydidae). No Australian bio-control agents introduced yet	*Z. acaciaphagus* destroys up to 84% seed crop	D Donnelly and M J Morris (unpublished data); Holmes and Rebelo (1988)
Acacia longifolia	Bud-galling wasp *Trichilogaster acaciaelongifoliae* Snout beetle *Melanterius ventralis*	*T. acaciaelongifoliae* reduces seeding and also causes growth suppression; *M. ventralis* attacks green developing seeds; the two agents together reduce seed production by about 99%	Dennill (1985) D Donnelly (unpublished data)
Acacia mearnsii	Several *Melanterius* species are potential bio-control agents; none have yet been released because of the perceived threat to commercial plantings of *A. mearnsii*	*Melanterius* spp. attack green developing seeds	Donnelly (1990)
Acacia melanoxylon	Snout beetle *Melanterius acaciae*	Attacks green developing seeds	D Donnelly (unpublished data)
Acacia saligna	Gall rust *Uromycladium tepperianum*	Galls reduce seed production and general vigour; heavily galled plants die	Morris (1987)
Hakea sericea	Snout beetle *Erytenna consputa* Moth *Carposina autologa* Fungus *Colletotrichum gloeosporioides*	Larvae destroy developing (*E. consputa*) and accumulated mature (*C. autologa*) fruits; the fungus reduces vigour and causes mortality	Morris (1982, 1989); Neser and Kluge (1985)
Leptospermum laevigatum	Seed-affecting organisms available, including a specialized nematode and a fly (Sphaerulariidae, Fergusoninidae) but not yet introduced	Not known	Neser and Kluge (1986)
Paraserianthus lophantha	Snout beetle *Melanterius servulus*	*M. servulus* attacks green developing seeds	D Donnelly (unpublished data)
Pinus pinaster	Pine woolly aphid *Pineus pini* (Homoptera: Adelgidae) (accidental introduction)	Causes cone deformation and stunting leading to reduced seed production	Zwolinski et al. (1989)
Pinus radiata	*Pineus pini* (accidental introduction)	Causes cone deformation and stunting leading to reduced seed production	Bruzas (1981)
Sesbania punicea	Snout weevils *Trichapion lativentre*, *Rhyssomatus marginatus* and *Neodiplogrammus quadrivittatus*	*T. lativentre* adults feed on leaves, flowers, and young pods and larvae develop in young flower buds; *R. marginatus* develops in immature pods; *N. quadrivittatus* larvae bore through mature branches and stems	Harris and Hoffmann (1985); Hoffmann and Moran (1988)

TABLE 14.4 The currently available options for the control of some important alien invasive plants in the fynbos biome. Specific information of bio-control agents is given in Table 14.3.

Species and salient life history attributes	Options	Focus of control measure	Comments / Constraints
Acacia cyclops	a) Bio-control		Not available at present.
— does not resprout	b) Cut-and-leave	Adult mortality and seed bank depletion	This species does not sprout and seed banks are relatively short-lived (Holmes 1989).
— relatively short-lived seed bank stored in soil	c) Cut-and-burn	Adult mortality and seed bank depletion	Fire destroys seeds near the soil surface.
— seeds not stimulated to germinate by fire	d) Burn standing	Adult mortality and seed bank depletion	Adult plants are killed. Follow-up weeding required; this is feasible because recruitment is not fire-coupled and because seed pools are relatively short-lived.
	e) Chemical	Chiefly seeding mortality	Non-target species adversely affected.
	f) Integrated	Long-term control	Use b, c, or d in combination with e where germination is prolific.
Acacia longifolia	a) Bio-control	Reduced seed production and vigour	Highly effective. Invasive potential greatly reduced.
— does not resprout	b) Cut-and-leave	Adult mortality	Adults killed but soil-stored seeds stimulated; follow-up weeding of seedlings necessary.
— relatively long-lived seed bank stored in soil	c) Cut-and-burn	Adult mortality and seed bank depletion	Adults killed but soil-stored seeds stimulated; follow-up weeding of seedlings necessary.
— seeds stimulated to germinate by fire	d) Burn standing	Adult mortality and seed bank depletion	
	e) Chemical	Seedling mortality Long-term control	Non-target species adversely affected.
	f) Integrated		Use b and a.
Acacia mearnsii	a) Bio-control	Reduced seed production	Effectiveness untested. Other forms of bio-control not considered due to commercial importance of species.
— resprouts	b) Cut-and-leave	Adult mortality	Ineffective; plant resprouts.
— relatively long-lived seed bank stored in soil	c) Cut-and-burn	Adult mortality	Ineffective. Adults resprout and germination of soil-stored seed stimulated.
— seeds stimulated to germinate by fire	d) Burn standing	Adult mortality	Ineffective. Adults resprout and germination of soil-stored seed stimulated.
	e) Chemical	Chiefly seedling mortality	Non-target species adversely affected.
	f) Integrated	Long-term control	Combine b, c, or d with e and a.

TABLE 14.4 continued.

Species and salient life history attributes	Options	Focus of control measure	Comments / Constraints
Acacia melanoxylon	a) Bio-control	—	Nothing available at present.
	b) Cut-and-leave	Adult mortality	Ineffective; plant resprouts.
— resprouts — relatively long-lived seed bank stored in soil	c) Cut-and-burn	Adult mortality and seed bank depletion	Ineffective. Adults resprout and germination of soil-stored seed stimulated.
— seeds stimulated to germinate by fire	d) Burn standing	Adult mortality and seed bank depletion	Ineffective. Adults resprout and germination of soil-stored seed stimulated.
	e) Chemical	Chiefly seeding mortality	Non-target species adversely affected.
	f) Integrated	Long-term control	Combine b, c, or d with e and a.
Acacia saligna	a) Bio-control	Reduced vigour	Introduced fungus slow to establish. Overall effectiveness unknown but potentially very effective.
— resprouts — relatively long-lived seed bank stored in soil	b) Cut-and-leave	—	Inappropriate. Adults resprout after felling and seed bank persists.
— seeds stimulated to germinate by fire	c) Cut-and-burn	Adult mortality and seed bank depletion	Ineffective. Adults resprout and germination of soil-stored seed stimulated.
	d) Burn standing	Adult mortality and seed bank depletion	Ineffective. Adults resprout and germination of soil-stored seed stimulated.
	e) Chemical	Chiefly seedling mortality	Non-target species adversely affected.
	f) Integrated	Long-term control	Combine b, c, or d with e and a.
Hakea gibbosa	a) Bio-control	Reduced seed production and vigour	Not highly effective.
— does not resprout	b) Cut-and-leave	Adult mortality	Ineffective; fire is needed to kill seedlings.
— seeds stored in heat-resistant follicles in the canopy	c) Cut-and-burn	Adult and seedling mortality	Adults felled, seeds germinate and are killed by fire. Highly effective but expensive.
— seed release stimulated by fire or felling	d) Burn standing	Adult mortality	Effective in some cases but usually results in dense stands of seedlings; standing dead trees make manual eradication of seedlings difficult. Seeds are released from standing shrubs and are dispersed to new sites by wind.
	e) Chemical	Adult mortality	Effective in low-density stands for selective control.
	f) Integrated	Long-term control	Combine c with a or e.
Hakea sericea	a) Bio-control	Seed production and vigour	Reduces invasion potential significantly in the long-term but existing stands need to be cleared.
— does not resprout	b) Cut-and-leave	Adult mortality	Fire is needed to kill seedlings.
— seeds stored in heat-resistant follicles in the canopy	c) Cut-and-burn	Adult and seedling mortality	As for *Hakea gibbosa*.
— seed release stimulated by fire or felling	d) Burn standing	Adult mortality	As for *Hakea gibbosa*.
	e) Chemical	Seedling mortality	Ineffective.
	f) Integrated	Long-term control	Combine c with a.

TABLE 14.4 continued.

Species and salient life history attributes	Options	Focus of control measure	Comments / Constraints
Hakea suaveolens	a) Bio-control	Seed production	None available at present.
	b) Cut-and-leave	Adult and seedling mortality	As for *Hakea gibbosa*.
— does not resprout	c) Cut-and-burn		As for *Hakea gibbosa*.
— seeds stored in heat-resistant follicles in the canopy	d) Burn standing	Adult mortality	As for *Hakea gibbosa*.
	e) Chemical	Seedling mortality	As for *Hakea gibbosa*.
— seed release stimulated by fire or felling	f) Integrated	Long-term control	Combine c with a.
Leptospermum laevigatum	a) Bio-control	—	None available at present.
	b) Cut-and-leave	Adult mortality	Not effective as seedling regeneration is prolific.
— does not resprout	c) Cut-and-burn	Adult and seedling mortality	Effectiveness has not been assessed.
— small seeds with no obvious storage mechanism	d) Burn standing	Adult mortality	Ineffective as seedling regeneration is prolific.
	e) Chemical	—	No information available.
	f) Integrated	Long-term control	Options limited by lack of effective control method.
Paraserianthus lophantha	a) Bio-control	Seed production	Has the potential to reduce invasion potential significantly in the long-term but existing stands need to be cleared.
— does not resprout	b) Cut-and-leave	Adult mortality	Satisfactory for low density stands.
— relatively long-lived seed bank stored in soil	c) Cut-and-burn	Adult and seedling mortality	This species grows along and in streams; fire is seldom practical.
— seeds stimulated to germinate by fire	d) Burn standing	(Adult mortality)	This species grows along and in streams; fire is seldom practical.
	e) Chemical	—	Untried.
	f) Integrated	Long-term control	a, b (and e?).
Pinus halepensis *P. pinaster* and *P. radiata*	a) Bio-control	Seed production	None available at present (but see Table 14.3). Problems because these are commercially important species.
	b) Cut-and-leave	Adult and seedling mortality	As for *Hakea gibbosa*.
— do not resprout	c) Cut-and-burn	Adult mortality	As for *Hakea gibbosa*.
— seeds stored in heat-resistant cones in the canopy	d) Burn standing	Adult mortality	Stem injections may be appropriate for control of scattered stands but seeds are released from standing trees.
— seed release stimulated by fire or felling	e) Chemical	Adult and seedling mortality	—
	f) Integrated	Long-term control	Combine c and a.
Sesbania punicea	a) Bio-control	Adult plants and seed production	Reduces vigour of adult trees and decreases invasive potential by reducing seed production.
— resprouts	b) Cut-and-leave	Adult mortality	Satisfactory for low density stands; felled plants should be stacked.
— relatively long-lived seed bank stored in the soil	c) Cut-and-burn	Adult and seedling mortality	This species grows along and in streams; fire is seldom practical.
	d) Burn standing	Adult mortality	

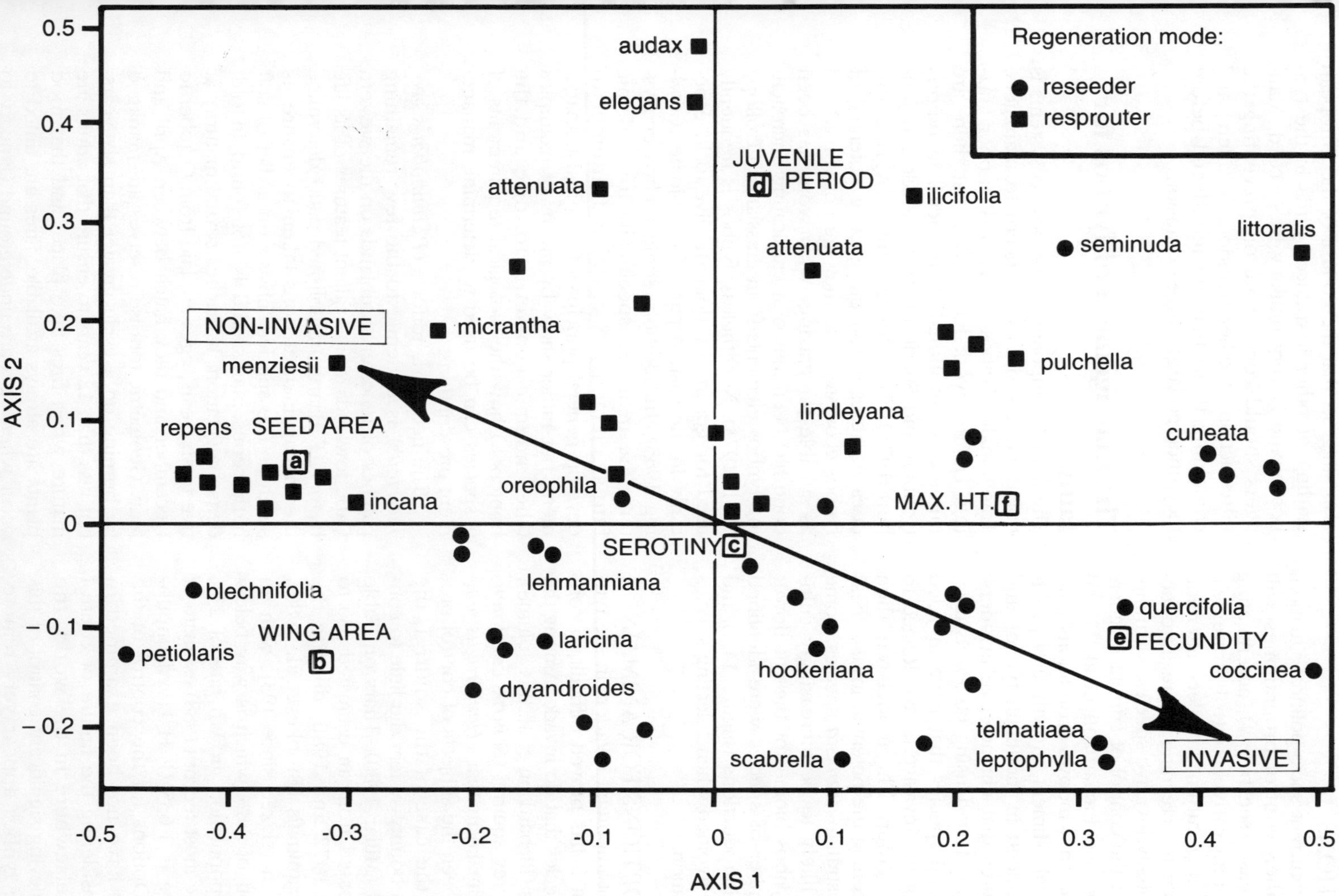

FIGURE 14.5 A simplified plot of the first two axes from a correspondence analysis of a matrix of six life history attributes (a to f) of 69 western Australian *Banksia* taxa. A trend of potential for invading mountain fynbos is defined (modified from Richardson et al. 1990). Tall serotinous shrubs with many small seeds per plant, short juvenile periods, and low fire tolerance (e.g. *B. coccinea*) were identified as high risk introductions. Low sprouting shrubs with few large seeds per plant and long juvenile periods (e.g. *B. menziesii*) are unlikely to become invasive in mountain fynbos. The model can be used for screening species that have good potential for the cut flower market.

SCREENING FUTURE INTRODUCTIONS

Because of the current socio-political situation, funding for alien control programmes will probably decrease (see below), accentuating the need for innovative control strategies. Besides controlling current invaders, attention should be given to screening future introductions for potentially invasive species. Meaningful priorities for the control of current invaders should also be set. Richardson et al. (1990) described invasion windows (conditions suitable for the establishment, growth, and proliferation of invaders) in mountain fynbos and defined attributes that characterize invaders. The scheme was tested using 60 *Pinus* taxa, including several species that have invaded fynbos. Most of the invasive taxa belonged to the 'high risk' group. The model was then applied in screening the potential invasiveness of western Australian *Banksia* species, some of which are likely to be increasingly cultivated in the fynbos biome for the cut flower market. A number of species were identified as 'high risk' introductions (Figure 14.5) and these could be avoided when selecting species for local cultivation.

CONTROL OPTIONS FOR ANIMALS

Animals are not important invaders in the fynbos biome, but it has proved difficult to control the few species that do invade. Among the faunal invaders (Richardson et al. this volume), the American grey squirrel (*Sciurus carolinensis*) and the Himalayan thar (*Hemitragus jemlahicus*) have been the subjects of control programmes. In the case of the squirrel, the imposition of a bounty system did little to contain its spread (Millar 1980). Thars on Table Mountain increased from an estimated 330 to 403 between 1972 and 1980, despite the culling of 629 animals (Brooke et al. 1986). Culling has been relaxed since 1981, with an average take-off of 2.8% which is way below the estimated population growth rate of 23% per annum. The most recent (1988) estimate of population size is 1 635 (P H Lloyd unpublished data). Options for the control of the Argentine ant have not been examined in detail but, considering the failure of control programmes elsewhere in the world, the chances of achieving significant control of this species in the fynbos appear very remote. Experience from other regions (Erhlich 1989) also suggests that the control of the European starling or other introduced birds in the fynbos biome is very unlikely to succeed. Managers should concentrate on preventing the introduction of other notorious invaders, and containing those that are introduced before they undergo large range expansions.

The management of fynbos for fauna

The faunal component of fynbos ecosystems is not often a subject of concern to managers. Traditionally, fynbos is regarded as poor grazing. Limited work has shown that there are numerous nutritional deficiencies in fynbos (Joubert and Stindt 1979a, b; Stindt and Joubert 1978). The Department of Agriculture does no research on the grazing potential of fynbos ecosystems, as they are not considered as suitable for grazing. What work has been done has been aimed at eradicating encroaching fynbos elements from grasslands (Trollope 1971, 1973). Nonetheless, fynbos is frequently used for grazing by domestic livestock, especially in the eastern part of the biome (Cowling 1983). In addition, some fynbos ecosystems are managed specifically for endemic faunal elements. We will not attempt to develop general principles for faunal management, but rather show by means of examples (the endemic geometric tortoise and the bontebok) how the biological requirements of a species can be used to determine management prescriptions.

The geometric tortoise (*Psammobates geometricus*) is endemic to the few remaining patches of renoster shrublands on the western Cape lowlands. Some salient features of its life history (E Baard unpublished data) determine the management actions taken to ensure its survival. The species subsists on a diet of soft herbaceous plants, and is not found in old, dense vegetation. It reaches sexual maturity at five to six years; eggs are laid from October to December and they hatch between April and May. Geometric tortoise reserves are managed by burning on a cycle greater than six years, but less than 12 years to ensure that adults are mature when fires take place, and that food plants are always available. Fires are timed to occur when eggs are underground (January to

March). The fire regime applied to ensure the survival of geometric tortoises is remarkably similar to that needed for serotinous *Protea* species. Grazing by domestic stock is sometimes considered as an alternative to burning as it opens up the vegetation (thus providing conditions favouring the tortoises' food plants) and does not kill any adults.

The bontebok (*Damaliscus dorcas* subsp. *dorcas*) occurred historically in renoster shrubland, most of which is now under cultivation. It is preserved in a few patches of fynbos, not its favoured habitat. Bontebok consume grasses (Beukes 1987) which are seldom a major component in fynbos in the western part of the biome (Cowling and Holmes this volume). In the Bontebok National Park, the vegetation is burnt on a short (four year) cycle to favour grass at the expense of shrubby elements (Novellie 1984; Verster 1989). The relatively frequent fire cycles applied to provide bontebok with food will undoubtedly lead to the loss of plant species. In addition, the application of grazing pressure far in excess of that under which the plants evolved will probably cause further attrition. For this reason, parts of the park are managed for floral conservation and are burnt on longer (10–12 year) cycles (Novellie 1984). However, the preservation of Africa's rarest antelope (Smithers 1983) is seen as justification for relatively frequent burning in limited areas of fynbos. Similar conflicts may potentially exist in the case of the Cape Mountain Zebra (*Equus zebra* subsp. *zebra*).

SCENARIOS

In previous sections we addressed management in fynbos ecosystems and attempted to clarify the scientific basis for thes e actions. But what of the future? It is clear that the biome faces increasing fragmentation of the natural landscape, invasion of these fragments by alien plants, a decline in funds for management, and climatic change. How can managers make rational decisions faced with such uncertainty? In this section we examine some scenarios in the biome and explore the implications for the management of fynbos.

Fragmentation

Although large areas of mountain fynbos remain intact, lowland fynbos has become fragmented into small remnants by agriculture, urbanization, and forestry (Rebelo this volume). This poses problems for conservation and management (Kruger 1977b). Biologists have become aware that the survival of species is linked to the area of habitat (Bond et al. 1988). Consequently, if large extensive areas of habitat become fragmented into small isolated parts, the local extinction of species may follow. Bond et al. (1988) showed that more than 75% of plant species had become extinct on isolated islands of fynbos in the Knysna Forest. The most vulnerable species were those most dependent on fire (sprouters and small, early successional shrubs). To reduce the level of extinctions, managers can either increase the area of natural land conserved, or ensure that isolated areas experience regular fires. The minimum area needed for plant conservation may vary greatly depending on circumstances, but some guidelines are beginning to emerge (Bond et al. 1988; Cowling and Bond 1991). Appro-priate fire regimes are of more immediate importance but there are often logistical and legal problems in burning small reserves, particularly near urban areas.

Besides changes in fire regime, other processes can also be influenced by habitat fragmentation. Some of these are ecological in nature. Few have been documented. Our own observations suggest that vertebrate densities are sometimes very high in refugia of natural veld leading to intense seed predation or herbivory; Argentine ants may invade more readily from cultivated land into small patches of veld; and pesticides blown or washed in from farm lands may disrupt plant-animal interactions.

Studies on extinction risks for specific taxa are still in their infancy. One demographic study (Bond 1989) predicts, surprisingly, that the species most vulnerable to extinction because of habitat fragmentation may also be the most common ones (e.g. *Protea repens* and *P. neriifolia*). Genetic processes are also influenced by population size as determined by habitat area (Franklin 1980; Lande 1988). In theory, managers can reduce the genetic effects of small population size by translocation. Although the problem has been recognized for fynbos (e.g. Hall 1987a), there is almost no work on the topic. Besides maintaining an awareness of the problem (see also Rebelo this volume), there is little the manager can do at present.

The inverse of fragmentation is that new corridors of habitat are opened up. With human-induced alteration of habitats, the fynbos biome has become less insular and many species have migrated into the fynbos from adjacent biomes. For example, 120 species of birds (35% of the current avifauna) have invaded the biome since 1850 (Macdonald 1991). We should expect more 'invaders' from adjacent regions as changes become more pervasive. For example, habitats such as forest patches, formally isolated in a 'sea' of fynbos, are now joined by corridors (plantations of alien trees) that permit diffusion of biota (e.g. Richardson 1989).

Funding

The management of fynbos ecosystems for nature conservation is currently dependent on state funding. As an example, the current (1990) level of funding for catchment conservation in the Western Cape Region is R7.5 million. The budget was increased by only R7 000 (0.1%) from the previous year — this with an annual inflation rate of around 15%. The money must also be used for additional tasks other than the prescribed burning and weed control programmes for which it was originally budgeted. There is an increased demand for recreational facilities, and as farming activities encroach further into mountain land, the demand for fire protection increases. Social upliftment is a priority in the region, and labour costs have increased at a rate several times greater than inflation. Eight years ago, managers were of the opinion that the alien weed problem would be brought under control within 15 years (Kluge and Richardson 1983); they no longer hold this view (R H Andrag personal communication). Changes in the levels of funding for weed control will affect the ultimate condition of fynbos ecosystems.

Four scenarios of funding, and their implication, provide a framework within which changes can be examined. In real terms, current levels of funding will allow prescribed burning to take place once every 15 years, but the cost of weed control programmes would mean that each management unit could only be cleared every second fire cycle (Scenario 1). Using the known rates of proliferation for *Pinus radiata* (Richardson and Brown 1986), alien cover would reach 34% in the second fire cycle. It would return to low levels again following clearing. In order to keep areas clear of invasives, funding would have to be increased to a level where clearing would be possible in conjunction with every prescribed fire (Scenario 2). Under such a scenario, it would be possible to keep the cover of aliens down to < 1%. A more likely scenario would be that funding would decrease in real terms, so that clearing would only be possible every third fire cycle (Scenario 3). Here, cover of aliens would increase to 34% in the second cycle, and to 90% in the third cycle. Should clearing of these infestations be attempted, they would be extremely costly compared to Scenario 1 or 2, due to the dense nature of the stands. Under a worst-case scenario (Scenario 4), no funding for the clearing of aliens would be provided. Alien invasions would increase to 100% cover after about four fire cycles, and would remain at these levels thereafter. The implications of these scenarios for species diversity, water yield, and fire hazard (Table 14.5) are discussed below. Such implications should be weighed carefully when decisions to cut back on alien control programmes are considered.

IMPLICATIONS FOR SPECIES DIVERSITY

In the fynbos biome, invasion by alien plants has resulted in a maximum of 26 extinctions (Hall and Veldhuis 1985). Whereas none of these extinctions can be attributed exclusively to alien plants (Richardson et al. 1989), a cascade of extinctions is virtually inevitable in the next few decades should Scenarios 3, 4, or probably even Scenario 1 prevail. Approximately 750 fynbos plant species (nine per cent of the flora) are currently at risk (Hall 1987b; Macdonald et al. 1989; Richardson et al. this volume). The list of extinct species will grow rapidly as an increasing number of communities are disrupted by invasions and, probably more importantly, as the time since the establishment of thickets of alien species increases, with concomitant attrition of seed banks of suppressed endemic species. Serotinous Proteaceae, small-leaved sprouting and myrmeco-

TABLE 14.5 The outcome of various funding scenarios and the implications for species diversity, water yields, and fire hazard in fynbos ecosystems. The projections for species diversity are based on data in Richardson et al. (1989), those for water yield on data in Van Wyk (1987), and those for fire hazard on data in Versfeld and Van Wilgen (1986).

Funding level	Outcome	Implications for species diversity	Effect on water yield	Effect on fire hazard
I Unchanged	Cover of aliens never exceeds 50%	Local extinction of sensitive species could occur	Reduced by 9%	Fuel loads increased by 50%
II Increased	Cover of aliens remains low (< 1%)	Current species diversity maintained	Little or no reduction	Little or no increase
III Decreased	Cover of aliens becomes dense (> 90%) at times	Local extinction of sensitive species likely; population sizes of common species reduced	Reduced by 21%	Fuel loads increased by 140%
IV Curtailed	Aliens form closed stands over large areas	Local extinction of most indigenous species	Reduced by 50%	Fuel loads increased by 330%

chorous woody shrubs, and large-leaved sprouting shrubs are especially susceptible to extinction (Richardson and Van Wilgen 1986b). A major effort and expenditure of funds are clearly required to prevent extinctions.

Control measures aimed at reducing the detrimental effects of alien plants on biotic diversity must:

- remove the existing dense stands with the least possible additional damage to surviving vegetation and soil-stored seed banks; and
- prevent the establishment of new dense stands.

Alien weed species that have commercial value are the largest problem, as continued propagation and dispersal will mean that control programmes can never be scaled down. We pose the question of whether agencies responsible for the cultivation of these species should not help to offset the costs incurred elsewhere in controlling the resultant invasions?

IMPLICATIONS FOR WATER YIELD

The conversion of fynbos shrublands to closed-canopy pine forests reduces stream flow by almost 50% (Van Wyk 1987). Similar reductions are likely for self-sown stands of pines and other species. For the various scenarios examined, reductions in stream flow will range from almost nothing (Scenario 2) to 50% (Scenario 4). The real costs of this, in terms of reduced water supplies to cities, industries, and agriculture are probably enormous. The question of whether we can afford to allow this to happen (through allowing funding to decrease) needs to be addressed urgently.

IMPLICATIONS FOR FIRE BEHAVIOUR

Fuel loads are greater in dense stands of aliens than in fynbos, but fires are generally more easily ignited in fynbos where there is an abundance of fine material in the herbaceous layer. Dense stands of alien plants burn only under extreme weather conditions, although such fires will be of much higher intensity than fires in uninvaded fynbos. Fires under these conditions are more difficult to contain and are potentially more damaging to ecosystems than fires in indigenous vegetation (Van Wilgen and Richardson 1985b). Under the scenarios examined here, the increase in fuel loads would range from almost nothing to more than 300% (Table 14.5). Fires that occur in stands with vastly increased fuel loads would be uncontrollable, and would result in damage both to the soil and to adjacent properties and crops.

Climatic change

Global climate change as a result of increases in carbon dioxide and other greenhouse gasses will affect all of the earth's ecosystems. Although the predictions for change are crude, they may have important implications for fynbos. Again, we can only examine potential scenarios of climatic change, and attempt to quantify their effects.

CHANGES TO THE FIRE REGIME

We selected some scenarios of likely climatic change (Table 14.6) and used these, together with a 10 year weather record from Swartboskloof (33°57'S, 18°55'E), to generate fire danger indices from the United States National Fire Danger Rating System (Deeming et al. 1978). The results are shown in Figure 14.6. Increases in temperature alone (Scenario 2), or in temperature and rainfall (Scenario 4) do not affect the likelihood of fires markedly. However, a seasonal shift in rainfall patterns will change the seasonal distribution of fires (Scenario 3). Exactly how plants will respond to these changes is not known. Scenario 3 shows that fires will most likely occur between March and July. Fynbos plants reach a minimum flowering activity in April, and this increases steadily through to a maximum in September (Kruger 1981). At times of active flowering (e.g. June and July), fires will destroy unripe seed crops and therefore reduce recruitment. Furthermore, for serotinous Proteaceae, the time between seed release after fire and germination could lengthen, causing high levels of seed predation (Johnson this volume). This could result in individual species becoming extinct (Figure 14.7).

CLIMATE AND PLANT SPECIES DISTRIBUTION

It is difficult to predict the effect of climate change on the distribution of plant species (Bond and Richardson 1990). We can, how-

TABLE 14.6 Scenarios of climatic change used to simulate changes to seasonal fire potential in the fynbos biome.

Climate change scenario	Temperature	Rainfall
1	No change	No change
2	Increase of 2°C in daily maximum temp	No change
3	Increase of 2°C in daily maximum temp	Seasonal shift forward by 4 months
4	Increase of 2°C in daily maximum and minimum temperature in summer and 3°C in winter	Increase of 15% in daily rainfall figures

ever, speculate on the primary influence of factors such as changed seasonal distribution of rainfall and changed mean annual temperatures on community structure. According to some predictions (R E Schulze personal communication), the western Cape will receive more rain in summer and annual mean temperatures will increase. This might favour invasion by summer-growing C_4 grasses from the east, at least in the lowlands where nutrients are not severely limiting (Cowling and Holmes this volume; Stock this volume). This could have a massive impact on community structure, directly if the grasses suppress geophytes and shrub seedlings in the vulnerable post-fire period, and indirectly by facilitating shorter fire rotations. Fire intensity would decline, reducing the abundance of species which require hot fires for regeneration (Bond et al. 1990). Since natural areas in the lowlands are already fragmented, extinction would probably exceed immigration, at least for woody elements, and the result would be reduced richness of fynbos components. Increased summer rain would alleviate soil moisture deficits and possibly favour the invasion of fynbos by forest species that are currently confined to riparian zones, although the increased abundance of grasses and the shorter fire rotations may counter such invasions (Cowling and Holmes this volume).

Plant responses to climate change are likely to be individualistic, especially in their reproduction. Increased temperatures, for example, could inhibit germination in several important fynbos taxa, with seeds stimulated by cold temperatures. For example, *Leucospermum cordifolium* has soil-stored seeds which germinate best when maximum temperatures exceed 24°C and minima are less than 9°C. Under current conditions, these temperature combinations occur in burnt, unshaded sites in autumn and spring (Brits 1986; Figure 14.7). Greenhouse warming of 3–4°C could cause delays of months in germination timing or outright germination failure (Figure 14.7), leading to the rapid population decline of this species. However, germination response is quite variable and we do not know the capacity for rapid selective change. An enormous amount of detailed physiological knowledge of reproductive cues would be needed to predict the potential threats to fynbos species. The most sensitive species would be long-lived nonsprouters. More effort is needed in determining the possible effects of climate change on the reproductive biology of such species.

Much more research is required before we can make useful predictions on the distribution of vegetation under different scenarios for climatic change. The problem is made worse by the complex interactions that regulate range limits (Richardson and Bond 1991). The limited information available suggests that

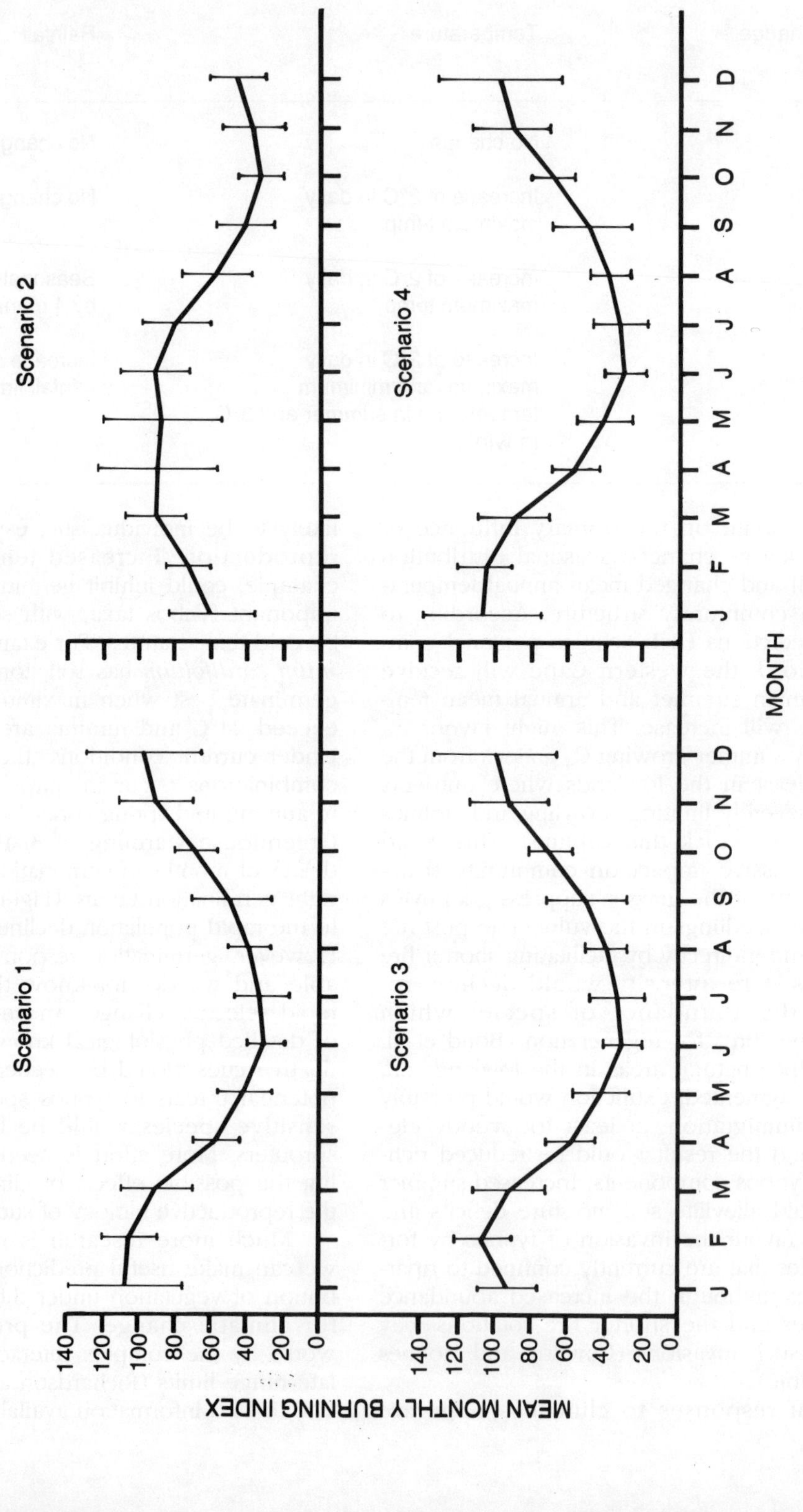

FIGURE 14.6 The annual cycles of mean monthly fire danger (burning index) under four scenarios of climatic change (Table 14.6) at a site in mountain fynbos.

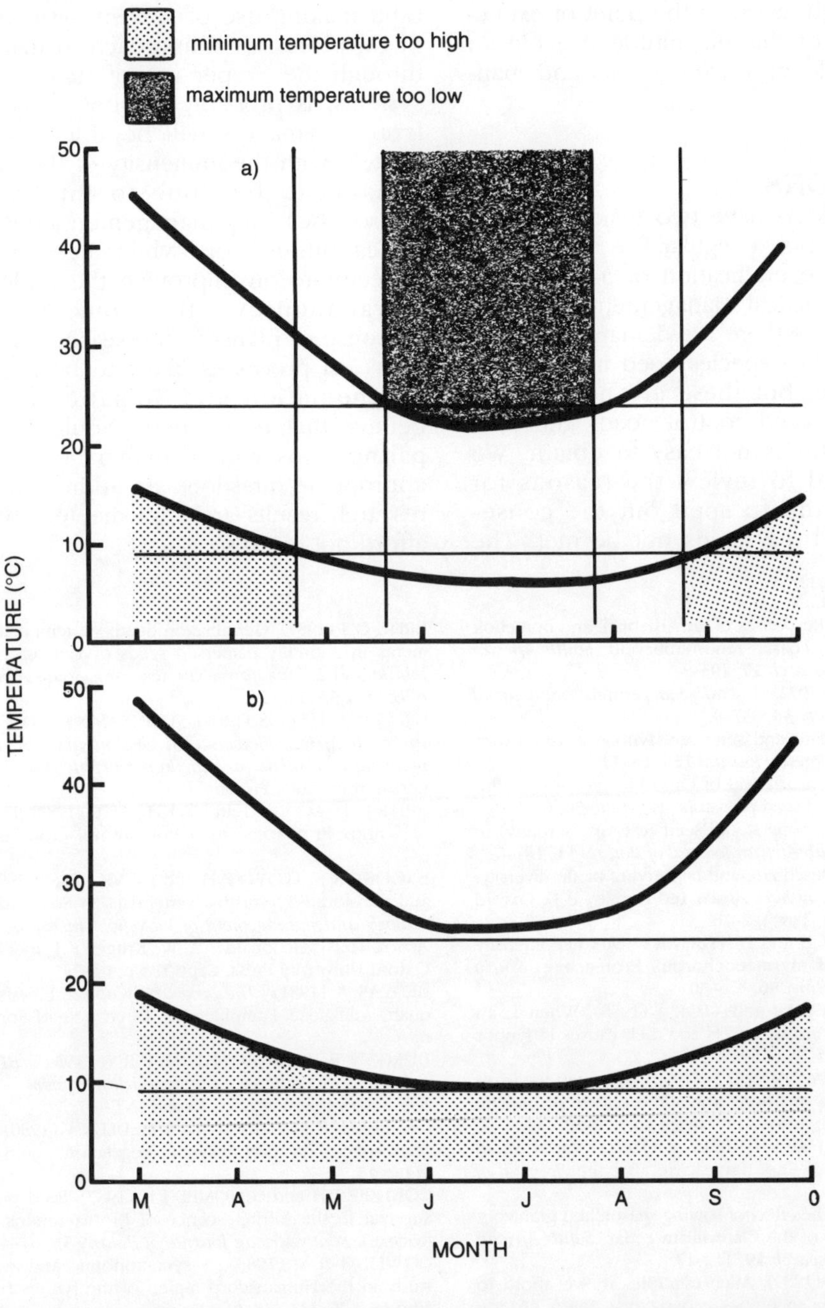

FIGURE 14.7 The potential effects of climatic warming on the seed germination of *Leucospermum cordifolium*, Proteaceae. (a) Optimum germination requires the simultaneous occurrence of minimum temperatures of < 9°C and maximum temperatures of > 24°C (Brits 1986). Under current conditions, these temperature combinations occur only after fire in autumn and spring (shaded areas on the temperature curves indicate periods and causes of inhibition at other times). (b) With a 4°C warming, seeds would not germinate. From data in Brits (1986).

the impact of climate change will be most marked in the lowlands, and that the seed-regenerating shrubs of the Proteaceae will be severely stressed, many to the point of extinction. Changes of this magnitude provide an enormous challenge to researchers and managers.

CONCLUSIONS

Fynbos ecosystems have two major management requirements: a regular fire regime and the need for the eradication of alien plants. Both require practical management solutions, and in both cases there are demanding problems. Many fynbos species need intense fires for persistence, but these are not easy to manage. Alien weed control needs generous funding, and this is not easy to obtain. We have attempted to review the reasons for these needs, and to spell out the consequences should the needs not be met. The chapter has highlighted the need to improve technical skills, such as making use of fire behaviour prediction models, and developing (and making use of) expert systems. A thorough, professional approach to management, through the proper use of such models and systems, is probably the only way in which fynbos managers will be able to cope adequately with the immensity of the task. Much remains to be done to integrate these approaches into management, but such an endeavour is worthwhile. Research should concentrate on improving the understanding of variation over the biome (such as the response of plants to fire season), and should focus on processes likely to be impacted by fragmentation and climatic change. We believe that researchers should form strong partnerships with managers to ensure that appropriate questions are addressed, and that research results are implemented. We cannot afford not to do so.

REFERENCES

BEUKES P C (1987). Responses of grey rhebuck and bontebok to controlled fires in coastal renosterbosveld. *South African Journal of Wildlife Research* **17**, 103–8.

BLOMMAERT K L J (1972). Buchu seed germination. *Journal of South African Botany* **38**, 237–9.

BOND W J (1980). Fire and senescent fynbos in the Swartberg. *South African Forestry Journal* **114**, 68–71.

BOND W J (1984). Fire survival of Cape Proteaceae — influence of fire season and seed predators. *Vegetatio* **56**, 65–74.

BOND W J (1985). Canopy stored seed reserves (serotiny) in Cape Proteaceae. *South African Journal of Botany* **51**, 181–6.

BOND W J (1989). Describing and conserving biotic diversity. In *Biotic diversity in southern Africa.* (ed Huntley B J), Oxford University Press, Cape Town, 2–18.

BOND W J, LE ROUX D and ERNTZEN R (1990). Fire intensity and regeneration of myrmecochorous Proteaceae. *South African Journal of Botany* **56**, 326–30.

BOND W J, MIDGLEY J J and VLOK J (1988). When is an island not an island? Insular effects and their causes in fynbos shrublands. *Oecologia* **77**, 515–21.

BOND W J and RICHARDSON D M (1990). What can we learn from extinctions and invasions about the effects of climate change? *South African Journal of Science* **86**, 429–33.

BOND W J, VLOK J and VIVIERS M (1984). Variation in seedling recruitment of Cape Proteaceae after fire. *Journal of Ecology* **72**, 209–21.

BOTHA S A (1989). The effect of sowing season and granivory on the establishment of the Clanwilliam cedar. *South African Journal of Wildlife Research* **19**, 112–17.

BREYTENBACH G J (1989). Alien control: can we afford to slash and burn hakea in fynbos ecosystems? *South African Forestry Journal* **151**, 6–16.

BRITS G J (1986). Influence of fluctuating temperatures and H_2O_2 treatment on germination of *Leucospermum cordifolium* and *Serruria florida* (Proteaceae) seeds. *South African Journal of Botany* **52**, 286–90.

BRITS G J (1987). Germination depth vs. temperature requirements in naturally dispersed seeds of *Leucospermum cordifolium* and *L. cuneiforme* (Proteaceae). *South African Journal of Botany* **53**, 119–24.

BRITS G J, JACOBS G and STEENKAMP J C (1986). *Farming in South Africa. Flowers and Ornamental shrubs B15. The pruning of proteas for cut flower production.* Horticultural Research Institute, Pretoria.

BRITS G J, JACOBS G and VOGTS M M (1983). Domestication of fynbos Proteaceae as a floricultural crop. *Bothalia* **14**, 642–6.

BROOKE R K, LLOYD P H and DE VILLIERS A L (1986). Alien and translocated terrestrial vertebrates in South Africa. In *The ecology and management of biological invasions in southern Africa.* (eds Macdonald I A W, Kruger F J and Ferrar A A). Oxford University Press, Cape Town, 63–74.

BRUZAS A (1981). *The pine wooly aphid,* Pineus pini (Macquart) Adelgidae. Pamphlet 273, Directorate of Forestry, Pretoria.

BURGAN R E and ROTHERMEL R C (1984). *BEHAVE: Fire behaviour prediction and fuel modelling system — fuel subsystem.* USDA Forest Service, Report INT-167.

CAMPBELL B M and VAN DER MEULEN F (1980). Patterns of plant species diversity in fynbos vegetation, South Africa. *Vegetatio* **43**, 43–7.

COETZEE J H and GILIOMEE J H (1987). Seed predation and survival in the infructescence of Protea repens (L) L (Proteaceae). *South African Journal of Botany* **53**, 61–4.

COWLING R M (1983). A syntaxonomic and synecological study in the Humansdorp region of the fynbos biome. *Bothalia* **15**, 175–227.

COWLING R M (1990). Farming fynbos — Reconciling conservation with exploitation. *UCT News Magazine* **17** (1), 8–10.

COWLING R M and BOND W J (1991). How small can reserves be? An empirical approach in Cape fynbos. *Biological Conservation* **58**.

COWLING R M, PIERCE S M and MOLL E J (1986). Conservation and utilization of South Coast Renosterveld, an endangered South African vegetation type. *Biological Conservation* **37**, 363–77.

DEEMING J E, BURGAN R E and COHEN J D (1978). The National Fire Danger Rating System — 1978. USDA Forest Service, General Technical Report INT-39.

DENNILL G B (1985). The effect of the gall wasp *Trichlogaster acaciaelongifoliae* (Hymenoptera: Pteromalidae) on the reproductive potential and vegetative growth of the weed *Acacia longifolia. Agriculture, Ecosystems and Environment* **14**, 53–61.

DONNELLY D (1990). Resolving conflicts of interest: the release of a biocontrol agent against stinkbean. *Plant Protection News* **19**, 1.

EHRLICH P R (1989). Attributes of invaders and the invading process: vertebrates. In *Biological invasions: a global synthesis.* (eds Drake J, Mooney H A, Di Castri F, Groves R H, Kruger F J, Rejmanek M and Williamson M). Wiley, Chichester, 315–28.

ELOFF J N and LIEDE S (1987). The viability of seed supplied to Botanical Society members by the National Botanic Gardens. *Veld & Flora* **73**, 2–9.

ESLER K J and COWLING R M (1990). Effects of density on the reproductive output of *Protea lepidocarpodendron. South African Journal of Botany* 56, 29–33.

FRANKLIN I R (1980). Evolutionary change in small populations. In *Conservation Biology: An Evolutionary-Ecological Perspective.* (eds Soule M E and Wilcox B A), Sinauer, Sunderland, Mass., 19–34.

GILL A M (1981). Coping with fire. In *The biology of Australian plants* (eds Pate J S and McComb A J), University of Western Australia Press, 65–87.

GREYLING T and DAVIS G W (eds) (1989). *The wildflower resource: commerce, conservation and research.* CSIR, Ecosystems Programme, Terrestrial Ecosystems Section, Occasional Report No. 40.

HALL A V (1987a). Gene flow in plant populations. In *A preliminary synthesis of pollination biology in the Cape flora* (ed Rebelo A G), *South African National Scientific Programmes Report* **141**, 193–211.

HALL A V (1987b). Threatened plants in the fynbos and karoo biomes, South Africa. *Biological Conservation* **40**, 11–28.

HALL A V and VELDHUIS H A (1985). South African Red Data Book: Plants — fynbos and karoo biomes. *South African National Scientific Programmes Report* **117**, CSIR, Pretoria.

HARRIS M S and HOFFMANN J H (1985). The weed *Sesbania punicea* (Leguminosae) in South Africa nipped in the bud by the weevil *Trichapion lativentre. Proceedings of the 6th international symposium on biological control of weeds,* Vancouver, 757–60.

HOFFMANN J H and MORAN V C (1988). The invasive weed *Sesbania punicea* in South Africa and prospects for its biological control. *South African Journal of Science* **84**, 740–2.

HOLMES P M (1989). A comparative study of the seed bank dynamics of two congeneric alien invasive species. Ph.D. thesis, University of Cape Town.

HOLMES P M and REBELO A G (1988). The occurrence of seed-feeding *Zulubius acaciaphagus* (Hemiptera, Alydidae) and its effects on *Acacia cyclops* seed germination and seed banks in South Africa. *South African Journal of Botany* **54**, 319–24.

JEFFERY D J, HOLMES P M and REBELO A G (1988). Effects of dry heat on seed germination in selected indigenous and alien legume species in South Africa. *South African Journal of Botany* **54**, 28–34.

JORDAAN P G (1945). Aantekeninge oor die voortplanting en brandperiodes van *Protea meliferra* Thunb. *Journal of South African Botany* **15**, 121–5.

JORDAAN P G (1965). Die invloed van n winterbrand op die voortplanting van vier soorte van die Proteaceae. *Tydskryf vir Natuurwetenskap* **5**, 27–31.

JORDAAN P G (1982). The influence of a fire in April on the reproduction of three species of the Proteaceae. *South African Journal of Botany* **48**, 1–4.

JOUBERT J G V and STINDT H W (1979a). The nutritive value of natural pastures in the district of Swellendam in the winter rainfall area of the Republic of South Africa. *Department of Agricultural Technical Services, Technical Communication* **156**, Pretoria.

JOUBERT J G V and STINDT H W (1979b). The nutritive value and general evaluation of natural pastures in the districts of Montagu, Robertson and Worcester in the winter rainfall area of the Republic of South Africa. *Department of Agricultural Technical Services, Technical Communication* **155**, Pretoria.

JUHNKE S R and FUGGLE R F (1987). Predicting weather for prescribed burns in the south-western Cape, Republic of South Africa. *South African Forestry Journal* **142**, 41–6.

KLUGE R L and RICHARDSON D M (1983). Progress in the fight against hakea. *Veld & Flora* **69**, 136–8.

KLUGE R L, ZIMMERMANN H G, CILLIERS C J and HARDING G B (1986). Integrated control for invasive alien weeds. In *The ecology and management of biological invasions in southern Africa.* (eds Macdonald I A W, Kruger F J and Ferrar A A), Oxford University Press, Cape Town, 295–303.

KRUGER F J (1977a). A preliminary account of aerial plant biomass in fynbos communities of the mediterranean-type climate zone of the Cape Province. *Bothalia* **12**, 301–7.

KRUGER F J (1977b). Ecological reserves in the Cape fynbos: toward a strategy for conservation. *South African Journal of Science* **73**, 81–5.

KRUGER F J (1981). Seasonal growth and flowering rhythms: South African Heathlands. In *Heathlands and related shrublands of the world. B. Analytical studies.* (ed Specht R L), Elsevier, Amsterdam, 1–4.

KRUGER F J (1987). Succession after fire in selected fynbos communities of the south-western Cape. Ph.D. thesis, University of Witwatersrand.

KRUGER F J and LAMB A J (1978). Conservation of the Kogelberg State Forest. Preliminary assessment of the effects of management from 1967 to 1978. Unpublished report, Jonkershoek Forestry Research Centre.

LANDE R (1988). Genetics and demography in biological conservation. *Science* **24**, 1 455–60.

LE MAITRE D C (1984). A short note on seed predation in *Watsonia pyramidata* (Andr.) Stapf in relation to season of burn. *Journal of South African Botany* **50**, 407–15.

LE MAITRE D C (1988). Effects of season of burn on the regeneration of two Proteaceae with soil-stored seed. *South African Journal of Botany* **54**, 575–80.

LE MAITRE D C (1990). The influence of seed ageing on the plant on seed germination in *Protea neriifolia* (Proteaceae). *South African Journal of Botany* **56**, 49–53.

LINDER H P (1990). The thatching reed of Albertinia. *Veld & Flora,* **76**, 86–9.

MACDONALD I A W (1991). Conservation implications of the invasion of southern Africa by alien organisms. PhD thesis, University of Cape Town.

MACDONALD I A W, LOOPE L L, USHER M B and HAMMAN O (1989). Wildlife conservation and the invasion of nature resources by introduced species: a global perspective. In *Biological invasions: a global perspective* (eds Drake J, Di Castri F, Groves R H, Kruger F J, Mooney H A, Rejmanek M and Williamson M). Wiley, Chichester, 215–55.

MALAN D G (1988). Harvesting and germination of *Helichrysum vestitum* seed. *Protea News* **8**, 4.

MANDERS P T (1985). The autecology of *Widdringtonia*

cederbergensis in relation to its conservation management. M.Sc. thesis, University of Cape Town.

MANDERS P T (1986). An assessment of the current status of the Clanwilliam cedar *Widdringtonia cedarbergensis* and the reasons for its decline. *South African Forestry Journal* **139**, 48–58.

MANDERS P T (1987). A transition matrix model of the population dynamics of the Clanwilliam cedar (*Widdringtonia cedarbergensis*) in natural stands subject to fire. *Forest Ecology and Management* **20**, 171–86.

MENTIS M T and WALKER R S (1989). An expert system for monitoring vegetation. *South African Journal of Science* **85**, 241–4.

MIDGLEY J J (1989). Season of burn of serotinous Proteaceae: a critical review and further data. *South African Journal of Botany* **55**, 165–70.

MIDGLEY J J and CLAYTON P (1990). Short term effects of an autumn fire on small mammal populations in southern Cape coastal mountain fynbos. *South African Forestry Journal* **153**, 27–30.

MIDGLEY J J and VLOK J (1986). Flowering patterns in Cape Proteaceae. *Acta Horticulturae* **185**, 273–6.

MILLAR J C B (1980). Aspects of the ecology of the American Grey Squirrel *Sciurus carolinensis* Gmelin in South Africa. M.Sc. thesis, University of Stellenbosch.

MORRIS M J (1982). Gummosis and die-back of *Hakea sericea* in South Africa. In *Proceedings of the Fourth National Weeds Conference of South Africa.* (eds Van de Venter H A and Mason M), Balkema, Cape Town, 51–4.

MORRIS M J (1987). Biology of the Acacia gall rust *Uromycladium tepperianum. Plant Pathology* **36**, 100–6.

MORRIS M J (1989). A method for controlling *Hakea sericea* Shrad. seedlings using the fungus *Colletotrichum gloeosporioides* (Penz.) Sacc. *Weed Research* **29**, 449–54.

MUSIL C F and DE WITT D M (1990). Post-fire regeneration in a sand plain lowland fynbos community. *South African Journal of Botany* **56**, 167–84.

MUSTART P J and COWLING R M (submitted). Impact of flower and cone harvesting on the seed banks of serotinous Proteaceae. *Journal of Applied Ecology.*

NESER S and KLUGE R L (1985). A seed-feeding insect showing promise in the control of a woody invasive plant: the weevil *Erytenna consputa* Pascoe on *Hakea sericea* (Proteaceae) in South Africa. In *Proceedings of the sixth international symposium on the biological control of weeds.* (ed Delfosse E S), Agricultura Canada, Vancouver, 805–9.

NESER S and KLUGE R L (1986). The importance of seed attaching insects in the biological control of invasive alien plants. In *The ecology and management of biological invasions in southern Africa.* (eds Macdonald I A W, Kruger F J and Ferrar A A). Oxford University Press, Cape Town, 285–9.

NOBLE I R (1987). The role of expert systems in vegetation science. *Vegetatio* **69**, 115–21.

NOBLE I R and SLATYER R O (1980). The use of vital attributes to predict successional changes in plant communities subject to recurrent disturbances. *Vegetatio* **43**, 5–21.

NOVELLIE P (1984). The fire regime at the Bontebok National Park: factors to be considered in applying rotational burning and vegetation monitoring. Typescript, National Parks Board.

PIERCE S M (1990). Pattern and process in south coast dune fynbos: population, community and landscape level studies. Ph.D. thesis, University of Cape Town.

PIERCE S M and COWLING R M (1991). Dynamics of soil-stored seed banks of six shrubs in fire-prone dune fynbos. *Journal of Ecology.*

REBELO A G (1987). Management implications. In *A preliminary synthesis of pollination biology in the Cape flora.* (ed Rebelo A G), *South African National Scientific Programmes Report* **141**, 193–211.

REBELO A G and HOLMES P M (1988). Commercial exploitation of *Brunia albiflora* (Bruniaceae) in South Africa. *Biological Conservation* **45**, 195–207.

RICHARDSON D M (1989). Colonization behaviour of the Sombre Bulbul at Jonkershoek, Stellenbosch. *Promerops* **189**, 7.

RICHARDSON D M and BOND W J (1991). Determinants of plant distribution: evidence from pine invasions. *American Naturalist* **137**, 639–68.

RICHARDSON D M and BROWN P J (1986). Invasion of mesic mountain fynbos by *Pinus radiata. South African Journal of Botany* **52**, 529–36.

RICHARDSON D M, COWLING R M and LE MAITRE D C (1990). Assessing the risk of invasive success in *Pinus* and *Banksia* in South African mountain fynbos. *Journal of Vegetation Science*, **1**, 629–42.

RICHARDSON D M, MACDONALD I A W and FORSYTH G G (1989). Reductions in plant species richness under stands of alien trees and shrubs in the fynbos biome. *South African Forestry Journal* **149**, 1–8.

RICHARDSON D M and VAN WILGEN B W (1986a). The effects of fire in felled *Hakea sericea* and natural fynbos and the implications for weed control in mountain catchments. *South African Forestry Journal* **139**, 4–14.

RICHARDSON D M and VAN WILGEN B W (1986b). Effects of thirty-five years of afforestation with *Pinus radiata* on the composition of mesic mountain fynbos near Stellenbosch. *South African Journal of Botany* **52**, 309–15.

ROTHERMEL R C (1972). A mathematical model for predicting fire spread in wildland fuels. *USDA Forest Service, Research Paper* INT-115.

SMITHERS R H N (1983). *The mammals of the southern African subregion.* University of Pretoria, Pretoria.

STINDT H W and JOUBERT J G V (1979). The nutritive value of natural pastures in the districts of Ladismith, Riversdale and Heidelberg in the winter rainfall area of the Republic of South Africa. *Department of Agricultural Technical Services Technical Communication* **154**, Pretoria.

TROLLOPE W S W (1971). Fire as a method of eradicating macchia vegetation in the Amatole Mountains of South Africa — experimental and field scale results. *Proceedings of the Tall Timbers Fire Ecology Conference* **11**, 99–120.

TROLLOPE W S W (1973). Fire as a method of controlling macchia (fynbos) vegetation on the Amatole Mountains of the eastern Cape. *Proceedings of the Grassland Society of South Africa* **8**, 35–41.

VAN DER MERWE P (1966). Die flora van Swartboskloof, Stellenbosch, en die herstel van soorte na 'n brand. *Annale van die Universiteit van Stellenbosch* **14**A, 691–736.

VAN WILGEN B W (1981). Some effects of fire frequency on fynbos plant community composition and structure at Jonkershoek, Stellenbosch. *South African Forestry Journal* **118**, 42–55.

VAN WILGEN B W (1982). Some effects of post-fire age on the above-ground biomass of fynbos (macchia) vegetation in South Africa. *Journal of Ecology* **70**, 217–25.

VAN WILGEN B W (1984). Adaptation of the United States Fire Danger Rating System to fynbos conditions. I. A fuel model for fire danger rating in the fynbos biome. *South African Forestry Journal* **129**, 61–5.

VAN WILGEN B W (1985). The derivation of fire hazard indices and burning prescriptions from climatic and ecological features of the fynbos biome. Ph.D. thesis, University of Cape Town.

VAN WILGEN B W (1987). Fire regimes in the fynbos biome. In *Disturbance and the dynamics of fynbos biome communities.* (eds Cowling R M, Le Maitre D C, McKenzie B, Prys-Jones

R P and Van Wilgen B W). *South African National Scientific Programmes Report* **135**, CSIR, Pretoria.
VAN WILGEN B W and BURGAN R E (1984). Adaptation of the United States Fire Danger Rating System to fynbos conditions. II. Historic fire danger in the fynbos biome. *South African Forestry Journal* **129**, 66–78.
VAN WILGEN B W and KRUGER F J (1981). Observations on the effects of fire in mountain fynbos at Zachariashoek, Paarl. *Journal of South African Botany* **47**, 195–212.
VAN WILGEN B W and LAMB A J (1986). The flower picking industry in relation to mountain catchment in the fynbos. *Acta Horticulturae* **185**, 181–8.
VAN WILGEN B W and MANDERS P T (1990). Adoption of United States fire technology in South Africa: a technology transfer exercise. *Technology Transfer in the South African forestry industry,* Southern African Institute of Forestry, Pretoria.
VAN WILGEN B W and RICHARDSON D M (1985a). Factors influencing burning by prescription in mountain fynbos catchment areas. *South African Forestry Journal* **134**, 22–32.
VAN WILGEN B W and RICHARDSON D M (1985b). The effect of alien shrub invasions on vegetation structure and fire behaviour in South African fynbos shrublands: a simulation study. *Journal of Applied Ecology* **22**, 955–66.
VAN WILGEN B W and VIVIERS M (1985). The effect of season of fire on serotinous Proteaceae in the western Cape and the implications for fynbos management. *South African Forestry Journal* **133**, 49–53.
VAN WILGEN B W, EVERSON C S and TROLLOPE W S W (1990a). Fire management in southern Africa: some examples of current objectives, practices and problems. In *Fire in the tropical biota.* (ed Goldammer J G) Ecological Studies **84**, Springer, Berlin, 179–215.
VAN WILGEN B W, HIGGINS K B and BELLSTEDT D U (1990b). The role of vegetation structure and fuel chemistry in excluding fire from forest patches in the fire-prone fynbos shrublands of South Africa. *Journal of Ecology* **78**, 210–22.
VAN WILGEN B W, LE MAITRE D C and KRUGER F J (1985). Fire behaviour in South African fynbos (macchia) vegetation and predictions from Rothermel's fire model. *Journal of Applied Ecology* **22**, 207–16.
VAN WYK D B (1987). Some effects of afforestation on streamflow in the Western Cape Province, South Africa. *Water S A* **13**, 31–6.
VERSFELD D B and VAN WILGEN B W (1986). Impacts of woody aliens on ecosystem properties. In *The ecology and control of biological invasions in South Africa.* (eds Macdonald I A W, Kruger F J and Ferrar A A) Oxford University Press, Cape Town, 239–46.
VERSTER M C (1989). Bewaringsplan, Bontebok Nasionale Park, Swellendam. Project report. Forestry Faculty, University of Stellenbosch.
VOGTS M (1982). *South Africa's Proteaceae: know them and grow them.* Struik, Cape Town.
WINTERBOTTOM J M (1968). A check list of the land and fresh water birds of the western Cape Province. *Annals of the South African Museum* **53**, 1–276.
ZWOLINSKI J B, GREY D C and MATHER J A (1989). Impact of pine woolly aphid, *Pineus pini* (Homoptera: Adelgidae) on cone development and seed production of *Pinus pinaster* in the southern Cape. *South African Forestry Journal* **148**, 1–6.

15

A Californian's view of fynbos

J E Keeley

Convergent evolution is a hypothesis that predicts similar environments will select for similar structures and functions in phylogenetically unrelated organisms. Remarkable similarities in the physiognomy of vegetations in mediterranean climate regions of the world have resulted in a great deal of interchange between scientists interested in evolutionary convergence. One ambition of mediterranean climate ecosystem comparisons has been the expectation of using information on ecosystem function in one region to make predictions in the other systems. The focus of this chapter is to consider to what extent conclusions drawn from Californian chaparral apply to the structure and function of South African fynbos.

Mooney (1977) provided a detailed comparison of Californian chaparral and Chilean matorral that revealed remarkable levels of convergence in both structure and function at the primary producer level. Non-convergence between vegetations on these two continents was attributable to subtle differences in the environments. Thus, we might conclude that deviations from convergence between California and South Africa should be attributable to deviations in ambient conditions. I suggest that much of the difference observed in vegetation structure and function between Californian chaparral and South African fynbos can be related, in large part, to differences between these two regions in stresses induced by deficits of water and inorganic nutrients.

FYNBOS VERSUS CHAPARRAL ENVIRONMENTS

Cody and Mooney (1978) summarized the major differences between chaparral and fynbos and pointed out that soil anomalies are undoubtedly an important factor accounting for much of the non-convergence between these mediterranean climate vegetations. The nutrient deficient nature of fynbos soils is well known, in particular the deficiency of nitrogen and phosphorous. To a southern Californian it was surprising to hear chaparral soils characterized as 'high fertility', relative to fynbos soils.

Differences in water stress between these regions, however, are potentially as important as soil nutrition in accounting for degrees of non-convergence. A comparison of fynbos and chaparral sites at similar latitudes and elevations reveals several indicators of more severe summer water stress in California (Table 15.1). At similar latitudes, elevations, and distances from the coast, southern Californian chaparral sites have substantially less precipitation and a greater proportion of it concentrated in winter. On most chaparral sites summer quarter precipitation seldom exceeds 1–4% and interior sites typically have summer temperatures above 30°C. Rutherford and Westfall (1986) compared the distribution of fynbos and other more arid vegetations relative to the percentage of winter-half rainfall and summer aridity (Figure 15.1). Californian chaparral sites all fall on the border between fynbos and the more arid succulent karoo vegetation. Sites with annual precipitation comparable to many fynbos sites are forested in California.

In addition to differences in precipitation, temperature extremes are more severe in chaparral relative to fynbos (Table 15.1). Although the temperature differences may not seem particularly great, Axelrod (1981) maintains that Holocene climate changes of only 1°C in mean temperature of the warmest or coldest month have had a profound impact on vegetation distributions. Another indication of greater extremes in California is the unpredictability of precipitation; for example,

TABLE 15.1 A climatic comparison between typical South African fynbos stations and southern Californian chaparral stations.

	South Africa			Southern California		
	Paarl	Ceres	Swartboskloof	La Mesa	Echo Valley	San Dimas
Latitude	33° 43'S	33° 22'S	34° 00'S	32° 46'N	32° 54'N	34° 12'N
Longitude	18° 57'E	19° 18'E	18° 57'E	117° 01'W	116° 39'W	117° 46'W
Elevation (m)	166	456	305	162	1 070	850
Distance to coast (km)	45	89	20	20	55	66
Precipitation:						
Annual mean (mm)	933	1 276	1 553	304	449	670
% wettest quarter	44	39	42	51	46	59
% driest quarter	9	7	11	2	4	1
Driest month (mm)	22	14	50	1	4	0
Mean temperature:						
Coolest month (°C)	6.2	2.9	6	12.2	1	2.6
Warmest month (°C)	29.7	29.9	27.6	22.3	32	30.7

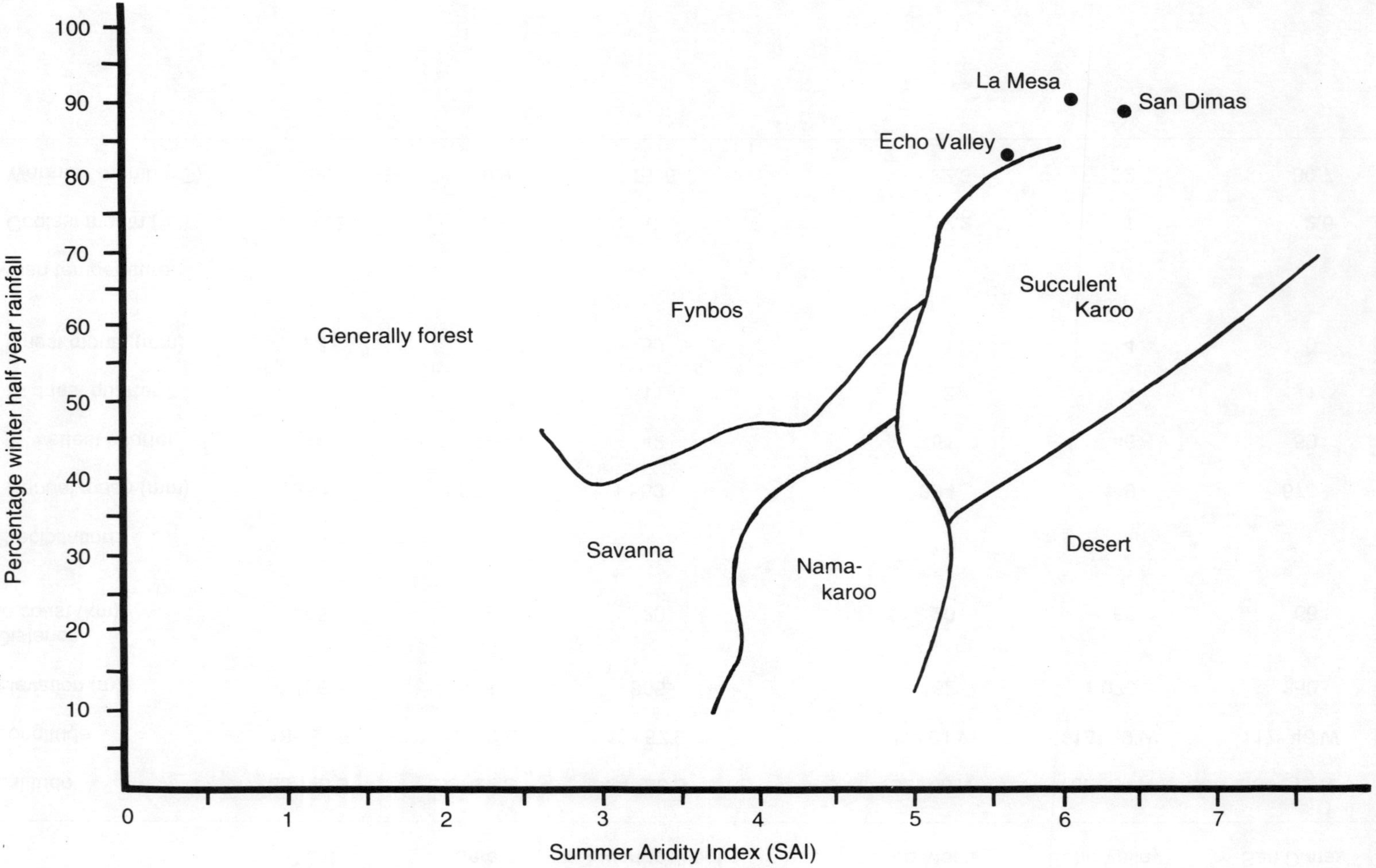

FIGURE 15.1 The distribution of the fynbos biome relative to the summer aridity index and the percentage of winter half year rainfall (redrawn from Rutherford and Westfall 1986). The relative position of southern Californian chaparral sites from Table 15.1 is indicated.

than 200% (Keeley and Keeley 1988) whereas most fynbos sites seldom exceed 100% (Le Maitre 1984). In many respects, fynbos could be described as having a much more equable climate.

In summary, abiotic stresses are different in these regions, and their potential for inducing structural and functional differences in the communities is a function of the time scale over which they operate. Water stress is a greater bottleneck to survival than nutrient stress; brief periods of several months of intense water stress may result in extirpation of a species from a site, whereas longer time scales may be required for nutrient stress to have a similar impact.

CHAPARRAL/FYNBOS COMPARISONS

Community structure

A comparison of community attributes of chaparral and fynbos is shown in Table 15.2.

STRATA

At a rather gross level, these vegetations are readily separable by the fact that chaparral sites are typically dominated by one to a few shrubs that form a mono-layer of overlapping branches often exceeding 100% ground surface cover and commonly 2.5–4 m high (Figure 15.2). This contrasts markedly with the multi-layer restioid-ericoid-proteoid mix in fynbos communities in which several growth forms share dominance and cover is generally less than 80% (Kruger 1979, 1985). Undisturbed mature chaparral has no component comparable to the restioid and ericoid strata.

Intense competition for scarce nutrients has been suggested as a factor leading to growth form diversity in fynbos (Tilman 1983). Another factor favouring low-growing, shallow-rooted restioid and ericoid growth forms would be the equable seasonal distribution of precipitation. In chaparral, brief intense storms concentrated in winter would, as horticulturists have long known, result in 'deep watering' which favours deep-rooted shrubs and trees.

BIOMASS

The multi-strata, more open nature of shrub distribution in fynbos is reflected in lower biomass relative to chaparral; mature fynbos sites are typically between 15 000 to 26 000 kg per ha (Kruger 1977; Rutherford 1978; Kathan 1981; Stock and Allsopp this volume) whereas mature chaparral is often double that amount (Schlesinger and Gill 1980; Ehleringer and Mooney 1982). Some fynbos sites — e.g. ones dominated by *Widdringtonia* (Cupressaceae) or other trees — may have biomass levels comparable to chaparral (Van Wilgen 1982).

SPECIES DIVERSITY

Species richness is significantly greater in fynbos than in chaparral and possibly greater, at a regional level, than the richest tropical rain forests (Cowling et al. this volume). This conclusion is, however, dependent upon the particular measure of diversity and the scale of focus (Bond 1989). In terms of landscape level comparisons, the Cape Floristic Region is markedly richer in species than the California Floristic Province (Cowling et al. this volume).

A precise comparison of community or alpha diversity is difficult because Californian ecologists have not collected the detailed information on species richness at different scales as is widely available for fynbos; the data on chaparral cited in Specht (1988) or in Naveh and Whittaker (1979) is incomplete by fynbos standards.

One of the complications in comparing alpha diversity in fynbos and chaparral is that species richness is usually reported for a single point in time. Species richness is often a function of time since fire and is greatest after fire (Campbell and Van der Meulen 1980; Keeley ct al. 1981; Cowling 1983). Several factors would tend to inflate, relative to chaparral, the species richness reported for fynbos. Due to the higher fire frequency (see below), most fynbos stands are younger than chaparral stands. In fynbos, diversity peaks in the second year and may remain at that level for many years (Kruger 1986). In fact, it appears that the species richness values for mature fynbos (Cowling et al. this volume) are not markedly lower than values for post-fire fyn-

TABLE 15.2 A comparison of Californian chaparral and South African fynbos, based on literature discussed in the text and personal observation.

	California		South Africa
Vegetation structure	Mono-layer		Multi-layer
Shrub dominants	Few species		Many species
Shrub coverage	80–150%		< 80%
Community diversity		<<	
Shrub structure and function			
Branching		>>	
Spinescence		>>	
Summer xylem water potentials	Low		Moderate
Foliage nutrients (N and P)		>>	
Sclerophyll index		<<	
Consumers			
Phytophagous		>?	
Granivory		<?	
Termites		<	
Shrub reproduction			
Flowering phenology	Largely spring		Spring and summer
Flower size	Small–medium		Medium–large
Dioecy	Rare		Frequent
Wind pollination	Rare		Frequent
Bird and mammal pollination	Rare–uncommon		Frequent
Insect pollination	Hymenoptera Lepidoptera		Coleoptera
Seed production	Water limited ?	>?	Pollinator limited ? Nutrient limited ?
Seed banks			
Transient	Many species		Few species
Persistent	Soil-stored		Canopy-stored and soil-stored
Seed dispersal			
Ornithochory	Frequent		Rare
Myrmecochory	Rare		Very common
Anemochory	Uncommon		Common
Telochory	Common		Common
Fire regime			
Frequency		<<	
Intensity		>	
Post-fire response			
Temporary flora	Annuals		Geophytes
Shrub resprouters	Facultative and obligate		Mostly facultative
Seeders	Facultative and obligate		Facultative and obligate
Plant resilience			
Short fire-free periods		<<	
Long fire-free periods		>>	

FIGURE 15.2 The typical single stratum of shrub dominants in Californian chaparral.

bos (reported in Specht 1988). This pattern would not be observed in chaparral. Due to the ephemeral nature of the chaparral post-fire flora, the decrease in species diversity is more rapid; peak species richness in chaparral occurs in the first post-fire year and declines rapidly over the next several years (Keeley et al. 1981; Davis et al. 1988). In chaparral more than half of the flora may be composed of these post-fire annuals, many of which are present between fires only as dormant seed; most studies fail to include these species. Even annuals that establish between fires are often not included due to their very ephemeral nature. Clearly, a detailed comparison of species diversity at different successional stages in chaparral and fynbos is needed.

Although precise numbers are lacking for a comparison of chaparral and fynbos, it is clear that species richness in fynbos is extraordinarily high, as is the diversity of theories on the causes. Nutrient stress in fynbos has been suggested as a mechanism for preventing dominance by any single species and selecting for a diversity of growth forms with competing modes of nutrient capture (Tilman 1983). This theory may account for differences in species richness between chaparral and fynbos. Habitats with low nutrients have low plant biomass and high light penetration (Tilman 1988). More nutrient-rich chaparral sites with higher plant biomass have greater light extinction by the shrub canopy, a factor potentially limiting understorey diversity. In fynbos, Campbell and Van der Meulen (1980) noted that after fire, species diversity declined the most in fynbos that attained the greater biomass, and this was linked to loss of species from the lower strata. Others have also noted a decline in understorey species richness as the overstorey proteoid dominance increased in fynbos (Cowling and Gxaba 1990; J Midgley personal communication).

In addition, greater summer moisture stress in chaparral may limit the diversity of successful physiological modes and increase the extirpation of species from a site. Testing this hypothesis is problematical. However, regional comparisons provide circumstantial evidence that such a factor may be important. Today species-rich pockets of palaeoendemics in California are positively correlated with warm season precipitation (Raven and Axelrod 1978), suggesting that increasing summer aridity, although contributing to accelerated speciation in some annual taxa, may have contributed to decreased species richness in other groups. More extreme winter and summer temperatures may also limit the options available to chaparral taxa. Axelrod (1973) maintains that a gradual decrease in equability of the Californian climate has resulted in a decrease in species richness during the Quaternary.

Nutrient stress, coupled with very frequent disturbance from fire, may maintain the fynbos in a non-equilibrium condition that prevents a few species from dominating a site such as in mature chaparral (Huston 1979; Tilman 1983; Cowling 1987). Cowling (1987) has suggested that fire has played an important role in disruptive selection leading to high species diversity in fynbos, although it is not clear if this could account for differences in species richness between fynbos and chaparral.

To summarize, in fynbos, the more equable climate with less stressful summer drought, reduces the potential abiotic weeding-out of species, and the nutrient-poor soils, by limiting biomass production and thus areal competition, reduce the potential biotic weeding-out of species. Although there are no community level tests of these ideas, regional patterns in California are consistent with these hypotheses. For example, in California, species richness increases with equability of the climate (Moldenke 1975; Richerson and Lum 1980) and the most diverse and endemic-rich chaparral communities are on oligotrophic Quaternary aeolian sands with a marked marine influence in central coastal California (Griffin 1978; Raven and Axelrod 1978; Axelrod 1982; Davis et al. 1988). Species richness on these sites is several times greater than areas of comparable size on more fertile sites under a less equable climate in the interior parts of the California Floristic Province (Richerson and Lum 1980).

Shrub structure and function

Relatively detailed structural comparisons of chaparral and fynbos have revealed marked differences (Cowling and Campbell 1980) and similarities (Cody and Mooney 1978).

MORPHOLOGY

The branching of the dominant woody species is distinctly greater in chaparral species, although the reasons for this are not obvious. Possibly nutrient allocation patterns are different in morphologies which grow from fewer more robust shoots, with lower surface to volume ratio, than in morphologies that proliferate many side shoots. Another possibility is that there are allometric explanations tied to the size of flowers, such as Bond and Midgley (1988) have suggested for *Leucadendron.*

DECIDUOUSNESS

Cowling and Campbell (1980) pointed out a marked difference in growth forms; low elevation summer-deciduous elements and high elevation winter-deciduous elements are largely lacking in fynbos. The greater summer rainfall and higher winter temperatures in the fynbos biome (Table 15.1) may play a role, although low nutrient soils would also select for evergreenness (Campbell and Werger 1988).

WATER POTENTIALS

Summer water stress is generally much greater in chaparral than in fynbos species (Stock et al. this volume). Dominant fynbos shrubs seldom reach xylem water potentials below -3 MPa (Miller et al. 1983; Miller et al. 1984; Moll and Sommerville 1985; Smith and Richardson 1990), whereas chaparral shrubs in summer routinely reach levels twice as low as these (Dunn et al. 1976; Miller and Poole 1979; Mooney and Miller 1985). Lack of intense drought stress is also suggested by the fact that fynbos shrubs reach zero stomatal conductance at substantially higher water potentials than observed for chaparral species (Miller 1985; Poole et al. 1981).

These differences may be attributable to greater total precipitation and greater summer rainfall coupled with lower summer temperatures on most fynbos sites. Richardson and Kruger (1990), however, suggest that sparse canopies of fynbos communities may be insufficient to dry the soils. If true, then the greater coverage typical of chaparral communities may account for a more rapid exhaustion of soil moisture leading to greater water stress. Richardson and Kruger's suggestion seems to contradict the prediction of Poole and Miller (1981) that communities should converge in total transpiration. However, Poole and Miller did not consider sites of radically different nutrient status.

Despite these differences, photosynthetic characteristics are remarkably similar between fynbos and chaparral taxa (Mooney et al. 1983; Van der Heyden and Lewis 1989; Von Willert et al. 1989; Stock et al. this volume).

FOLIAGE NUTRIENTS

Foliage nutrients, in particular N and P, are generally substantially lower in fynbos than in chaparral species (Rundel 1988). This results in much higher sclerophyll indices (lignin + cellulose/protein) in fynbos; 391–1 885 with 50% of the taxa reported as over 1 000 in fynbos, whereas in chaparral reports range from 266–688 with 50% less than 500 (Specht 1988). These differences are undoubtedly related to the greater nitrogen use efficiency of fynbos taxa, which is to be expected in plants from nutrient-poor sites (Rundel 1982; Field et al. 1983).

Additionally, the more equable climate may also contribute to a higher sclerophyll index in fynbos species. Favourable soil moisture conditions during the summer and higher air temperatures during the winter (Table 15.1) allow fynbos shrubs to maintain positive carbon balances throughout the year (Stock et al. this volume).

It is of interest that the level of foliage nutrients in chaparral from oligotrophic soils in central coastal California is more similar to fynbos than to chaparral on other sites (Rundel 1988).

Consumers

Based on differences in the sclerophyll index, the foliage of fynbos plants could be expected to be of lower nutritive value and thus I would predict that foliage herbivores would play a greater role in chaparral.

Although precise data on primary consumers are unavailable to test this hypothesis, Johnson (this volume) concludes that certain herbivore guilds are poorly developed in fynbos. However, the degree of similarity in the foraging ecology of birds in fynbos and chaparral (Cody and Mooney 1978) indicates a strong degree of convergence in the trophic and foraging levels of these secondary consumers, suggesting the primary consumer faunas may be similar. On the other hand, spines are developed to a much greater degree in chaparral (Cowling and Campbell 1980) and this would be expected if chaparral had evolved under greater selective pressure by browsing animals. Much more comparative work between these two regions is needed before this problem is adequately addressed.

Fynbos communities are markedly different from chaparral with respect to termite faunas. Although numbers are lacking, certainly termite presence is far more impressive in fynbos, as evidenced by the numbers of termite mounds. Possibly the more equable climate, higher winter temperatures, and higher summer moisture favour termite faunas. Also, fynbos is rich in fine twigs and stems produced by the restioid and ericoid layers, and these resources would be more readily available to termites than the bulky wood products available in chaparral.

Reproductive biology

PHENOLOGY

Phenological studies suggest that flowering is more synchronized to spring in chaparral, although a few species do extend into the summer e.g. *Adenostoma sparsifolium*, *Heteromeles arbutifolia* (Cody and Mooney 1978). Also, in chaparral there are many taxa that flower (prior to growth) in late winter and early spring on 'old wood' from buds initiated in the previous growing season e.g. all *Ceanothus*, *Arctostaphylos*, and *Rhus* spp. (not *Malosma*). It has been suggested that this pattern allows these plants to speed up their spring flowering and possibly 'beat' the summer drought (Keeley and Keeley 1988).

Fynbos taxa flower over a more extended duration well into summer (Pierce 1984; Moll 1987) probably because of the greater xylem water potentials. Spreading out the duration of flowering over the year may be an important means of reducing competition for pollinators and could be a factor promoting species richness. Summer flowering and fruiting in myrmecochorous species (see below) may have been selected in order to disperse seeds at the time of maximum ant activity (Pierce 1984). Le Maitre (cited in Pierce 1984) notes that *Leucospermum* species flower on the previous year's twigs prior to vegetative growth, although these taxa typically flower in summer, and, therefore, this phenological pattern does not result in a speeding up of flowering, as is the case with Californian taxa that flower on old growth.

Although no published data are available, it seems to me that seed maturation is a much more extended process in fynbos species than in chaparral. One advantage of this would be to increase the period of time for nutrient capture, a luxury allowed by the better summer moisture conditions in fynbos.

POLLINATION

Differences in pollinating strategies are apparent. Most striking is the high incidence of anemophily (wind pollination) in the fynbos flora (Johnson this volume). In chaparral this syndrome is absent; perhaps the more closed nature of chaparral, resulting from less structural variation, may hinder the free movement of pollen. Also, wind is not nearly as predictable as in the Cape during the summer. Mammal pollination is present in fynbos but missing from chaparral. Furthermore, bird pollinators are less common and not responsible for the pollination of any major shrub species in chaparral. Although phylogenetic constraints cannot be ruled out, it is possible that the shorter duration of flower availability in chaparral may represent a bottleneck in resources that large animals are unable to survive.

Additionally, for the same reasons that fynbos shrubs have a high sclerophyll index, it is expected that carbon-rich products, e.g. nectar and flowers, are more readily available for pollinators than nutrient-rich products such as pollen (see Le Maitre and Midgley this volume). Consequently, larger pollinators, which require a carbon-rich nectar reward, may be selected for, whereas pollen-gatherers such as bees would play a subsidiary role. To me, it seemed as though bees were much less common in fynbos than in chaparral and apparently there is some data to support this observation (see Johnson this volume). In place of bees, flower-feeding Coleoptera seemed to be much more common in fynbos than in chaparral, and this would be expected if carbon-rich compounds are more readily available. One consequence of this guild of fynbos pollinators may be that it has selected for larger flowers than are typical of chaparral species.

POLLINATION LIMITATION

Although there is little evidence (see Johnson this volume), there are reasons for suspecting that seed production in fynbos is more pollinator-limited than in chaparral; in the latter vegetation, there is good reason to believe that seed set is more often water-limited (Keeley and Keeley 1988). One phenomenon that needs closer scrutiny is the very high number of unfilled non-viable seeds reported for several fynbos species (see references in Pierce 1990). I also observed this in several taxa I collected. If true, it could reflect pollinator limitation although other reasons are possible e.g. nutrient limitations and inbreeding depression. Alternatively, a strategy of unpredictable resource availability may have selected this characteristic in order to disguise good seeds from predators or satiate them with empty seeds.

DIOECY

Dioecy is markedly more common in fynbos than in chaparral (Le Maitre and Midgley this volume). This is true both in the percentage of the flora and the percentage cover of vegetation at a site; dioecious species, e.g. *Garrya* spp., are uncommon in chaparral, whereas dioecious *Leucadendron* spp. are dominant on many fynbos sites. Curiously, *Garrya* spp. and the dioecious *Simmondsia chinesis* are more common on the arid margin of chaparral, extending into desert environments.

Theories attempting to account for dioecy are widely published and none is overly compelling for fynbos. Selective pressure for dioecy may increase in environments where there are barriers to outcrossing. Perhaps in fynbos there are indirect barriers to outcrossing such as limited seed set; *vis-à-vis* if outcrossed seeds constitute a smaller proportion of the potential seed pool and few seeds ever mature, this may result in a proportionately greater adverse impact on outcrossed seeds. In the case of the chaparral shrub *Simmondsia*, the barrier may be the precarious situation of successfully maturing a particularly large 'nut' in a rather arid environment.

Alternatively, as Midgley (1987) and others have suggested, there may be nutrient limitations that would select for dioecy. By eliminating the male function, female shrubs are able to devote all of the nutrients sequestered by their roots to seed production. This argument may also account for the incidence of dioecy on nutrient-deficient sites in chaparral.

Seed dispersal

Seed dispersal patterns are markedly different in fynbos and chaparral.

ORNITHOCHORY

Bird-dispersed fruits are relatively common in chaparral but much rarer in fynbos where they are present in only a few taxa (e.g. *Rhus*, *Heeria*, and *Myrica*) (see Le Maitre and Midgley this volume; Johnson this volume). In chaparral species, this mode of dispersal is coupled with other reproductive characteristics (Keeley 1991a); seeds are non-dormant at dispersal, thus bird-dispersed taxa do not accumulate a persistent seed bank and consequently do not have post-fire seedling recruitment. In chaparral, ornithochory does not seem to be important as a means of dispersing seeds to gaps, since successful seedling recruitment in these species is generally confined to shaded understoreys of very old (> 50 yrs.) chaparral (Keeley 1991b). Seedlings of these species are more drought-sensitive than other chaparral species and it has been suggested that establishment requires shady sites with a well-developed soil litter layer, conditions that have selected for larger seeds which necessitate ornithochory. The lack of old fynbos sites (see below), except perhaps along river courses, may have selected against ornithochorous species.

Interestingly, many forest species in the Cape are bird-dispersed and share many of the same characteristics described above for such chaparral taxa (Manders 1990, 1991).

MYRMECOCHORY

California has at most three myrmecochorous, or ant-dispersed, plant species (Keeley and Keeley 1988), whereas approximately 2 500 fynbos species are myrmecochorous (Breytenbach 1988). Hypotheses on the selective advantage to myrmecochory are:

Dispersal to reduce intra- or inter-specific competition

I suggest this is an unlikely explanation for the high incidence of myrmecochory in fynbos because it does not result in particularly widespread dispersal and also concentrates seeds much more than other forms of dispersal. If the hypothesis were true, one would predict a higher incidence of myrmecochory in chaparral where higher shrub cover would generate greater shrub competition than in fynbos.

Burial to avoid destruction by fire

This hypothesis seems unlikely because soil-stored seeds of chaparral species are not obviously better protected against fire than seeds of fynbos species, and fire intensities are greater in chaparral than in fynbos (B W van Wilgen personal communication).

Dispersal to nutrient-rich microsites

Empirical data fail to support this hypothesis (Westoby et al. 1982; Bond and Stock 1989).

A dispersal alternative to ornithochory on phosphorous-poor soils

One observation that weakens this hypothesis (Milewski 1982) is that most myrmecochorous species are obligate reseeders recruiting post-fire from persistent seed banks; in other mediterranean climate regions, this mode of regeneration is never ornithochorous (Keeley 1991a). Apparently the 'character syndrome' associated with ornithochory is not compatible with post-fire seedling recruitment. Thus, myrmecochory should not be viewed as a potential disperal alternative to ornithochory.

Rapid removal and burial to reduce predation

This is supported from studies in fynbos (Bond and Breytenbach 1985). However, in order for this hypothesis to have much explanatory power, it must account for why predatory pressure is greater in fynbos than in chaparral. I support the suggestion made by others that the high incidence of myrmecochory in fynbos is tied to the nutrient-deficient soils (e.g. Breytenbach 1987) and suggest it has the following consequences.

One result of nutrient-deficient soils is that they give rise to foliage with a very high sclerophyll index, making leaves a less desirable food source than other plant parts such as seeds (e.g. see Johnson this volume). Seed predation would be intensified by the fact that low nutrient soils would make seedling establishment more precarious and select for seeds that are nutrient sinks (Jongens-Roberts and Mitchell 1986; Stock et al. 1990; Stock and Allsopp this volume). Since, within the 'herbivore guild', seeds are one of the few high quality nutrient (N and P) sources for consumers, granivory may constitute a greater selective pressure in fynbos than in chaparral. Low nutrient fynbos soils may also generate lower seed production over that observed for chaparral shrubs (Le Maitre and Midgley this volume) and possibly more empty seeds than typical of chaparral species; both of which could exacerbate the predation intensity in fynbos and increase the selective pressure for myrmecochory.

Myrmecochory reduces seed predation in two ways. The obvious means is by collecting and burying seeds, thus making them unavailable to seed predators. The less obvious means is through the competitive displacement of seed predators i.e. myrmecochorous ants competitively displace harvester ants in much the same way harvester ants displace other seed-consuming guilds e.g. Brown and Davidson (1977). In effect, myrmecochorous ants are being 'bribed' with a carbon-rich reward for not destroying the nutrient-rich seed. The non-destructive dispersal of myrmecochorous seeds by normally predaceous harvester ants (Bullock 1974) suggests they are bribable and is consistent with the presumed evolution of this behaviour from a seed-harvesting ancestor (Holldobler and Wilson 1990).

Several observations are consistent with the predator-avoidance hypothesis. One is the rarity of myrmecochory in post-fire flowering geophytes (Le Maitre and Midgley this volume). Most of these species sprout and flower profusely after fire (Le Maitre and Midgley this volume) and since most seeds are dispersed at the end of the first post-fire growing season, I would speculate that the main seedling recruitment stage is in subsequent years after fire, a period when animal predation is low (Breytenbach 1987). Another observation supporting the predation hypothesis is the rarity of myrmecochorous species in renoster shrubland, a vegetation juxtaposed with fynbos but with a soil nutrient status more similar to chaparral (Specht 1988). Also, consistent with this hypothesis is the high number of myrmecochorous species on nutrient-poor sites in Australia (Bond and Slingsby 1983; Westoby et al. 1982).

It has been suggested that the presence of vertebrate predators in chaparral, and depauperate small mammal fauna on myrmecochorous-rich sites in Australia, is evidence against this hypothesis (Midgley 1987). These facts alone are insufficient to disregard the hypothesis since the loss of vertebrate predators from the 'granivorous guild' is readily compensated for by increased invertebrate predation (Brown and Davidson 1977).

Post-fire regeneration

FIRE REGIME

One environmental factor that is less readily quantifiable is fire regime. There are several reasons for believing that fires are much more frequent in fynbos. Fynbos stands over 20 years of age are uncommon and areas over 40 years are essentially non-existent. In fact, fynbos stands have a 90–100% probability of burning before they reach the age of 25 years (Kruger 1983; Van Wilgen and Van Hensbergen 1991). This contrasts with Californian chaparral where many stands are over 20 years old and sites over 50 years are present throughout the region (Black 1987; Keeley 1991b). One reason for this difference is that prescribed burning is practised to a greater degree in South Africa and wildfires are only suppressed if they are an immediate threat (B W van Wilgen personal communication). Also, even though most fires in both regions are anthropogenic, there is evidence that natural lightning-caused fires may be more important in the Cape (Kruger 1979; Van Wilgen 1981; Horne 1981; Keeley 1982; Le Maitre and Midgley this volume).

One observation that suggests a greater fire frequency in fynbos is the much more rapid developmental rates of the dominant fynbos species. The juvenile period in many fynbos shrubs is less than seven years (Kruger 1986; Van Wilgen et al. this volume) whereas in many chaparral species it may be substantially greater than that (personal observation).

One factor that could increase fire frequency is the relentless summer southeasterly winds in the Cape that make suppression more precarious and play a significant role in the fynbos fire regime (Van Wilgen 1981). Another factor is the difference in vegetation structure between chaparral and fynbos, which generates very different fuel characteristics. Finer fuels generated by restioids and ericoids make fynbos susceptible to burning at any time of the year, possibly contributing to higher fire frequency.

Historically, fire frequency may have been increased by human occupation and this factor has been present much longer in fynbos than in chaparral (Deacon 1986).

Due to a combination of higher fire frequencies and extensive coverage of restioid and ericoid growth forms, fire intensities are lower in fynbos (Van Wilgen et al. 1985).

Both fynbos and chaparral are resilient to fire and illustrate a remarkable degree of convergence in some aspects of regeneration, and marked differences in other aspects.

TEMPORARY POST-FIRE FLORA

The post-fire flora in fynbos is dominated by geophytes, whereas in chaparral it is dominated by annuals (Kruger 1983). Certainly part of the explanation is that bulbs and corms are a means of sequestering valuable nutrients (Le Maitre and Midgley this volume). Additionally, other factors could be involved. In chaparral, the longer intervals between fires may be important; for example, seed banks are a reliable mode of surviving a long interval between fires, and thus the annual habit may be more compatible with longer, unpredictable fire-free periods. In fynbos, geophytes can persist for many more years after fire due to the more open nature of the vegetation. As a result of the closed canopy in chaparral, geophytes that sprout between fires are less likely to find adequate light and are susceptible to animal predation under the chaparral. Also, the greater summer water stress may make survival of perennials less likely in post-fire chaparral and this may have played a role in selecting for the annual habit. This is supported by the fact that in arid fynbos on the edge of karoo (in a climate more typical of chaparral e.g. Figure 15.1), annuals become more important (Kruger 1979).

Although annuals dominate the post-fire chaparral environment, geophytes are present and flowering is largely restricted to the first post-fire year, as in fynbos geophytes. One marked difference from fynbos is the depth of bulb or corm burial. Although comparative detailed data are largely lacking, my observation was that geophyte bulbs and corms were much more easily extracted from the soil than is the case for chaparral species. In general, the bulbs and corms of fynbos geophytes are buried between 5 and 10 cm (J Vlok personal communication; personal observation), whereas in chaparral they are nearly always greater than 20 cm in depth and often more than 30 cm (personal observation). During the dry summer months, this greater depth of burial, coupled with the finer textured chaparral soils, makes excavation nearly impossible. There are two reasons why Californian geophyte bulbs are buried so deeply: to reduce predation from animals and to reduce dessication during the summer. In sandy fynbos soils, burial may not be a viable means of escaping predation and thus we might expect more species to have evolved toxins that deter predation (see Johnson this volume). Also, the high predictability of measurable precipitation in fynbos during the summer may reduce the dessication of bulbs and corms near the soil surface.

SHRUB REGENERATION

Shrub regeneration in fynbos and chaparral is remarkably similar in the importance of lignotuberous resprouters and non-sprouting seeders (Le Maitre and Midgley this volume). One noteworthy difference, however, is in the preponderance of fynbos species that maintain a seed bank on the plant in serotinous fruits (Bond 1985) versus the preponderance of soil-stored seeds in chaparral. Several factors may be involved. Nutrient-deficient soils may have selected for larger seeds, and if, as argued above, seed predation pressure is greater in fynbos, storing seeds on the plant may be safer. This strategy is apparently quite successful under the relatively frequent and predictable fire regime in the Cape. However, if the fire-free interval exceeds the lifespan of the plant, this could mean localized extinction since the seeds of serotinous species will not survive more than a year following dispersal (Midgley 1987) and successful seedling recruitment between fires is precarious (an important exception, deserving further study, is on semiarid interior fynbos sites where some proteoid taxa have an uneven age structure suggesting continuous recruitment; W Bond personal communication; personal observation). Serotiny is a less viable mode in chaparral due to longer intervals between fire; the few serotinous species are often restricted to nutrient-poor soils where growth rates, and plant longevities, are significantly greater than on more fertile soils (Zedler et al. 1984).

SEED GERMINATION

A remarkable degree of convergence is evident in the seed germination patterns in chaparral and fynbos (e.g. Table 15.3). Particularly striking is the presence of charred wood-stimulated species in both fire-prone regions. These species arise after fire from a dormant seed bank that persists between fires. Other species in both regions, which accumulate a persistent seed bank between fires, have seeds that are stimulated to germinate by heat shock (Table 15.3). Additionally, in both regions geophytes have similar life histories; they resprout immediately after fire from bulbs or corms but not from seed. Due to substantial carbon reserves, these sprouts are capable of flowering in profusion in the first post-fire year. Consequently, seedling recruitment in such species is likely restricted to the second or third post-fire year and thus is not cued directly to any fire-related stimulus. In both regions such species have seeds that germinate readily under adequate temperature and moisture conditions (Table 15.3).

CONCLUSIONS

The convergence of many aspects of fynbos and chaparral supports the hypothesis that similar climates select for similar plant structures and functions. The lack of convergence focused on in this chapter may be explained by regional differences in climate, soil characteristics, and disturbance regimes. Thus, it may be more appropriate to consider the hypothesis that similar *environments* will select for similar structures and functions. Since the environment consists of an n-dimensional hypervolume, it may not be possible to test this hypothesis, and attempts to do so invariably lead to equivocal conclusions (e.g. Barbour and Minnich 1990). In many respects testing the notion of ecosystem convergence is like early attempts to test Gause's theorem that no two species could occupy the same niche. Numerous papers purportedly falsified this hypothesis until Hutchinson's seminal paper (1957) defined niche in such a way as to make it no longer possible to test the competitive exclusion principle. Today it is taken as axiomatic that no two species occupy the same niche and ecologists follow a more productive line of enquiry and focus on the question: 'How dissimilar do species have to be in order to coexist in stable equilibrium?' Likewise, I suggest it be taken as axiomatic that 'similar environments will select for similar characteristics' (within phylogenetic constraints e.g. Peet 1978). Thus, a productive line of enquiry for mediterranean ecologists would be to address the question: 'How dissimilar do environments have to be in order to generate differences in structure and function between mediterranean regions?'

TABLE 15.3 A comparison of seed germination response between species from fynbos and chaparral (Keeley l991a; J Keeley and W J Bond unpublished data). Methods are as described in the publication; heating treatments were slightly different between fynbos and chaparral experiments and are indicated; n = 3 dishes of 50 seeds each; percentages within the same row with the same superscript are not significantly different at P > 0.05. F = fynbos; C = chaparral.

Species Family	Life form	Percentage germination: Control	Heated F: 80°C/30 min C: 80°C/120 min	Heated 115°C/5 min 120°C/5 min	Charred wood added
Fynbos species					
Non-refractory seeds:					
Wachendorfia paniculata (Haemodoraceae)	Geo-	80^{a}	72^{a}	0	80^{a}
Geissorhiza sp. (Iridaceae)	Geo-	74^{a}	77^{a}	0	60^{a}
Heat shock-stimulated seeds:					
Phylica ericoides (Rhamnaceae)	Phano-	9^{a}	44	84	11^{a}
Hermannia alnifolia (Sterculiaceae)	Chamae-	0^{a}	17	67	1^{a}
Charred wood-stimulated seeds:					
Pharnaceum elongatum (Aizoaceae)	Chamae-	0^{a}	4^{a}	1^{a}	47
Nemesia cf *lucida* (Scrophulariaceae)	Thero-	7^{a}	9^{a}	2^{a}	84
Chaparral species					
Non-refractory seeds:					
Marah macrocarpus (Cucurbitaceae)	Geo-	73^{a}	19	25	89^{a}
Zigadenus fremontii (Liliaceae)	Geo-	84^{a}	71^{a}	9	83^{a}
Heat shock-stimulated seeds:					
Ceanothus leucodermis (Rhamnaceae)	Phano-	3^{a}	29	64	7^{a}
Camissonia hirtella (Onagraceae)	Chamae-	30^{a}	49	66	26^{a}
Charred wood-stimulated seeds:					
Eriophyllum confertiflorum (Asteraceae)	Chamae-	4^{a}	6^{a}	4^{a}	52
Emmenanthe penduliflora (Hydrophyllaceae)	Thero-	5^{a}	6^{a}	1^{a}	67

ACKNOWLEDGEMENTS

I thank W J Bond, D C le Maitre, S and R M Cowling, P Manders, J J Midgley, E Moll, W D Stock, and B W van Wilgen, as well as all the other authors of this volume who contributed through their writing, discussions, field trips, and hospitality to my understanding of fynbos. Field assistance from M Keeley is much appreciated. This research was funded by the South African Foundation for Research Development and Occidental College.

REFERENCES

AXELROD D I (1973). History of the mediterranean ecosystem in California. In *Mediterranean ecosystems: origin and structure.* (eds Di Castri F and Mooney H A) Springer-Verlag New York, 225–77.

AXELROD D I (1981). Holocene climatic changes in relation to vegetation disjunction and speciation. *American Naturalist* **117**, 847–70.

AXELROD D I (1982). Age and origin of the Monterey endemic area. *Madrono* **29**, 127–47.

BARBOUR M G and MINNICH R A (1990). The myth of chaparral convergence. *Israel Journal of Botany* **39**, 453–63.

BLACK C H (1987). Biomass nitrogen and phosphorus accumulation over a southern California fire cycle chronosequence. In *Plant response to stress: functional analysis in mediterranean ecoystems.* (eds Tenhunen J D, Catarino F M, Lange O L and Oechel W C) Springer-Verlag, New York, 445–58.

BOND W J (1985). Canopy-stored seed reserves (serotiny) in Cape Proteaceae. *South African Journal of Botany* **51**, 181–6.

BOND W J (1989). Describing and conserving biotic diversity. In *Biotic diversity in southern Africa: concepts and conservation.* (ed Huntley B J) Oxford University Press, Cape Town, 2–18.

BOND W J and BREYTENBACH G J (1985). Ants rodents and seed predation in Proteaceae. *South African Journal of Zoology* **20**, 150–4.

BOND W J and MIDGLEY J (1988). Allometry and sexual differences in leaf size. *American Naturalist* **131**, 901–10.

BOND W J and SLINGSBY P (1983). Seed dispersal in Cape shrublands and its evolutionary implications. *South African Journal of Science* **79**, 231–3.

BOND W J and STOCK W D (1989). The costs of leaving home: ants disperse myrmecochorous seeds to low nutrient sites. *Oecologia* **81**, 412–17.

BREYTENBACH G J (1987). Small mammal dynamics in relation to fire. In *Disturbance and the dynamics of fynbos biome communities.* (eds Cowling R M, Le Maitre D C, McKenzie B, Prys-Jones R P and Van Wilgen B W) *South African National Scientific Programmes Report* **135**. CSIR, Pretoria.

BREYTENBACH G J (1988). Why are myrmecochorous plants limited to fynbos (macchia) vegetation types? *South African Forestry Journal* **144**, 3–5.

BROWN J H and DAVIDSON D W (1977). Competition between seed-eating rodents and ants in desert ecosystems. *Science* **196**, 880–2.

BULLOCK S H (1974). Seed dispersal of *Dendromecon* by the seed predator *Pogonomyrmex. Madrono* **22**, 378–9.

CAMPBELL B M and VAN DER MEULEN F (1980). Patterns of plant species diversity in fynbos vegetation, South Africa. *Vegetatio* **43**, 43–7.

CAMPBELL B M and WERGER M J A (1988). Plant form in the mountains of the Cape South Africa. *Journal of Ecology* **76**, 637–53.

CODY M L and MOONEY H A (1978). Convergence versus nonconvergence in mediterranean-climate ecosystems. *Annual Review of Ecology and Systematics* **9**, 265–321.

COWLING R M (1983). Diversity relations in Cape shrublands and other vegetation in the southeastern Cape, South Africa. *Vegetatio* **54**, 103–27.

COWLING R M (1987). Fire and its role in coexistence and speciation in Gondwanan shrublands. *South African Journal of Science* **83**, 106–12.

COWLING R M and CAMPBELL B M (1980). Convergence in vegetation structure in the mediterranean communities of California, Chile and South Africa. *Vegetatio* **43**, 191–7.

COWLING R M and GXABA T (1990). Effects of a fynbos overstory shrub on understory community structure: implications for the maintenance of community-wide species richness. *South African Journal of Ecology* **1**, 1–7.

DAVIS F W, HICKSON D E and ODION D C (1988). Composition of maritime chaparral related to fire history and soil at Burton Mesa, Santa Barbara County California. *Madrono* **35**, 169–95.

DEACON J (1986). Human settlement in South Africa and archaeological evidence for alien plants and animals. In *The ecology and management of biological invasions in southern Africa.* (eds Macdonald I A W, Kruger F J and Ferrar A A) Oxford University Press, Cape Town, 3–19.

DUNN E L, SHROPSHIRE F M, SONG L C and MOONEY H A (1976). The water factor in convergent evolution in mediterranean-type vegetation. In *Water and plant life.* (eds Lange O L, Kappen L and Schultze E D) Springer-Verlag New York, 492–505.

EHLERINGER J and MOONEY H A (1982). Productivity of desert and mediterranean-climate plants. In *Physiological plant ecology IV. Ecosystem processes: mineral cycling productivity and man's influence.* (eds Lange O L, Nobel P S, Osmond C B and Ziegler H) Springer-Verlag, New York, 206–31.

FIELD D, MERINO J and MOONEY H A (1983). Compromises between water-use efficiency and nitrogen-use efficiency in five species of California evergreens. *Oecologia* **60**, 384–9.

GRIFFIN J R (1978). Maritime chaparral and endemic shrubs of the Monterey Bay Region California. *Madrono* **25**, 65–81.

HOLLDOBLER B and WILSON E O (1990). *The ants.* Belknap Press of Harvard University Press, Cambridge MA.

HOOVER R F (1970). *The vascular plants of San Luis Obispo County California.* University of California Press, Los Angeles.

HORNE I P (1981). The frequency of veld fires in the Groot Swartberg Mountain Catchment Area, Cape Province. *South African Forestry Journal* **118**, 56–60.

HUSTON M (1979). A general hypothesis of species diversity. *American Naturalist* **113**, 81–101.

HUTCHINSON G E (1957). Concluding remarks. *Cold Spring Harbor Symposium on Quantitative Biology* **22**, 415–27.

JONGENS-ROBERTS S M and MITCHELL D T (1986). The distribution of dry mass and phosphorus in an evergreen fynbos shrub species *Leucospermum parile* (Salisb. from J Knight) Sweet (Proteaceae) at different stages of development. *New Phytologist* **103**, 669–83.

KATHAN L (1981). A study of certain ecological aspects pertaining to a *Leucadendron laureolum* community at the Silver Mine Nature Reserve, South Africa. M.Sc. thesis, University of Cape Town.

KEELEY J E (1982). Distribution of lightning- and man-caused wildfires in California. In *Proceedings of the international symposium on the dynamics and management of mediterranean type ecosystems.* (eds Conrad C E and Oechel W C) *General Technical Report* **PSW-58**. USDA Forest Service, Pacific Southwest Forest and Range Experiment Station, Berkeley, 431–7.

KEELEY J E (1991a). Seed germination and life history syndromes in the California chaparral. *Botanical Review* **57**, 81–116.

KEELEY J E (1991b). Seedling recruitment and age structure of

California chaparral in the long absence of fire. *Ecology* (submitted).
KEELEY J E (in press). Seed germination patterns in fire-prone mediterranean climate regions. In *Mediterranean ecosystems of the Pacific Basin: similarities and differences.* (eds Kalin Arroyo M, Fox M and Zedler P H) Kluwer Academic Press.
KEELEY J E and KEELEY S C (1988). Chaparral. In *North American terrestrial vegetation.* (eds Barbour M G and Billings W D) Cambridge University Press, New York, 165–207.
KEELEY S C, KEELEY J E, HUTCHINSON S M and JOHNSON A W (1981). Postfire succession of the herbaceous flora in southern California chaparral. *Ecology* **62**, 1 608–21.
KRUGER F J (1977). A preliminary account of aerial plant biomass in fynbos communities of the mediterranean-type climate zone of the Cape Province. *Bothalia* **12**, 301–7.
KRUGER F J (1979). South African heathlands. In *Heathlands and related shrublands.* (ed Specht R L) Elsevier, New York, 19–80.
KRUGER F J (1983). Plant community diversity and dynamics in relation to fire. In *Mediterranean-type ecosystems: the role nutrients.* (eds Kruger F J, Mitchell D T and Jarvis J U M) Springer-Verlag, New York, 446–72.
KRUGER F J (1985). Climatic and vegetation patterns in the mediterranean zone of the southwestern Cape Province, South Africa. *Actualits Botaniques* **131**, 213–25.
KRUGER F J (1986). Succession after fire in selected fynbos communities of the south-western Cape. Ph.D. thesis, University of Witwatersrand.
LE MAITRE D C (1984). Aspects of the structure and phenology of two fynbos communities. M.Sc. thesis, University of Cape Town.
MANDERS P T (1990). Soil seed banks and post-fire seed deposition across a forest-fynbos ecotone in the Cape Province. *Journal of Vegetation Science* **1**, 491–8.
MANDERS P T (1991). The relationships between forest and mountain fynbos communities in the southwestern Cape Province of South Africa. Ph.D. thesis, University of Cape Town.
MIDGLEY J J (1987). Aspects of the evolutionary biology of the Proteaceae, with emphasis on the genus *Leucadendron* and its phylogeny. Ph.D. thesis, University of Cape Town.
MILEWSKI A V (1982). The occurrence of seeds and fruits taken by ants versus birds in mediterranean Australia and southern Africa, in relation to the availability of soil potassium. *Journal of Biogeography* **9**, 505–16.
MILLER J M (1985). Plant water relations along a rainfall gradient, between the succulent karoo and mesic mountain fynbos, in the Cedarberg Mountains near Clanwilliam, South Africa. M.Sc. thesis, University of Cape Town.
MILLER J M, MILLER P C and MILLER P M (1984). Leaf conductances and xylem pressure potentials in fynbos plant species. *South African Journal of Science* **80**, 381–5.
MILLER P C, MILLER J M and MILLER P M (1983). Seasonal progression of plant water relations in fynbos in the western Cape Province, South Africa. *Oecologia* **56**, 392–6.
MILLER P C and POOLE D K (1979). Patterns of water use by shrubs in southern California. *Forest Science* **25**, 84–98.
MOLDENKE A R (1975). Niche specialization and species diversity along a California transect. *Oecologia* **21**, 219–42.
MOLL E J (1987). Review of some new concepts in 'fynbos' ecology. In *Papers on the prehistory of the Western Cape, South Africa.* (eds Parkington J and Hall M) *BAR International Series* **332**, 120–31.
MOLL E J and SOMMERVILLE J E M (1985). Seasonal xylem pressure potentials of two South African coastal fynbos species in three soil types. *South African Journal of Botany* **51**, 187–93.
MOONEY H A (1977). *Convergent evolution in Chile and California mediterranean climate ecoystems.* Dowden, Hutchinson & Ross, Inc., Stroudsburg, Pennsylvania.
MOONEY H A, FIELD C, GULMON S L, RUNDEL P and KRUGER F J (1983). Photosynthetic characteristic of South African sclerophylls. *Oecologia* **58**, 398–401.
MOONEY H A and MILLER P C (1985). Chaparral. In *Physiological ecology of North American plant communities.* (eds Chabot B F and Mooney H A) Chapman and Hall, New York, 213–31.
NAVEH Z and WHITTAKER R H (1979). Structural and floristic diversity of shrublands and woodlands in northern Israel and other mediterranean areas. *Vegetatio* **41**, 171–90.
PEET R K (1978). Ecosystem convergence. *American Naturalist* **112**, 441–4.
PIERCE S M (1984). A synthesis of plant phenology in the fynbos biome. *South African National Scientific Programmes Report* **88**. CSIR, Pretoria.
PIERCE S M (1990). Pattern and process in south coast dune fynbos: population, community and landscape level studies. Ph.D. thesis, University of Cape Town.
POOLE D K and MILLER P C (1981). The distribution of plant water stress and vegetation characteristics in southern California chaparral. *American Midland Naturalist* **105**, 32–43.
POOLE D K, ROBERTS S W and MILLER P C (1981). Water utilization. In *Resource use by chaparral and matorral: a comparison of vegetation function in two mediterranean type ecosystems.* (ed Miller P C) Springer-Verlag, New York, 123–49.
RAVEN P H and AXELROD D I (1978). Origin and relationships of the California flora. *University of California Publications in Botany* **72**, 1–134.
RICHERSON P J and LUM K-L (1980). Patterns of plant species diversity in California: relation to weather and topography. *American Naturalist* **116**, 504–36.
RICHARDSON D M and KRUGER F J (1990). Water relations and photosynthetic characteristics of selected trees and shrubs of riparian and hillslope habitats in the south-western Cape Province, South Africa. *South African Journal of Botany* **56**, 214–25.
RUNDEL P W (1982). Nitrogen use efficiency in mediterranean-climate shrubs of California and Chile. *Oecologia* **55**, 409–13.
RUNDEL P W (1988). Leaf structure and nutrition in mediterranean-climate sclerophylls. In *Mediterranean-type ecosystems: a data source book.* (ed Specht R L) Kluwer Academic Publishers, Boston, 157–67.
RUTHERFORD M C (1978). Karoo-fynbos biomass along an elevational gradient in the western Cape. *Bothalia* **12**, 555–60.
RUTHERFORD M C and WESTFALL R H (1986). Biomes of Southern Africa — an objective categorization. *Memoirs of the Botanical Survey of South Africa* **54**, 1–98.
SCHLESINGER W H and GILL D S (1980). Biomass, production, and changes in the availability of light, water, and nutrients during the development of pure stands of the chaparral shrub, *Ceanothus megacarpus*, after fire. *Ecology* **61**, 781–9.
SMITH R E and RICHARDSON D M (1990). Comparative post-fire water relations of selected reseeding and resprouting fynbos plants in the Jonkershoek Valley, Cape Province, South Africa. *South African Journal of Botany* **56**, 683–94.
SPECHT R L (ed) (1988). *Mediterranean-type ecosystems: a data source book.* Kluwer Academic Publishers, Boston.
STOCK W D, PATE J S and DELFS J (1990). Influence of seed size and quality on seedling development under low nutrient conditions in five Australian and South African members of the Proteaceae. *Journal of Ecology* **78**, 1 005–20.
TILMAN D (1983). Some thoughts on resource competition and diversity in plant communities. In *Mediterranean-type ecosystems, the role of nutrients.* (eds Kruger F J, Mitchell D T and Jarvis J U M) Springer-Verlag, New York, 322–36.
TILMAN D (1988). *Plant strategies and the dynamics and struc-*

ture of plant communities. Princeton University Press, Princeton, New Jersey.
VAN DER HEYDEN F and LEWIS O A M (1989). Seasonal variation in photosynthetic capacity with respect to plant water status of five species of the mediterranean climate region of South Africa. *South African Journal of Botany* **55**, 509–15.
VAN WILGEN B W (1981). An analysis of fires and associated weather factors in mountain fynbos areas of the south-western Cape. *South African Forestry Journal* **119**, 29–34.
VAN WILGEN B W (1982). Some effects of post-fire age on the above-ground plant biomass of fynbos (macchia) vegetation in South Africa. *Journal of Ecology* **70**, 217–25.
VAN WILGEN B W, LE MAITRE D C and KRUGER F J (1985). Fire behaviour in South African fynbos (macchia) vegetation and predictions from Rothermel's fire model. *Journal of Applied Ecology* **22**, 207–16.
VAN WILGEN B W and VAN HENSBERGEN H J (1991). Fuel properties of vegetation in Swartboskloof. In *Fire in South African mountain fynbos: ecosystem, community and species response at Swartboskloof.* (eds Van Wilgen B W, Richardson D M, Kruger F J and Andrag R H) Springer-Verlag, Heidelberg.
VON WILLERT D J, HERPPICH M AND MILLER J M (1989). Photosynthetic characteristics and leaf water relations of mountain fynbos vegetation in the Cedarberg area (South Africa). *South African Journal of Botany* **55**, 288–98.
WESTOBY M, RICE B, SHELLY J M, HAIG D and KOHEN J L (1982). Plant's use of ants for dispersal at West Head New South Wales. In *Ant-plant interactions in Australia.* (ed Buckley R C) Dr. W. Junk, The Hague, 75–87.
ZEDLER P H, GAUTIER C R and JACKS P (1984). Edaphic restriction of *Cupressus forbesii* (Tecate cypress) in southern California, USA — a hypothesis. In *Being alive on land.* (eds Margaris N S, Arianoustou-Farragitaki M and Oechel W C) Dr. W. Junk, The Hague, 237–43.

Index